Computational Methods for Fluid Dynamics

Joel H. Ferziger · Milovan Perić ·
Robert L. Street

Computational Methods
for Fluid Dynamics

Fourth Edition

 Springer

Joel H. Ferziger
Department of Mechanical Engineering
Stanford University
Stanford, CA, USA

Robert L. Street
Department of Civil and Environmental
Engineering
Stanford University
Stanford, CA, USA

Milovan Perić
University of Duisburg-Essen
Duisburg, Germany

ISBN 978-3-319-99691-2   ISBN 978-3-319-99693-6   (eBook)
https://doi.org/10.1007/978-3-319-99693-6

This Springer imprint is published by the registered company Springer Nature Switzerland AG
The registered company address is: Gewerbestrasse 11, 6330 Cham, Switzerland

Preface

Computational fluid dynamics, commonly known by the acronym 'CFD', continues to have significant expansion. There are many software packages available that solve fluid flow problems; thousands of engineers are using them across a broad range of industries and research areas. The market is growing apparently at a rate of around 15% each year. CFD codes are accepted nowadays as design tools in many industries and are used not only to solve problems but also to help in designing and optimizing various products and as a vehicle for research. While the user-friendliness of commercial CFD-tools has greatly increased since the first edition of this book appeared in 1996, for their efficient and reliable application it is still necessary that the user has a solid background in both fluid mechanics and CFD-methods. We assume that our readers are familiar with theoretical fluid mechanics, so we try to provide useful information on the other component—computational methods for fluid dynamics.

The book is based on material offered by the authors in the past in courses at Stanford University, the University of Erlangen-Nürnberg and the Technical University of Hamburg-Harburg, as well as in a number of short courses. It reflects the authors' experience in developing numerical methods, writing CFD codes and using them to solve engineering and geophysical problems. Many of the codes used in the examples, from the simple ones involving rectangular grids to the ones using non-orthogonal grids and multigrid methods, are available to interested readers; the information on how to access them via the Internet is given in the appendix. These codes illustrate some of the methods described in the book; they can be extended and adapted to the solution of many fluid mechanical problems. Students should try to modify them (e.g., to implement different boundary conditions, interpolation schemes, differentiation and integration approximations, etc.). This is important as one does not really know a method until she or he has programmed and/or run it. We have learned that many researchers have used these codes in the past as the basis for their research projects.

The finite volume method is favored in this book, although finite difference methods are described in what we hope is sufficient detail. Finite element methods are not covered in detail as a number of books on that subject already exist.

The basic ideas of each topic are described in such a way that they can be understood by the reader; where possible, we have avoided lengthy mathematical analysis. Usually, a general description of an idea or method is followed by a more detailed description (including the necessary equations) of one or two numerical schemes representative of the better methods of the type; other possible approaches and extensions are briefly described. We have tried to emphasize common elements of methods rather than their differences and to provide the basis upon which variants can be built.

We have placed considerable emphasis on the need to estimate numerical errors; almost all examples in this book are accompanied with error analysis. Although it is possible for a *qualitatively incorrect* solution of a problem to look reasonable (it may even be a good solution of another problem), the consequences of accepting it may be severe. On the other hand, sometimes a solution of a relatively low accuracy can be of value if treated with care. Industrial users of commercial codes need to learn to judge the quality of the results before believing them. Likewise, researchers have the same challenge. We hope that this book will contribute to the awareness that numerical solutions are always approximate and need to be properly assessed.

We have tried to cover a cross-section of modern approaches, including arbitrary polyhedral and overlapping grids, multigrid methods and parallel computing, methods for moving grids and free surface flows, direct and large-eddy simulation of turbulence, etc. Obviously, we could not cover all these topics in detail, but we hope that the information contained herein will provide the reader with a useful general knowledge of the subject; those interested in a more detailed study of a particular topic will find recommendations for further reading.

The long time between the previous and the current edition of this book was caused by the sudden passing of Joel H. Ferziger in 2004. Although the remaining co-author of the previous edition found an excellent partner for continuing the project in Bob Street, for various reasons (but mostly the lack of time), it took a while before this new edition was finished. The new co-author has brought a new expertise, and the time passed also required that most chapters be significantly revised. Notably, the former Chap. 7 dealing with methods for solving the Navier-Stokes equations has been completely re-written and broken-up into two chapters. Fractional-step methods, which are widely used for large-eddy simulations, have been described in more detail and a new, implicit version has been derived. New codes based on the fractional-step method have been added to the set which can be downloaded by readers from the web-site created specially for that purpose (www.cfd-peric.de). Most examples in later chapters have been re-computed using commercial software; the simulation files for these examples with some instructions can also be downloaded from the above web-site.

While we have invested every effort to avoid typing, spelling, and other errors, no doubt some remain to be found by readers. We will appreciate your notifying us of any mistakes you might find, as well as your comments and suggestions for improvement of future editions of the book. For that purpose, the authors' electronic mail addresses are given below. Corrections, as well as additional extended reports on some of the examples will become available for download at the above web-site.

We also hope that colleagues whose work has not been referenced will forgive us, because any omissions are unintentional.

We have to thank all our present and former students, colleagues, and friends, who helped us in one way or another to finish this work; the complete list of names is too long to present here. Names that we cannot avoid mentioning include (in alphabetic order) Drs. Steven Armfield, David Briggs, Fotini (Tina) Katapodes Chow, Ismet Demirdžić, Gene Golub, Sylvain Lardeau, Željko Lilek, Samir Muzaferija, Joseph Oliger, Eberhard Schreck, Volker Seidl, and Kishan Shah. The help provided by those people who created and made available TeX, LaTeX, Linux, Xfig, Gnuplot, and other tools which made our job easier is also greatly appreciated. Special thanks to Rafael Ritterbusch who provided the fluid-structure interaction example in Chap. 13.

Our families gave us a tremendous support during this endeavor; our special thanks go to Eva Ferziger, Anna James, Robinson and Kerstin Perić and Norma Street.

The initial collaboration between geographically distant colleagues was made possible by grants and fellowships from the Alexander von Humboldt Foundation (to JHF) and the Deutsche Forschungsgemeinschaft (German National Research Foundation, to MP). Without their support, this work would never have come into existence and we cannot express sufficient thanks to them. One of the authors (MP) is especially indebted to the late Peter S. MacDonald, former president of CD-adapco, for his support, and to managers at Siemens (Jean-Claude Ercollanely, Deryl Sneider, and Sven Enger), who provided both support and the software Simcenter STAR-CCM+ to create examples in Chaps. 9–13[1]. RLS is deeply appreciative of the opportunity to contribute to the continuation of the work of his great friend and research colleague, Joel Ferziger.

Duisburg, Germany   Milovan Perić
                                                                milovan.peric@t-online.de
Stanford, USA   Joel H. Ferziger
Stanford, USA   Robert L. Street
                                                                  street@stanford.edu

---

[1]Examples in Sects. 9.12.2, 10.3.5, 10.3.8, 12.2.2, 12.5.2, 12.6.4, and all of Chap. 13 (except where another source is explicitly named) were simulated and the images were created with Simcenter STAR-CCM+, a trademark or registered trademark of Siemens Industry Software NV and any of its affiliates.

Contents

Acronyms

1D	One-dimensional
2D	Two-dimensional
3D	Three-dimensional
ADI	Alternating direction implicit
ALM	Actuator line model
BDS	Backward difference scheme
CDS	Central difference scheme
CFD	Computational fluid dynamics
CG	Conjugate gradient method
CGSTAB	CG stabilized
CM	Control mass
CV	Control volume
CVFEM	Control-volume-based finite element method
DDES	Delayed detached-eddy simulation
DES	Detached-eddy simulation
DNS	Direct numerical simulation
EARSM	Explicit algebraic Reynolds-stress model
EB	Elliptic blending
ENO	Essentially non-oscillatory
FAS	Full approximation scheme
FD	Finite difference
FDS	Forward difference scheme
FE	Finite elements
FFT	Fast Fourier transform
FMG	Full multigrid method
FV	Finite volume
GC	Global communication
GS	Gauss-Seidel method
ICCG	CG preconditioned by incomplete Cholesky method
IDDES	Improved delayed detached-eddy simulation

IFSM	Implicit fractional-step method
ILES	Implicit large-eddy simulation
ILU	Incomplete lower-upper decomposition
LC	Local communication
LES	Large-eddy simulation
LU	Lower-upper decomposition
MAC	Marker-and-cell
MG	Multigrid
MPI	Message-passing interface
ODE	Ordinary differential equation
PDE	Partial differential equation
PVM	Parallel virtual machine
RANS	Reynolds averaged Navier-Stokes
rms	Root mean square
rpm	Revolutions per minute
RSFS	Resolved sub-filter scale
RSM	Reynolds-stress model
SBL	Stable boundary layer
SCL	Space conservation law
SFS	Sub-filter scale
SGS	Subgrid scale
SIP	Strongly implicit procedure
SOR	Successive over-relaxation
SST	Shear stress transport
TDMA	Tridiagonal matrix algorithm
TRANS	Transient RANS
TVD	Total variation diminishing
UDS	Upwind difference scheme
URANS	Unsteady RANS
VLES	Very-large-eddy simulation
VOF	Volume-of-fluid

Chapter 1
Basic Concepts of Fluid Flow

1.1 Introduction

Fluids are substances whose molecular structure offers no resistance to external shear forces: even the smallest force causes *deformation* of a fluid particle. Although a significant distinction exists between *liquids* and *gases*, both types of fluids obey the same laws of motion. In most cases of interest, a fluid can be regarded as a *continuum*, i.e., a continuous substance.

Fluid flow is caused by the action of externally applied forces. Common driving forces include pressure differences, gravity, shear, rotation, and surface tension. They can be classified as *surface forces* (e.g., the shear force due to wind blowing above the ocean or pressure and shear forces created by a movement of a rigid wall relative to the fluid) and *body forces* (e.g., gravity and forces induced by rotation).

While all fluids behave similarly under action of forces, their *macroscopic properties* differ considerably. These properties must be known if one is to study fluid motion; the most important properties of simple fluids are the *density* and *viscosity*. Others, such as *Prandtl number, specific heat*, and *surface tension* affect fluid flows only under certain conditions, e.g., when there are large temperature differences. Fluid properties are functions of other physical variables (e.g., temperature and pressure); although it is possible to estimate some of them from statistical mechanics or kinetic theory, they are usually obtained by laboratory measurement.

Fluid mechanics is a very broad field. A small library of books would be required to cover all of the topics that could be included in it. In this book we shall be interested mainly in flows of interest to engineers but even that is a very broad area (ranging, for example, from wind turbines to gas turbines, from nano-scale to Airbus-scale, and from HVAC to human blood flow). However, we can try to classify the types of problems that may be encountered. A more mathematical, but less complete, version of this scheme will be found in Sect. 1.8.

The speed of a flow affects its properties in a number of ways. At low enough speeds, the inertia of the fluid may be ignored and we have *creeping flow*. This regime is of importance in flows containing small particles (suspensions), in flows

© Springer Nature Switzerland AG 2020
J. H. Ferziger et al., *Computational Methods for Fluid Dynamics*,
https://doi.org/10.1007/978-3-319-99693-6_1

through porous media or in narrow passages (coating techniques, micro-devices). As the speed is increased, inertia becomes important but each fluid particle follows a smooth trajectory; the flow is then said to be *laminar*. Further increases in speed may lead to instability that eventually produces a more random type of flow that is called *turbulent*; the process of laminar-turbulent *transition* is an important area in its own right. Finally, the ratio of the flow speed to the speed of sound in the fluid (the *Mach number*) determines whether exchange between kinetic energy of the motion and internal degrees of freedom needs to be considered. For small Mach numbers, $Ma < 0.3$, the flow may be considered *incompressible*; otherwise, it is *compressible*. If $Ma < 1$, the flow is called *subsonic*; when $Ma > 1$, the flow is *supersonic* and shock waves are possible. Finally, for $Ma > 5$, the compression may create high enough temperatures to change the chemical nature of the fluid; such flows are called *hypersonic*. These distinctions affect the mathematical nature of the problem and therefore the solution method. Note that we call the flow compressible or incompressible depending on the Mach number, even though compressibility is a property of the fluid. This is common terminology because the flow of a compressible fluid at low Mach number is essentially incompressible.

It is common now for engineers to deal with geophysical flows, e.g., in the ocean and atmosphere. There, the fluid density responds to pressure so that the fluid is effectively compressible in many cases, even in the absence of motion. However, except in issues dealing with the deep ocean, the speed of sound in sea water is very large and sea water can be taken as incompressible even though its density depends on the ocean temperature and salt concentration. The atmosphere is quite different. There the pressure and air density decrease exponentially with altitude so the fluid may need to be treated as compressible, except perhaps in the atmospheric boundary layer near the earth's surface.

In many flows, the effects of viscosity are important only near walls, so that the flow in the largest part of the domain can be considered as *inviscid*. In the fluids we treat in this book, Newton's law of viscosity is a good approximation and it will be used exclusively. Fluids obeying Newton's law are called *Newtonian*; *non-Newtonian* fluids are important for some engineering applications but are not treated here.

Many other phenomena affect fluid flow. These include temperature differences which lead to *heat transfer* and density differences which give rise to *buoyancy*. They, and differences in concentration of solutes, may affect flows significantly or, even be the sole cause of the flow. Phase changes (boiling, condensation, melting and solidification), when they occur, always lead to important modifications of the flow and give rise to *multiphase* flow. Variation of other properties such as viscosity, surface tension etc. may also play an important role in determining the nature of the flow. With only a few exceptions, these effects will not be considered in this book.

In this chapter the basic equations governing fluid flow and associated phenomena will be presented in several forms: (i) a coordinate-free form, which can be specialized to various coordinate systems, (ii) an integral form for a finite control volume, which serves as starting point for an important class of numerical methods, and (iii) a differential (tensor) form in a Cartesian reference frame, which is the basis for another important approach. The basic conservation principles and laws used to derive these equations will only be briefly summarized here; more detailed derivations can be found in a number of standard texts on fluid mechanics (e.g., Bird et al. 2006; White 2010). It is assumed that the reader is somewhat familiar with the physics of fluid flow and related phenomena, so we shall concentrate on techniques for the numerical solution of the governing equations.

1.2 Conservation Principles

Conservation laws can be derived by considering a given quantity of matter or *control mass* (CM) and its *extensive* properties, such as mass, momentum and energy. This approach is used to study the dynamics of solid bodies, where the CM (sometimes called the *system*) is easily identified. In fluid flows, however, it is difficult to follow a parcel of matter. It is more convenient to deal with the flow within a certain spatial region we call a *control volume* (CV), rather than in a parcel of matter which quickly passes through the region of interest. This method of analysis is called the *control volume approach*.

We shall be concerned primarily with two extensive properties, mass and momentum. The conservation equations for these and other properties have common terms which will be considered first.

The conservation law for an extensive property relates the rate of change of the amount of that property in a given control mass to externally determined effects. For mass, which is neither created nor destroyed in the flows of engineering interest, the conservation equation can be written:

$$\frac{dm}{dt} = 0 \,. \tag{1.1}$$

On the other hand, momentum can be changed by the action of forces and its conservation equation is Newton's second law of motion:

$$\frac{d(m\mathbf{v})}{dt} = \sum \mathbf{f} \,, \tag{1.2}$$

where t stands for time, m for mass, $\mathbf{v}$ for the velocity, and $\mathbf{f}$ for forces acting on the control mass.[1]

We shall transform these laws into a control volume form that will be used throughout this book. The fundamental variables will be *intensive* rather than extensive properties; the former are properties which are independent of the amount of matter considered. Examples are density ρ (mass per unit volume) and velocity $\mathbf{v}$ (momentum per unit mass).

If ϕ is any conserved intensive property (for mass conservation, $\phi = 1$; for momentum conservation, $\phi = \mathbf{v}$; for conservation of a scalar, ϕ represents the con-

---

[1] Bold symbols, e.g., $\mathbf{v}$ or $\mathbf{f}$ are vectors with three components in the context of this book.

served property per unit mass), then the corresponding extensive property Φ can be expressed as:

$$\Phi = \int_{V_{CM}} \rho\phi \, dV \,, \tag{1.3}$$

where V_{CM} stands for volume occupied by the CM. Using this definition, the left hand side of each conservation equation for a control volume can be written[2]:

$$\frac{d}{dt} \int_{V_{CM}} \rho\phi \, dV = \frac{d}{dt} \int_{V_{CV}} \rho\phi \, dV + \int_{S_{CV}} \rho\phi \, (\mathbf{v} - \mathbf{v_s}) \cdot \mathbf{n} \, dS \,, \tag{1.4}$$

where V_{CV} is the CV volume, S_{CV} is the surface enclosing CV, $\mathbf{n}$ is the unit vector orthogonal to S_{CV} and directed outwards, $\mathbf{v}$ is the fluid velocity and $\mathbf{v_s}$ is the velocity with which the CV surface is moving. For a fixed CV, which we shall be considering most of the time, $\mathbf{v_s} = \mathbf{0}$ and the first derivative on the right hand side becomes a local (partial) derivative. This equation states that the rate of change of the amount of the property in the control mass, Φ, is the rate of change of the property within the control volume plus the net flux of it through the CV boundary due to fluid motion relative to CV boundary. The last term is usually called the *convection* (or sometimes, advection) flux of ϕ through the CV boundary. If the CV moves so that its boundary coincides with the boundary of a control mass, then $\mathbf{v} = \mathbf{v_s}$ and this term will be zero as required.

A detailed derivation of this equation is given in many textbooks on fluid dynamics (e.g., in Bird et al. 2006; Street et al. 1996; Pritchard 2010) and will not be repeated here. The mass, momentum and scalar conservation equations will be presented in the next three sections. For convenience, a fixed CV will be considered; V represents the CV volume and S its surface.

1.3 Mass Conservation

The integral form of the mass conservation (continuity) equation follows directly from the control volume equation, by setting $\phi = 1$:

$$\frac{\partial}{\partial t} \int_V \rho \, dV + \int_S \rho\mathbf{v} \cdot \mathbf{n} \, dS = 0 \,. \tag{1.5}$$

By applying the Gauss' divergence theorem to the convection term, we can transform the surface integral into a volume integral. Allowing the control volume to become infinitesimally small leads to a differential coordinate-free form of the continuity equation:

---

[2]This equation is often called the *control volume equation* or the *Reynolds' transport theorem*.

$$\frac{\partial \rho}{\partial t} + \nabla \cdot (\rho \mathbf{v}) = 0 . \tag{1.6}$$

This form can be transformed into a form specific to a given coordinate system by providing the expression for the divergence operator in that system. Expressions for common coordinate systems such as the Cartesian, cylindrical and spherical systems can be found in many textbooks (e.g., Bird et al. 2006); expressions applicable to general non-orthogonal coordinate systems are given, e.g., in Aris (1990) or Chen et al. (2004a). We present below the Cartesian form in both tensor and expanded notation. Here and throughout this book we shall adopt the Einstein convention that whenever the same index appears twice in any term, summation over the range of that index is implied:

$$\frac{\partial \rho}{\partial t} + \frac{\partial (\rho u_i)}{\partial x_i} = \frac{\partial \rho}{\partial t} + \frac{\partial (\rho u_x)}{\partial x} + \frac{\partial (\rho u_y)}{\partial y} + \frac{\partial (\rho u_z)}{\partial z} = 0 , \tag{1.7}$$

where x_i ($i = 1, 2, 3$) or (x, y, z) are the Cartesian coordinates and u_i or (u_x, u_y, u_z) are the Cartesian components of the velocity vector $\mathbf{v}$. The conservation equations in Cartesian form are often used and this will be the case in this work. Differential conservation equations in non-orthogonal coordinates will be presented in Chap. 9.

1.4 Momentum Conservation

There are several ways of deriving the momentum conservation equation. One approach is to use the control volume method described in Sect. 1.2; in this method, one uses Eqs. (1.2) and (1.4) and replaces ϕ by $\mathbf{v}$, e.g., for a fixed fluid-containing volume of space:

$$\frac{\partial}{\partial t} \int_V \rho \mathbf{v} \, dV + \int_S \rho \mathbf{v} \mathbf{v} \cdot \mathbf{n} \, dS = \sum \mathbf{f} . \tag{1.8}$$

To express the right hand side in terms of intensive properties, one has to consider the forces which may act on the fluid in a CV:

- surface forces (pressure, normal and shear stresses, surface tension etc.);
- body forces (gravity, centrifugal and Coriolis forces, electromagnetic forces, etc.).

The surface forces due to pressure and stresses are, from the molecular point of view, the microscopic momentum fluxes across a surface. If these fluxes cannot be written in terms of the properties whose conservation the equations govern (density and velocity), the system of equations is not closed; that is, there are fewer equations than dependent variables and solution is not possible. This possibility can be avoided by making certain assumptions. The simplest assumption is that the fluid is Newtonian; fortunately, the Newtonian model applies to many actual fluids.

For Newtonian fluids, the stress tensor T, which is the molecular rate of transport of momentum, can be written:

$$\mathsf{T} = -\left(p + \frac{2}{3}\mu\,\nabla\cdot\mathbf{v}\right)\mathsf{I} + 2\mu\mathsf{D}\;, \tag{1.9}$$

where μ is the dynamic viscosity, I is the unit tensor, p is the static pressure and D is the rate of strain (deformation) tensor:

$$\mathsf{D} = \frac{1}{2}\left[\nabla\mathbf{v} + (\nabla\mathbf{v})^T\right]\;. \tag{1.10}$$

These two equations may be written, in index notation in Cartesian coordinates, as follows:

$$T_{ij} = -\left(p + \frac{2}{3}\mu\,\frac{\partial u_j}{\partial x_j}\right)\delta_{ij} + 2\mu D_{ij}\;, \tag{1.11}$$

$$D_{ij} = \frac{1}{2}\left(\frac{\partial u_i}{\partial x_j} + \frac{\partial u_j}{\partial x_i}\right)\;, \tag{1.12}$$

where δ_{ij} is Kronecker symbol ($\delta_{ij} = 1$ if $i = j$ and $\delta_{ij} = 0$ otherwise). For incompressible flows, the second term in the brackets in Eq. (1.11) is zero by virtue of the continuity equation. The following notation is often used in literature to describe the viscous part of the stress tensor:

$$\tau_{ij} = 2\mu D_{ij} - \frac{2}{3}\mu\delta_{ij}\,\nabla\cdot\mathbf{v}\;. \tag{1.13}$$

For non-Newtonian fluids, the relation between the stress tensor and the velocity is often defined by a set of partial differential equations and the total problem is far more complicated; see, e.g., Bird and Wiest (1995). For the class of non-Newtonian fluids which are described using the same kind of constitutive relation as above, but only require a variable viscosity (typically a non-linear function of velocity gradients and temperature) or a stress model comparable to the Reynolds stress models described in Chap. 10, the same methods used for Newtonian fluids can be used.[3] However, different types of non-Newtonian fluids require different constitutive equations as seen in Bird and Wiest (1995); these may, in turn, require specialized solution methods. This subject is complex and is just briefly touched on in Chap. 13.

   With the body forces (per unit mass) being represented by $\mathbf{b}$, the integral form of the momentum conservation equation becomes:

$$\frac{\partial}{\partial t}\int_V \rho\mathbf{v}\,\mathrm{d}V + \int_S \rho\mathbf{v}\mathbf{v}\cdot\mathbf{n}\,\mathrm{d}S = \int_S \mathsf{T}\cdot\mathbf{n}\,\mathrm{d}S + \int_V \rho\mathbf{b}\,\mathrm{d}V\;. \tag{1.14}$$

A coordinate-free vector form of the momentum conservation equation (1.14) is readily obtained by applying Gauss' divergence theorem to the convection and diffusion

---

[3]For example, blood can be treated as Newtonian at high shear rates (Tokuda et al. 2008), but with a variable viscosity in other cases (Perktold and Rappitsch 1995).

flux terms:

$$\frac{\partial(\rho \mathbf{v})}{\partial t} + \nabla \cdot (\rho \mathbf{vv}) = \nabla \cdot \mathsf{T} + \rho \mathbf{b} . \tag{1.15}$$

The corresponding equation for the ith Cartesian component is:

$$\frac{\partial(\rho u_i)}{\partial t} + \nabla \cdot (\rho u_i \mathbf{v}) = \nabla \cdot \mathbf{t}_i + \rho b_i . \tag{1.16}$$

Because momentum is a vector quantity, the convection and diffusion fluxes of it through a CV boundary are the scalar products of second rank tensors ($\rho \mathbf{vv}$ and T) with the surface vector $\mathbf{n}\, \mathrm{d}S$. The integral form of the above equations is:

$$\frac{\partial}{\partial t} \int_V \rho u_i \, \mathrm{d}V + \int_S \rho u_i \mathbf{v} \cdot \mathbf{n} \, \mathrm{d}S = \int_S \mathbf{t}_i \cdot \mathbf{n} \, \mathrm{d}S + \int_V \rho b_i \, \mathrm{d}V , \tag{1.17}$$

where (see Eqs. (1.9) and (1.10)):

$$\mathbf{t}_i = \mu \nabla u_i + \mu (\nabla \mathbf{v})^T \cdot \mathbf{i}_i - \left(p + \frac{2}{3}\mu \nabla \cdot \mathbf{v} \right) \mathbf{i}_i = \tau_{ij} \mathbf{i}_j - p \mathbf{i}_i . \tag{1.18}$$

Here b_i stands for the ith component of the body force, superscript ^{T} means transpose and $\mathbf{i}_i$ is the Cartesian unit vector in the direction of the coordinate x_i. In Cartesian coordinates one can write the above expression as:

$$\mathbf{t}_i = \mu \left(\frac{\partial u_i}{\partial x_j} + \frac{\partial u_j}{\partial x_i} \right) \mathbf{i}_j - \left(p + \frac{2}{3}\mu \frac{\partial u_j}{\partial x_j} \right) \mathbf{i}_i . \tag{1.19}$$

A vector field can be represented in a number of different ways. The basis vectors in terms of which the vector is defined may be local or global. In curvilinear coordinate systems, which are often required when the boundaries are complex (see Chap. 9) one may choose either a covariant or a contravariant basis, see Fig. 1.1. The former expresses a vector in terms of its components along the local coordinates; the latter uses the projections normal to coordinate surfaces. In a Cartesian system, the two become identical. Also, the basis vectors may be dimensionless or dimensional. Including all of these options, over 70 different forms of the momentum equations are possible. Mathematically, all are equivalent; from the numerical point of view, some are more difficult to deal with than others.

The momentum equations are said to be in "strong conservation form" if all terms have the form of the divergence of a vector or tensor. This is possible for the component form of the equations only when components in fixed directions are used. A coordinate-oriented vector component turns with the coordinate direction and an "apparent force" is required to produce the turning; these forces are non-conservative in the sense defined above. For example, in cylindrical coordinates the radial and circumferential directions change so the components of a spatially constant vector (e.g., a uniform velocity field) vary with r and θ and are singular at

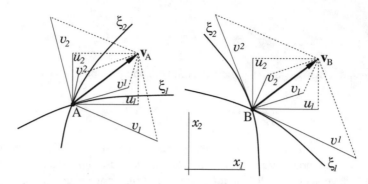

Fig. 1.1 Representation of a vector through different components: u_i—Cartesian components; v^i—contravariant components; v_i—covariant components [$\mathbf{v}_A = \mathbf{v}_B$, $(u_i)_A = (u_i)_B$, $(v^i)_A \neq (v^i)_B$, $(v_i)_A \neq (v_i)_B$]

the coordinate origin. To account for this, the equations in terms of these components contain centrifugal and Coriolis force terms.

Figure 1.1 shows a vector **v** and its contravariant, covariant and Cartesian components. Obviously, the contravariant and covariant components change as the base vectors change even though the vector **v** remains constant. We shall discuss the effect of the choice of velocity components on numerical solution methods in Chap. 9.

The strong conservation form of the equations, when used together with a finite volume method, automatically insures global momentum conservation in the calculation. This is an important property of the conservation equations and its preservation in the numerical solution is equally important. Retention of this property can help to insure that the numerical method will not diverge during the solution and may be regarded as a kind of "realizability".

For some flows it is advantageous to resolve the momentum in spatially variable directions. For example, the velocity in a line vortex has only one component u_θ in cylindrical coordinates but two components in Cartesian coordinates. Axisymmetric flow without swirl is two-dimensional (2D) when analyzed in a polar-cylindrical coordinate frame, but three-dimensional (3D) when a Cartesian frame is used. Some numerical techniques that use non-orthogonal coordinates require use of contravariant velocity components. The equations then contain so-called "curvature terms", which are hard to compute accurately because they contain second derivatives of the coordinate transformations that are difficult to approximate.

Throughout this book we shall work with velocity vectors and stress tensors in terms of their Cartesian components, and we shall use conservative form of the Cartesian momentum equations.

Equation (1.16) is in strong conservation form. A non-conservative form of this equation can be obtained by employing the continuity equation; because

$$\nabla \cdot (\rho \mathbf{v} u_i) = u_i \, \nabla \cdot (\rho \mathbf{v}) + \rho \mathbf{v} \cdot \nabla u_i \ ,$$

it follows that:

$$\rho \frac{\partial u_i}{\partial t} + \rho \mathbf{v} \cdot \nabla u_i = \nabla \cdot \mathbf{t}_i + \rho b_i \ . \tag{1.20}$$

The pressure term contained in $\mathbf{t}_i$ can also be written as

$$\nabla \cdot (p \, \mathbf{i}_i) = \nabla p \cdot \mathbf{i}_i \ .$$

The pressure gradient is then regarded as a body force; this amounts to non-conservative treatment of the pressure term. The non-conservative form of equations is often used in finite difference methods, as it is somewhat simpler. In the limit of a very fine grid, all equation forms and numerical solution methods give the same solution; however, on coarse grids the non-conservative form introduces additional errors which may become important.

If the expression for the viscous part of the stress tensor, Eq. (1.13), is substituted into Eq. (1.16) written in index notation and for Cartesian coordinates, and if gravity is the only body force, one has:

$$\frac{\partial (\rho u_i)}{\partial t} + \frac{\partial (\rho u_j u_i)}{\partial x_j} = \frac{\partial \tau_{ij}}{\partial x_j} - \frac{\partial p}{\partial x_i} + \rho g_i \ , \tag{1.21}$$

where g_i is the component of the gravitational acceleration $\mathbf{g}$ in the direction of the Cartesian coordinate x_i. For the case of constant density and gravity, the term $\rho \mathbf{g}$ can be written as $\nabla (\rho \mathbf{g} \cdot \mathbf{r})$, where $\mathbf{r}$ is the position vector, $\mathbf{r} = x_i \mathbf{i}_i$ (usually, gravity is assumed to act in the negative z-direction, i.e., $\mathbf{g} = g_z \mathbf{k}$, g_z being negative; in this case $\mathbf{g} \cdot \mathbf{r} = g_z z$). Then $-\rho g_z z$ is the hydrostatic pressure, and it is convenient—and for numerical solution more efficient—to define $\tilde{p} = p - \rho g_z z$ as the *head* and use it in place of the pressure. The term ρg_i then disappears from the above equation. If the actual pressure is needed, one only has to add $\rho g_z z$ to $\tilde{p}$.

Because only the gradient of the pressure appears in the equation, the absolute value of the pressure is not important except in compressible flows (including atmospheric and some ocean flows) and in flows with a free surface exposed to the atmosphere.

In variable density flows (the variation of gravity can be neglected in all flows considered in this book), one can split the ρg_i term into two parts: $\rho_0 g_i + (\rho - \rho_0) g_i$, where ρ_0 is a reference density. The first part can then be included with pressure and if the density variation is retained only in the gravitational term, we have the *Boussinesq* approximation, see Sect. 1.7.

1.5 Conservation of Scalar Quantities

The integral form of the equation describing conservation of a scalar quantity, ϕ, is analogous to the previous equations and reads:

$$\frac{\partial}{\partial t} \int_V \rho\phi \; dV + \int_S \rho\phi\mathbf{v} \cdot \mathbf{n} \; dS = \sum f_\phi \, , \tag{1.22}$$

where f_ϕ represents transport of ϕ by mechanisms other than convection and any sources or sinks of the scalar. Diffusion transport is always present (even in stagnant fluids), and it is usually described by a gradient approximation, e.g., *Fourier's law* for heat diffusion and *Fick's law* for mass diffusion:

$$f_\phi^d = \int_S \Gamma \nabla\phi \cdot \mathbf{n} \; dS \, , \tag{1.23}$$

where Γ is the diffusivity for the quantity ϕ. An example is the energy equation which, for most engineering flows, can be written:

$$\frac{\partial}{\partial t} \int_V \rho h \; dV + \int_S \rho h\mathbf{v} \cdot \mathbf{n} \; dS = \int_S k \nabla T \cdot \mathbf{n} \; dS +$$
$$\int_V (\mathbf{v} \cdot \nabla p + \mathsf{S} : \nabla\mathbf{v}) \; dV + \frac{\partial}{\partial t} \int_V p \; dV \, , \tag{1.24}$$

where $h = p/\rho + e$ is the enthalpy per unit mass or specific enthalpy, which is a measure of the total energy of the system, and e is the internal energy. In addition, T is the temperature, k is the thermal conductivity, $k = \mu c_p/\text{Pr}$, and S is the viscous part of the stress tensor, $\mathsf{S} = \mathsf{T} + p\mathsf{I}$. Pr is the Prandtl number, which is defined as the ratio of momentum diffusivity to heat diffusivity, and c_p is the specific heat at constant pressure. The source term represents work done by pressure and viscous forces; it may be neglected in incompressible flows. Further simplification is achieved by considering a fluid with constant specific heat, in which case a convection/diffusion equation for the temperature results:

$$\frac{\partial}{\partial t} \int_V \rho T \; dV + \int_S \rho T\mathbf{v} \cdot \mathbf{n} \; dS = \int_S \frac{\mu}{\text{Pr}} \nabla T \cdot \mathbf{n} \; dS \, . \tag{1.25}$$

Species concentration equations have the same form, with T replaced by the concentration c and Pr replaced by Sc, the Schmidt number, which then is the ratio of the momentum diffusivity to the species diffusivity.

It is useful to write the conservation equations in a general form, as all of the above equations have common terms. The discretization and analysis can then be carried out in a general manner; when necessary, terms peculiar to an equation can be handled separately.

The integral form of the generic conservation equation follows directly from Eqs. (1.22) and (1.23):

$$\frac{\partial}{\partial t} \int_V \rho\phi \; dV + \int_S \rho\phi\mathbf{v} \cdot \mathbf{n} \; dS = \int_S \Gamma \nabla\phi \cdot \mathbf{n} \; dS + \int_V q_\phi \; dV \, , \tag{1.26}$$

where q_ϕ is the source or sink of ϕ. The coordinate-free vector form of this equation is:

$$\frac{\partial(\rho\phi)}{\partial t} + \nabla \cdot (\rho\phi\mathbf{v}) = \nabla \cdot (\Gamma \nabla\phi) + q_\phi . \qquad (1.27)$$

In Cartesian coordinates and tensor notation, the differential form of the generic conservation equation is:

$$\frac{\partial(\rho\phi)}{\partial t} + \frac{\partial(\rho u_j \phi)}{\partial x_j} = \frac{\partial}{\partial x_j}\left(\Gamma \frac{\partial\phi}{\partial x_j}\right) + q_\phi . \qquad (1.28)$$

Numerical methods will first be described for this generic conservation equation. Special features of the continuity and momentum equations (which are usually called *Navier–Stokes equations*) will be described afterwards as an extension of the methods for the generic equation.

1.6 Dimensionless Form of Equations

Experimental studies of flows are often carried out on models, and the results are displayed in dimensionless form, thus allowing scaling to real flow conditions. The same approach can be undertaken in numerical studies as well. The governing equations can be transformed to dimensionless form by using appropriate normalization. For example, velocities can be normalized by a reference velocity v_0, spatial coordinates by a reference length L_0, time by some reference time t_0, pressure by ρv_0^2, and temperature by some reference temperature difference $T_1 - T_0$. The dimensionless variables are then:

$$t^* = \frac{t}{t_0} ; \quad x_i^* = \frac{x_i}{L_0} ; \quad u_i^* = \frac{u_i}{v_0} ; \quad p^* = \frac{p}{\rho v_0^2} ; \quad T^* = \frac{T - T_0}{T_1 - T_0} .$$

If the fluid properties are constant, the continuity, momentum and temperature equations are, in dimensionless form:

$$\frac{\partial u_i^*}{\partial x_i^*} = 0 , \qquad (1.29)$$

$$\mathrm{St}\frac{\partial u_i^*}{\partial t^*} + \frac{\partial(u_j^* u_i^*)}{\partial x_j^*} = \frac{1}{\mathrm{Re}}\frac{\partial^2 u_i^*}{\partial x_j^{*2}} - \frac{\partial p^*}{\partial x_i^*} + \frac{1}{\mathrm{Fr}^2}\gamma_i , \qquad (1.30)$$

$$\mathrm{St}\frac{\partial T^*}{\partial t^*} + \frac{\partial(u_j^* T^*)}{\partial x_j^*} = \frac{1}{\mathrm{Re}\,\mathrm{Pr}}\frac{\partial^2 T^*}{\partial x_j^{*2}} . \qquad (1.31)$$

The following dimensionless numbers appear in the equations:

$$\text{St} = \frac{L_0}{v_0 t_0} \; ; \quad \text{Re} = \frac{\rho v_0 L_0}{\mu} \; ; \quad \text{Fr} = \frac{v_0}{\sqrt{L_0 g}} \; , \tag{1.32}$$

which are called Strouhal, Reynolds, and Froude number, respectively. γ_i is the component of the normalized gravitational acceleration vector in the x_i direction.

For natural convection flows, the Boussinesq approximation is often used, in which case the last term in the momentum equations becomes:

$$\frac{\text{Ra}}{\text{Re}^2 \, \text{Pr}} T^* \gamma_i \; ,$$

where Ra is the Rayleigh number, defined as:

$$\text{Ra} = \frac{\rho^2 g \beta (T_1 - T_0) L_0^3}{\mu^2} \, \text{Pr} = \text{Gr} \, \text{Pr} \; . \tag{1.33}$$

Here Gr is another dimensionless number called the Grashof number and β is the coefficient of thermal expansion.

The choice of the normalization quantities is obvious in simple flows; v_0 is the mean velocity and L_0 is a geometric length scale; T_0 and T_1 are the cold and hot wall temperatures. If the geometry is complicated, the fluid properties are not constant, or the boundary conditions are unsteady, the number of dimensionless parameters needed to describe a flow can become very large and dimensionless equations may no longer be useful.

The dimensionless equations are useful for analytical studies and for determining the relative importance of various terms in the equations. They show, for example, that steady flow in a channel or pipe depends only on the Reynolds number; however, if the geometry changes, the flow will also be influenced by the shape of boundary. Because we are interested in computing flows in complex geometries, we shall use the dimensional form of transport equations throughout this book.

1.7 Simplified Mathematical Models

The conservation equations for mass and momentum are more complex than they appear. They are non-linear, coupled, and difficult to solve. It is difficult to prove by the existing mathematical tools that a unique solution exists for particular boundary conditions. Experience shows that the Navier–Stokes equations describe the flow of a Newtonian fluid accurately. Only in a small number of cases—mostly fully developed flows in simple geometries, e.g., in pipes, between parallel plates etc.—is it possible to obtain an analytical solution of the Navier–Stokes equations. These flows are important for studying the fundamentals of fluid dynamics, but their practical relevance is limited.

In all cases in which such a solution is possible, many terms in the equations are zero. For other flows some terms are unimportant and we may neglect them; this simplification introduces an error. In most cases, even the simplified equations cannot be solved analytically; one has to use numerical methods. The computing effort may be much smaller than for the full equations, which is a justification for simplifications. We list below some flow types for which the equations of motion can be simplified.

1.7.1 Incompressible Flow

The conservation equations for mass and momentum presented in Sects. 1.3 and 1.4 are the most general ones; they assume that all fluid and flow properties vary in space and time. In many applications the fluid density may be assumed constant. This is true not only for flows of liquids, whose compressibility may indeed be neglected in most cases, but also for gases if the Mach number is below 0.3. Such flows are said to be incompressible.[4] If the flow is also *isothermal*, the viscosity is also constant. In that case the mass and momentum conservation equations (1.6) and (1.16) reduce to:

$$\nabla \cdot \mathbf{v} = 0 , \tag{1.34}$$

$$\frac{\partial u_i}{\partial t} + \nabla \cdot (u_i \mathbf{v}) = \nabla \cdot (\nu \nabla u_i) - \frac{1}{\rho} \nabla \cdot (p \, \mathbf{i}_i) + b_i , \tag{1.35}$$

where $\nu = \mu/\rho$ is the kinematic viscosity. This simplification is generally not of a great value, as the equations are hardly any simpler to solve. However, it does help in numerical solution.

1.7.2 Inviscid (Euler) Flow

In flows far from solid surfaces, the effects of viscosity are usually very small. If viscous effects are neglected altogether, i.e., if we assume that the stress tensor reduces to $\mathsf{T} = -p\mathsf{I}$, the Navier–Stokes equations reduce to the Euler equations. The continuity equation is identical to (1.6), and the momentum equations are:

$$\frac{\partial(\rho u_i)}{\partial t} + \nabla \cdot (\rho u_i \mathbf{v}) = -\nabla \cdot (p \, \mathbf{i}_i) + \rho b_i . \tag{1.36}$$

---

[4]Under certain circumstances, e.g., very high pressure or in the deep ocean, the compressibility of liquids needs to be accounted for. Likewise, as noted in Sect. 1.1, in simulating the atmosphere, the compressible version of the flow equations might need to be used even though the Mach number is very small.

Because the fluid is assumed to be inviscid, it cannot stick to walls and slip is possible at solid boundaries. The Euler equations are often used to study compressible flows at high Mach numbers. At high velocities, the Reynolds number is very high and viscous and turbulence effects are important only in a small region near the walls. These flows are often well predicted using the Euler equations.

Although the Euler equations are not easy to solve, the fact that no boundary layers near the walls need to be resolved allows the use of coarser grids. However, as Hirsch (2007) reports in his Introduction, the evolution of solution methods and computer power has made it possible for full three-dimensional Navier–Stokes simulations of flow over entire aircraft, ships, vehicles, etc., and through multistage compressors and pumps, etc., since the mid-1990s. Full Euler simulations have been possible since the early 1980s. Today, engineers employ the most efficient tool for the task, and the Euler equations are still part of the essential tool set (see, e.g., Wie et al. 2010).

There are many methods designed to solve the compressible Euler equations. Some of them will be briefly described in Chap. 11. More details on these methods can be found in books by Hirsch (2007), Fletcher (1991), Knight (2006) and Tannehill et al. (1977), among others. The solution methods described in this book can also be used to solve the compressible Euler equations and, as we shall see in Chap. 11, they perform as well as the special methods designed for compressible flows.

1.7.3 Potential Flow

One of the simplest flow models is potential flow. The fluid is assumed to be inviscid (as in the Euler equations); however, an additional condition is imposed on the flow—the velocity field must be irrotational, i.e.:

$$\text{rot } \mathbf{v} = 0 \,. \tag{1.37}$$

From this condition it follows that there exists a *velocity potential* Φ, such that the velocity vector can be defined as $\mathbf{v} = -\nabla\Phi$. The continuity equation for an incompressible flow, $\nabla \cdot \mathbf{v} = 0$, then becomes a Laplace equation for the potential Φ:

$$\nabla \cdot (\nabla\Phi) = 0 \,. \tag{1.38}$$

The momentum equation can then be integrated to give the Bernoulli equation, an algebraic equation that can be solved once the potential is known. Potential flows are therefore described by the scalar Laplace equation. The latter cannot be solved analytically for arbitrary geometries, although there are simple analytical solutions (uniform flow, source, sink, vortex), which can also be combined to create more complicated flows, e.g., flow around a cylinder.

For each velocity potential Φ one can also define the corresponding *streamfunction* Ψ. The velocity vectors are tangential to streamlines (lines of constant streamfunction); the streamlines are orthogonal to lines of constant potential, so these families of lines form an orthogonal *flow net*.

Occasionally, potential flows are not very realistic. For example, potential theory applied to flow around a body leads to D'Alembert's paradox, i.e., the body experiences neither drag nor lift in a potential flow (see, e.g., Street et al. 1996, or Kundu and Cohen 2008). However, potential flow theory has many applications in porous media flow and computational methods based on potential flow theory are used in many domains, e.g., shipbuilding (for wave resistance, propeller performance, motion of floating bodies, etc.). Numerical methods used to predict potential flows are usually based on the boundary element approach or panel methods (Hess 1990, and Kim et al. 2018); there are also special methods developed for specific applications. These will not be covered in this book, but interested readers may find relevant information in Wrobel (2002) or the journal *Engineering analysis with boundary elements*.

1.7.4 Creeping (Stokes) Flow

When the flow velocity is very small, the fluid is very viscous, or the geometric dimensions are very small (i.e., when the Reynolds number is small), the convection (inertial) terms in the Navier–Stokes equations play a minor role and can be neglected (see the dimensionless form of the momentum equation, Eq. (1.30)). The flow is then dominated by the viscous, pressure, and body forces and is called *creeping flow*. If the fluid properties can be considered constant, the momentum equations become linear; they are usually called the *Stokes equations*. Due to the low velocities the unsteady term can also be neglected, a substantial simplification. The continuity equation is identical to Eq. (1.34), while the momentum equations become:

$$\nabla \cdot (\mu \, \nabla u_i) - \frac{1}{\rho} \nabla \cdot (p \, \mathbf{i}_i) + b_i = 0 \,. \tag{1.39}$$

Creeping flows are found in porous media, coating technology, micro-devices, etc.

1.7.5 Boussinesq Approximation

In flows accompanied by heat transfer, the fluid properties are normally functions of temperature. The variations may be small and yet be the cause of the fluid motion. If the density variation is not large, one may treat the density as constant in the unsteady and convection terms, and treat it as variable only in the gravitational term. This is called the *Boussinesq approximation*. One usually assumes that the density varies linearly with temperature. If one includes the effect of the body force on the mean density in the pressure term as described in Sect. 1.4, the remaining term can be expressed as:

$$(\rho - \rho_0)g_i = -\rho_0 g_i \beta (T - T_0) , \tag{1.40}$$

where β is the coefficient of volumetric expansion. This approximation introduces errors of the order of 1% if the temperature differences are below, e.g., 2° for water and 15° for air. The error may be more substantial when temperature differences are larger; the solution may even be qualitatively wrong (for an example, see Bückle and Perić 1992).

1.7.6 Boundary Layer Approximation

When the flow has a predominant direction (i.e., there is no reversed flow or recirculation) and the variation of the geometry is gradual, the flow is mainly influenced by what happened upstream. Examples are flows in channels and pipes and flows over plane or mildly curved solid walls. Such flows are called *thin shear layer* or *boundary layer flows*. The Navier–Stokes equations can be simplified for such flows as follows:

- diffusion of momentum in the principal flow direction is much smaller than convective transport and can be neglected;
- the velocity component in the main flow direction is much larger than the components in other directions;
- the pressure gradient across the flow is much smaller than in the principal flow direction.

The two-dimensional boundary layer equations reduce to:

$$\frac{\partial(\rho u_1)}{\partial t} + \frac{\partial(\rho u_1 u_1)}{\partial x_1} + \frac{\partial(\rho u_2 u_1)}{\partial x_2} = \mu \frac{\partial^2 u_1}{\partial x_2^2} - \frac{\partial p}{\partial x_1} , \tag{1.41}$$

which must be solved together with the continuity equation; the equation for the momentum normal to the principal flow direction reduces to $\partial p / \partial x_2 = 0$. The pressure as a function of x_1 must be supplied by a calculation of the flow exterior to the boundary layer—which is usually assumed to be potential flow, so the boundary layer equations themselves are not a complete description of the flow. The simplified equations can be solved by using marching techniques similar to those used to solve ordinary differential equations with initial conditions. These techniques see considerable use in aerodynamics. The methods are very efficient but can be applied only to flows without separation.

1.7.7 Modeling of Complex Flow Phenomena

Many flows of practical interest are difficult to describe exactly mathematically, let alone solve exactly. These flows include turbulence, combustion, and multiphase flow, and are very important. Because exact description is often impracticable, one usually uses semi-empirical models to represent these phenomena. Examples are turbulence models (which will be treated in some detail in Chap. 10), combustion models, multiphase models, etc. These models, as well as the above-mentioned simplifications affect the accuracy of the solution. The errors introduced by the various approximations may either augment or cancel each other; therefore, care is needed when drawing conclusions from calculations in which models are used. Due to the importance of various kinds of errors in numerical solutions we shall devote a lot of attention to this topic. The error types will be defined and described as they are encountered.

1.8 Mathematical Classification of Flows

Quasi-linear second-order partial differential equations in two independent variables can be divided into three types: hyperbolic, parabolic, and elliptic. This distinction is based on the nature of the characteristics, curves along which information about the solution is carried. Every equation of this type has two sets of characteristics (see, e.g., Street 1973).

In the hyperbolic case, the characteristics are real and distinct. This means that information propagates at finite speeds in two sets of directions. In general, the information propagation is in a particular direction so that one datum needs to be given at an initial point on each characteristic; the two sets of characteristics therefore demand two initial conditions. If there are lateral boundaries, usually only one condition is required at each point because one characteristic is carrying information out of the domain and one is carrying information in. There are, however, exceptions to this rule.

In parabolic equations the characteristics degenerate to a single real set. Consequently, only one initial condition is normally required. At lateral boundaries one condition is needed at each point.

Finally, in the elliptic case, there are no real characteristics; the two sets of characteristics are complex (imaginary) and distinct. As a consequence, there are no special directions of information propagation. Indeed, information travels essentially equally well in all directions. Generally, one boundary condition is required at each point on the boundary and the domain of solution is usually closed although part of the domain may extend to infinity. Unsteady problems are never elliptic.

These differences in the nature of the equations are reflected in the methods used to solve them. It is an important general rule that numerical methods should respect the properties of the equations they are solving.

The Navier–Stokes equations are a system of non-linear second-order equations in four independent variables. Consequently the classification scheme does not apply directly to them. Nonetheless, the Navier–Stokes equations do possess many of the properties outlined above and the many of the ideas used in solving second-order equations in two independent variables are applicable to them but care must be exercised.

1.8.1   Hyperbolic Flows

To begin, consider the case of unsteady inviscid compressible flow. A compressible fluid can support sound and shock waves and it is not surprising that these equations have essentially hyperbolic character. Most of the methods used to solve these equations are based on the idea that the equations are hyperbolic and, given sufficient care, they work quite well; these are the methods referred to briefly above.

For steady compressible flows, the character depends on the speed of the flow. If the flow is supersonic, the equations are hyperbolic while the equations for subsonic flow are essentially elliptic. This leads to a difficulty that we shall discuss further below.

It should be noted however, that the equations for a viscous compressible flow are still more complicated. Their character is a mixture of elements of all of the types mentioned above; they do not fit well into the classification scheme and numerical methods for them are more difficult to construct.

1.8.2   Parabolic Flows

The boundary layer approximation described briefly above leads to a set of equations that have essentially parabolic character. Information travels only downstream in these equations and they may be solved using methods that are appropriate for parabolic equations.

Note, however, that the boundary layer equations require specification of a pressure that is usually obtained by solving a potential flow problem. Subsonic potential flows are governed by elliptic equations (in the incompressible limit the Laplace equation suffices) so the overall problem actually has a mixed parabolic-elliptic character.

1.8.3   Elliptic Flows

When a flow has a region of recirculation i.e., flow in a sense opposite to the principal direction of flow, information may travel upstream as well as downstream. As a result, one cannot apply conditions only at the upstream end of the flow. The problem then acquires elliptic character. This situation occurs in subsonic (including incompressible) flows and makes solution of the equations a very difficult task.

It should be noted that unsteady incompressible flows actually have a combination of elliptic and parabolic character. The former comes from the fact that information travels in both directions in space while the latter results from the fact that information can only flow forward in time. Problems of this kind are called incompletely parabolic.

1.8.4 Mixed Flow Types

As we have just seen, it is possible for a single flow to be described by equations that are not purely of one type. Another important example occurs in steady transonic flows, that is, steady compressible flows that contain both supersonic and subsonic regions. The supersonic regions are hyperbolic in character while the subsonic regions are elliptic. Consequently, it may be necessary to change the method of approximating the equations as a function of the nature of the local flow. To make matters even worse, the regions cannot be determined prior to solving the equations.

1.9 Plan of This Book

This book contains thirteen chapters. We now give a brief summary of the remaining twelve chapters.

In Chap. 2 an introduction to numerical solution methods is given. The advantages and disadvantages of numerical methods are discussed and the possibilities and limitations of the computational approach are outlined. This is followed by a description of the components of a numerical solution method and their properties. Finally, a brief description of basic computational methods (finite difference, finite volume and finite element) is given.

In Chap. 3 finite difference (FD) methods are described. Here we present methods of approximating first, second, and mixed derivatives, using Taylor series expansion and polynomial fitting. Derivation of higher-order methods, and treatment of nonlinear terms and boundaries is discussed. Attention is also paid to the effects of grid non-uniformity on truncation error and to the estimation of discretization errors. Spectral methods are also briefly described here.

In Chap. 4 the finite volume (FV) method is described, including the approximation of surface and volume integrals and the use of interpolation to obtain variable values and derivatives at locations other than cell centers. Development of higher-order schemes and simplification of the resulting algebraic equations using the deferred-correction approach is also described. Special attention is paid to the analysis of discretization errors caused by interpolation and integral approximations. Finally, implementation of the various boundary conditions is discussed.

Applications of basic FD and FV methods are described and their use is demonstrated in Chaps. 3 and 4 for structured Cartesian grids. This restriction allows us to separate the issues connected with geometric complexity from the concepts behind discretization techniques. The treatment of complex geometries is introduced later, in Chap. 9.

In Chap. 5 we describe methods of solving the algebraic equation systems resulting from discretization. Direct methods are briefly described, but the major part of the chapter is devoted to iterative solution techniques. Incomplete lower-upper decomposition, conjugate gradient and multigrid methods are given special attention. Approaches to solving coupled and non-linear systems are also described, including the issues of under-relaxation and convergence criteria.

Chapter 6 is devoted to methods of time integration. First, the methods of solving ordinary differential equations are described, including basic methods, predictor-corrector and multipoint methods, and Runge–Kutta methods. The application of these methods to the unsteady transport equations is described next, including analysis of stability and accuracy.

The complexity of the Navier–Stokes equations and special features for incompressible flows are considered in Chaps. 7 and 8. The staggered and colocated variable arrangements, the pressure equation, and the pressure-velocity coupling for incompressible flows using the fractional-step and SIMPLE algorithms are described in detail. Other approaches (PISO algorithm, streamfunction-vorticity, artificial compressibility) are also described. The solution method for staggered and colocated Cartesian grids is described in sufficient detail to enable writing of a computer code; such codes are available on the Internet. Finally, some illustrative examples of steady and unsteady laminar flows computed using provided codes based on the fractional-step and SIMPLE algorithm are presented and discussed, including evaluation of iteration and discretization errors.

Chapter 9 is devoted to the treatment of complex geometries. The choices of grid type, grid generation approaches in complex geometries, grid properties, velocity components and variable arrangements are discussed. FD and FV methods are revisited, and the features special to complex geometries (like non-orthogonal, block-structured and unstructured grids, non-conformal grid interfaces, control volumes of arbitrary shape, overlapping grids, etc.) are discussed. Special attention is paid to the pressure-correction equation and boundary conditions. Again, some illustrative examples of steady and unsteady, two- and three-dimensional laminar flows computed using provided codes based on the fractional-step and SIMPLE algorithms are presented and discussed; evaluation of discretization errors and comparison of results obtained using different grid types (trimmed Cartesian and arbitrary polyhedral) are also included.

Chapter 10 deals with computation of turbulent flows. We discuss the nature of turbulence and three methods for its simulation: direct and large-eddy simulation and methods based on Reynolds-averaged Navier–Stokes equations. Some widely used models in the latter two approaches are described, including details related to boundary conditions. Examples of application of these approaches, including comparison of their performance, are presented.

In Chap. 11 compressible flows are considered. Methods designed for compressible flows are briefly discussed. The extension of pressure-correction approaches based on the fractional-step method and the SIMPLE algorithm for incompressible flows to compressible flows is described. Methods for dealing with shocks (e.g., grid adaptation, total-variation-diminishing and essentially-non-oscillating schemes) are also discussed. Boundary conditions for various types of compressible flows (subsonic, transonic and supersonic) are described. Finally, application examples are presented and discussed.

Chapter 12 is devoted to accuracy and efficiency improvement. The increased efficiency provided by multigrid algorithms is described first, followed by examples. Adaptive grid methods and local grid refinement are the subject of another section. Finally, parallelization is discussed. Special attention is paid to parallel processing for implicit methods based on domain decomposition in space and time, and to analysis of the efficiency of parallel processing. Example calculations are used to demonstrate these points.

Finally, in Chap. 13 some special issues are considered. These include conjugate heat transfer, flows with free surfaces, the treatment of moving boundaries which require moving grids, simulation of cavitation and fluid-structure interaction. Special effects in flows with heat and mass transfer, two phases and chemical reactions are briefly discussed.

We end this introductory chapter with a short note. Computational fluid dynamics (CFD) may be regarded as a sub-field of either fluid dynamics or numerical analysis. Competence in CFD requires that the practitioner has a fairly solid background in both areas. Poor results have been produced by individuals who are experts in one area but regarded the other as unnecessary. We hope the reader will take note of this and acquire the necessary background.

Chapter 2
Introduction to Numerical Methods

2.1 Approaches to Solving Problems in Fluid Dynamics

As the first chapter stated, the equations of fluid mechanics—which have been known for over a century—are solvable for only a limited number of flows. The known solutions are extremely useful in helping to understand fluid flow but rarely can they be used directly in engineering analysis or design. The engineer has traditionally been forced to use other approaches.

In the most common approach, simplifications of the equations are used. These are usually based on a combination of approximations and dimensional analysis; empirical input is almost always required. For example, dimensional analysis shows that the drag force on an object can be represented by:

$$F_D = C_D S \rho v^2 , \tag{2.1}$$

where S is the frontal area presented to the flow by the body, v is the flow velocity and ρ is the density of the fluid; the parameter C_D is called the drag coefficient. It is a function of the other non-dimensional parameters of the problem and is nearly always obtained by correlating experimental data. This approach is very successful when the system can be described by one or two parameters so applications to complex geometries (which can only be described by many parameters) are ruled out.

A related approach is arrived at by noting that for many flows non-dimensionalization of the Navier–Stokes equations leaves the Reynolds number as the only independent parameter. If the body shape is held fixed, one can get the desired results from an experiment on a scaled model with that shape. The desired Reynolds number is achieved by careful selection of the fluid and the flow parameters or by extrapolation in Reynolds number; the latter can be dangerous. These approaches are very valuable and are the primary methods of practical engineering design even today.

The problem is that many flows require several dimensionless parameters for their specification and it may be impossible to set up an experiment which correctly scales the actual flow. Examples are flows around aircraft or ships. In order to achieve the

© Springer Nature Switzerland AG 2020
J. H. Ferziger et al., *Computational Methods for Fluid Dynamics*,
https://doi.org/10.1007/978-3-319-99693-6_2

same Reynolds number with smaller models, fluid velocity has to be increased. For aircraft, this may give too high a Mach number if the same fluid (air) is used; one tries to find a fluid which allows matching of both parameters. For ships, the issue is to match both the Reynolds and Froude numbers, which is nearly impossible.

In other cases, experiments are very difficult if not impossible. For example, the measuring equipment might disturb the flow or the flow may be inaccessible (e.g., flow of a liquid silicon in a crystal growth apparatus). Some quantities are simply not measurable with present techniques or can be measured only with an insufficient accuracy.

Experiments are an efficient means of measuring global parameters, like the drag, lift, pressure drop, or heat transfer coefficient. In many cases, details are important; it may be essential to know whether flow separation occurs or whether the wall temperature exceeds some limit. As technological improvement and competition require more careful optimization of designs or, when new high-technology applications demand prediction of flows for which the database is insufficient, experimental development may be too costly and/or time consuming. Finding a reasonable alternative is essential.

An alternative—or at least a complementary method—came with the birth of electronic computers. Although many of the key ideas for numerical solution methods for partial differential equations were established more than a century ago, they were of little use before computers appeared. The performance-to-cost ratio of computers has increased at a spectacular rate since the 1950s and shows no sign of slowing down. While the first computers built in the 1950s performed only a few hundred operations per second, as of June 2017, the 1st ranked computer on the TOP500 list (https://www.top500.org) had a measured performance peak of 93 Pflops (petaflops $= 10^{15}$ floating point operations per second); it has over 10^6 processing cores and the memory size is 1.3 PB (petabytes $= 10^{15}$ bytes). Even laptop computers have multicore chips as well as multiple processors, and the GPUs can also be used to accomplish massive parallel computations (Thibault and Senocak 2009; Senocak and Jacobsen 2010). The ability to store data has also increased dramatically: hard discs with ten gigabyte ($10\,GB = 10^{10}$ bytes or characters) capacity could be found only on supercomputers twenty years ago—now laptop computers have 1 TB hard disks. Back-up disks the size of smart phones hold $500\,GB$ or more. A machine that cost millions of dollars, filled a large room, and required a permanent maintenance and operating staff in 1980 is now available on a laptop! It is difficult to predict what will happen in the future, but further significant increases in both computing speed and memory of affordable computers are certain.

It requires little imagination to understand that computers make the study of fluid flow easier and more effective. Once this power of computers had been recognized, interest in numerical techniques increased dramatically. This field is known as *computational fluid dynamics* (CFD). Contained within it are many sub-specialties. CFD has evolved over the decades from a specialized research area into a powerful tool, included in university curricula, used in virtually every industry and employed by researchers to study the very nature of fluid flows.

We shall discuss in detail only a small subset of methods for solving the equations describing fluid flow and related phenomena; others will be briefly mentioned and references for other literature will be given where appropriate.

2.2 What is CFD?

As we have seen in Chap. 1, flows and related phenomena can be described by partial differential (or integro-differential) equations, which cannot be solved analytically except in special cases. To obtain an approximate solution numerically, we have to use a *discretization method* which approximates the differential equations by a system of algebraic equations, which can then be solved on a computer. The approximations are applied to small domains in space and/or time so the numerical solution provides results at *discrete locations* in space and time. Much as the accuracy of experimental data depends on the quality of the tools used, the accuracy of numerical solutions is dependent on the quality of the discretizations used.

Contained within the broad field of computational fluid dynamics are activities that cover the range from the automation of well-established engineering design methods to the use of detailed solutions of the Navier–Stokes equations as substitutes for experimental research into the nature of complex flows. At one end, one can purchase design packages for pipe systems that solve problems in a few seconds or minutes on personal computers or workstations. On the other, there are codes that may require hundreds of hours on the largest super-computers. The range is as large as the field of fluid mechanics itself, making it impossible to cover all of CFD in a single work. Also, the field is evolving so rapidly that we run the risk of becoming out of date in a short time.

We shall not deal with automated simple methods in this book. The basis for them is covered in elementary textbooks and undergraduate courses and the available program packages are relatively easy to understand and to use.

We shall be concerned with methods designed to solve the equations of fluid motion in two or three dimensions. These are the methods used in non-standard applications, by which we mean applications for which solutions (or, at least, good approximations) cannot be found in textbooks or handbooks. While these methods have been used in high-technology engineering (for example, aeronautics and astronautics) from the very beginning, they are being used more frequently in fields of engineering where the geometry is complicated or some important feature (such as the prediction of the concentration of a pollutant) cannot be dealt with by standard methods.

CFD has found its way into mechanical, process, chemical, civil, and environmental engineering, and is a major component of all aspects of atmospheric science—from weather forecasting to climate change. Optimization in these areas can produce large savings in equipment and energy costs, yield improved prediction of floods and storms, and lead to reduction of environmental pollution.

2.3 Possibilities and Limitations of Numerical Methods

We have already noted some problems associated with experimental work. Some of these problems are easily dealt with in CFD. For example, if we want to simulate the flow around a moving car in a wind tunnel, we need to fix the car model and blow air at it—but the floor has to move at the air speed, which is difficult to do. It is not difficult to do in a numerical simulation. Other types of boundary conditions are easily prescribed in computations; for example, temperature or opaqueness of the fluid pose no problem. If we solve the unsteady three-dimensional Navier–Stokes equations accurately (as in direct simulation of turbulence), we obtain a complete data set from which any quantity of physical significance can be derived.

This sounds to good to be true. Indeed, these advantages of CFD are conditional on being able to solve the Navier–Stokes equations accurately, which is extremely difficult for most flows of engineering interest. We shall see in Chap. 10 why obtaining accurate numerical solutions of the Navier–Stokes equations for high Reynolds number flows is so difficult.

If we are unable to obtain *accurate* solutions for all flows, we have to determine what we can produce and learn to analyze and judge the results. First of all, we have to bear in mind that numerical results are always *approximate*. There are reasons for differences between computed results and 'reality', i.e., errors arise from each part of the process used to produce numerical solutions:

- The differential equations may contain approximations or idealizations, as discussed in Sect. 1.7;
- Approximations are made in the discretization process;
- In solving the discretized equations, iterative methods are used. Unless they are run for a very long time, the exact solution of the discretized equations is not produced.

When the governing equations are known accurately (e.g., the Navier–Stokes equations for incompressible Newtonian fluids), solutions of any desired accuracy can be achieved in principle. However, for many phenomena (e.g., turbulence, combustion, and multiphase flow) the exact equations are either not available or numerical solution is not feasible. This makes introduction of models a necessity. Even if we solve the equations exactly, the solution would not be a correct representation of reality. In order to *validate* the models, we have to rely on experimental data. Even when the exact treatment is possible, models are often needed to reduce the cost.

Discretization errors can be reduced by using more accurate interpolation or approximations or by applying the approximations to smaller regions but this usually increases the time and cost of obtaining the solution. Compromise is usually needed. We shall present some schemes in detail but shall also point out ways of creating more accurate approximations.

Compromises are also needed in solving the discretized equations. Direct solvers, which obtain accurate solutions, are seldom used, because they are too costly. Iterative methods are more common but the errors due to stopping the iteration process too soon need to be taken into account.

Errors and their estimation will be emphasized throughout this book. We shall present error estimates for many examples; the need to analyze and estimate numerical errors can not be overemphasized.

Visualization of numerical solutions using vector, contour or other kinds of plots or movies (videos) of unsteady flows is important for the interpretation of results. It is far and away the most effective means of interpreting the huge amount of data produced by a calculation. However, there is the danger that an erroneous solution may look good but may not correspond to the actual boundary conditions, fluid properties, etc.! The authors have encountered incorrect numerically-produced flow features that could be and have been interpreted as physical phenomena. Industrial users of commercial CFD codes should especially be careful, as the optimism of salesmen is legendary. Wonderful color pictures make a great impression, but are of no value if they are not quantitatively correct. Results must be examined very critically before they are believed.

2.4 Components of a Numerical Solution Method

Considering that this book is meant not only for users of commercial codes but also for young researchers developing new codes, we shall present the important ingredients of a numerical solution method here. More details will be presented in the following chapters.

2.4.1 Mathematical Model

The starting point of any numerical method is the mathematical model, i.e., the set of partial differential or integro-differential equations and boundary conditions. Some sets of equations used for flow prediction were presented in Chap. 1. One chooses an appropriate model for the target application (incompressible, inviscid, turbulent; two- or three-dimensional, etc.). As already mentioned, this model may include simplifications of the exact conservation laws. A solution method is usually designed for a particular set of equations. Trying to produce a general purpose solution method, i.e., one which is applicable to all flows, is impractical, if not impossible and, as with most general purpose tools, they are usually not optimum for any one application.

2.4.2 Discretization Method

After selecting the mathematical model, one has to choose a suitable discretization method, i.e., a method of approximating the differential equations by a system of algebraic equations for the variables at some set of discrete locations in space and time.

There are many approaches, but the most important of which are: finite difference (FD), finite volume (FV) and finite element (FE) methods. Important features of these three kinds of discretization methods are described at the end of this chapter. Other methods, like spectral schemes, boundary element methods, and lattice-Boltzmann methods are used in CFD but their use is limited to special classes of problems.

Each type of method yields the same solution if the grid is very fine. However, some methods are more suitable to some classes of problems than others. The preference is often determined by the attitude of the developer. We shall discuss the pros and cons of the various methods later.

2.4.3 Coordinate and Basis Vector Systems

It was mentioned in Chap. 1 that the conservation equations can be written in many different forms, depending on the coordinate system and the basis vectors used. For example one can select Cartesian, cylindrical, spherical, curvilinear orthogonal or non-orthogonal coordinate systems, which may be fixed or moving. The choice depends on the target flow, and may influence the discretization method and grid type to be used.

One also has to select the basis in which vectors and tensors will be defined (fixed or variable, covariant or contravariant, etc.). Depending on this choice, the velocity vector and stress tensor can be expressed in terms of, e.g., Cartesian, covariant or contravariant, physical or non-physical coordinate-oriented components. In this book we shall use Cartesian components exclusively for reasons explained in Chap. 9.

2.4.4 Numerical Grid

The discrete locations at which the variables are to be calculated are defined by the numerical grid which is essentially a discrete representation of the geometric domain on which the problem is to be solved. It divides the solution domain into a finite number of subdomains (elements, control volumes, etc.). Some of the options available are the following:

- *Structured (regular) grid*—Regular or structured grids consist of families of grid lines with the property that members of a single family do not cross each other and cross each member of the other families only once. This allows the lines of a given set to be numbered consecutively. The position of any grid point (or control volume) within the domain is uniquely identified by a set of two (in 2D) or three (in 3D) indexes, e.g., (i, j, k).
  This is the simplest grid structure, because it is logically equivalent to a Cartesian grid. Each point has four nearest neighbors in two dimensions and six in three dimensions; one of the indexes of each neighbor of point P (indexes i, j, k) differs

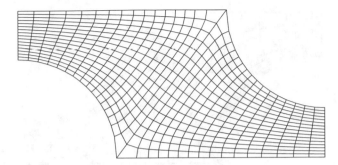

Fig. 2.1 Example of a 2D, structured, non-orthogonal grid, designed for calculation of flow in a symmetric segment of a staggered tube bank

by ±1 from the corresponding index of P. An example of a structured 2D grid is shown in Fig. 2.1. This neighbor connectivity simplifies programming and the matrix of the algebraic equation system has a regular structure, which can be exploited in developing a solution technique. Indeed, there is a large number of efficient solvers applicable only to structured grids (see Chap. 5). The disadvantage of structured grids is that they can be used only for geometrically simple solution domains. Another disadvantage is that it may be difficult to control the distribution of the grid points: concentration of points in one region for reasons of accuracy produces unnecessarily small spacing in other parts of the solution domain and a waste of resources. This problem is exaggerated in 3D geometries. The long thin cells may also affect the convergence adversely.

Structured grids may be of H-, O-, or C-type; the names are derived from the topology of solution domain boundaries. Figure 2.1 shows an H-type grid which, when mapped onto a rectangle, has distinct east, west, north, and south boundaries. In an O-grid, two opposite boundaries connect to each other (e.g., east to west, or south to north). An example is a grid around a circular cylinder (see Fig. 9.22): if the cylinder wall is the south boundary and the outer edge is the north boundary, then the west and east boundaries are merged to create endless grid lines around cylinder (they may or may not be circular). The counting for index i starts at an arbitrary radial line which represents the interface between the west and the east boundaries. The grid may or may not be conformal at this interface; how such interfaces are treated will be described in Chap. 9.

In a C-grid, one boundary falls partially onto itself. An example is a grid around an airfoil: one set of grid lines wraps around the foil, while the other set is (nearly) orthogonal to it. For example, if the airfoil wall is the south boundary, than it is extended from trailing edge up to some distance behind the foil; the outlet is the west boundary below and the east boundary above the double south line extending from trailing edge, while the north boundary covers the bottom, left and top sides of the solution domain. As for the O-grid, the interface between two parts of the south boundary which are in contact may be either conformal or non-conformal.

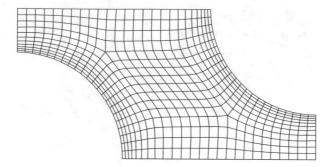

Fig. 2.2 Example of a 2D block-structured grid with conformal interfaces

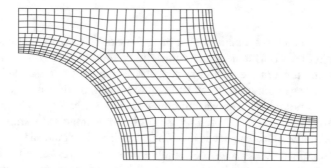

Fig. 2.3 Example of a 2D block-structured grid with non-conformal interfaces

- *Block-structured grid*—In a block-structured grid, there is a two (or more) level subdivision of solution domain. On the coarse level, there are blocks which are relatively large segments of the domain; their structure may be irregular and they may or may not overlap. On the fine level (within each block) a structured grid is defined. Special treatment is necessary at block interfaces. Some methods of this kind are described in Chap. 9.

  In Fig. 2.2 a block-structured grid with conformal interfaces is shown; it is designed for the same flow and geometry used in Fig. 2.1. Block interfaces were initially regular surfaces (cylinder or plane segments), but with some smoothing operations designed to improve grid quality, they became curved surfaces. With a good design of block structure, one can create a grid with good quality—but it takes time. Indeed, it is not unusual that the generation of a good quality grid of this type for moderately complicated geometries takes a week or two of engineer's time.

  In Fig. 2.3 a block-structured grid with non-conformal interfaces is shown; it is similar to the grid from Fig. 2.2, except that the grid is finer in blocks around tubes. This kind of grid is more flexible than the previous ones, as it allows use of finer grids in regions, where greater resolution is required (e.g., around tubes where one wants to capture heat transfer accurately). The non-conformal interface can be treated in a fully conservative manner or by using "hanging nodes", as will be

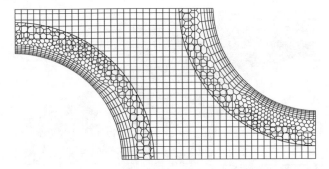

Fig. 2.4   A composite 2D grid, designed for the same flow and geometry used in Fig. 2.1

discussed in Chap. 9. Here grid smoothing can only be applied within each block; the interfaces usually retain their original shape, as can be seen by comparing grids in Figs. 2.2 and 2.3. The programming is more difficult than for grid types described above. Solvers for structured grids can be applied block-wise, and more complex flow domains can be treated with these grids. Local refinement is possible block-wise (i.e., the grid may be refined in some blocks).

Block-structured grids with overlapping blocks are sometimes called *composite* or *Chimera* grids. One such grid is shown in Fig. 2.4. In the overlap region, boundary conditions for one block are obtained by interpolating the solution from the other (overlapped) block. The disadvantage of these grids is that conservation is not easily enforced at block boundaries. The advantages of this approach are that complex domains are dealt with more easily and it can be used to follow moving bodies: one block is attached to the body and moves with it, while a stagnant grid covers the surroundings.

Methods that use overlapping structured grids have been developed by many authors in late 1980s and early 1990s (Tu and Fuchs 1992; Perng and Street 1991; Hinatsu and Ferziger 1991; Zang and Street 1995; Hubbard and Chen 1994, 1995). Several semi-commercial codes use this approach, e.g., SHIPFLOW, CFDShip-Iowa, and OVERFLOW (NASA). A strong interest in this approach has emerged again around the turn of the century, particularly for handling flows around moving bodies. One such method has been developed by Hadžić (2005) and a version applicable to arbitrary polyhedral grids is available in the commercial software STAR-CCM+. It will be described in more detail in Chap. 9.

- *Unstructured grids*—For very complex geometries, the most flexible type of grid is one which can fit an arbitrary solution domain boundary. In principle, such grids could be used with any discretization scheme, but they are best adapted to the finite volume and finite element approaches. The elements or control volumes may have any shape; nor is there a restriction on the number of neighbor elements or nodes. In practice, grids made of triangles, quadrilaterals or arbitrary polygons in 2D, and tetrahedra, hexahedra or arbitrary polyhedra in 3D are most often used. Three examples of typical unstructured grids with prism layers along walls are shown in Fig. 2.5.

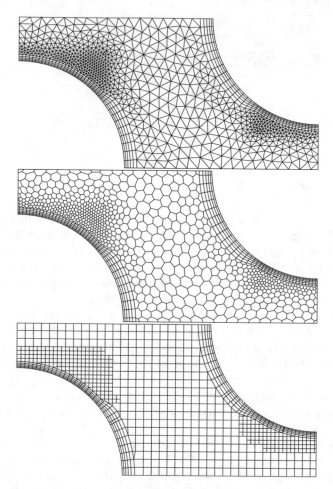

Fig. 2.5 Three examples of unstructured grids: tetrahedral (upper), polyhedral (middle) and trimmed hexahedral (lower) with prism layers along walls and local grid refinement

Recently, grids made of arbitrary polyhedral control volumes have become popular, because they have better properties than terahedral grids and are easier to generate automatically than unstructured grids consisting of hexahedra. Such grids can be generated automatically by commercial grid-generation tools. If desired, the grid can be optimized for orthogonality, the aspect ratio is easily controlled, and the grid may be easily locally refined. The advantage of flexibility is offset by the disadvantage of the irregularity of the data structure. Node locations and neighbor connections need be specified explicitly. The matrix of the algebraic equation system no longer has regular, diagonal structure; the band width needs to be reduced by reordering of the points. The solvers for the algebraic equation systems are usually slower than those for structured grids due to the use of indirect addressing. This, for example, makes the use of graphics processing units (GPUs) for computation less efficient than is the case with structured grids; the same is true for the case of vector processors.

Unstructured grids are usually used with finite-element methods and, increasingly, with finite-volume methods. Computer codes for unstructured grids are more flexible. They need not be changed when the grid is locally refined, or when elements or control volumes of different shapes are used.

With the availability of automatic mesh generation methods for complex geometries, unstructured grids have become the rule rather than exception in industry. In order to retain the benefits of structured grids in near-wall region, most modern grid generation tools can also create prism layers along boundaries. The grid is then unstructured along the wall, but is layered (structured) and almost orthogonal in wall-normal direction, thus allowing a more accurate approximation of the boundary layer. The finite-volume method presented in this book is applicable to unstructured grids; more details will be given in Chap. 9.

Methods of grid generation will not be covered in detail in this book. Grid properties and some basic grid generation methods are discussed briefly in Chap. 9. There is a vast literature devoted to generation of block-structured and unstructured grids and interested reader is referred to books by Thompson et al. (1985) and Arcilla et al. (1991). However, many methods used in commercial grid generation and optimization tools are not described in public literature; grid generation is to some extent an art and many steps used to handle special situations cannot be described mathematically. The issues of grid quality—which is especially important in the case of unstructured grids—will be handled in Chap. 12.

2.4.5 Finite Approximations

Following the choice of grid type, one has to select the approximations to be used in the discretization process. In a finite difference method, approximations for the derivatives at the grid points have to be selected. In a finite volume method, one has to select the methods of approximating surface and volume integrals. In a finite element method, one has to choose the shape functions (elements) and weighting functions.

There are many possibilities to choose from; some of those most often used are presented in this book, some are simply mentioned and many more can be created. The choice influences the accuracy of the approximation. It also affects the difficulty of developing the solution method, coding it, debugging it, and the speed of the code. More accurate approximations involve more nodes and give fuller coefficient matrices. The increased memory demand may require using coarser grids, partially offsetting the advantage of higher accuracy. A compromise between simplicity, ease of implementation, accuracy and computational efficiency has to be made. The second-order methods presented in this book were selected with this compromise in mind.

2.4.6   Solution Method

Discretization yields a large system of non-linear algebraic equations. The method of solution depends on the problem. For unsteady flows, methods based on those used for initial value problems for ordinary differential equations (marching in time) are used. At each time step an elliptic problem has to be solved. Steady flow problems are usually solved by pseudo-time marching or an equivalent iteration scheme. Because the equations are non-linear, an iteration scheme is used to solve them. These methods use successive linearization of the equations and the resulting linear systems are almost always solved by iterative techniques. The choice of solver depends on the grid type and the number of nodes involved in each algebraic equation. Some solvers will be presented in Chap. 5.

2.4.7   Convergence Criteria

Finally, one needs to set the convergence criteria for the iterative method. Usually, there are two levels of iterations: inner iterations, within which the linear equations are solved, and outer iterations, that deal with the non-linearity and coupling of the equations. Deciding when to stop the iterative process on each level is important, from both the accuracy and efficiency points of view. These issues are dealt with in Chaps. 5 and 12.

2.5   Properties of Numerical Solution Methods

The solution method should have certain properties. In most cases, it is not possible to analyze the complete solution method. One analyzes the components of the method; if the components do not possess the desired properties, neither will the complete method but the reverse is not necessarily true. The most important properties are summarized below.

2.5.1   Consistency

The discretization should become exact as the grid spacing tends to zero. The difference between the discretized equation and the exact one is called the *truncation error*. It is usually estimated by replacing all the nodal values in the discrete approximation by a Taylor series expansion about a single point. As a result one recovers the original differential equation plus a remainder, which represents the truncation error. For a method to be *consistent*, the truncation error must become zero when the mesh spacing $\Delta t \to 0$ and/or $\Delta x_i \to 0$. Truncation error is usually proportional to a power of the grid spacing Δx_i and/or the time step Δt. If the most important term is proportional to $(\Delta x)^n$ or $(\Delta t)^n$ we call the method an nth-order approximation; $n > 0$ is required for consistency. Ideally, all terms should be discretized with approximations of the same order of accuracy; however, some terms (e.g., convection terms in high Reynolds number flows or diffusion terms in low Reynolds number flows) may be dominant in a particular flow and it may be reasonable to treat them with more accuracy than the others.

Some discretization methods lead to truncation errors which are functions of the ratio of Δx_i to Δt or vice versa. In such a case the consistency requirement is only conditionally fulfilled: Δx_i and Δt must be reduced in a way that allows the appropriate ratio to go to zero. In the next two chapters we shall demonstrate consistency for several discretization schemes.

Even if the approximations are consistent, it does not necessarily mean that the solution of the discretized equation system will become the exact solution of the differential equation in the limit of small step size. For this to happen, the solution method has to be *stable*; this is defined below.

2.5.2 Stability

A numerical solution method is said to be stable if it does not magnify the errors that appear in the course of numerical solution process. For temporal problems, stability guarantees that the method produces a bounded solution whenever the solution of the exact equation is bounded. For iterative methods, a stable method is one that does not diverge. Stability can be difficult to investigate, especially when boundary conditions and non-linearities are present. For this reason, it is common to investigate the stability of a method for linear problems with constant coefficients without boundary conditions. Experience shows that the results obtained in this way can often be applied to more complex problems but there are notable exceptions.

The most widely used approach to studying stability of numerical schemes is the von Neumann's method. We shall describe it briefly for one scheme in Chap. 6. Most of the schemes to be described in this book have been analyzed for stability and we shall state the important result when describing each scheme. However, when solving complicated, non-linear and coupled equations with complicated boundary conditions, there are few stability results so we may have to rely on experience and intuition. Many solution schemes require that the time step be smaller than a certain limit or that under-relaxation be used. We shall discuss these issues and give guidelines for selecting time step size and values of under-relaxation parameters in Chaps. 6, 7 and 8.

2.5.3   Convergence

A numerical method is said to be convergent if the solution of the discretized equations tends to the exact solution of the differential equation as the grid spacing tends to zero. For linear initial value problems, the *Lax equivalence theorem* (Richtmyer and Morton 1967, or Street 1973) states that "given a properly posed linear initial value problem and a finite difference approximation to it that satisfies the consistency condition, stability is the necessary and sufficient condition for convergence". Obviously, a consistent scheme is useless unless the solution method converges.

For non-linear problems which are strongly influenced by boundary conditions, the stability and convergence of a method are difficult to demonstrate. Therefore convergence is usually checked using numerical experiments, i.e., repeating the calculation on a series of successively refined grids. If the method is stable and if all approximations used in the discretization process are consistent, we usually find that the solution does converge to a *grid-independent solution*. For sufficiently small grid sizes, the rate of convergence is governed by the order of principal truncation error component. This allows us to estimate the error in the solution. We shall describe this in detail in Chaps. 3 and 5.

2.5.4   Conservation

Because the equations to be solved are conservation laws, the numerical scheme should also—on both a local and a global basis—respect these laws. This means that, at steady state and in the absence of sources, the amount of a conserved quantity leaving a closed volume is equal to the amount entering that volume. If the strong conservation form of equations and a finite volume method are used, this is guaranteed for each individual control volume and for the solution domain as a whole. Other discretization methods can be made conservative if care is taken in the choice of approximations. The treatment of source or sink terms should be consistent so that the total source or sink in the domain is equal to the net flux of the conserved quantity through the boundaries.

This is an important property of the solution method, because it imposes a constraint on the solution error. If the conservation of mass, momentum and energy are ensured, the error can only improperly distribute these quantities over the solution domain. Non-conservative schemes can produce artificial sources and sinks, changing the balance both locally and globally. However, non-conservative schemes can be consistent and stable and therefore lead to correct solutions in the limit of very fine grids. The errors due to non-conservation are in most cases appreciable only on relatively coarse grids. The problem is that it is difficult to know on which grid are these errors small enough. Conservative schemes are therefore preferred.

2.5.5 *Boundedness*

Numerical solutions should lie within proper bounds. Physically non-negative quantities (like density, kinetic energy of turbulence) must always be positive; other quantities, such as concentration, must lie between 0% and 100%. In the absence of sources, some equations (e.g., the heat equation for the temperature when no heat sources are present) require that the minimum and maximum values of the variable be found on the boundaries of the domain. These conditions should be inherited by the numerical approximation.

Boundedness is difficult to guarantee. We shall show later on that only some first-order schemes guarantee this property. All higher-order schemes can produce unbounded solutions; fortunately, this usually happens only on grids that are too coarse, so a solution with undershoots and overshoots is usually an indication that the errors in the solution are large and the grid needs some refinement (at least locally). The problem is that schemes prone to producing unbounded solutions may have stability and convergence problems. These methods should be avoided, if possible.

2.5.6 *Realizability*

Models of phenomena which are too complex to treat directly (for example, turbulence, combustion, or multiphase flow) should be designed to guarantee physically realistic solutions. This is not a numerical issue *per se* but models that are not realizable may result in unphysical solutions or cause numerical methods to diverge. We shall not deal with these issues in this book, but if one wants to implement a model in a CFD code, one has to be careful about this property.

2.5.7 *Accuracy*

Numerical solutions of fluid flow and heat transfer problems are only *approximate solutions*. In addition to the errors that might be introduced in the course of the development of the solution algorithm, in programming or setting up the boundary conditions, numerical solutions always include three kinds of systematic errors:

- *Modeling errors*, which are defined as the difference between the actual flow and the exact solution of the mathematical model;
- *Discretization errors*, defined as the difference between the exact solution of the conservation equations and the exact solution of the algebraic system of equations obtained by discretizing these equations, and
- *Iteration errors*, defined as the difference between the iterative and exact solutions of the algebraic equations systems.

Iteration errors are often called *convergence errors* (which was the case in the earlier editions of this book). However, the term *convergence* is used not only in conjunction with error reduction in iterative solution methods, but is also (quite appropriately) often associated with the convergence of numerical solutions towards a grid-independent solution, in which case it is closely linked to discretization error. To avoid confusion, we shall adhere to the above definition of errors and, when discussing issues of convergence, always indicate which type of convergence we are talking about.

It is important to be aware of the existence of these errors, and even more to try to distinguish one from another. Various errors may cancel each other, so that sometimes a solution obtained on a coarse grid may agree better with the experiment than a solution on a finer grid—which, by definition, should be more accurate.

Modeling errors depend on the assumptions made in deriving the transport equations for the variables. They may be considered negligible when laminar flows are investigated, because the Navier–Stokes equations represent a sufficiently accurate model of the flow. However, for turbulent flows, two-phase flows, combustion etc., the modeling errors may be very large—the exact solution of the model equations may be *qualitatively* wrong. Modeling errors are also introduced by simplifying the geometry of the solution domain, by simplifying boundary conditions etc. These errors are not known *a priori*; they can only be evaluated by comparing solutions in which the discretization and iteration errors are negligible with accurate experimental data or with data obtained by more accurate models (e.g., data from direct simulation of turbulence, etc.). It is essential to control and estimate the iteration and discretization errors before the models of physical phenomena (like turbulence models) can be judged.

We mentioned above that discretization approximations introduce errors which decrease as the grid is refined, and that the order of the approximation is a measure of accuracy. However, on a given grid, methods of the same order may produce solution errors which differ by as much as an order of magnitude. This is because the order only tells us the *rate* at which the error decreases as the mesh spacing is reduced—it gives no information about the error on a single grid. We shall show how discretization errors can be estimated in the next chapter.

Errors due to iterative solution and round-off are easier to control; we shall see how this can be done in Chap. 5, where iterative solution methods are introduced.

There are many solution schemes and the developer of a CFD code may have a difficult time deciding which one to adopt. The ultimate goal is to obtain desired accuracy with least effort, or the maximum accuracy with the available resources. Each time we describe a particular scheme we shall point out its advantages or disadvantages with respect to these criteria.

2.6 Discretization Approaches

2.6.1 Finite Difference Method

This is the oldest method for numerical solution of PDE's, believed to have been introduced by Euler in the 18th century. It is also the easiest method to use for simple geometries.

The starting point is the conservation equation in differential form. The solution domain is covered by a grid. At each grid point, the differential equation is approximated by replacing the partial derivatives by approximations in terms of the nodal values of the functions. The result is one algebraic equation per grid node, in which the variable value at that and a certain number of neighbor nodes appear as unknowns.

In principle, the FD method can be applied to any grid type. However, in all applications of the FD method known to the authors, it has been applied to structured grids. The grid lines serve as local coordinate lines.

Taylor series expansion or polynomial fitting is used to obtain approximations to the first and second derivatives of the variables with respect to the coordinates. When necessary, these methods are also used to obtain variable values at locations other than grid nodes (interpolation). The most widely used methods of approximating derivatives by finite differences are described in Chap. 3.

On structured grids, the FD method is very simple and effective. It is especially easy to obtain higher-order schemes on regular grids; some will be mentioned in Chap. 3. The disadvantage of FD methods is that the conservation is not enforced unless special care is taken. Also, the restriction to simple geometries is a significant disadvantage in complex flows.

2.6.2 Finite Volume Method

The FV method uses the integral form of the conservation equations as its starting point. The solution domain is subdivided into a finite number of contiguous control volumes (CVs), and the conservation equations are applied to each CV. At the centroid of each CV lies a computational node at which the variable values are to be calculated. Interpolation is used to express variable values at the CV surface in terms of the nodal (CV-center) values. Surface and volume integrals are approximated using suitable quadrature formulas. As a result, one obtains an algebraic equation for each CV, in which a number of neighbor nodal values appear.

The FV method can accommodate any type of grid, so it is suitable for complex geometries. The grid defines only the control volume boundaries and need not be related to a coordinate system. The method is conservative by construction, so long as surface integrals (which represent convection and diffusion fluxes) are the same for the CVs sharing the boundary.

The FV approach is perhaps the simplest to understand and to program. All terms that need be approximated have physical meaning which is why it is popular with engineers.

The disadvantage of FV methods compared to FD-schemes is that methods of order higher than second are more difficult to develop in 3D. This is due to the fact that the FV-approach requires three levels of approximation: interpolation, differentiation, and integration. We shall give a detailed description of the FV method in Chap. 4; it is the most used method in this book.

2.6.3   Finite Element Method

The FE-method is similar to the FV-method in many ways. The domain is broken into a set of discrete volumes or finite elements that are generally unstructured; in 2D, they are usually triangles or quadrilaterals, while in 3D tetrahedra or hexahedra are most often used. The distinguishing feature of FE-methods is that the equations are multiplied by a *weight function* before they are integrated over the entire domain. In the simplest FE-methods, the solution is approximated by a linear shape function within each element in a way that guarantees continuity of the solution across element boundaries. Such a function can be constructed from its values at the corners of the elements. The weight function is usually of the same form.

This approximation is then substituted into the weighted integral of the conservation law and the equations to be solved are derived by requiring the derivative of the integral with respect to each nodal value to be zero; this corresponds to selecting the best solution within the set of allowed functions (the one with minimum residual). The result is a set of non-linear algebraic equations.

An important advantage of finite element methods is the ability to deal with arbitrary geometries; there is an extensive literature devoted to the construction of grids for finite element methods. The grids are easily refined; each element is simply subdivided. Finite element methods are relatively easy to analyze mathematically and can be shown to have optimality properties for certain types of equations. The principal drawback, which is shared by any method that uses unstructured grids, is that the matrices of the linearized equations are not as well structured as those for regular grids making it more difficult to find efficient solution methods. For more details on finite element methods and their application to the Navier–Stokes equations, see books by Oden (2006), Zienkiewicz et al. (2005), Donea and Huerta (2003), Glowinski and Pironneau (1992) or Fletcher (1991).

A hybrid method called *control-volume-based finite element method* (CVFEM) should also be mentioned. In it, shape functions are used to describe the variation of the variables over an element. Control volumes are formed around each node by joining the centroids of the elements. The conservation equations in integral form are applied to these CVs in the same way as in the finite volume method. The fluxes through CV boundaries and the source terms are calculated element-wise. We shall give a short description of this approach in Chap. 9.

Chapter 3
Finite Difference Methods

3.1 Introduction

As was mentioned in Chap. 1, all conservation equations have similar structure and may be regarded as special cases of a generic transport equation, Eq. (1.26), (1.27) or (1.28). For this reason, we shall treat only a single, generic conservation equation in this and the following chapters. It will be used to demonstrate discretization methods for the terms which are common to all conservation equations (convection, diffusion, and sources). The special features of the Navier–Stokes equations and techniques for solving coupled non-linear problems will be introduced later. Also, for the time being, the unsteady term will be dropped so we consider only time-independent problems.

For simplicity, we shall use only Cartesian grids at this point. The equation we shall deal with is:

$$\frac{\partial(\rho u_j \phi)}{\partial x_j} = \frac{\partial}{\partial x_j}\left(\Gamma\frac{\partial \phi}{\partial x_j}\right) + q_\phi \, . \tag{3.1}$$

We shall assume that ρ, u_j, Γ and q_ϕ are known. This may not be the case because the velocity may not have been computed yet and the properties of the fluid may depend on the temperature and, if turbulence models are used, on the velocity field as well. As we shall see, the iterative schemes used to solve these equations treat ϕ as the only unknown; all other variables are fixed at their values determined on the previous iteration so regarding these as known is a reasonable approach.

The special features of non-orthogonal and unstructured grids will be discussed in Chap. 9. Furthermore, of the many possible discretization techniques, only a selected few which illustrate the main ideas will be described; others may be found in the literature cited.

© Springer Nature Switzerland AG 2020
J. H. Ferziger et al., *Computational Methods for Fluid Dynamics*,
https://doi.org/10.1007/978-3-319-99693-6_3

3.2 Basic Concept

The first step in obtaining a numerical solution is to discretize the geometric domain—
i.e., a numerical grid must be defined. In finite difference (FD) discretization methods
the grid is usually locally structured, i.e., each grid node may be considered the
origin of a local coordinate system, whose axes coincide with grid lines. This also
implies that two grid lines belonging to the same family, say ξ_1, do not intersect,
and that any pair of grid lines belonging to different families, say $\xi_1 = \text{const.}$ and
$\xi_2 = \text{const.}$, intersect only once. In three dimensions, three grid lines intersect at each
node; none of these lines intersect each other at any other point. Figure 3.1 shows
examples of one-dimensional (1D) and two-dimensional (2D) Cartesian grids used
in FD methods.

Each node is uniquely identified by a set of indexes, which are the indexes of the
grid lines that intersect at it, (i, j) in 2D and (i, j, k) in 3D. The neighbor nodes are
defined by increasing or reducing one of the indexes by unity.

The generic scalar conservation equation in differential form, (3.1), serves as the
starting point for FD methods. As it is linear in ϕ, it will be approximated by a
system of linear algebraic equations, in which the variable values at the grid nodes
are the unknowns. The solution of this system approximates the solution to the partial
differential equation (PDE).

Each node thus has one unknown variable value associated with it and must
provide one algebraic equation. The latter is a relation between the variable value
at that node and those at some of the neighboring nodes. It is obtained by replacing
each term of the PDE at the particular node by a finite-difference approximation.
Of course, the numbers of equations and unknowns must be equal. At boundary

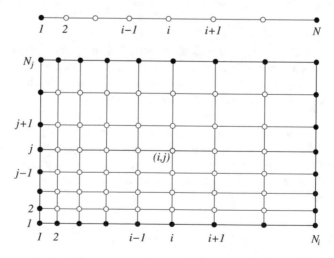

Fig. 3.1 An example of a 1D (above) and 2D (below) Cartesian grid for FD methods (full symbols
denote boundary nodes and open symbols denote internal computational nodes)

Fig. 3.2 On the definition of a derivative and its approximations

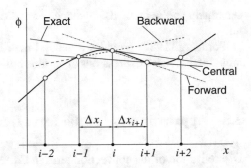

nodes where variable values are given (Dirichlet conditions), no equation is needed. When the boundary conditions involve derivatives (as in Neumann conditions), the boundary condition must be discretized to contribute an equation to the set that must be solved.

The idea behind finite-difference approximations is borrowed directly from the definition of a derivative:

$$\left(\frac{\partial \phi}{\partial x}\right)_{x_i} = \lim_{\Delta x \to 0} \frac{\phi(x_i + \Delta x) - \phi(x_i)}{\Delta x} . \tag{3.2}$$

A geometrical interpretation is shown in Fig. 3.2 to which we shall refer frequently. The first derivative $\partial \phi / \partial x$ at a point is the slope of the tangent to the curve $\phi(x)$ at that point, the line marked 'Exact' in the figure. Its slope can be approximated by the slope of a line passing through two nearby points on the curve. The dotted line shows approximation by a *forward difference*; the derivative at x_i is approximated by the slope of a line passing through the point x_i and another point at $x_i + \Delta x$. The dashed line illustrates approximation by *backward difference* for which the second point is $x_i - \Delta x$. The line labeled 'Central' represents approximation by a *central difference*: it uses the slope of a line passing through two points lying on opposite sides of the point at which the derivative is approximated.

It is obvious from Fig. 3.2 that some approximations are better than others. The line for the central-difference approximation has a slope very close to the slope of the exact line; if the function $\phi(x)$ were a second-order polynomial and the points were equally spaced in x-direction, the slopes would match exactly.

It is also obvious from Fig. 3.2 that the quality of the approximation improves when the additional points are close to x_i, i.e., as the grid is refined, the approximation improves. The approximations shown in Fig. 3.2 are a few of many possibilities; the following sections outline the principal approaches to deriving approximations for the first and second derivatives.

In the following two sections, only the one-dimensional case is considered. The coordinate may be either Cartesian or curvilinear, the difference is of little importance here. In multidimensional finite differences, each coordinate is usually treated separately so the methods developed here are readily adapted to higher dimensionality.

Fornberg (1988) provides a general method for difference formulas and useful tables of derivative expressions of various orders of derivative and orders of accuracy.

3.3 Approximation of the First Derivative

Discretization of the convection term in Eq. (3.1) requires the approximation of the first derivative, $\partial(\rho u \phi)/\partial x$. We shall now describe some approaches to approximation of the first derivative of a generic variable ϕ; the methods can be applied to the first derivative of any quantity.

In the previous section, one means of deriving approximations to the first derivative was presented. There are more systematic approaches that are better suited to the derivation of more accurate approximations; some of these will be described later.

3.3.1 Taylor Series Expansion

Any continuous differentiable function $\phi(x)$ can, in the vicinity of x_i, be expressed as a Taylor series:

$$\phi(x) = \phi(x_i) + (x - x_i)\left(\frac{\partial \phi}{\partial x}\right)_i + \frac{(x - x_i)^2}{2!}\left(\frac{\partial^2 \phi}{\partial x^2}\right)_i +$$
$$\frac{(x - x_i)^3}{3!}\left(\frac{\partial^3 \phi}{\partial x^3}\right)_i + \cdots + \frac{(x - x_i)^n}{n!}\left(\frac{\partial^n \phi}{\partial x^n}\right)_i + H, \quad (3.3)$$

where H means "higher-order terms". By replacing x by x_{i+1} or x_{i-1} in this equation, one obtains expressions for the variable values at these points in terms of the variable and its derivatives at x_i. This can be extended to any other point near x_i, for example, x_{i+2} and x_{i-2}.

Using these expansions, one can obtain approximate expressions for the first and higher derivatives at point x_i in terms of the function values at neighboring points. For example, using Eq. (3.3) for ϕ at x_{i+1}, we can show that:

$$\left(\frac{\partial \phi}{\partial x}\right)_i = \frac{\phi_{i+1} - \phi_i}{x_{i+1} - x_i} - \frac{x_{i+1} - x_i}{2}\left(\frac{\partial^2 \phi}{\partial x^2}\right)_i - \frac{(x_{i+1} - x_i)^2}{6}\left(\frac{\partial^3 \phi}{\partial x^3}\right)_i + H. \quad (3.4)$$

Another expression may be derived using the series expression (3.3) at x_{i-1}:

$$\left(\frac{\partial \phi}{\partial x}\right)_i = \frac{\phi_i - \phi_{i-1}}{x_i - x_{i-1}} + \frac{x_i - x_{i-1}}{2}\left(\frac{\partial^2 \phi}{\partial x^2}\right)_i - \frac{(x_i - x_{i-1})^2}{6}\left(\frac{\partial^3 \phi}{\partial x^3}\right)_i + H. \quad (3.5)$$

Still another expression may be obtained by using Eq. (3.3) at both x_{i-1} and x_{i+1}:

$$\left(\frac{\partial \phi}{\partial x}\right)_i = \frac{\phi_{i+1} - \phi_{i-1}}{x_{i+1} - x_{i-1}} - \frac{(x_{i+1} - x_i)^2 - (x_i - x_{i-1})^2}{2(x_{i+1} - x_{i-1})} \left(\frac{\partial^2 \phi}{\partial x^2}\right)_i -$$
$$\frac{(x_{i+1} - x_i)^3 + (x_i - x_{i-1})^3}{6(x_{i+1} - x_{i-1})} \left(\frac{\partial^3 \phi}{\partial x^3}\right)_i + H . \qquad (3.6)$$

All three of these expressions are *exact* if all terms on the right-hand side are retained. Because the higher-order derivatives are unknown, these expressions are not of great value as they stand. However, if the distance between the grid points i.e., $x_i - x_{i-1}$ and $x_{i+1} - x_i$ is small, the higher-order terms will be small except in the unusual situation in which the higher derivatives are locally very large. Ignoring the latter possibility, *approximations* to the first derivative result from truncating each of the series after the first terms on the right-hand sides:

$$\left(\frac{\partial \phi}{\partial x}\right)_i \approx \frac{\phi_{i+1} - \phi_i}{x_{i+1} - x_i} ; \qquad (3.7)$$

$$\left(\frac{\partial \phi}{\partial x}\right)_i \approx \frac{\phi_i - \phi_{i-1}}{x_i - x_{i-1}} ; \qquad (3.8)$$

$$\left(\frac{\partial \phi}{\partial x}\right)_i \approx \frac{\phi_{i+1} - \phi_{i-1}}{x_{i+1} - x_{i-1}} . \qquad (3.9)$$

These are the forward- (FDS), backward- (BDS), and central-difference (CDS) schemes mentioned earlier, respectively. The terms that were deleted from the right-hand sides are called the *truncation errors;* they measure the accuracy of the approximation and determine the rate at which the error decreases as the spacing between points is reduced. In particular, the first truncated term is usually the principal source of error.

The truncation error is the sum of products of a power of the spacing between the points and a higher-order derivative at the point $x = x_i$:

$$\epsilon_\tau = (\Delta x)^m \alpha_{m+1} + (\Delta x)^{m+1} \alpha_{m+2} + \cdots + (\Delta x)^n \alpha_{n+1} , \qquad (3.10)$$

where Δx is the spacing between the points (assumed all equal for the present) and the α's are higher-order derivatives multiplied by constant factors. From Eq. (3.10) we see that the terms containing higher powers of Δx are smaller for small spacing so that the leading term (the one with the smallest exponent) is the dominant one. As Δx is reduced, the above approximations converge to the exact derivatives with an error proportional to $(\Delta x)^m$, where m is the exponent of the leading truncation error term. The order of an approximation indicates how fast the error is *reduced* when the grid is *refined*; it does not indicate the absolute magnitude of the error. The error is thus reduced by a factor of two, four, eight or sixteen for first-, second-, third- or fourth-order approximations, respectively. It should be remembered that this rule is valid only for *sufficiently small spacings;* the definition of 'small enough' depends on the profile of the function $\phi(x)$.

Equations (3.7) and (3.8) represent first-order approximations, irrespective of whether the grid spacing is uniform or non-uniform, because the leading term in the truncation error is proportional to grid spacing (see Eqs. (3.4) and (3.5)). The leading term for the expression (3.9) vanishes when the grid spacing is uniform; the remaining leading term is proportional to grid spacing squared and the scheme is thus second-order accurate. A more detailed discussion of the effects of grid non-uniformity on truncation errors will be given in Sect. 3.3.4.

3.3.2 Polynomial Fitting

An alternative way of obtaining approximations for the derivatives is to fit the function to an interpolation curve and differentiate the resulting curve. For example, if piece-wise linear interpolation is used, we obtain the FDS or BDS approximations, depending on whether the second point lies to the left or the right of point x_i.

Fitting a parabola to the data at points x_{i-1}, x_i, and x_{i+1}, and computing the first derivative at x_i from the interpolant, we obtain:

$$\left(\frac{\partial\phi}{\partial x}\right)_i = \frac{\phi_{i+1}(\Delta x_i)^2 - \phi_{i-1}(\Delta x_{i+1})^2 + \phi_i[(\Delta x_{i+1})^2 - (\Delta x_i)^2]}{\Delta x_{i+1}\Delta x_i(\Delta x_i + \Delta x_{i+1})} , \qquad (3.11)$$

where $\Delta x_i = x_i - x_{i-1}$. This approximation has a second-order truncation error on any grid. An identical second-order approximation can be obtained using the Taylor series approach by eliminating the term containing the second derivative in Eq. (3.6); see Eq. (3.26). For uniform spacing, the above expression reduces to the CDS approximation given in Eq. (3.9).

Other polynomials, splines, etc. can be used as interpolants and then to approximate the derivative. In general, approximation of the first derivative possesses a truncation error of the same order as the degree of the polynomial used to approximate the function. We give below two third-order approximations obtained by fitting a cubic polynomial to four points and a fourth-order approximation obtained by fitting a polynomial of degree four to five points on a uniform grid:

$$\left(\frac{\partial\phi}{\partial x}\right)_i = \frac{2\phi_{i+1} + 3\phi_i - 6\phi_{i-1} + \phi_{i-2}}{6\Delta x} + \mathcal{O}\left((\Delta x)^3\right) ; \qquad (3.12)$$

$$\left(\frac{\partial\phi}{\partial x}\right)_i = \frac{-\phi_{i+2} + 6\phi_{i+1} - 3\phi_i - 2\phi_{i-1}}{6\Delta x} + \mathcal{O}\left((\Delta x)^3\right) ; \qquad (3.13)$$

$$\left(\frac{\partial \phi}{\partial x}\right)_i = \frac{-\phi_{i+2} + 8\,\phi_{i+1} - 8\,\phi_{i-1} + \phi_{i-2}}{12\,\Delta x} + \mathcal{O}\big((\Delta x)^4\big)\,. \tag{3.14}$$

The above approximations are third-order BDS, third-order FDS, and fourth-order CDS schemes, respectively. On non-uniform grids, the coefficients in the above expressions become functions of grid expansion ratios.

In the case of FDS and BDS, the major contribution to the approximation comes from one side. In convection-dominated problems, BDS is sometimes used when flow is locally from node x_{i-1} to x_i and FDS when the flow is in the negative direction. Such methods are called *upwind schemes* (UDS). First-order upwind schemes are very inaccurate; their truncation error has the effect of a false diffusion (i.e., the solution corresponds to a larger diffusion coefficient, which is sometimes much larger than the actual diffusivity). Higher-order upwind schemes are more accurate, but one can usually implement a CDS of higher order with less effort, because it is not necessary to check the flow direction (see above expressions).

We have demonstrated only one-dimensional polynomial fitting here; a similar approach can be used together with any type of *shape function* or interpolant in one, two, or three dimensions. The only constraint is the obvious one that the number of grid points used to compute the coefficients of the shape function must equal the number of available coefficients. This approach is attractive when irregular grids are used, because it allows the possibility of avoiding the use of coordinate transformations; see Sect. 9.5.

3.3.3 Compact Schemes

For uniformly spaced grids, many special schemes can be derived. Among these are compact schemes (Lele 1992; Mahesh 1998) and the spectral methods described later. Here, only Padé schemes will be described.

Compact schemes can be derived through the use of polynomial fitting. However, instead of using only the variable values at computational nodes to derive the coefficients of the polynomial, one also uses values of the derivatives at some of the points. We will use this idea to derive a fourth-order Padé scheme. The objective is to use information from near-neighbor points only; this makes solution of the resulting equations simpler and reduces the difficulty of finding approximations near the domain boundaries. In the particular schemes described here, we will use the variable values at nodes i, $i+1$, and $i-1$, and the first derivatives at nodes $i+1$ and $i-1$, to obtain an approximation for the first derivative at the node i. To this end, a polynomial of degree four is defined in the vicinity of node i:

$$\phi = a_0 + a_1(x - x_i) + a_2(x - x_i)^2 + a_3(x - x_i)^3 + a_4(x - x_i)^4\,. \tag{3.15}$$

The coefficients $a_0, \ldots, a_4$ can be found by fitting the above polynomial to the three variable and two derivative values. However, as we are interested only in the first derivative at the node i, we only need to compute the coefficient a_1. Differentiating Eq. (3.15), we have:

$$\frac{\partial \phi}{\partial x} = a_1 + 2a_2(x - x_i) + 3a_3(x - x_i)^2 + 4a_4(x - x_i)^4 \tag{3.16}$$

so that

$$\left(\frac{\partial \phi}{\partial x}\right)_i = a_1 . \tag{3.17}$$

By writing Eq. (3.15) for $x = x_i$, $x = x_{i+1}$, and $x = x_{i-1}$, and Eq. (3.16) for $x = x_{i+1}$ and $x = x_{i-1}$, we obtain after some rearrangement:

$$\left(\frac{\partial \phi}{\partial x}\right)_i = -\frac{1}{4}\left(\frac{\partial \phi}{\partial x}\right)_{i+1} - \frac{1}{4}\left(\frac{\partial \phi}{\partial x}\right)_{i-1} + \frac{3}{4}\frac{\phi_{i+1} - \phi_{i-1}}{\Delta x} . \tag{3.18}$$

A polynomial of degree six can be used if the variable values at nodes $i + 2$ and $i - 2$ are added and one of degree eight can be employed if the derivatives at these two nodes are also used. An equation like Eq. (3.18) may be written at each point. The complete set of equations is actually a tridiagonal system of equations for the derivatives at the grid points. To compute the derivatives, this system has to be solved.

A family of compact centered approximations of up to sixth order can be written:

$$\alpha\left(\frac{\partial \phi}{\partial x}\right)_{i+1} + \left(\frac{\partial \phi}{\partial x}\right)_i + \alpha\left(\frac{\partial \phi}{\partial x}\right)_{i-1} = \beta\frac{\phi_{i+1} - \phi_{i-1}}{2\,\Delta x} + \gamma\frac{\phi_{i+2} - \phi_{i-2}}{4\,\Delta x} . \tag{3.19}$$

Depending on the choice of parameters α, β, and γ, the second- and fourth-order CDS, and fourth- and sixth-order Padé schemes are obtained; the parameters and the corresponding truncation errors are listed in Table 3.1.

Obviously, for the same order of approximation, Padé schemes use fewer computational nodes and thus have more compact computational molecules than central-difference approximations. If the variable values at all grid points were known, we

Table 3.1 Compact schemes: the parameters and truncation errors

Scheme	Truncation error	α	β	γ
CDS-2	$\dfrac{(\Delta x)^2}{3!}\dfrac{\partial^3 \phi}{\partial x^3}$	0	1	0
CDS-4	$\dfrac{13(\Delta x)^4}{3\cdot 3!}\dfrac{\partial^5 \phi}{\partial x^5}$	0	$\dfrac{4}{3}$	$-\dfrac{1}{3}$
Padé-4	$\dfrac{(\Delta x)^4}{5!}\dfrac{\partial^5 \phi}{\partial x^5}$	$\dfrac{1}{4}$	$\dfrac{3}{2}$	0
Padé-6	$\dfrac{4(\Delta x)^6}{7!}\dfrac{\partial^7 \phi}{\partial x^7}$	$\dfrac{1}{3}$	$\dfrac{14}{9}$	$\dfrac{1}{9}$

can compute the derivatives at all nodes on a grid line by solving the tridiagonal system (see Chap. 5 for details on how this can be done). We shall see, in Sect. 5.6, that these schemes can also be applied in implicit methods. This issue will be addressed again in Sect. 3.7.

The schemes derived here are only a few of the possibilities; extensions to higher order and multi-dimensional approximations are possible. It is also possible to derive schemes for non-uniform grids but the coefficients are particular to the grid (see, e.g., Gamet et al. 1999).

3.3.4 Non-uniform Grids

Because the truncation error depends not only on the grid spacing but also on the derivatives of the variable, we cannot achieve a uniform distribution of discretization error on a uniform grid. We therefore need to use a non-uniform grid. The idea is to use a smaller Δx in regions where the derivatives of the function are large and a larger Δx in regions where the function is smooth. In this way, it should be possible to spread the error nearly uniformly over the domain, thus obtaining a better solution for a given number of grid points. In this section, we will discuss the accuracy of finite-difference approximations on non-uniform grids.

In some approximations, the leading term in the truncation error expression becomes zero when the spacing of the points is uniform, i.e., $x_{i+1} - x_i = x_i - x_{i-1} = \Delta x$. This is the case for the CDS approximation, see Eq. (3.6). Even though different approximations are formally of the same order for non-uniform spacing, they do not have the same truncation error. Moreover, the rate at which the error decreases when the grid is refined does not deteriorate when CDS is applied to non-uniform grids, as we shall now show.

To demonstrate this point, on which there is some confusion in the literature, note that the truncation error for the CDS is (compare Eqs. (3.9) and (3.6)):

$$\epsilon_\tau = -\frac{(\Delta x_{i+1})^2 - (\Delta x_i)^2}{2\,(\Delta x_{i+1} + \Delta x_i)}\left(\frac{\partial^2 \phi}{\partial x^2}\right)_i - \frac{(\Delta x_{i+1})^3 + (\Delta x_i)^3}{6\,(\Delta x_{i+1} + \Delta x_i)}\left(\frac{\partial^3 \phi}{\partial x^3}\right)_i + H\,,$$

(3.20)

where we have used the notation (see Fig. 3.2):

$$\Delta x_{i+1} = x_{i+1} - x_i\,,\quad \Delta x_i = x_i - x_{i-1}\,.$$

The leading term is proportional to Δx, but becomes zero when $\Delta x_{i+1} = \Delta x_i$. This means that the more non-uniform the mesh spacing, the larger the error.

Let us assume that the grid expands or contracts with a constant factor r_e. This is called a *compound interest grid*; for it:

$$\Delta x_{i+1} = r_e \Delta x_i\,.$$

(3.21)

In this case, the leading truncation error term for the CDS can be rewritten:

$$\epsilon_\tau \approx \frac{(1 - r_e)\Delta x_i}{2}\left(\frac{\partial^2\phi}{\partial x^2}\right)_i .$$

(3.22)

The leading error term of the first-order FDS or BDS schemes is:

$$\epsilon_\tau \approx \frac{\Delta x_i}{2}\left(\frac{\partial^2\phi}{\partial x^2}\right)_i .$$

When r_e is close to unity, the first-order truncation error of the CDS is substantially smaller than the BDS error.

Now let us see what happens when the grid is refined. We consider two possibilities: halving the spacing between two coarse grid points and inserting new points so that the fine grid also has a constant ratio of spacings.

In the first case, the spacing is uniform around the new points, and the expansion factor r_e at the old points remains the same as on the coarse grid. If the refinement is repeated several times, we obtain a grid which is uniform everywhere except near the coarsest grid points. At this stage, at all grid points except those belonging to the coarsest grid, the spacing is uniform and the leading error term in the CDS vanishes. After some refinements, the number of points at which the spacing is non-uniform will be small. Therefore, the global error will decrease just a bit more slowly than in a true second-order scheme.

In the second case the expansion factor of the fine grid is smaller than on the coarse grid. Simple arithmetic shows that

$$r_{e,h} = \sqrt{r_{e,2h}} ,$$

(3.23)

where h represents the refined grid and $2h$, the coarse grid. Let us consider a node common to both grids; the ratio of the leading truncation error term at node i on the two grids is (see Eq. (3.22)):

$$r_\tau = \frac{(1 - r_e)_{2h}\,(\Delta x_i)_{2h}}{(1 - r_e)_h\,(\Delta x_i)_h} .$$

(3.24)

The following relation holds between the mesh spacing on the two grids (see Fig. 3.3):

$$(\Delta x_i)_{2h} = (\Delta x_i)_h + (\Delta x_{i-1})_h = (r_e + 1)_h(\Delta x_{i-1})_h .$$

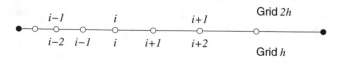

Fig. 3.3 Refinement of a non-uniform grid which expands by a constant factor r_e

When these are inserted in Eq. (3.24), taking into account Eq. (3.23), it follows that the first-order truncation error of the CDS is reduced by a factor

$$r_\tau = \frac{(1 + r_{e,h})^2}{r_{e,h}} \tag{3.25}$$

when the grid is refined. This factor has the value 4 when $r_e = 1$, i.e., when the grid is uniform. When $r_e > 1$ (expanding grid) or $r_e < 1$ (contracting grid), this factor is $r_\tau > 4$, which means that the error due to the first-order term decreases faster than the second-order error term! Because, in this method, $r_e \to 1$ as the grid is refined, the convergence becomes asymptotically second order. This will be demonstrated in the examples presented later.

A similar analysis can be performed for any scheme with the same conclusion: systematic refinement of non-uniform grids gives a rate of reduction of truncation error that has the same order as for a uniform grid.

For a given number of grid points, smaller errors are almost always obtained with non-uniform spacing. This is their purpose. However, for the grid to do its job, the user must know where smaller spacing is needed or an automatic means of grid adaptation to the solution needs to be used. An experienced user can identify regions that require fine grids; see Chap. 12 for a discussion of this issue. Methods which provide automatic error-guided grid refinement will also be presented there. It should be emphasized that grid generation becomes more difficult as the dimension of the problem is increased. Indeed, the generation of effective grids remains one of the most difficult problems in computational fluid dynamics.

Higher-order approximations of the first derivative can be obtained by using more points to eliminate more of the truncation error terms in the above expressions. For example, using ϕ_{i-1} to obtain an expression for the second derivative at x_i and substituting this expression in Eq. (3.6), we obtain the following second-order approximation (whose leading truncation error term—which is also given—is proportional to the grid spacing squared, for any grid):

$$\left(\frac{\partial \phi}{\partial x}\right)_i = \frac{\phi_{i+1}(\Delta x_i)^2 - \phi_{i-1}(\Delta x_{i+1})^2 + \phi_i[(\Delta x_{i+1})^2 - (\Delta x_i)^2]}{\Delta x_{i+1} \Delta x_i (\Delta x_i + \Delta x_{i+1})} -$$
$$\frac{\Delta x_{i+1} \Delta x_i}{6} \left(\frac{\partial^3 \phi}{\partial x^3}\right)_i + H . \tag{3.26}$$

For equispaced grids this reduces to the simple form given by Eq. (3.9).

3.4 Approximation of the Second Derivative

Second derivatives appear in the diffusive terms, see Eq. (3.1). To estimate the second derivative at a point, one may use the approximation for the first derivative twice. This is the only approach possible when the fluid properties are variable, because we need the derivative of the product of diffusion coefficient and the first derivative. We next consider approximations to the second derivative; application to the diffusion term in the conservation equation will be discussed later.

Geometrically, the second derivative is the slope of the line tangent to the curve representing the first derivative, see Fig. 3.2. By inserting approximations for the first derivatives at locations x_{i+1} and x_i, an approximation for the second derivative is obtained:

$$\left(\frac{\partial^2 \phi}{\partial x^2}\right)_i \approx \frac{\left(\frac{\partial \phi}{\partial x}\right)_{i+1} - \left(\frac{\partial \phi}{\partial x}\right)_i}{x_{i+1} - x_i}. \tag{3.27}$$

All such approximations involve data from at least three points.

In the above equation, the outer derivative was estimated by FDS. For inner derivatives one may use a different approximation, e.g., BDS; this results in the following expression:

$$\left(\frac{\partial^2 \phi}{\partial x^2}\right)_i = \frac{\phi_{i+1}(x_i - x_{i-1}) + \phi_{i-1}(x_{i+1} - x_i) - \phi_i(x_{i+1} - x_{i-1})}{(x_{i+1} - x_i)^2(x_i - x_{i-1})}. \tag{3.28}$$

One could also use the CDS approach which requires the first derivative at at x_{i-1} and x_{i+1}. A better choice is to evaluate $\partial \phi / \partial x$ at points halfway between x_i and x_{i+1} and x_i and x_{i-1}. The CDS approximations for these first derivatives are:

$$\left(\frac{\partial \phi}{\partial x}\right)_{i+\frac{1}{2}} \approx \frac{\phi_{i+1} - \phi_i}{x_{i+1} - x_i} \quad \text{and} \quad \left(\frac{\partial \phi}{\partial x}\right)_{i-\frac{1}{2}} \approx \frac{\phi_i - \phi_{i-1}}{x_i - x_{i-1}},$$

respectively. The resulting expression for the second derivative is:

$$\left(\frac{\partial^2 \phi}{\partial x^2}\right)_i \approx \frac{\left(\frac{\partial \phi}{\partial x}\right)_{i+\frac{1}{2}} - \left(\frac{\partial \phi}{\partial x}\right)_{i-\frac{1}{2}}}{\frac{1}{2}(x_{i+1} - x_{i-1})} \approx$$

$$\frac{\phi_{i+1}(x_i - x_{i-1}) + \phi_{i-1}(x_{i+1} - x_i) - \phi_i(x_{i+1} - x_{i-1})}{\frac{1}{2}(x_{i+1} - x_{i-1})(x_{i+1} - x_i)(x_i - x_{i-1})}. \tag{3.29}$$

For equidistant spacing of the points, expressions (3.28) and (3.29) become:

$$\left(\frac{\partial^2 \phi}{\partial x^2}\right)_i \approx \frac{\phi_{i+1} + \phi_{i-1} - 2\phi_i}{(\Delta x)^2}. \tag{3.30}$$

Taylor series expansion offers another way of deriving approximations to the second derivative. Using the series (3.6) at x_{i-1} and x_{i+1}, one can re-derive Eq. (3.28) with an explicit expression for the error:

$$\left(\frac{\partial^2 \phi}{\partial x^2}\right)_i = \frac{\phi_{i+1}(x_i - x_{i-1}) + \phi_{i-1}(x_{i+1} - x_i) - \phi_i(x_{i+1} - x_{i-1})}{\frac{1}{2}(x_{i+1} - x_{i-1})(x_{i+1} - x_i)(x_i - x_{i-1})} -$$

$$\frac{(x_{i+1} - x_i) - (x_i - x_{i-1})}{3}\left(\frac{\partial^3 \phi}{\partial x^3}\right)_i + H . \tag{3.31}$$

The leading truncation error term is first order but vanishes when the spacing between the points is uniform, making the approximation second-order accurate. However, even when the grid is non-uniform, the argument given in Sect. 3.3.4 shows that the truncation error is reduced in a second-order manner when the grid is refined. When a compound interest grid is used, the error decreases in the same way as for the CDS approximation of the first derivative, see Eq. (3.25).

Higher-order approximations for the second derivative can be obtained by including more data points, say x_{i-2} or x_{i+2}.

Finally, one can use interpolation to fit a polynomial of degree n through $n + 1$ data points. From that interpolation, approximations to all derivatives up to the nth can be obtained by differentiation. Using quadratic interpolation on three points leads to the formulas given above. Approaches like those described in Sect. 3.3.3 can also be extended to the second derivative.

In general, the truncation error of the approximation to the second derivative is the degree of the interpolating polynomial minus one (first order for parabolas, second order for cubics, etc.). One order is gained when the spacing is uniform and even-order polynomials are used. For example, a polynomial of degree four fit through five points leads to a fourth-order approximation on uniform grids:

$$\left(\frac{\partial^2 \phi}{\partial x^2}\right)_i = \frac{-\phi_{i+2} + 16\,\phi_{i+1} - 30\,\phi_i + 16\,\phi_{i-1} - \phi_{i-2}}{12(\Delta x)^2} + \mathcal{O}\big((\Delta x)^4\big) . \tag{3.32}$$

One can also use approximations of the second derivative to increase the accuracy of approximations to the first derivative. For example, using the FDS expression for the first derivative, Eq. (3.4), keeping just two terms on the right-hand side, and using the CDS expression (3.29) for the second derivative, results in the following expression for the first derivative:

$$\left(\frac{\partial \phi}{\partial x}\right)_i \approx \frac{\phi_{i+1}(\Delta x_i)^2 - \phi_{i-1}(\Delta x_{i+1})^2 + \phi_i[(\Delta x_{i+1})^2 - (\Delta x_i)^2]}{\Delta x_{i+1}\Delta x_i(\Delta x_i + \Delta x_{i+1})} . \tag{3.33}$$

This expression possesses a second-order truncation error on any grid and reduces to the standard CDS expression for the first derivative on uniform grids. This approximation is identical to Eq. (3.26). In a similar way, one can upgrade any approximation by eliminating the derivative in the leading truncation error term. Higher-order approximations always involve more nodes, yielding more complex equations to solve and more complicated treatment of boundary conditions so a trade-off has to be made. Second-order approximations usually offer a good combination of ease of use, accuracy, and cost-effectiveness in engineering applications. Schemes of third and fourth order offer higher accuracy for a given number of points when the grid is sufficiently fine but are more difficult to use. Methods of still higher order are used only in special cases.

For the conservative form of the diffusion term (3.1), one has to approximate the inner first derivative $\partial\phi/\partial x$ first, multiply it by Γ and differentiate the product again. As shown above, one does not have to use the same approximation for the inner and outer derivatives.

The most often used approach is a second-order, central-difference approximation; the inner derivative is approximated at points midway between nodes, and then a central difference with a grid size Δx is used. One obtains:

$$
\left[\frac{\partial}{\partial x}\left(\Gamma\frac{\partial\phi}{\partial x}\right)\right]_i \approx \frac{\left(\Gamma\frac{\partial\phi}{\partial x}\right)_{i+\frac{1}{2}} - \left(\Gamma\frac{\partial\phi}{\partial x}\right)_{i-\frac{1}{2}}}{\frac{1}{2}(x_{i+1}-x_{i-1})} \approx
$$

$$
\frac{\Gamma_{i+\frac{1}{2}}\dfrac{\phi_{i+1}-\phi_i}{x_{i+1}-x_i} - \Gamma_{i-\frac{1}{2}}\dfrac{\phi_i-\phi_{i-1}}{x_i-x_{i-1}}}{\frac{1}{2}(x_{i+1}-x_{i-1})}. \tag{3.34}
$$

Other approximations are easily obtained using different approximations for the inner and outer first derivatives; any of the approximations presented in the previous section can be used.

3.5 Approximation of Mixed Derivatives

Mixed derivatives occur only when the transport equations are expressed in non-orthogonal coordinate systems; see Chap. 9 for an example. The mixed derivative, $\partial^2\phi/\partial x\partial y$ may be treated by combining the one-dimensional approximations as was described above for the second derivative. One can write:

$$
\frac{\partial^2\phi}{\partial x\partial y} = \frac{\partial}{\partial x}\left(\frac{\partial\phi}{\partial y}\right). \tag{3.35}
$$

The mixed second derivative at (x_i, y_j) can be estimated using CDS by first evaluating the first derivative with respect to y at (x_{i+1}, y_j) and (x_{i-1}, y_j) and then evaluating the first derivative of this new function with respect to x, in the manner described above.

The order of differentiation can be changed; the numerical approximation may depend on the order. Although this may seem a drawback, it really poses no problem. All that is required is that the numerical approximation becomes exact in the limit of infinitesimal grid size. The difference in the solutions obtained with two approximations is due to the discretization errors being different.

3.6 Approximation of Other Terms

3.6.1 Non-differentiated Terms

In the scalar conservation equation there may be terms—which we have lumped together into the source term q_ϕ—which do not contain derivatives; these also have to be evaluated. In the FD method, only the values at the nodes are normally needed. If the non-differentiated terms involve the dependent variable, they may be expressed in terms of the nodal value of the variable. Care is needed when the dependence is non-linear. The treatment of these terms depends on the equation and further discussion is put off until Chaps. 5, 7 and 8.

3.6.2 Differentiated Terms Near Boundaries

A problem does arise when higher-order approximations of the derivatives are used; as they require data at more than three points, approximations at interior nodes may demand data at points beyond the boundary. It may then be necessary to use different approximations for the derivatives at points close to boundary; usually these are of lower order than the approximations used deeper in the interior and may be one-sided differences. For example, from a cubic fit to the boundary value and three inner points, Eq. (3.13) may be derived for the first derivative at the next-to-boundary point. Fitting a fourth-order polynomial through the boundary and four inner points, the following approximation for the first derivative results at $x = x_2$, the first interior point:

$$\left(\frac{\partial \phi}{\partial x}\right)_2 = \frac{-\phi_5 + 6\phi_4 + 18\phi_3 + 10\phi_2 - 33\phi_1}{60\,\Delta x} + \mathcal{O}\big((\Delta x)^4\big). \qquad (3.36)$$

Approximation of the second derivative using the same polynomial gives:

$$\left(\frac{\partial^2 \phi}{\partial x^2}\right)_2 = \frac{-21\phi_5 + 96\phi_4 + 18\phi_3 - 240\phi_2 + 147\phi_1}{180\,(\Delta x)^2} + \mathcal{O}\big((\Delta x)^3\big). \qquad (3.37)$$

In a similar way one can derive one-sided derivative approximations of any order at the first interior grid point which use the variable values at that point, the boundary point, and a certain number of interior grid points.

3.7   Implementation of Boundary Conditions

A finite-difference approximation to the partial differential equation is required at every interior grid point. To render the solution unique, the continuous problem requires information about the solution at the domain boundaries. Generally, the value of the variable at the boundary (Dirichlet boundary conditions) or its gradient in a particular direction (usually normal to the boundary—Neumann boundary conditions) or a linear combination of the two quantities is given.

Boundary conditions can be implemented in various ways. In one approach, transport equations are always solved at interior grid points only; if boundary values are not known at the boundary (Neumann or Robin condition), the discretized boundary condition is used to express the boundary value through values at interior points and thus eliminate it as unknown. In another approach, ghost points are used outside of the boundary and transport equations are solved for the unknown boundary values, while the values at ghost points are obtained from discretized boundary condition.

3.7.1   Implementation of Boundary Conditions Using Internal Grid Points

If the variable value is known at some boundary point, then there is no need to solve for it. In all FD-equations which contain data at these points, the known values are used and nothing more is necessary.

If the gradient is prescribed at the boundary, a suitable FD-approximation for it (it must be a one-sided approximation if only internal grid points are employed) can be used to compute the boundary value of the variable. If, for example, a zero gradient in the normal direction is prescribed, a simple FDS-approximation leads to (see Fig. 3.1):

$$\left(\frac{\partial \phi}{\partial x}\right)_1 = 0 \quad \Rightarrow \quad \frac{\phi_2 - \phi_1}{x_2 - x_1} = 0 \, , \tag{3.38}$$

which gives $\phi_1 = \phi_2$, allowing the boundary value to be replaced by the value at the node next to boundary and eliminated as an unknown. This is a first-order approximation; higher-order approximations can be obtained by using polynomial fits of a higher degree.

From a parabolic fit to the boundary and two inner points, the following second-order approximation, valid on any grid, is obtained for the first derivative at the boundary:

$$\left(\frac{\partial \phi}{\partial x}\right)_1 \approx \frac{-\phi_3(x_2 - x_1)^2 + \phi_2(x_3 - x_1)^2 - \phi_1[(x_3 - x_1)^2 - (x_2 - x_1)^2]}{(x_2 - x_1)(x_3 - x_1)(x_3 - x_2)} \, .$$

On a uniform grid this expression reduces to:

$$\left(\frac{\partial \phi}{\partial x}\right)_1 \approx \frac{-\phi_3 + 4\phi_2 - 3\phi_1}{2\Delta x} \quad \Rightarrow \quad \phi_1 = \frac{4}{3}\phi_2 - \frac{1}{3}\phi_3 - 2\Delta x \left(\frac{\partial \phi}{\partial x}\right)_1. \quad (3.39)$$

Now the boundary value appearing in any discretized term at interior grid nodes is replaced by a combination of nodes at points 2 and 3 and the specified boundary gradient. This requires adaptation of the elements of the coefficient matrix for nodes next to boundary (and possibly at the next layer of nodes, if higher-order approximations are used), but the system of equations to be solved remains always the same.

A third-order approximation on equispaced grids is obtained from a cubic fit to four points:

$$\left(\frac{\partial \phi}{\partial x}\right)_1 \approx \frac{2\phi_4 - 9\phi_3 + 18\phi_2 - 11\phi_1}{6\Delta x}. \quad (3.40)$$

Sometimes one needs to calculate first derivative normal to boundary at points at which the boundary value of the variable is given (for example, to calculate heat flux through an isothermal surface). In this case, any of the one-sided approximations given above are suitable. The accuracy of the result depends not only on the approximation used, but also on the accuracy of the values at interior points. It is sensible to use approximations of the same order for both purposes. See Table 3 of Fornberg (1988) for a list of expressions for one-sided derivatives of order up to four and various accuracies.

When the compact schemes described in Sect. 3.3.3 are used, one has to provide both the variable value and the derivative at boundary nodes. Usually, one of these is known and the other must be computed using information from the interior. For example, a one-sided approximation to the derivative at the boundary node, like Eq. (3.40), can be employed when the variable value is prescribed. On the other hand, polynomial interpolation can be used to compute the boundary value if the derivative is known. From a cubic fit to four points the following expression is obtained for the boundary value:

$$\phi_1 = \frac{18\phi_2 - 9\phi_3 + 2\phi_4}{11} - \frac{6\Delta x}{11}\left(\frac{\partial \phi}{\partial x}\right)_1. \quad (3.41)$$

Approximations of lower or higher order can be obtained in a similar way.

3.7.2 Implementation of Boundary Conditions Using Ghost Points

As an alternative to the strategy of the previous section, one can implement derivative-based boundary conditions using central differences and ghost points, i.e., points that

Fig. 3.4 An example of a
1D FD-grid showing the
ghost node at $i = 0$ (an
extension of 1D-grid in
Fig. 3.1)

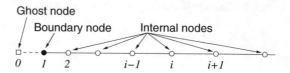

are outside the problem boundary. An immediate result is an increase in accuracy
compared to Eq. (3.38), for example. Again, it is enough to discuss the implementation of a flux boundary condition, because the value of a variable is known on the
boundary for the Dirichlet boundary condition.

If, for example, a boundary condition of the third kind, i.e., a *Robin* condition, is
prescribed in the normal direction, then a simple CDS-approximation leads to:

$$\left(\frac{\partial \phi}{\partial x} + c\phi \right)_1 = 0 \quad \Rightarrow \quad \frac{\phi_2 - \phi_0}{2(x_2 - x_1)} + c\phi_1 = 0 \,, \tag{3.42}$$

which gives $\phi_0 = \phi_2 + 2(x_2 - x_1)c\phi_1$. From Fig. 3.4 we see that node 0 is outside of
the computational domain; however, the derivative term is centered so this approximation of the boundary condition is second-order accurate.

There are two ways to use this approach. First, write the difference equation for
the node 1 and then replace ϕ_0 with $\phi_2 + 2(x_2 - x_1)c\phi_1$. This means that transport
equations are solved for the boundary nodes as well where the Neumann or Robin
boundary conditions are imposed, but all of the unknowns are within the computational domain. Alternatively, the values of the variables at the ghost points can be
regarded as unknowns and an equation equivalent to Eq. (3.42) added to the equation
set wherever the derivative boundary condition is imposed. This alters the equation
structure, e.g., away from being tridiagonal in some cases where direct matrix inversion is used; this approach is not favored then. However, in iterative solution methods,
this second approach is easily programmed and efficient.

3.8 The Algebraic Equation System

A finite-difference approximation provides an algebraic equation at each grid node; it
contains the variable value at that node as well as values at neighboring nodes. If the
differential equation is non-linear, the approximation will contain some non-linear
terms. The numerical solution process will then require linearization; methods for
solving these equations will be discussed in Chap. 5. For now, we consider only the
linear case. The methods described are applicable in the non-linear case as well. For
the linear case, the result of discretization is a system of linear algebraic equations
of the form:

$$A_P \phi_P + \sum_l A_l \phi_l = Q_P \,, \tag{3.43}$$

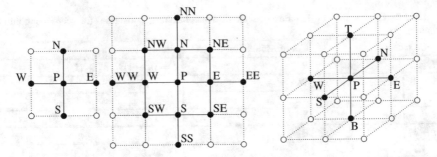

Fig. 3.5 Examples of computational molecules in 2D and 3D

where P denotes the node at which the partial differential equation is approximated and index l runs over the neighbor nodes involved in finite-difference approximations. The node P and its neighbors form the so-called *computational molecule*; two examples, which result from second and third-order approximations, are shown in Fig. 3.5.[1] The coefficients A_l depend on geometrical quantities, fluid properties and, for non-linear equations, the variable values themselves. Q_P contains all the terms which do not contain unknown variable values; it is presumed known.

The numbers of equations and unknowns must be equal, i.e., there has to be one equation for each grid node. Thus we have a large set of linear algebraic equations, which must be solved numerically. This system is *sparse*, meaning that each equation contains only a few unknowns. The system can be written in matrix notation as follows:

$$A\boldsymbol{\phi} = \mathbf{Q} , \tag{3.44}$$

where A is the square sparse coefficient matrix, $\boldsymbol{\phi}$ is a vector (or column matrix) containing the variable values at the grid nodes, and $\mathbf{Q}$ is the vector containing the terms on the right-hand side of Eq. (3.43).

The structure of matrix A depends on the ordering of variables in the vector $\boldsymbol{\phi}$. For structured grids, if the variables are labeled starting at a corner and traversing line after line in a regular manner (*lexicographic ordering*), the matrix has a polydiagonal structure. For the case of a five-point computational molecule, all the non-zero coefficients lie on the main diagonal, the two neighboring diagonals, and two other diagonals removed by N positions from the main diagonal, where N is the number of nodes in one direction. All other coefficients are zero. This structure allows use of efficient iterative solvers.

Throughout this book we shall, for the sake of definiteness, order the entries in vector $\boldsymbol{\phi}$ starting at the southwest corner of the domain, proceeding northwards along each grid line and then eastward across the domain (in three-dimensional cases we shall start at the bottom computational surface and proceed on each horizontal plane in the manner just described, and then go from bottom to top). The variables are

---

[1]For example, an equation such as Eq. (3.43) and the pictured molecules would be generated by discretization of the *Poisson equation* $\nabla \cdot (\nabla \Phi) = f$.

Table 3.2 Conversion of grid indexes to one-dimensional storage locations for vectors or column matrices

Grid location	Compass notation	Storage location
i, j, k	P	$l = (k-1)N_j N_i + (i-1)N_j + j$
$i-1, j, k$	W	$l - N_j$
$i, j-1, k$	S	$l - 1$
$i, j+1, k$	N	$l + 1$
$i+1, j, k$	E	$l + N_j$
$i, j, k-1$	B	$l - N_i N_j$
$i, j, k+1$	T	$l + N_i N_j$

normally stored in computers in one-dimensional arrays. The conversion between the grid locations, compass notation, and storage locations is indicated in Table 3.2.

Because the matrix A is sparse, it does not make sense to store it as a two- (in 2D) or three-dimensional (in 3D) array in computer memory (this is standard practice for full matrices). In 2D, storing the elements of each non-zero diagonal in a separate array of dimension $1 \times N_i N_j$, where N_i and N_j are the numbers of grid points in the two coordinate directions, requires only $5 N_i N_j$ words of storage; full array storage would require $N_i^2 N_j^2$ words of storage. In 3D, the numbers are $7 N_i N_j N_k$ and $N_i^2 N_j^2 N_k^2$, respectively. The difference is sufficiently large that the diagonal-storage scheme may allow the problem to be kept in main memory when the full-array scheme does not.

If the nodal values are referenced using the grid indexes, say $\phi_{i,j}$ in 2D, they look like matrix elements or components of a tensor. Because they are actually components of a vector $\boldsymbol{\phi}$, they should have only the single index indicated in Table 3.2.

The linearized algebraic equations in two dimensions can now be written in the form:

$$A_{l,l-N_j}\phi_{l-N_j} + A_{l,l-1}\phi_{l-1} + A_{l,l}\phi_l +$$
$$A_{l,l+1}\phi_{l+1} + A_{l,l+N_j}\phi_{l+N_j} = Q_l . \tag{3.45}$$

As noted above, it makes little sense to store the matrix as an array. If, instead, the diagonals are kept in separate arrays, it is better to give each diagonal a separate name. Because each diagonal represents the connection to the variable at a node that lies in a particular direction with respect to the central node, we shall call them A_{W}, A_{S}, A_{P}, A_{N} and A_{E}; their locations in the matrix for a grid with 5×5 internal nodes are shown in Fig. 3.6. With this ordering of points, each node is identified with an index l, which is also the relative storage location. In this notation the Eq. (3.45) can be written

$$A_{\mathrm{W}}\phi_{\mathrm{W}} + A_{\mathrm{S}}\phi_{\mathrm{S}} + A_{\mathrm{P}}\phi_{\mathrm{P}} + A_{\mathrm{N}}\phi_{\mathrm{N}} + A_{\mathrm{E}}\phi_{\mathrm{E}} = Q_{\mathrm{P}} , \tag{3.46}$$

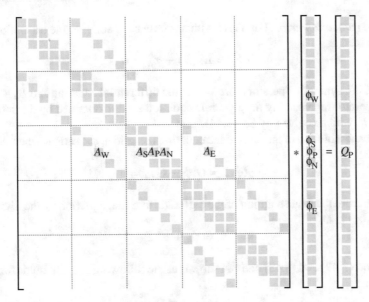

Fig. 3.6 Structure of the matrix for a five-point computational molecule (non-zero entries in the coefficient matrix on five diagonals are shaded; each horizontal set of boxes corresponds to one grid line)

where the index l, which indicated rows in Eq. (3.45), is understood, and the index indicating column or location in the vector has been replaced by the corresponding letter. We shall use this shorthand notation from now on. When necessary for clarity, the index will be inserted. A similar treatment applies to three-dimensional problems.

For block-structured and composite grids, this structure is preserved within each block, and the solvers for regular structured grids may be used. This is discussed further in Chap. 5.

For unstructured grids, the coefficient matrix remains sparse, but it no longer has banded structure. For a 2D grid of quadrilaterals and approximations that use only the four nearest neighbor nodes, there are only five non-zero coefficients in any column or row. The main diagonal is full and the other non-zero coefficients lie within a certain range of the main diagonal but not necessarily on definite diagonals. A different type of iterative solver must be used for such matrices; they will be discussed in Chap. 5. The storage scheme for unstructured grids will be introduced in Chap. 9, because such grids are used mostly in complex geometries with the FV-method.

3.9 Discretization Errors

Because the discretized equations represent approximations to the differential equation, the exact solution of the latter, which we shall denote by Φ, does not satisfy the difference equation. The imbalance, which is due to truncation of the Taylor series, is called *truncation error*. For a grid with a reference spacing h, the truncation error τ_h is defined as:

$$\mathcal{L}(\Phi) = L_h(\Phi) + \tau_h = 0 , \tag{3.47}$$

where $\mathcal{L}$ is a symbolic operator representing the differential equation and L_h is a symbolic operator representing the algebraic equation system obtained by discretization on grid h, which is given by Eq. (3.44).

The exact solution of the discretized equations on grid h, ϕ_h, satisfies the following equation:

$$L_h(\phi_h) = (A\phi - Q)_h = 0 . \tag{3.48}$$

It differs from the exact solution of the partial differential equation by the *discretization error*, ϵ_h^d, i.e.:

$$\Phi = \phi_h + \epsilon_h^d . \tag{3.49}$$

From Eqs. (3.47) and (3.48) one can show that the following relation holds for linear problems:

$$L_h(\epsilon_h^d) = -\tau_h . \tag{3.50}$$

This equation states that the truncation error acts as a source of the discretization error, which is convected and diffused by the operator L_h. Exact analysis is not possible for non-linear equations, but we expect similar behavior; in any case, if the error is small enough, we can locally linearize about the exact solution and what we will say in this section is valid. Information about the magnitude and distribution of the truncation error can be used as a guide for grid refinement and can help achieve the goal of having the same level of the discretization error everywhere in the solution domain. However, as the exact solution Φ is not known, the truncation error cannot be calculated exactly. An approximation to it may be obtained by using a solution from another (finer or coarser) grid. The estimate of the truncation error thus obtained is not always accurate but it serves the purpose of pointing to regions that have large errors and need finer grids.

For sufficiently fine grids, the truncation error (and the discretization error as well) is proportional to the leading term in the Taylor series:

$$\epsilon_h^d \approx \alpha h^p + H , \tag{3.51}$$

where H stands for higher-order terms and α depends on the derivatives at the given point but is independent of h. The discretization error can be estimated from the difference between solutions obtained on systematically refined (or coarsened) grids. Considering that the exact solution may be expressed as (see Eq. (3.49)):

$$\Phi = \phi_h + \alpha h^p + H = \phi_{2h} + \alpha(2h)^p + H , \tag{3.52}$$

the exponent p, which is the order of the scheme, may be estimated as follows:

$$p = \frac{\log\left(\dfrac{\phi_{2h} - \phi_{4h}}{\phi_h - \phi_{2h}}\right)}{\log 2}. \tag{3.53}$$

From Eq. (3.52) it also follows that the discretization error on grid h can be approximated by:

$$\epsilon_h^d \approx \frac{\phi_h - \phi_{2h}}{2^p - 1}. \tag{3.54}$$

If the ratio of the grid sizes on successive grids is not two, the factor 2 in the last two equations needs to be replaced by that ratio (see Roache 1994, for details on error estimates when the grid is not systematically refined or coarsened).

When solutions on several grids are available, one can obtain an approximation of Φ which is more accurate than the solution ϕ_h on the finest grid by adding the error estimate (3.54) to ϕ_h; this method is known as *Richardson extrapolation*, (Richardson 1910). It is simple and, when the convergence is monotonic, accurate. When a number of solutions are available, the process can be repeated to improve the accuracy further.

We have shown above that it is the rate at which the error is reduced when the grid is refined that matters, not the formal order of the scheme as defined by the leading term in the truncation error. Equation (3.53) takes this into account and returns the correct exponent p. This estimate of the order of a scheme is also a useful tool in code validation. If a method should be, say, second-order accurate but Eq. (3.53) finds that it is only first-order accurate, there is probably an error in the code.

The order of convergence estimated using Eq. (3.53) is valid only when the convergence is monotonic. Monotonic convergence can be expected only on sufficiently fine grids. We shall show in the examples that the error dependence on grid size may be irregular when the grid is coarse. Therefore, care should be taken when comparing solutions on two grids; when convergence is not monotonic, solutions on two consecutive grids may not differ much even though the errors are not small. A third grid is necessary to assure that the solution is really converged. Also, when the solution is not smooth, the error estimates obtained with Taylor series approximations may be misleading. For example, in simulations of turbulent flows, the solution varies on a wide range of scales and the order of the solution method may not be a good indicator of solution quality. In Sect. 3.11 it will be shown that the error of a fourth-order scheme may not be much smaller than of a second-order scheme for these types of simulations.

3.10 Finite-Difference Example

In this example we solve the steady 1D convection-diffusion equation with Dirichlet boundary conditions at both ends. The aim is to demonstrate the properties of the FD discretization technique for a simple problem which has an analytic solution.

Fig. 3.7 Boundary
conditions and solution
profiles for the 1D problem
as a function of the Peclet
number

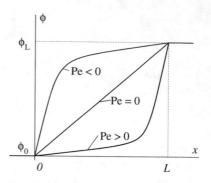

The equation to be solved reads (see Eq. (1.28)):

$$\frac{\partial(\rho u \phi)}{\partial x} = \frac{\partial}{\partial x}\left(\Gamma \frac{\partial \phi}{\partial x}\right) , \qquad (3.55)$$

with the boundary conditions: $\phi = \phi_0$ at $x = 0$, $\phi = \phi_L$ at $x = L$, see Fig. 3.7; the partial derivatives may be replaced by ordinary derivatives in this case. The density ρ and the velocity u are assumed constant. This problem has the exact solution:

$$\phi = \phi_0 + \frac{e^{x Pe/L} - 1}{e^{Pe} - 1}(\phi_L - \phi_0) . \qquad (3.56)$$

Here Pe is the Peclet number, defined as:

$$Pe = \frac{\rho u L}{\Gamma} . \qquad (3.57)$$

Because it is so simple, this problem is often used as a test of numerical methods, including both discretization and solution schemes. Physically, it represents a situation in which convection is balanced by diffusion in the streamwise direction. There are few actual flows in which this balance plays an important role. Normally, convection is balanced by either a pressure gradient or diffusion in the direction normal to the flow. In the literature, one finds many methods that were developed for Eq. (3.55) and then applied to the Navier–Stokes equations. The results are often very poor and most of these methods are best avoided. Indeed, use of this problem as a test case has probably produced more poor choices of method than any other in the field. Despite these difficulties, we shall consider this problem as some of the issues it raises are worthy of attention.

Let us consider the case $u \geq 0$ and $\phi_0 < \phi_L$; other situations are easily dealt with. In the case of small velocity ($u \approx 0$) or large diffusivity Γ, the Peclet number tends to zero and convection can be neglected; the solution is then linear in x. When the Peclet number is large, ϕ grows slowly with x and then suddenly rises to ϕ_L over a short distance close to $x = L$. The sudden change in the gradient of ϕ provides a severe test of the discretization method.

We shall discretize Eq. (3.55) using FD schemes which use the three-point computational molecule. The resulting algebraic equation at node i reads:

$$A_P^i \phi_i + A_E^i \phi_{i+1} + A_W^i \phi_{i-1} = Q_i \ . \tag{3.58}$$

It is common practice to discretize the diffusion term using CDS; thus, for the outer derivative, we have:

$$-\left[\frac{\partial}{\partial x}\left(\Gamma\frac{\partial\phi}{\partial x}\right)\right]_i \approx -\frac{\left(\Gamma\frac{\partial\phi}{\partial x}\right)_{i+\frac{1}{2}} - \left(\Gamma\frac{\partial\phi}{\partial x}\right)_{i-\frac{1}{2}}}{\frac{1}{2}(x_{i+1} - x_{i-1})} \ . \tag{3.59}$$

The CDS-approximations of the inner derivatives are:

$$\left(\Gamma\frac{\partial\phi}{\partial x}\right)_{i+\frac{1}{2}} \approx \Gamma\frac{\phi_{i+1} - \phi_i}{x_{i+1} - x_i} \ ; \quad \left(\Gamma\frac{\partial\phi}{\partial x}\right)_{i-\frac{1}{2}} \approx \Gamma\frac{\phi_i - \phi_{i-1}}{x_i - x_{i-1}} \ . \tag{3.60}$$

The contributions of the diffusion term to the coefficients of the algebraic equation (3.58) are thus:

$$A_E^d = -\frac{2\,\Gamma}{(x_{i+1} - x_{i-1})(x_{i+1} - x_i)} \ ;$$

$$A_W^d = -\frac{2\,\Gamma}{(x_{i+1} - x_{i-1})(x_i - x_{i-1})} \ ;$$

$$A_P^d = -(A_E^d + A_W^d) \ .$$

If the convection term is discretized using first-order upwind differences (UDS—FDS or BDS, depending on the flow direction), we have:

$$\left[\frac{\partial(\rho u\phi)}{\partial x}\right]_i \approx \begin{cases} \rho u\dfrac{\phi_i - \phi_{i-1}}{x_i - x_{i-1}} \ , & \text{if } u > 0 \ ; \\[2mm] \rho u\dfrac{\phi_{i+1} - \phi_i}{x_{i+1} - x_i} \ , & \text{if } u < 0 \ . \end{cases} \tag{3.61}$$

This leads to the following contributions to the coefficients of Eq. (3.58):

$$A_E^c = \frac{\min(\rho u, 0)}{x_{i+1} - x_i} \ ; \quad A_W^c = -\frac{\max(\rho u, 0)}{x_i - x_{i-1}} \ ;$$

$$A_P^c = -(A_E^c + A_W^c) \ .$$

Either A_E^c or A_W^c is zero, depending on the flow direction.

The CDS-approximation leads to:

$$\left[\frac{\partial(\rho u \phi)}{\partial x}\right]_i \approx \rho u \frac{\phi_{i+1} - \phi_{i-1}}{x_{i+1} - x_{i-1}} . \tag{3.62}$$

The CDS-contributions to the coefficients of Eq. (3.58) are:

$$A_E^c = \frac{\rho u}{x_{i+1} - x_{i-1}} ; \qquad A_W^c = -\frac{\rho u}{x_{i+1} - x_{i-1}} ;$$

$$A_P^c = -(A_E^c + A_W^c) = 0 .$$

The total coefficients are the sums of the convection and diffusion contributions, A^c and A^d.

The values of ϕ at boundary nodes are specified: $\phi_1 = \phi_0$ and $\phi_N = \phi_L$, where N is the number of nodes including the two at the boundaries. This means that, for the node at $i = 2$, the term $A_W^2 \phi_1$ can be calculated and added to Q_2, the right-hand side, and we set the coefficient A_W^2 in that equation to zero. Analogously, we add the product $A_E^{N-1} \phi_N$ for the node $i = N - 1$ to Q_{N-1} and set the coefficient $A_E^{N-1} = 0$.

The resulting tridiagonal system is easily solved. We shall only discuss the solutions here; the solver used to obtain them will be introduced in Chap. 5.

In order to demonstrate the false diffusion associated with UDS and the possibility of oscillations when using CDS, we shall consider the case with Pe $= 50$ ($L = 1.0$, $\rho = 1.0$, $u = 1.0$, $\Gamma = 0.02$, $\phi_0 = 0$ and $\phi_L = 1.0$). We start with results obtained using a uniform grid with 11 nodes (10 equal subdivisions). The profiles of $\phi(x)$ obtained using CDS and UDS for convection and CDS for diffusion terms are shown in Fig. 3.8.

The UDS-solution is obviously over-diffusive; it corresponds to the exact solution for Pe ≈ 18 (instead of 50). The false diffusion is stronger than the true diffusion! On the other hand, the CDS-solution exhibits severe oscillations. The oscillations are due to the sudden change of gradient in ϕ at the last two points. The local Peclet number based on mesh spacing, $Pe_\Delta = \rho u \Delta x / \Gamma$, is equal to 5 at every node.

If the grid is refined, the CDS-oscillations are reduced, but they are still present when 21 points are used. After the second refinement (41 grid nodes), the CDS-solution is oscillation-free and very accurate, see Fig. 3.9. The accuracy of the UDS-solution has also been improved by grid refinement, but it is still substantially in error for $x > 0.8$.

The CDS-oscillations depend on the value of the local Peclet number. It can be shown that no oscillations occur if the local Peclet number is $Pe_\Delta \leq 2$ at every grid node (see Patankar 1980). This is a sufficient, but not necessary condition for boundedness of CDS-solution. The so-called *hybrid scheme* (Spalding 1972) was designed to switch from CDS to UDS at any node at which $Pe_\Delta \geq 2$. This is too restrictive and reduces the accuracy. Oscillations appear only when the solution changes rapidly in a region of high local Peclet number.

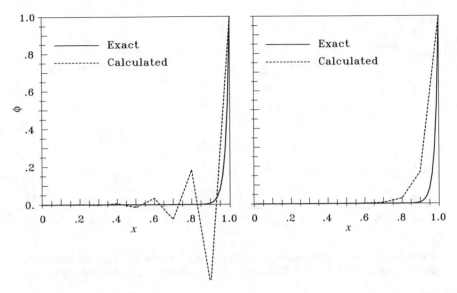

Fig. 3.8 Solution of the 1D convection-diffusion equation at Pe = 50 using CDS (left) and UDS (right) for convection terms and a uniform grid with 11 nodes

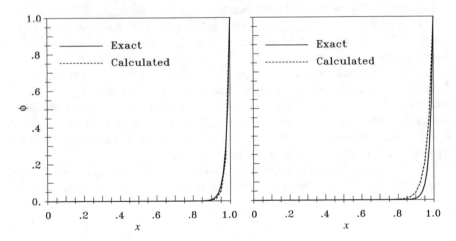

Fig. 3.9 Solution of the 1D convection-diffusion equation at Pe = 50 using CDS (left) and UDS (right) for convection terms and a uniform grid with 41 nodes

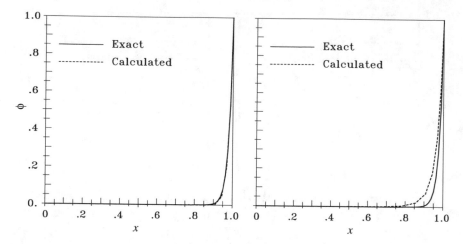

Fig. 3.10 Solution of the 1D convection-diffusion equation at Pe = 50 using CDS (left) and UDS (right) for convection terms and a non-uniform grid with 11 nodes (grid dense near right end)

In order to demonstrate this, we repeat the calculation using a non-uniform grid with 11 nodes. The smallest and the largest mesh spacings are $\Delta x_{\min} = x_N - x_{N-1} = 0.0125$ and $\Delta x_{\max} = x_2 - x_1 = 0.31$, corresponding to an expansion factor $r_e = 0.7$, see Eq. (3.21). The minimum local Peclet number is thus $\text{Pe}_{\Delta,\min} = 0.625$ near right boundary, and the maximum is $\text{Pe}_{\Delta,\max} = 15.5$ near left boundary. The local Peclet number is thus smaller than 2 in the region where ϕ undergoes strong change and is large in the region of nearly constant ϕ. The calculated profiles on this grid using the CDS and UDS schemes are shown in Fig. 3.10. There are no visible oscillations in the CDS solution. Moreover, it is as accurate as the solution on a uniform grid with four times as many nodes. The accuracy of the UDS solution has also been improved by using a non-uniform grid but is still unacceptable.

Because this problem has an analytical solution, Eq. (3.56), we can calculate the error in the numerical solution directly. The following average error is used as a measure:

$$\epsilon = \frac{\sum_i |\phi_i^{\text{exact}} - \phi_i|}{N} . \tag{3.63}$$

The problem was solved using both CDS and UDS and both uniform and non-uniform grids with up to 321 nodes. The average error is plotted as a function of average mesh spacing in Fig. 3.11. The UDS-error asymptotically approaches the slope expected of a first-order scheme. The CDS shows, from the second grid onwards, the slope expected of a second-order scheme: the error is reduced by two orders of magnitude when the grid spacing is reduced one order of magnitude.

Fig. 3.11 Average error in
the solution of the 1D
convection-diffusion
equation for Pe = 50 as a
function of the average mesh
spacing

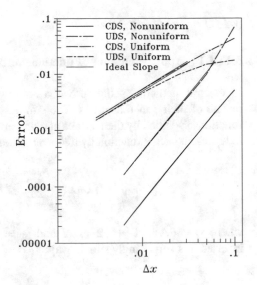

This example clearly shows that the solution on a non-uniform grid converges in the same way as the solution on a uniform grid, even though the truncation error contains a first-order term, as explained in Sect. 3.3.4. For the CDS, the average error on a non-uniform grid is almost an order of magnitude smaller than on a uniform grid with the same number of grid nodes. This is due to the fact that the mesh spacing is small where the error would be large. That Fig. 3.11 indicates larger error for UDS on a non-uniform than on a uniform grid is due to the fact that large errors at a few nodes on a uniform grid have a small effect on the average; maximum nodal error is much larger on uniform than on non-uniform grids, as can be seen by examining Figs. 3.8 and 3.10.

For related examples, see the last section of the next chapter.

3.11   An Introduction to Spectral Methods

Spectral methods are a class of methods less suited for general-purpose CFD codes than FV and FE-methods, but they are important in many applications (e.g. global and climate simulation codes for the ocean and the atmosphere (Washington and Parkinson 2005), high resolution mesoscale modeling of the atmospheric boundary layer (Moeng 1984), and simulation of turbulence (Moin and Kim 1982)). Some spectral methods are briefly described here. For a more complete description of them, see the books by Canuto et al. (2006 and 2007), Boyd (2001), Durran (2010), and Moin (2010).

3.11.1 Tools for Analysis

3.11.1.1 Approximation of Functions and Derivatives

In spectral methods, spatial derivatives are evaluated with the aid of Fourier series or one of their generalizations. The simplest spectral method deals with periodic functions specified by their values at a uniformly spaced set of points. It is possible to represent such a function by a *discrete* Fourier series:

$$f(x_i) = \sum_{q=-N/2}^{N/2-1} \hat{f}(k_q) \, e^{ik_q x_i} \,, \tag{3.64}$$

where $x_i = i \, \Delta x$, $i = 1, 2, \ldots N$ and $k_q = 2\pi q/\Delta x \, N$. Equation (3.64) can be inverted in a surprisingly simple way:

$$\hat{f}(k_q) = \frac{1}{N} \sum_{i=1}^{N} f(x_i) \, e^{-ik_q x_i} \,, \tag{3.65}$$

as can be proven by using the well-known formula for the summation of geometric series. The set of values of q is somewhat arbitrary; changing the index from q to $q \pm lN$, where l is an integer, produces no change in the value of $e^{\pm ik_q x_i}$ at the grid points. This property is known as *aliasing*; aliasing is a common and important source of error in numerical solutions of non-linear differential equations, including ones that do not use spectral methods. We note it again in Sect. 10.3.4.3 in relation to an application of a spectral code to turbulence simulation.

What makes these series useful is that Eq. (3.64) can be used to interpolate $f(x)$. We simply replace the discrete variable x_i by the continuous variable x; $f(x)$ is then defined for all x, not just the grid points. Now the choice of the range of q becomes very important. Different sets of q produce different interpolants; the best choice is the set which gives the smoothest interpolant, which is the one used in Eq. (3.64). (The set $-N/2 + 1, \ldots, N/2$ is as good a choice as the one selected.) Having defined the interpolant, we can differentiate it to produce a Fourier series for the derivative:

$$\frac{df}{dx} = \sum_{q=-N/2}^{N/2-1} ik_q \hat{f}(k_q) \, e^{ik_q x} \,, \tag{3.66}$$

which shows that the Fourier coefficient of df/dx is $ik_q \hat{f}(k_q)$. This provides a method of evaluating the derivative:

- Given $f(x_i)$, use Eq. (3.65) to compute its Fourier coefficients $\hat{f}(k_q)$;
- Compute the Fourier coefficients of $g = d \, f/dx$; $\quad \hat{g}(k_q) = i \, k_q \hat{f}(k_q)$;
- Evaluate the series (3.66) to obtain $g = df/dx$ at the grid points.

Several points need to be noted.

- The method is easily generalized to higher derivatives; for example, the Fourier coefficient of $d^2 f/dx^2$ is $-k_q^2 \hat{f}(k_q)$.
- The error in the computed derivative decreases exponentially with N when the number of grid points N is large if $f(x)$ is periodic in x. This makes spectral methods much more accurate than finite difference methods for large N; however, for small N, this may not be the case. The definition of 'large' depends on the function.
- The cost of computing the Fourier coefficients using Eq. (3.65) and/or the inverse using Eq. (3.64), if done in the most obvious manner, scales as N^2. This would be prohibitively expensive; the method is made practical by the existence of a fast method of computing Fourier transform (FFT) for which the cost is proportional to $N \log_2 N$.

To obtain the advantages of this particular spectral method, the function must be periodic and the grid points uniformly spaced. These conditions can be relaxed by using functions other than complex exponentials but any change in geometry or boundary conditions requires a considerable change in the method, making spectral methods relatively inflexible. For the problems to which they are ideally suited (for example, the simulation of turbulence in geometrically simple domains), they are unsurpassed.

3.11.1.2 Another View of Discretization Error

Spectral methods are as useful for providing another way of looking at truncation errors as they are as computational methods on their own. So long as we deal with periodic functions, the series (3.64) represents the function and we may approximate its derivative by any method we choose. In particular, we can use the exact spectral method of the example above or a finite-difference approximation. Any of these methods can be applied term-by-term to the series so it is sufficient to consider differentiation of e^{ikx}. The exact result is ike^{ikx}. On the other hand, if we apply the central-difference operator of Eq. (3.9) to this function we find:

$$\frac{\delta e^{ikx}}{\delta x} = \frac{e^{ik(x+\Delta x)} - e^{ik(x-\Delta x)}}{2\Delta x} = i\,\frac{\sin(k\,\Delta x)}{\Delta x}\,e^{ikx} = ik_{\mathrm{eff}}\,e^{ikx}\,, \qquad (3.67)$$

where k_{eff} is called the *effective wavenumber*[2] because using the finite difference approximation is equivalent to replacing the exact wavenumber k by k_{eff}. Similar expressions can be derived for other schemes; for example, the fourth-order CDS, Eq. (3.14), leads to:

$$k_{\mathrm{eff}} = \frac{\sin(k\,\Delta x)}{3\Delta x}\,[4 - \cos(k\,\Delta x)]\,. \qquad (3.68)$$

---

[2]It is called *modified wavenumber* by some authors.

Fig. 3.12 Effective
wavenumber for the second
and fourth-order
CDS-approximation of the
first derivative, normalized
by $k_{max} = \pi / \Delta x$

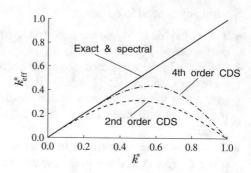

For low wavenumber (corresponding to smooth functions), the effective wavenumber of the CDS-approximation can be expanded in a Taylor series:

$$k_{eff} = \frac{\sin(k \, \Delta x)}{\Delta x} = k - \frac{k^3 (\Delta x)^2}{6} , \tag{3.69}$$

which shows the second-order nature of the approximation for small k and small Δx. However, in any computation, wavenumbers up to $k_{max} = \pi / \Delta x$ may be encountered (see Eq. (3.64)). The magnitude of a given Fourier coefficient depends on the function whose derivatives are being approximated; smooth functions have small high wavenumber components but rapidly varying functions give Fourier coefficients that decrease slowly with wavenumber.

In Fig. 3.12 the effective wavenumbers of the second and fourth-order CDS scheme, normalized by k_{max}, are shown as functions of the normalized wavenumber $k^* = k / k_{max}$. Both schemes give a poor approximation if the wavenumber is larger than half the maximum value. More wavenumbers are included as the grid is refined. In the limit of small spacings, the function is smooth relative to the grid, only the small wavenumbers have large coefficients, and accurate results may be expected. Alternatively, the better the modified wavenumber approximates the actual wavenumber, the more accurate the result.

If we are solving a problem with a solution that is not very smooth, the order of the discretization method may no longer be a good indicator of its accuracy. One needs to be very careful about claims that a particular scheme is accurate because the method used is of high order. The result is accurate only if there are enough nodes per wavelength of a highest wavenumber in the solution.

Spectral methods yield an error that decreases more rapidly than any power of the grid size as the latter goes to zero. This is often cited as an advantage of the method. However, this behavior is obtained only when enough points are used (the definition of 'enough' depends on the function). For small numbers of grid points, spectral methods may actually yield larger errors than finite difference methods.

Finally, we note that the effective wavenumber of the upwind-difference method is:

$$k_{\text{eff}} = \frac{1 - e^{-ik\,\Delta x}}{i\Delta x} \tag{3.70}$$

and is complex; this in an indication of the diffusive or dissipative nature of this approximation. The former was manifested in the UDS-example in the previous section. In Sect. 6.3, we will see that, when used in an unsteady differential equation, this upwind approximation is dissipative, e.g., causing a propagating wave's amplitude to decay unnaturally and rapidly with time.

3.11.2 Solution of Differential Equations

Here we examine two well-known spectral methods for solving differential equations (see Boyd 2001; Durran 2010; or Moin 2010). They are based on expansion of the unknown solution in a series of terms, e.g., Eq. (3.64), as are many other spectral methods (Canuto et al. 2007), including spectral element, patching collocation, spectral discontinuous Galerkin, etc.

The first formulation is called the *weak form* of the differential equation because the weighted integral of the equation over the solution domain is satisfied, but the differential equation is not satisfied at every point. Galerkin formulations would be of this type. The second formulation is the *strong form* of the differential equation and the equation is required to be satisfied at each point of the domain. An important variant of this formulation is called spectral collocation or the *pseudospectral method*. In this case, while the solution is represented by a truncated expansion, the differential equation is satisfied at a finite set of grid points in the domain. This method is used in fluid mechanics, e.g., for global circulation models and turbulence modeling in the atmospheric boundary layer (Fox and Orszag 1973; Moeng 1984; Pekurovsky et al. 2006; Sullivan and Patton 2011), because the derivatives of the nonlinear terms in the Navier–Stokes or Euler equations can be evaluated in physical space, which is efficient and less costly than doing the evaluation in spectral space.

For this development, consider the 1D equation for the diffusion of a pollutant in a narrow basin. The pollutant flows into the basin unevenly over its length and is extracted at each end so that the concentration there remains zero. If the diffusivity of the pollutant is D_x and the pollutant concentration ϕ, then the following boundary-value problem can be posed on a channel of length π:

$$D_x \frac{\partial^2 \phi}{\partial x^2} + A(x) = 0, \ 0 < x < \pi, \tag{3.71}$$

with

$$\phi(0) = 0 \text{ and } \phi(\pi) = 0. \tag{3.72}$$

Here $A(x)$ represents the influx of pollutant along the channel and is, in this one-dimensional problem, represented by a source term rather than inlet over a boundary (as would be the case in a multi-dimensional problem).

3.11.2.1 Using the Weak Form and Fourier Series

The approach used here is formally called the *Fourier–Galerkin* method (Canuto et al. 2006; Boyd 2001). This method is based on two key principles, namely, the ability of a complete set of *basis* functions to accurately represent any arbitrary, but reasonable, function and the orthogonality concept (Street 1973; Boyd 2001; Moin 2010). The basic idea then is to represent the unknown variable with a series composed of the basis functions, multiplied by unknown coefficients [cf., Eq. (3.64)]. Then, orthogonality of the basis functions is used by multiplying a given equation, e.g., (3.71) above, by a set of *test* functions and integrating over the domain of the problem. In the Fourier–Galerkin method used here the basis and test functions are the same.

For the current problem, posed in Eqs. (3.71) and (3.72), the boundary conditions are homogeneous Dirichlet, not periodic, conditions, so it is convenient to expand the solution using a trigonometric sine function, as the half-range expansion (i.e., on $0 < x < \pi$):

$$\phi^N(x) = \sum_{q=1}^{N} \hat{\phi}_q \sin(qx) . \tag{3.73}$$

The superscript N just indicates that the solution is represented by a partial (finite) sum of N elements. Recall that $e^{ikx} = \cos(kx) + i \sin(kx)$ and note that the elements in the partial sum each satisfy the boundary conditions of the problem. In general the partial sum $\phi^N(x)$ will not satisfy the governing Eq. (3.71). Thus, there would be a *residual* at each point:

$$R(\phi^N) = \frac{\partial^2 \phi^N}{\partial x^2} + \frac{A(x)}{D_x} \neq 0 . \tag{3.74}$$

This residual is equal to the error in the solution estimate provided by the partial sum. There are a number of ways of minimizing this residual (see, e.g., Boyd 2001; or Durran 2010), including minimization by a least-squares technique or use of weighted residuals, which is the method illustrated here. In our weighted-residual method, we ask that the weighted average of the equation be zero; this is a *weak solution* and will allow us to determine the unknown coefficients $\hat{\phi}_q$.

Now substitution of the partial sum (3.73) into Eq. (3.71), multiplying the entire equation by the weighting function $\sin(jx)$, and integrating over the domain yields:

$$\int_0^\pi \left(\frac{\partial^2 \phi^N}{\partial x^2} + \frac{A(x)}{D_x} \right) \sin(jx) \, dx = 0 . \tag{3.75}$$

This equation must be satisfied for each $j = 1, 2, \ldots N$. At this point, because the source term $A(x)/D_x$ is presumed known but may take an arbitrary form, it is necessary to expand the source in terms of the sine-series as:

$$\frac{A(x)}{D_x} = \sum_{q=1}^{N} \hat{a}_q \, \sin(qx) \,. \tag{3.76}$$

As we shall see below, the form of the source term affects the solution, especially if the source has discontinuities.

Inserting the partial sums in Eq. (3.75) and rearranging produces

$$\sum_{q=1}^{N} (-q^2 \hat{\phi}_q + \hat{a}_q) \int_0^{\pi} \sin(qx) \, \sin(jx) \, \mathrm{d}x = 0 \,. \tag{3.77}$$

However, because the chosen expansion and weighting functions are the same and are orthogonal, i. e.,

$$\int_0^{\pi} \sin(qx) \, \sin(jx) \, \mathrm{d}x = \begin{cases} \pi/2 & \text{if } q = j \\ 0 & \text{if } q \neq j \end{cases} \tag{3.78}$$

we obtain:

$$- j^2 \hat{\phi}_j + \hat{a}_j = 0 \quad \text{or} \quad \hat{\phi}_j = \frac{1}{j^2} \hat{a}_j, \quad \text{for all } j \,. \tag{3.79}$$

Multiplying Eq. (3.76) by weighting functions, integrating over the domain and using orthogonality again yields

$$\hat{a}_j = \frac{2}{\pi} \int_0^{\pi} \frac{A(x)}{D_x} \sin(jx) \, \mathrm{d}x \,. \tag{3.80}$$

The final solution then is

$$\phi^N(x) = \frac{2}{\pi} \sum_{q=1}^{N} \frac{1}{q^2} \left(\int_0^{\pi} \frac{A(x)}{D_x} \sin(qx) \, \mathrm{d}x \right) \sin(qx) \,. \tag{3.81}$$

The partial sum (3.81) is an approximate (*weak*) spectral solution for Eq. (3.71), satisfying the boundary conditions (3.72). As $N \to \infty$ in the partial sum, readers will recognize this solution becomes the standard Fourier series solution in this simple case. The references cited above deal with more complex situations and other basis and test functions.

Figure 3.13 shows the result of choosing a pollution input at the center of the domain. The narrow step input approximates a delta function so the solution converges slowly in the neighborhood of the source, i.e., the rate at which the coefficients

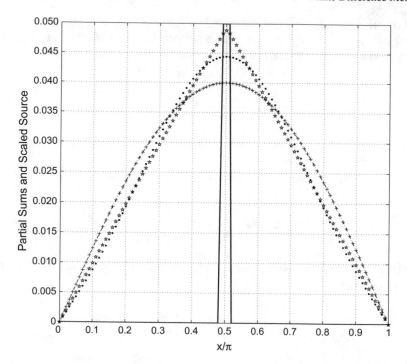

Fig. 3.13 Solution of pollution-input problem for source of unit strength located at x between $(0.49 - 0.51)\pi$: Shown are the source/20 (solid line) and solution using 1 (+), 2 $(- \cdot -)$, 3 (•) and 512 (⋆) terms of the partial expansion

$\hat{\phi}_q$ decrease with increasing q is smaller near the rapid changes in the source term. Accordingly, more terms are needed there to achieve a desired level of accuracy. Outside that area the exact solution is linear. Also in this case, the even-numbered terms (2,4,6, ...) in the partial expansion are zero as is seen by partial sums with 1 and 2 terms being identical in the figure. A 100 term expansion is already essentially identical to the exact solution.

3.11.2.2 Using the Strong Form and Collocation with Fourier Series

For the strong form, pseudo-spectral method, the procedure is rather straightforward. Given the problem posed by Eqs. (3.71) and (3.72) and the partial expansion (3.73), we simply require Eq. (3.71) be satisfied at $N - 1$ equally-spaced (grid or node) points in the domain.[3] Because the partial sum already satisfies the boundary conditions, no special account of them is necessary; see Boyd (2001) for a discussion of

---

[3]Chapter 3 of Boyd (2001) points out that this pseudo-spectral constraint can be obtained by using the Dirac delta-function, $\delta(x - x_i)$ as a test function in the weighted-residual method described above.

cases where the boundary conditions must be added in as equations in the final set obtained here.

For the Fourier series approximation, the points in the domain can be equally spaced as $\delta x_j = \pi/N$, for $j = 1, 2, 3, \ldots, N-1$ so $x_j = \pi j/N$. However, it is clear that the differential equation with the partial expansion entered can only be satisfied at $N-1$ points so Eq. (3.73) is modified to read:

$$\phi^{N-1}(x) = \sum_{q=1}^{N-1} \hat{\phi}_q \sin(qx) . \qquad (3.82)$$

What is required is that

$$\left(\frac{\partial^2 \phi^{N-1}}{\partial x^2} + \frac{A(x)}{D_x} \right) = 0 \quad \text{at} \quad x_j = \pi \, j/N, \quad \text{for} \quad j = 1, 2, 3, \ldots, N-1.$$
$$(3.83)$$

Using Eq. (3.82) yields

$$\sum_{q=1}^{N-1} (q^2 \hat{\phi}_q \sin(qx_j)) = \frac{A(x_j)}{D_x} , \qquad (3.84)$$

which yields $N-1$ equations for $\hat{\phi}_q$ in the form of an $(N-1) \times (N-1)$ coefficient matrix. For $N = 4$, $x_j = \pi \, j/4$ so

$$A = \begin{pmatrix} \sin(x_1) & 4\sin(2x_1) & 9\sin(3x_1) \\ \sin(x_2) & 4\sin(2x_2) & 9\sin(3x_2) \\ \sin(x_3) & 4\sin(2x_3) & 9\sin(3x_3) \end{pmatrix} ,$$

$$Q = \begin{pmatrix} A(x_1)/D_x \\ A(x_2)/D_x \\ A(x_3)/D_x \end{pmatrix} ,$$

$$\hat{\phi} = \begin{pmatrix} \hat{\phi}_1 \\ \hat{\phi}_2 \\ \hat{\phi}_3 \end{pmatrix} .$$

The solution is obtained by solving the equation set (see Sect. 5.2)

$$A \hat{\phi} = Q \qquad (3.85)$$

which can be generated for any value of N.

Applying the collocation method to the problem solved by the Fourier–Galerkin method above yields similar results. When the average error according to Eq. (3.63) and normalized by the maximum value of ϕ in the domain is examined in Fig. 3.14

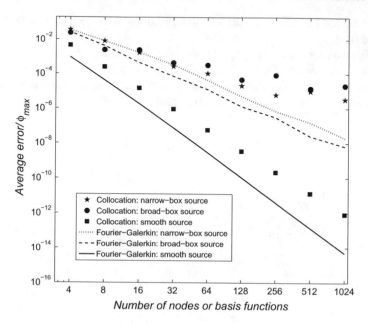

Fig. 3.14 Error in Fourier–Galerkin or collocation methods as a function of number of basis functions or nodes and shape of pollution source

for the weak and strong solutions as a function of the number of basis functions used (or equivalently for the collocation method, the number of nodes or grid points used), a significant difference in the errors is revealed.

We have used three different source shapes here to illustrate the effects this has on the errors. The narrow-box source is the one shown on Fig. 3.13 with a constant non-zero value only in the range $0.49 \geq x/\pi \geq 0.51$; the broad box has a constant non-zero source from $0.45 \geq x/\pi \geq 0.55$. The source and accordingly the second derivative of ϕ are discontinuous at these edges. The Fourier–Galerkin weak solution integrates across these discontinuities and requires the differential equation to be satisfied only on average, not point-wise. Examination of the partial sum for the second derivative of the solution shows (Fig. 3.15) that it exhibits Gibbs phenomenon (Ferziger 1998; Gibbs 1898, 1899) at the discontinuities, i.e., the value of partial sum over- and undershoots the correct value (it wiggles!) because the series does not converge there. However, the first derivative of the solution is continuous, so this problem does not occur there, and the solution itself is not infected. Indeed, because the solution is based on analytical integration of the source term, the mass of source material is conserved at all times. If one does the integrations numerically, this would introduce some error that is dependent on the spacing of the integration nodes and the integration method.

The collocation method sees the discontinuities explicitly because we force the differential equation to be satisfied on node points straddling the discontinuities, and the source is not as accurately represented. This causes a persistent error in the

Fig. 3.15 Gibbs phenomenon in 2nd derivative of Fourier–Galerkin partial sum solution for pollution source of Fig. 3.13: 1024 terms

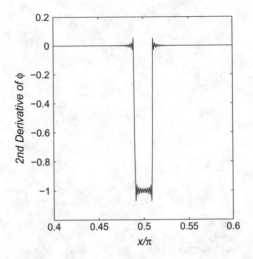

amount of source material that is present. This error is dependent on the grid spacing and its relationship to the area of non-zero source that is prescribed. Normalizing the source volume so as to conserve mass overall reduces the error.

Changing the width of the box-shaped source makes some, but not a large, difference on the reduction of error with larger number of nodes or basis functions. However, the impact of the source shape is clearly revealed in Fig. 3.14 by the use of the smooth source $A(x)/D_x = 1 - \cos(2x)$. Now the error decreases rapidly for both methods as the number of terms or nodes is increased.

Finally, given the different means of determining the partial expansion coefficients in the weak and strong methods, it is important to note that *the expansion coefficients from the two methods may not be the same, even though the basis functions are the same.*

The application of the collocation method is straightforward in general, but the presence of nonlinear terms (as would appear in the Navier–Stokes equations) requires special treatment which is beyond the scope of this introduction (see Boyd 2001; Moin 2010; Canuto et al. 2006, 2007).

Chapter 4
Finite Volume Methods

4.1 Introduction

As in the previous chapter, we consider only the generic conservation equation for a quantity ϕ and assume that the velocity field and all fluid properties are known. The finite volume method uses the integral form of the conservation equation as the starting point:

$$\int_S \rho \phi \mathbf{v} \cdot \mathbf{n} \, \mathrm{d}S = \int_S \Gamma \, \nabla \phi \cdot \mathbf{n} \, \mathrm{d}S + \int_V q_\phi \, \mathrm{d}V \ . \tag{4.1}$$

The solution domain is subdivided into a finite number of small control volumes (CVs) by a grid which, in contrast to the finite difference (FD) method, defines the control volume boundaries, not the computational nodes. For the sake of simplicity we shall demonstrate the method using Cartesian grids; complex geometries are treated in Chap. 9.

The usual approach is to define CVs by a suitable grid and assign the computational node to the CV center. However, one could as well (for structured grids) define the nodal locations first and construct CVs around them, so that CV faces lie midway between nodes; see Fig. 4.1. Nodes on which boundary conditions are applied are shown as full symbols in this figure.

The advantage of the first approach is that the nodal value represents the mean over the CV volume to higher accuracy (second order) than in the second approach, because the node is located at the centroid of the CV. The advantage of the second approach is that CDS approximations of derivatives at CV faces are more accurate when the face is midway between two nodes. The first variant is used more often and will be adopted in this book.

There are several other specialized variants of FV-type methods (cell-vertex schemes, dual-grid schemes etc.); some of these will be described later in this chapter and in Chap. 9. Here we shall describe just the basic method. The discretization prin-

© Springer Nature Switzerland AG 2020
J. H. Ferziger et al., *Computational Methods for Fluid Dynamics*,
https://doi.org/10.1007/978-3-319-99693-6_4

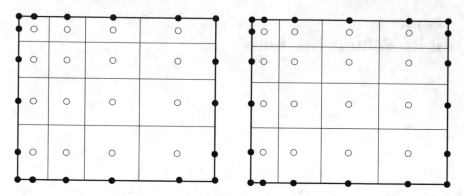

Fig. 4.1 Types of FV grids: nodes centered in CVs (left) and CV faces centered between nodes (right)

ciples are the same for all variants—one only has to take into account the relation between the various locations within the integration volume.

The integral conservation equation (4.1) applies to each CV, as well as to the solution domain as a whole. If we sum equations for all CVs, we obtain the global conservation equation, because surface integrals over inner CV faces cancel out. Thus global conservation is built into the method and this provides one of its principal advantages.

To obtain an algebraic equation for a particular CV, the surface and volume integrals need to be approximated using quadrature formulas. *Depending on the approximations used, the resulting equations may or may not be those obtained from the FD method.*

4.2   Approximation of Surface Integrals

In Figs. 4.2 and 4.3, typical 2D and 3D Cartesian control volumes are shown together with the notation we shall use. The CV surface consists of four (in 2D) or six (in 3D) plane faces, denoted by lower-case letters corresponding to their direction (e, w, n, s, t, and b) with respect to the central node (P). The 2D case can be regarded as a special case of the 3D one in which the dependent variables are independent of z. In this chapter we shall deal mostly with 2D grids; the extension to 3D problems is straightforward.

The net flux through the CV boundary is the sum of integrals over the four (in 2D) or six (in 3D) CV faces:

$$\int_S f \, \mathrm{d}S = \sum_k \int_{S_k} f \, \mathrm{d}S \, , \tag{4.2}$$

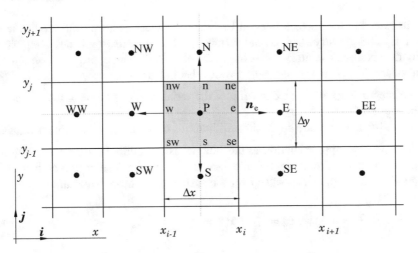

Fig. 4.2 A typical CV and the notation used for a Cartesian 2D grid

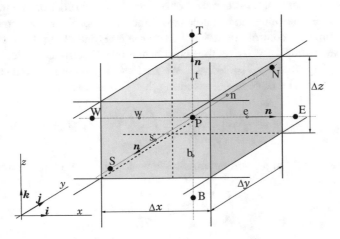

Fig. 4.3 A typical CV and the notation used for a Cartesian 3D grid

where f is the component of the convection ($\rho\phi\mathbf{v} \cdot \mathbf{n}$) or diffusion ($\Gamma\nabla\phi \cdot \mathbf{n}$) flux vector in the direction normal to CV face. As the velocity field and the fluid properties are assumed known, the only unknown is ϕ. If the velocity field is not known, we have a more complex problem involving non-linear coupled equations; we shall deal with it in Chaps. 7 and 8.

For maintenance of conservation, it is important that CVs do not overlap; each CV face is unique to the two CVs which lie on either side of it.

In what follows, only a typical CV face, the one labeled 'e' in Fig. 4.2 will be considered; analogous expressions may be derived for all faces by making appropriate index substitutions.

To calculate the surface integral in Eq. (4.2) exactly, one would need to know the integrand f everywhere on the surface S_e. This information is not available, as only the nodal (CV center) values of ϕ are calculated so an approximation must be introduced. This is best done using two levels of approximation:

- the integral is approximated in terms of the variable values at one or more locations on the cell face;
- the cell-face values are approximated in terms of the nodal (CV center) values.

The simplest approximation to the integral is the midpoint rule: the integral is approximated as a product of the integrand at the cell-face center (which is itself an approximation to the mean value over the surface) and the cell-face area:

$$F_e = \int_{S_e} f \, dS = \overline{f}_e S_e \approx f_e S_e \,. \tag{4.3}$$

This approximation of the integral—provided the value of f at location 'e' is known—is of second-order accuracy. Let us show how the order of integral approximation can be determined in a simple two-dimensional problem. We consider east side of a Cartesian control volume and therefore integrate along the y-coordinate. For the sake of simplicity, we set the coordinate origin at cell-face center "e"; therefore $y_{se} = -\Delta y/2$ and $y_{ne} = +\Delta y/2$. We first use Taylor-series expansion of the integrand f around face centroid, assuming that f_e is known:

$$f = f_e + \left(\frac{\partial f}{\partial y}\right)_e y + \left(\frac{\partial^2 f}{\partial y^2}\right)_e \frac{y^2}{2} + H \,, \tag{4.4}$$

where H stands for "higher-order terms". We want to integrate the function f along east face, thus:

$$\int_{S_e} f \, dS = \int_{-\Delta y/2}^{\Delta y/2} f \, dy \,. \tag{4.5}$$

By inserting expression (4.4) into Eq. (4.5) we obtain:

$$\int_{-\Delta y/2}^{\Delta y/2} f \, dy = \left[f_e y + \left(\frac{\partial f}{\partial y}\right)_e \frac{y^2}{2} + \left(\frac{\partial^2 f}{\partial y^2}\right)_e \frac{y^3}{6} + H \right]_{-\Delta y/2}^{+\Delta y/2} \,. \tag{4.6}$$

The term involving the first derivative of f drops out and we thus obtain:

$$\int_{-\Delta y/2}^{\Delta y/2} f \, dy = f_e \Delta y + \left(\frac{\partial^2 f}{\partial y^2}\right)_e \frac{(\Delta y)^3}{24} + H \,. \tag{4.7}$$

The first term on the right-hand side is the midpoint-rule approximation; the second term is the leading term from the truncation error, which is usually the largest. The local error is proportional to $(\Delta y)^3$, but we have to consider that integration needs to be performed over a finite area that includes n segments, where $n = Y/\Delta y$, with

Y being the total integration distance. The total error is thus proportional to $(\Delta y)^2$, because the local error is incurred n times.

Because the value of f is not available at the cell face center 'e', it has to be obtained by interpolation. In order to preserve the second-order accuracy of the midpoint-rule approximation of the surface integral, the value of f_e has to be computed with at least second-order accuracy. We shall present some widely used approximations in Sect. 4.4.

Another second-order approximation of the surface integral in 2D is the trapezoid rule, which leads to:

$$F_e = \int_{S_e} f \ dS \approx \frac{S_e}{2} \left(f_{ne} + f_{se} \right) . \tag{4.8}$$

In this case we need to evaluate the flux vector components at the CV corners. Using the same approach presented above for the midpoint-rule approximation but performing the Taylor-series expansion around location "se" and assuming that f_{ne} and f_{se} are known, one can show that the local truncation error for the trapezoid rule is twice as large as for the midpoint rule but of the opposite sign:

$$\int_{-\Delta y/2}^{\Delta y/2} f \ dy = \frac{f_{ne} + f_{se}}{2} \Delta y - \left(\frac{\partial^2 f}{\partial y^2} \right)_{se} \frac{(\Delta y)^3}{12} + H . \tag{4.9}$$

Note, however, that the reference location is now the face corner, as the midpoint value is not used. Again, interpolation from CV-centroids to face corners must be at least second-order accurate to retain the second order of the trapezoid rule integral approximation.

For higher-order approximation of the surface integrals, the flux vector components must be evaluated at more than two locations. A fourth-order approximation is Simpson's rule, which estimates the integral over S_e as:

$$F_e = \int_{S_e} f \ dS \approx \frac{S_e}{6} \left(f_{ne} + 4 f_e + f_{se} \right) . \tag{4.10}$$

Here the values of f are needed at three locations: the cell face center 'e' and the two corners, 'ne' and 'se'. In order to retain the fourth-order accuracy these values have to be obtained by interpolation of the nodal values at least as accurate as Simpson's rule. Cubic polynomials are suitable, as shown below.

In 3D, the midpoint rule is again the simplest second-order approximation. Higher-order approximations, which require the integrand at locations other than cell-face center (e.g., corners and centers of edges) are possible, but they are more difficult to implement. One possibility is mentioned in the following section.

If the variation of f is assumed to have some particular simple shape (e.g., an interpolation polynomial), the integration is easy. The accuracy of the approximation then depends on the order of shape functions.

4.3 Approximation of Volume Integrals

Some terms in the transport equations require integration over the volume of a CV. The simplest second-order accurate approximation is to replace the volume integral by the product of the mean value of the integrand and the CV volume and approximate the former as the value at the CV center:

$$Q_P = \int_V q \; dV = \overline{q} \, \Delta V \approx q_P \, \Delta V \, , \qquad (4.11)$$

where q_P stands for the value of q at the CV center. This quantity is easily calculated; because all variables are available at node P, no interpolation is necessary. The above approximation becomes exact if q is either constant or varies linearly within the CV; otherwise, it contains a second-order error, as is easily shown.

An approximation of higher order requires the values of q at more locations than just the center. These values have to be obtained by interpolating nodal values or, equivalently, by using shape functions.

In 2D the volume integral becomes an area integral. A fourth-order approximation can be obtained by using the bi-quadratic shape function:

$$q(x, y) = a_0 + a_1 x + a_2 y + a_3 x^2 + a_4 y^2 + a_5 xy + a_6 x^2 y + a_7 xy^2 + a_8 x^2 y^2 \, . \qquad (4.12)$$

The nine coefficients are obtained by fitting the function to the values of q at nine locations ('nw', 'w', 'sw', 'n', P, 's', 'ne', 'e' and 'se', see Fig. 4.2). The integral can then be evaluated. In 2D the integration gives (for Cartesian grids):

$$Q_P = \int_V q \; dV \approx \Delta x \, \Delta y \left[a_0 + \frac{a_3}{12} (\Delta x)^2 + \frac{a_4}{12} (\Delta y)^2 + \frac{a_8}{144} (\Delta x)^2 (\Delta y)^2 \right] . \qquad (4.13)$$

Only four coefficients need to be determined, but they depend on the values of q at all nine locations listed above. On a uniform Cartesian grid we obtain:

$$Q_P = \frac{\Delta x \, \Delta y}{36} \left(16 \, q_P + 4 \, q_s + 4 \, q_n + 4 \, q_w + 4 \, q_e + q_{se} + q_{sw} + q_{ne} + q_{nw} \right) . \qquad (4.14)$$

Because only the value at P is available, interpolation has to be used to obtain q at the other locations. It has to be at least fourth-order accurate to retain the accuracy of the integral approximation. Some possibilities will be described in the next section.

The above fourth-order approximation of the volume integral in 2D can be used to approximate the surface integrals in 3D. Higher-order approximations of volume integrals in 3D are more complex, but can be found using the same techniques.

4.4 Interpolation and Differentiation Practices

The approximations to the integrals require the values of variables at locations other than computational nodes (CV centers). The integrand, denoted in the previous sections by f, involves the product of several variables and/or variable gradients at those locations: $f^c = \rho\phi\mathbf{v} \cdot \mathbf{n}$ for the convection flux and $f^d = \Gamma\nabla\phi \cdot \mathbf{n}$ for the diffusion flux. We assume that the velocity field and the fluid properties ρ and Γ are known at all locations. To calculate the convection and diffusion fluxes, the value of ϕ and its gradient normal to the cell face at one or more locations on the CV surface are needed. Volume integrals of the source terms may also require these values. They have to be expressed in terms of the nodal values by interpolation.

Numerous possibilities are available for calculating ϕ and its gradient at the cell face. Indeed, Waterson and Deconinck (2007) have reviewed the performance of, literally, dozens of schemes in the finite volume context. Commercial codes generally implement a number of different schemes and give advice on their use and accuracy. Here, our strategy is to explore in some detail a few that are commonly used and then to summarize and extend our view to more general schemes including κ methods, flux limiters, and *total variation diminishing* (TVD) schemes.[1]

4.4.1 Upwind Interpolation (UDS)

Approximating ϕ_e by its value at the node upstream of 'e' is equivalent to using a backward- or forward-difference approximation for the first derivative (depending on the flow direction), hence the name *upwind-differencing scheme* (UDS) for this approximation. In UDS, ϕ_e is approximated as:

$$\phi_e = \begin{cases} \phi_P & \text{if } (\mathbf{v} \cdot \mathbf{n})_e > 0 ; \\ \phi_E & \text{if } (\mathbf{v} \cdot \mathbf{n})_e < 0 . \end{cases} \qquad (4.15)$$

This is the only approximation that unconditionally satisfies the boundedness criterion, i.e., it will never yield oscillatory solutions. However, it achieves this by being *numerically diffusive*. This was shown in the preceding chapter and will be shown again below.

Taylor-series expansion about P gives (for a Cartesian grid and $(\mathbf{v} \cdot \mathbf{n})_e > 0$):

$$\phi_e = \phi_P + (x_e - x_P)\left(\frac{\partial\phi}{\partial x}\right)_P + \frac{(x_e - x_P)^2}{2}\left(\frac{\partial^2\phi}{\partial x^2}\right)_P + H , \qquad (4.16)$$

---

[1]In this section we are looking at schemes to define ϕ and/or its derivative. In Sect. 11.3, we will examine so-called *flux-corrected transport* (FCT).

where H denotes higher-order terms. The UDS approximation retains only the first term on the right-hand side, so it is a first-order scheme. Its leading truncation error term is diffusive, i.e., it resembles a diffusion flux:

$$f_e^d = \Gamma_e \left(\frac{\partial \phi}{\partial x} \right)_e . \tag{4.17}$$

The coefficient of numerical, artificial, or false diffusion (it goes by various uncomplimentary names!) is $\Gamma_e^{num} = (\rho u)_e \Delta x / 2$. This numerical diffusion is magnified in multidimensional problems if the flow is oblique to the grid; the truncation error then produces diffusion in the direction normal to the flow as well as in the streamwise direction, a particularly serious type of error. Peaks or rapid variations in the variables will be smeared out and, because the rate of error reduction is only first order, very fine grids are required to obtain accurate solutions.

4.4.2 Linear Interpolation (CDS)

Another straightforward approximation for the value at the CV-face center is linear interpolation between the two nearest nodes. At location 'e' on a Cartesian grid we have (see Figs. 4.2 and 4.3):

$$\phi_e = \phi_E \lambda_e + \phi_P (1 - \lambda_e) , \tag{4.18}$$

where the linear interpolation factor λ_e is defined as:

$$\lambda_e = \frac{x_e - x_P}{x_E - x_P} . \tag{4.19}$$

Equation (4.18) is second-order accurate as can be shown by using the Taylor-series expansion of ϕ_E about the point x_P to eliminate the first derivative in Eq. (4.16). The result is:

$$\phi_e = \phi_E \lambda_e + \phi_P (1 - \lambda_e) - \frac{(x_e - x_P)(x_E - x_e)}{2} \left(\frac{\partial^2 \phi}{\partial x^2} \right)_P + H . \tag{4.20}$$

The leading truncation error term is proportional to the square of the grid spacing, on uniform or non-uniform grids.

As with all approximations of order higher than one, this scheme may produce oscillatory solutions. This is the simplest second-order scheme and is the one most widely used. It corresponds to the central-difference approximation of the first derivative in FD methods; hence the acronym CDS.

The assumption of a linear profile between the P and E nodes also offers the simplest approximation of the gradient, which is needed for the evaluation of diffusion fluxes:

$$\left(\frac{\partial \phi}{\partial x}\right)_e \approx \frac{\phi_E - \phi_P}{x_E - x_P} . \tag{4.21}$$

By using Taylor-series expansion around ϕ_e one can show that truncation error of the above approximation is:

$$\epsilon_\tau = \frac{(x_e - x_P)^2 - (x_E - x_e)^2}{2\,(x_E - x_P)}\left(\frac{\partial^2\phi}{\partial x^2}\right)_e -$$
$$\frac{(x_e - x_P)^3 + (x_E - x_e)^3}{6\,(x_E - x_P)}\left(\frac{\partial^3\phi}{\partial x^3}\right)_e + H . \tag{4.22}$$

When the location 'e' is midway between P and E (for example on a uniform grid), the approximation is of second-order accuracy, because the first term on the right-hand side vanishes and the leading error term is then proportional to $(\Delta x)^2$. When the grid is non-uniform, the leading error term is proportional to the product of Δx and the grid expansion factor minus unity. In spite of the formal first-order accuracy, the error reduction when the grid is refined is similar to that of a second-order approximation even on non-uniform grids. See Sect. 3.3.4 for a detailed explanation of this behavior.

4.4.3 Quadratic Upwind Interpolation (QUICK)

The next logical improvement is to approximate the variable profile between P and E by a parabola rather than a straight line. To construct a parabola we need to use data at one more point; in accord with the nature of convection, the third point is taken on the upstream side, i.e., W if the flow is from P to E (i.e., $u_x > 0$) or EE if $u_x < 0$, see Fig. 4.2. We thus obtain:

$$\phi_e = \phi_U + g_1(\phi_D - \phi_U) + g_2(\phi_U - \phi_{UU}) , \tag{4.23}$$

where D, U, and UU denote the downstream, the first upstream, and the second upstream node, respectively (E, P, and W or P, E, and EE, depending on the flow direction). The coefficients g_1 and g_2 can be expressed in terms of the nodal coordinates by:

$$g_1 = \frac{(x_e - x_U)(x_e - x_{UU})}{(x_D - x_U)(x_D - x_{UU})} ; \quad g_2 = \frac{(x_e - x_U)(x_D - x_e)}{(x_U - x_{UU})(x_D - x_{UU})} .$$

For uniform grids, the coefficients of the three nodal values involved in the interpolation turn out to be: $\frac{3}{8}$ for the downstream point, $\frac{6}{8}$ for the first upstream node and $-\frac{1}{8}$ for the second upstream node. This scheme is somewhat more complex than the CDS scheme: it extends the computational molecule one more node in each direction (in 2D, the nodes EE, WW, NN and SS are included), and, on non-orthogonal and/or non-uniform grids, the expressions for the coefficients g_i are not simple. Leonard

(1979) made this scheme popular and gave it the name QUICK (Quadratic Upwind Interpolation for Convective Kinematics).

This quadratic interpolation scheme has a third-order truncation error on both uniform and non-uniform grids. This can be shown by eliminating the second derivative from Eq. (4.20) using ϕ_W, which, on a uniform Cartesian grid with $u_x > 0$, leads to:

$$\phi_e = \frac{6}{8}\phi_P + \frac{3}{8}\phi_E - \frac{1}{8}\phi_W - \frac{3(\Delta x)^3}{48}\left(\frac{\partial^3\phi}{\partial x^3}\right)_P + H \ . \tag{4.24}$$

The first three terms on the right-hand side represent the QUICK approximation, while the last term is the principal truncation error. When this interpolation scheme is used in conjunction with the midpoint-rule approximation of the surface integral, the overall approximation is, however, still of second-order accuracy (the accuracy of the quadrature approximation). Although the QUICK approximation is slightly more accurate than CDS, both schemes converge asymptotically in a second-order manner and the differences are rarely large.

4.4.4 Higher-Order Schemes

Interpolation of order higher than third makes sense only if the integrals are approximated using higher-order formulas. If one uses Simpson's rule in 2D for surface integrals, one has to interpolate with polynomials of at least degree three in order to retain the fourth-order accuracy of the quadrature approximation, which leads to interpolation errors of fourth order. For example, by fitting a polynomial

$$\phi(x) = a_0 + a_1 x + a_2 x^2 + a_3 x^3 \tag{4.25}$$

through the values of ϕ at four nodes (two on either side of 'e': W, P, E and EE), one can determine the four coefficients a_i and find ϕ_e as a function of the nodal values. For a uniform Cartesian grid, the following expression is obtained:

$$\phi_e = \frac{9\phi_P + 9\phi_E - \phi_W - \phi_{EE}}{16} \ . \tag{4.26}$$

The same polynomial can be used to determine the derivative; we need only to differentiate it once to obtain:

$$\left(\frac{\partial\phi}{\partial x}\right)_e = a_1 + 2 a_2 x + 3 a_3 x^2 \ , \tag{4.27}$$

which, on a uniform Cartesian grid, produces:

$$\left(\frac{\partial\phi}{\partial x}\right)_e = \frac{27\phi_E - 27\phi_P + \phi_W - \phi_{EE}}{24\,\Delta x} \ . \tag{4.28}$$

The above approximation is sometimes called *fourth-order CDS*. Of course, both polynomials of higher degree and/or multi-dimensional polynomials can be used. Cubic splines, which ensure continuity of the interpolation function and its first two derivatives across the solution domain, can also be used (at some increase in cost).

Once the values of the variable and its derivative are obtained at the cell-face centers, one can interpolate on the cell faces to obtain values at the CV corners. This is not difficult to use with explicit methods but the fourth-order scheme based on Simpson's rule and polynomial interpolation produces too large a computational molecule for implicit treatment. One can avoid this complexity by using the *deferred-correction approach* described in Sect. 5.6.

Another approach is to use the techniques employed to derive the compact (Padé) schemes in FD methods. For example, one can obtain the coefficients of the polynomial (4.25) by fitting it to the variable values and first derivatives at the two nodes on either side of the cell face. For a uniform Cartesian grid, the following expression for ϕ_e results:

$$\phi_e = \frac{\phi_P + \phi_E}{2} + \frac{\Delta x}{8} \left[\left(\frac{\partial \phi}{\partial x} \right)_P - \left(\frac{\partial \phi}{\partial x} \right)_E \right] . \tag{4.29}$$

The first term on the right-hand side of the above equation represents second-order approximation by linear interpolation; the second term represents an approximation of the leading truncation error term for linear interpolation, see Eq. (4.20), in which the second derivative is approximated by CDS.

The problem is that the derivatives at nodes P and E are not known and must themselves be approximated. However, even if we approximate the first derivatives by a second-order CDS, i.e.:

$$\left(\frac{\partial \phi}{\partial x} \right)_P = \frac{\phi_E - \phi_W}{2 \Delta x} \; ; \quad \left(\frac{\partial \phi}{\partial x} \right)_E = \frac{\phi_{EE} - \phi_P}{2 \Delta x} ,$$

the resulting approximation of the cell-face value retains the fourth-order accuracy of the polynomial:

$$\phi_e = \frac{\phi_P + \phi_E}{2} + \frac{\phi_P + \phi_E - \phi_W - \phi_{EE}}{16} + \mathcal{O}(\Delta x)^4 . \tag{4.30}$$

This expression is identical to Eq. (4.26).

If we use as data the variable values on either side of the cell face and the derivative on the upstream side, we can fit a parabola. This leads to an approximation equivalent to the QUICK scheme described above:

$$\phi_e = \frac{3}{4} \phi_U + \frac{1}{4} \phi_D + \frac{\Delta x}{4} \left(\frac{\partial \phi}{\partial x} \right)_U \tag{4.31}$$

The same approach can be used to obtain an approximation of the derivative at the cell-face center; from the derivative of the polynomial (4.25) we obtain:

$$\left(\frac{\partial \phi}{\partial x}\right)_{\mathrm{e}} = \frac{\phi_{\mathrm{E}} - \phi_{\mathrm{P}}}{\Delta x} + \frac{\phi_{\mathrm{E}} - \phi_{\mathrm{P}}}{2\,\Delta x} - \frac{1}{4}\left[\left(\frac{\partial \phi}{\partial x}\right)_{\mathrm{P}} + \left(\frac{\partial \phi}{\partial x}\right)_{\mathrm{E}}\right]. \qquad (4.32)$$

Obviously, the first term on the right-hand side is the second-order CDS approximation. The remaining terms represent a correction which increases the accuracy.

The problem with approximations (4.29), (4.31) and (4.32) is that they contain first derivatives at CV centers, which are not known. Although we can replace these by second-order approximations expressed in terms of the nodal variable values without destroying their order of accuracy, the resulting computational molecules will be much larger than we would like them to be. For example, in 2D, using Simpson's rule and fourth-order polynomial interpolation, we find that each flux depends on 15 nodal values and the algebraic equation for one CV involves 25 values. The solution of the resulting equation system would be very expensive (see Chap. 5).

A way around this problem lies in the deferred-correction approach, that will be described in Sect. 5.6.

One should bear in mind that a higher-order approximation does not necessarily guarantee a more accurate solution on *any* single grid; high accuracy is achieved only when the grid is fine *enough* to capture all of the essential details of the solution; at what grid size this happens can be determined only by systematic grid refinement.

4.4.5 Other Schemes

A large number of approximations to the convection fluxes have been proposed; it is beyond the scope of this book to discuss all of them. The approach used above can be used to derive nearly all of them. We describe a few of them here and in the next section.

One can approximate ϕ_{e} by linear extrapolation from two upstream nodes, leading to the so called *linear upwind scheme* (LUDS). This scheme is of second-order accuracy but, as it is more complex than CDS and can produce unbounded solutions, the latter is a better choice.

Another approach, proposed by Raithby (1976), is to extrapolate from the upwind side, but along a streamline rather than a grid line (*skew upwind schemes*). First- and second-order schemes corresponding to the upwind and linear upwind schemes have been proposed. They have better accuracy than schemes based on extrapolation along grid lines. However, these schemes are very complex (there are many possible directions of flow) and a lot of interpolation is required. Because these schemes may produce oscillatory solutions when the grid is not sufficiently fine and are difficult to program, they have not found widespread use.

It is also possible to blend two or more different approximations (see, e.g., Sect. 4.4.6). One example that saw a great deal of use in the 1970s and 1980s (and is still used in some commercial codes) is the hybrid scheme of Spalding (1972), which switches between UDS and CDS, depending on the *local* value of the Peclet number (e.g., switch to UDS for local $\mathrm{Pe}_\Delta > 2$). Other researchers have proposed blending of lower and higher-order schemes to avoid unphysical oscillations, especially for compressible flows with shocks. Some of these ideas are covered in the next section and will be mentioned in Chap. 11 as well. Blending may be used to improve the rate of convergence of some iterative solvers, as we shall show below.

4.4.6 A General Strategy, TVD Schemes, and Flux Limiters

The schemes presented above range widely in their accuracy and ability to generate well-behaved, i.e., bounded, results. In particular, as noted, only the UDS scheme is guaranteed to yield solutions without oscillations. Waterson and Deconinck (2007) presented a comprehensive review of bounded, higher-order convection schemes. Here, we briefly outline the κ scheme for classifying linear models, *Total variation diminishing* (TVD) schemes, and the concept of flux-limiter schemes. Although we do not explicitly deal with unsteady problems until Chap. 6, it is convenient to consider here the impact of time on the spatial discretization schemes, because in Chap. 6, the focus is on the time-stepping schemes themselves. In what follows, a uniform grid and face spacing is assumed; it is straightforward to derive expressions for non-uniform grids.

A number of the schemes described above can be cast into a general framework, called κ schemes (Van Leer 1985; Waterson and Deconinck 2007). We follow the pattern laid out in the latter, in part to facilitate the reader using that review as a source for choosing appropriate schemes. In the context of Figs. 4.2 and 4.3, the κ scheme can be written, for the cell-face variable, ϕ_e, on a uniform grid with the velocity in the positive direction:

$$\phi_e = \phi_P + \left\{ \frac{1+\kappa}{4} (\phi_E - \phi_P) + \frac{1-\kappa}{4} (\phi_P - \phi_W) \right\}, \tag{4.33}$$

where the "lead" scheme is the upwind-differencing scheme (4.15) and the added term is an "anti-diffusive" term to counteract the diffusive and upwind-biased UDS. κ assigns the proportion of each scheme that is active as shown in Table 4.1.

Table 4.1 κ scheme examples for linear convection equation in the FVM

Scheme	$-1 \leq \kappa \leq 1$	Expression for ϕ_e	Notes
CDS	1	$\frac{1}{2}(\phi_P + \phi_E)$	2nd-order accurate; cf. Eq. (4.18) for $\lambda_e = \frac{1}{2}$
QUICK	$\frac{1}{2}$	$\frac{6}{8}\phi_P + \frac{3}{8}\phi_E - \frac{1}{8}\phi_W$	2nd-order accurate; cf., Eq. (4.24)
LUI	-1	$\frac{3}{2}\phi_P - \frac{1}{2}\phi_W$	2nd-order accurate; fully upwind
CUI	$\frac{1}{3}$	$\frac{5}{6}\phi_P + \frac{2}{6}\phi_E - \frac{1}{6}\phi_W$	2nd-order accurate; best κ scheme in Waterson-Deconinck 2007 tests

In a computer code, one needs to program just the general Eq. (4.33) to be able to utilize any of them by prescribing κ. Waterson and Deconinck (2007) summarized tests with the linear scalar convection equation and showed that the *modified differential equation* for steady, one-dimensional convection could be written in terms of the κ schemes as:

$$
\left(u\frac{\partial\phi}{\partial x}\right)_P = -\frac{1}{12}(3\kappa-1)u(\Delta x)^2\left(\frac{\partial^3\phi}{\partial x^3}\right)_P +
$$
$$
\frac{1}{8}(\kappa-1)u(\Delta x)^3\left(\frac{\partial^4\phi}{\partial x^4}\right)_P + H
$$

(4.34)

for the CV centered around point P. Note that the convection term on the left-hand side of the above equation is written using the non-conservative, differential form of the momentum equation, see Eq. (1.20), to be consistent with the notation used by Waterson and Deconinck (2007). It represents the difference between convection fluxes at the 'e' and 'w' faces.

The modified equation is obtained by using appropriate interpolation formulas to express the FV form of the differential equation and then expanding the resulting expression in Taylor series about point P (Warming and Hyett 1974; Fletcher 1991 (Sect. 9.2); Ferziger 1998). The result is the original equation plus other terms that are implicitly buried in the FV computational formula. This modified equation method is useful because (1) it can be applied to nonlinear schemes also, and (2) from the results one can understand the consequences of adopting a particular scheme. For example, in this case, there may be both third and fourth derivative terms present, depending on the value of κ. The third-order term is dispersive, meaning that in the unsteady case, various components of the solution move at different speeds, thereby dispersing an initial wave form incorrectly.[2] The fourth-order term is diffusive and would result in dissipation or degradation of a solution. Thus, for the CDS scheme, where $\kappa = 1$, the method is dispersive, but not diffusive. On the other hand, for the cubic-upwind scheme (CUI), $\kappa = \frac{1}{3}$, the method is more accurate and diffusive, but not dispersive.

While, generally speaking, the above approaches are normally adequate, there are specific issues associated, for example, with convection of scalars where oscillations yielding negative density or salinity are non-physical, or with velocities near regions of rapid change, viz., near extrema such as shocks. Thus, it is useful to create techniques whose results are bounded in some way. Of the techniques described above only UDS is guaranteed to yield bounded and/or monotonic behavior. The role of *total variation diminishing* (TVD) schemes and *flux limiters* is to provide bounded solutions.

Consider the convection of a scalar ϕ in one dimension as a function of time in the absence of any sources or sinks of material (e.g., the time rate of change of the flux $\left(u\frac{\partial\phi}{\partial x}\right)_P$ through the CV). Given that the concentration ϕ is bounded

---

[2]N.B.: This is *numerical* dispersion; some cases, e.g., nonlinear processes and waves in the ocean, may exhibit real *physical* dispersion.

initially, the values at later times must also be bounded and wiggles that might lead to the non-physical conditions should not occur. Harten (1983) (see Hirsch 2007, or Durran 2010) was the first to quantify a means of achieving the needed boundedness in a scheme: *the total variation of the scalar must be non-increasing over time* (the literature on this topic since that time is abundant as a perusal of the above references will confirm). Thus, if the total variation of the quantity ϕ is defined at time n by:

$$TV(\phi^n) = \sum_k |\phi_k^n - \phi_{k-1}^n|, \tag{4.35}$$

where k is a grid point index, the criteria would be

$$TV(\phi^{n+1}) \le TV(\phi^n) \tag{4.36}$$

i.e., the total variation at time $n + 1$ should be less than or equal to the total variation at time n. The effect of this is to limit the flux of the conserved quantity into a control volume to a level that will not produce a local maximum or minimum of the profile of that quantity in that control volume. Notwithstanding the definition, this constraint has become known as *total variation diminishing* or TVD in the literature (Durran 2010).

The question then becomes how to meet the test of Eq. (4.36)? Noted earlier, and demonstrated by Godunov in 1959 (see Roe 1986, or Hirsch 2007), among linear schemes, only first-order schemes can guarantee to meet the test. This motivated the search for non-linear schemes. Among the most successful have been the *flux limiters*; they are "simple functions which define the convection scheme based on a ratio of local gradients in the solution field" (Waterson and Deconinck 2007). Roe (1986) notes that one strategy is to begin with a scheme that meets the test, viz., UDS, and add to that nonlinear terms to improve the accuracy. We follow that here [in parallel to our approach in the κ scheme (Eq. (4.33)) where the "lead" term was also UDS], writing:

$$\phi_e = \phi_P + \frac{1}{2}\Psi(r)(\phi_P - \phi_W), \tag{4.37}$$

where

$$r = \frac{\frac{1}{\Delta x}(\phi_E - \phi_P)}{\frac{1}{\Delta x}(\phi_P - \phi_W)} = \frac{\phi_E - \phi_P}{\phi_P - \phi_W}. \tag{4.38}$$

Note that by writing the function r as shown in Eq. (4.38), we can see that it is actually the ratio of the centered to upstream derivatives. The weighting function $\Psi(r)$ can be chosen so that the total scheme is TVD, particularly being oscillation-free near rapid changes of the variable. A feature of such a scheme is that while it is second-order accurate when the variable changes are smooth, it reduces to first order at local extrema. There is a vast literature on the various methods of achieving this behavior, as reported, e.g., in Waterson and Deconinck (2007), Hirsch (2007), Sweby (1984, 1985), Yang and Przekwas (1992), and Jakobsen (2003). The basic

Table 4.2 $\Psi(r)$ for various schemes

Scheme	Expression for $\Psi(r)$
Upwind	0
CDS	r
MUSCL	$\max\left[0, \min\left(2, 2r, \dfrac{1+r}{2}\right)\right]$
OSPRE	$\dfrac{3r(r+1)}{2(r^2+r+1)}$
H-CUI	$\dfrac{3(r+|r|)}{2(r+2)}$
Van Leer Harmonic	$\dfrac{r+|r|}{1+r}$
MINMOD	$\max[0, \min(r, 1)]$
Superbee	$\max[0, \min(2r, 1), \min(r, 2)]$

idea in choosing the limiter $\Psi(r)$ is to maximize the impact of the "anti-diffusive", higher-order term within the TVD constraints. Hirsch (2007, Sect. 8.3.4) gives an example of how the appropriate limiter region can be chosen and displayed on the Sweby *flux-limiter diagram*. The general result for the scheme (4.37) to be TVD is

$$0 \le \Psi(r) \le \min(2r, 2), \quad r \ge 0 \quad \text{and} \quad \Psi(r) = 0, \quad r \le 0. \tag{4.39}$$

Waterson and Deconinck (2007) reported on tests with about 20 versions of flux limiters, among which MUSCL (Van Leer 1977) gave the best performance. A small set of the limiter functions is listed in Table 4.2 for reference.

If $\Psi(r) = 0$ the scheme reduces to the upwind scheme, which is first-order accurate and TVD, and if $\Psi(r) = r$ it is the CDS scheme, which is second-order accurate and *not* TVD. Of the TVD schemes shown, the first four (MUSCL, OSPRE, H-CUI and Van Leer Harmonic) were the highest scoring (listed in descending order of score) in the Waterson and Deconinck (2007) test report and are essentially second-order accurate. The bottom two are commonly used, but had much lower test scores and were significantly less than second-order accurate. Fringer et al. (2005), while using some of the TVD limiters shown in the table, demonstrated that it may be useful to change limiters during calculation in order to preserve particular behavior; the change was precipitated by a test on the behavior of the potential energy of the flow in their cases.

As an aside we note the similarity of Eqs. (4.37) and (4.33) and can deduce that for the κ schemes (Waterson and Deconinck 2007):

$$\Psi(r) = \frac{(1+\kappa)}{2} r + \frac{(1-\kappa)}{2}.$$

The schemes that have been described in this section on interpolation for generating face values of variables are useful for steady cases and for unsteady cases (irrespective of the time-stepping scheme used). Although schemes for two and three space dimensions can be derived, typically the one-dimensional schemes are extended to multiple dimensions by using the one-dimensional scheme in each direction. If the transporting velocity in the convection term varies in space, these schemes must be re-derived by using the actual face fluxes in each term because the transporting velocity may be different, e.g., on the e and w, n and s, or t and b faces of the CV in Figs. 4.2 and 4.3.

In addition, as noted above, many flow fields include oscillations and/or sharp gradients, which require some action to prevent problems in the numerical solution. We have discussed some methods for dealing with this, but other schemes exist which we have not discussed. One family of methods is called "essentially non-oscillating" (ENO) and "weighted essentially non-oscillating" (WENO) schemes, see Sect. 11.3. Durran (2010), who provides an in-depth discussion of these methods, cites "their ability to maintain genuinely high-order accuracy in the vicinity of smooth maxima and minima." Because they are used in geophysical applications, there is an active current literature, e.g., Gottlieb et al. (2006), Wang et al. (2016), and Li and Xing (1967), the last including a number of illustrative examples.

4.5 Implementation of Boundary Conditions

Each CV provides one algebraic equation. Volume integrals are calculated in the same way for every CV, but fluxes through CV faces coinciding with the domain boundary require special treatment. These boundary fluxes must either be known, or be expressed as a combination of interior values and boundary data. They should not introduce additional unknowns, because the number of equations is equal to the number of cells and thus only cell-center values can be treated as unknowns. Because there are no nodes outside the boundary, these approximations are usually based on one-sided differences or extrapolations.

However, as we showed in Sect. 3.7.2, it is possible to implement derivative-based boundary conditions with central differences and employ ghost points outside of the domain. The external points are typically then incorporated into the differential equation applied at the boundary, using the boundary condition relationship between the ghost point and the first internal point in the case of the Neumann boundary condition. We shall not do that in this book.

Usually, convection fluxes are prescribed at the inflow boundary. Convection fluxes are zero at impermeable walls and symmetry planes, and are usually assumed to be independent of the coordinate normal to an outflow boundary; in this case, upwind approximations can be used. Diffusion fluxes are sometimes specified at a wall, e.g., specified heat flux (including the special case of an adiabatic surface with zero heat flux) or boundary values of variables are prescribed. In such a case the diffusion fluxes are evaluated using one-sided approximations for normal gradients as outlined in Sect. 3.7. If the gradient itself is specified, it is used to calculate the flux, and an approximation for the flux in terms of nodal values can be used to calculate the boundary value of the variable. This will be demonstrated in an example below.

4.6 The Algebraic Equation System

By summing all the flux approximations and source terms, we produce an algebraic equation which relates the variable value at the center of the CV to the values at several neighbor CVs. The numbers of equations and unknowns are both equal to the number of CVs so the system is well-posed. The algebraic equation for a particular CV has the form (3.43), and the system of equations for the whole solution domain has the matrix form given by Eq. (3.44). When the ordering scheme of Sect. 3.8 is used, the matrix A has the form shown in Fig. 3.6. This is true only for structured grids with quadrilateral or hexahedral CVs; for other geometries, the matrix structure will be more complex (see Chap. 9 for details) but it will always be sparse. The maximum number of elements in any row is equal to the number of near neighbors for second-order approximations. For higher-order approximations, it depends on the number of neighbors used in the scheme.

4.7 Examples

In order to demonstrate the FV method and to display some of the properties of the discretization methods presented above, we present three examples.

4.7.1 Testing the Order of FV-Approximations

Because there are misconceptions regarding the order of FV-approximations in the literature, results of some representative tests are presented here. A uniform 2D Cartesian grid is considered, with 6 levels of refinement (see Fig. 4.4, where the three coarsest grids are shown). For the sake of simplicity, two analytical functions defining the variation of the variable ϕ in two dimensions are considered:

$$\phi = -2x + 3x^2 - 7x^3 + x^4 + 5y^4 \quad \text{and} \quad \phi = \cos(x) + \cos(y) . \qquad (4.40)$$

The coarsest grid has the spacing $\Delta x = \Delta y = 1$; it is then five times refined, so that on the finest grid 32 faces correspond to a single face on the coarsest grid. We test both the accuracy of various interpolation approximations to compute the variable value at the face centroid, and the accuracy of surface integral approximation for convection flux. Interpolation approximations are evaluated only for a single

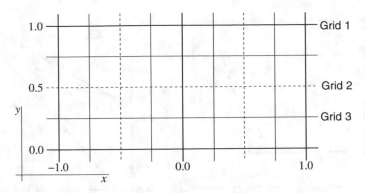

Fig. 4.4 Sketch of three coarsest grids used to test the order of interpolation and integration

location, while integral approximations are evaluated for a surface corresponding to one coarse grid face.

We first consider the accuracy of 2nd-order CDS (labeled CDS2) and 4th-order CDS (labeled CDS4) interpolation, as defined in Eqs. (4.18) and (4.26). Nodal values of the variable are the exact function values obtained by inserting node coordinates into expressions from Eq. (4.40). The interpolated value for the face centroid at $x = 0$ and $y = 0.5$ from each grid is compared with the exact value from Eq. (4.40). Figure 4.5 shows the variation of interpolated value and the error as the grid is refined for the polynomial and cosine function. Note that, in case of the polynomial function, CDS2 overestimates and CDS4 underestimates the exact value. Both approximations converge with expected order towards the exact function value at the specified location as the grid spacing is successively halved. The CDS4-approximation is more accurate on all grids; the error on the finest grid is 4 orders of magnitude lower.

Next the accuracy of integral approximations by midpoint and Simpson's rule is considered, as defined in Eqs. (4.3) and (4.10). In both cases three variants of variable values at integration points (face centroid and face corners) are used: approximations by CDS2 and CDS4, and the exact value from analytical expressions given by Eq. (4.40). Results for midpoint rule and the two functions (polynomial and cosine) are presented in Figs. 4.6. The results are quite surprising: in the case of polynomial function, CDS2 overestimates and CDS4 underestimates the integral, but the error in integral is an order of magnitude lower for CDS2 than for CDS4 or exact midpoint variable value. In the case of cosine function, CDS2 underestimates and CDS4 (and the exact face centroid value) overestimate the integral, and the error is only about 3 times larger for CDS2 than for CDS4.

In any case, 2nd-order convergence is obtained in all cases. These tests confirm that using an interpolation scheme which is much more accurate does not necessarily mean the integral approximation will also be as much more accurate: here CDS4 on fine grids produced 4 orders of magnitude lower interpolation errors, but the error of the integral approximation is only about 3 times lower for the cosine function, while 10 times higher for the polynomial function!

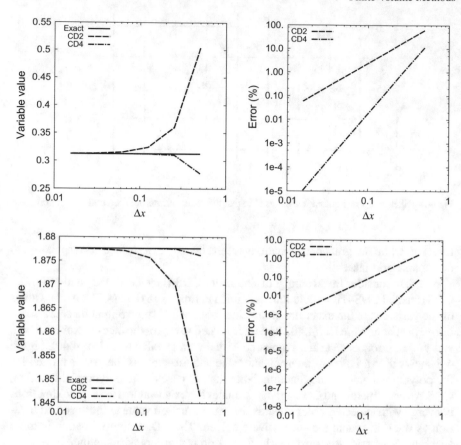

Fig. 4.5 Variation of interpolated face value from 2nd-order CDS and 4th-order CDS (left) and the associated error (right) as the grid spacing is reduced by halving, for the polynomial (upper) and cosine function (lower)

Especially it is clear that the order of interpolation does not affect the order of midpoint rule approximation, as long as it is a second-order approximation or better (even the exact variable value at the face centroid makes no big difference). Thus, solution methods that use schemes like QUICK or CDS4 only interpolate with order higher than second, but the overall order of approximation cannot be better than second, as long as midpoint rule is used for integral approximation.

Finally, integral approximations from Simpson's rule are investigated. Again, variable values at the three integration points (face centroid and the two corners) are obtained either exactly from specified analytical functions, or from CDS2 or CDS4 interpolations. Figure 4.7 shows the variation of integral value and the error as the grid is refined for the polynomial and cosine function. For both functions, CDS4 leads to under-estimated integrals, while CDS2 leads to overestimation for the polynomial function and underestimation for the cosine function. Here the use

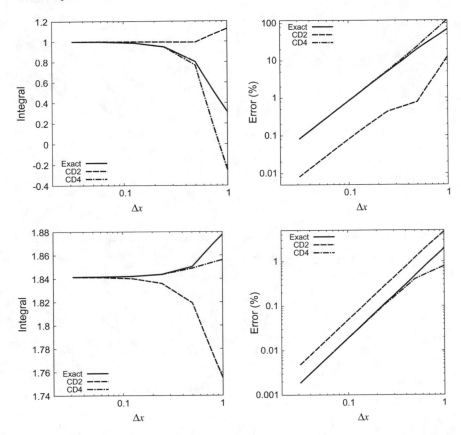

Fig. 4.6 Variation of surface integral in midpoint-rule approximation when the variable value at face centroid is exact or obtained from CDS2 or CDS4 interpolation (left) and the associated error (right) as the grid spacing is reduced by halving, for the polynomial (upper) and cosine function (lower)

of exact function values at integration points leads to the lowest errors (by more than an order of magnitude on all grids). CDS4 always leads to lower integral errors than CDS2; at the coarsest grid the difference is not that large, but on the finest grid the difference is more than 3 orders of magnitude.

These results also show that the order of integral computation equals the lowest order of involved approximations (interpolation and integral approximation). Interpolation using CDS2 leads to integral approximation by Simpson's rule becoming only a 2nd-order approximation. If the 1st-order upwind scheme was used for interpolation, the overall order would be first. However, one should also note that, when CDS2 is used for interpolation, errors are more than an order of magnitude lower when Simpson's rule is used to approximate the integral than in the case of midpoint rule, although both approximations of the integral are then of second order (see lines labeled CD2 in Figs. 4.6 and 4.7).

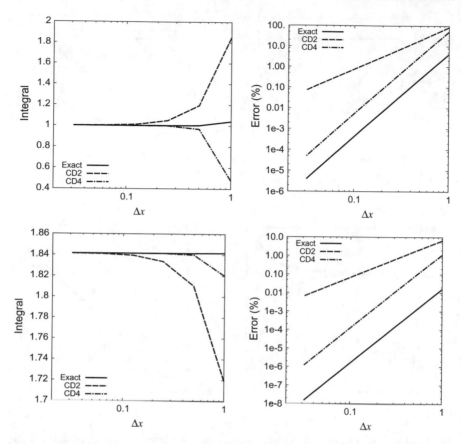

Fig. 4.7 Variation of surface integral in Simpson's rule approximation when the variable value at face centroid is exact or obtained from CDS2 or CDS4 interpolation (left) and the associated error (right) as the grid spacing is reduced by halving, for the polynomial (upper) and cosine function (lower)

The same conclusions would result if we compared 2nd and 4th-order approximations of the first derivative at the cell face, needed to compute the diffusion flux, and different surface integral approximations.

The aim of this exercise was to demonstrate that the accuracy of the FV method depends on three factors: (i) interpolation to locations other than cell centroids, (ii) approximation of derivatives (for diffusion fluxes or source terms), and (iii) approximation of integrals. For optimum results one needs a good balance between these three kinds of approximations; making one much more accurate than the others does not necessarily improve the overall result.

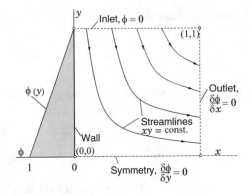

Fig. 4.8 Geometry and boundary conditions for the scalar transport in a stagnation point flow

4.7.2 Scalar Transport in a Known Velocity Field

Consider the problem, illustrated in Fig. 4.8, of transport of a scalar quantity in a known velocity field. The latter is given by $u_x = x$ and $u_y = -y$, which represents the inviscid flow near a stagnation point. The streamlines are the lines $xy = $ const. and change direction with respect to the Cartesian grid. On the other hand, on any cell face the normal velocity component is constant so the error in the approximation of the convection flux depends only on the approximation used for ϕ_e. This aids in the analysis of the accuracy.

The scalar transport equation to be solved reads:

$$\int_S \rho \phi \mathbf{v} \cdot \mathbf{n} \, dS = \int_S \Gamma \nabla \phi \cdot \mathbf{n} \, dS , \qquad (4.41)$$

and the following boundary conditions are to be applied:

- $\phi = 0$ along the north (inlet) boundary;
- Linear variation of ϕ from $\phi = 0$ at $y = 1$ to $\phi = 1$ at $y = 0$ along the west boundary;
- Symmetry condition (zero gradient normal to boundary) on the south boundary;
- Zero gradient in the flow direction at outlet (east) boundary.

The geometry and the flow field are sketched in Fig. 4.8. We give the details of discretization for the 'e' face.

The convection flux will be evaluated using the midpoint rule and either UDS or CDS interpolation. We express the convection flux as a product of the mass flux and the mean face value of ϕ:

$$F_e^c = \int_{S_e} \rho \phi \mathbf{v} \cdot \mathbf{n} \, dS \approx \dot{m}_e \phi_e , \qquad (4.42)$$

where $\dot{m}_e$ is the mass flux through the 'e' face:

$$\dot{m}_e = \int_{S_e} \rho \mathbf{v} \cdot \mathbf{n} \, dS = (\rho u_x)_e \Delta y \, . \tag{4.43}$$

Expression (4.43) is *exact* on any grid, because the velocity $u_{x,e}$ is constant along the face. The flux approximation is then:

$$F_e^c = \begin{cases} \max(\dot{m}_e, 0.) \, \phi_P + \min(\dot{m}_e, 0.) \, \phi_E & \text{for UDS} , \\ \dot{m}_e(1 - \lambda_e) \, \phi_P + \dot{m}_e \lambda_e \, \phi_E & \text{for CDS} . \end{cases} \tag{4.44}$$

The linear interpolation coefficient λ_e is defined by Eq. (4.19). Analogous expressions for the fluxes through the other CV faces are obtained by simply rotating the face 'e' around P until it collapses onto the particular face and substituting indexes. Note that $\mathbf{n}$ at each face points outwards, i.e., from cell center P to the center of the neighbor cell, cf. Fig. 4.2. For example, at the face 'w' we obtain:

$$F_w^c = \begin{cases} \max(\dot{m}_w, 0.) \, \phi_P + \min(\dot{m}_w, 0.) \, \phi_W & \text{for UDS} , \\ \dot{m}_w(1 - \lambda_w) \, \phi_P + \dot{m}_w \lambda_w \, \phi_W & \text{for CDS} , \end{cases} \tag{4.45}$$

and

$$\lambda_w = \frac{x_w - x_P}{x_W - x_P} \, . \tag{4.46}$$

The following contributions from convection fluxes to coefficients in the algebraic equation for the case of UDS are obtained:

$$\begin{aligned} A_E^c &= \min(\dot{m}_e, 0.) \, ; & A_W^c &= \min(\dot{m}_w, 0.) \, , \\ A_N^c &= \min(\dot{m}_n, 0.) \, ; & A_S^c &= \min(\dot{m}_s, 0.) \, , \\ A_P^c &= -(A_E^c + A_W^c + A_N^c + A_S^c) \, . \end{aligned} \tag{4.47}$$

For the CDS case, the coefficients are:

$$\begin{aligned} A_E^c &= \dot{m}_e \lambda_e \, ; & A_W^c &= \dot{m}_w \lambda_w \, , \\ A_N^c &= \dot{m}_n \lambda_n \, ; & A_S^c &= \dot{m}_s \lambda_s \, , \\ A_P^c &= -(A_E^c + A_W^c + A_N^c + A_S^c) \, . \end{aligned} \tag{4.48}$$

The expression for A_P^c follows from the continuity condition:

$$\dot{m}_e + \dot{m}_w + \dot{m}_n + \dot{m}_s = 0$$

which is satisfied by the velocity field. Note that $\dot{m}_w$ and λ_w for the CV centered around node P are equal to $-\dot{m}_e$ and $1 - \lambda_e$ for the CV centered around node W, respectively. In a computer code the mass fluxes and interpolation factors are therefore calculated once and stored as $\dot{m}_e$, $\dot{m}_n$ and λ_e, λ_n for each CV.

The diffusion flux integral is evaluated using the midpoint rule and CDS approximation of the normal derivative; this is the simplest and most widely used approximation:

$$F_e^d = \int_{S_e} \Gamma \, \nabla \phi \cdot \mathbf{n} \, \mathrm{d}S \approx \left(\Gamma \frac{\partial \phi}{\partial x} \right)_e \Delta y = \frac{\Gamma \, \Delta y}{x_E - x_P} (\phi_E - \phi_P) \,. \qquad (4.49)$$

Note that $x_E = \frac{1}{2}(x_{i+1} + x_i)$ and $x_P = \frac{1}{2}(x_i + x_{i-1})$, see Fig. 4.2. The diffusion coefficient Γ is assumed constant; if not, it could be interpolated linearly between the nodal values at P and E. The contribution of the diffusion term to the coefficients of the algebraic equation are:

$$A_E^d = -\frac{\Gamma \, \Delta y}{x_E - x_P} \;; \qquad A_W^d = -\frac{\Gamma \, \Delta y}{x_P - x_W} \,,$$

$$A_N^d = -\frac{\Gamma \, \Delta x}{y_N - y_P} \;; \qquad A_S^d = -\frac{\Gamma \, \Delta x}{y_P - y_S} \,, \qquad (4.50)$$

$$A_P^d = -(A_E^d + A_W^d + A_N^d + A_S^d) \,.$$

With same approximations applied to other CV faces, the integral equation becomes:

$$A_W \phi_W + A_S \phi_S + A_P \phi_P + A_N \phi_N + A_E \phi_E = Q_P \,, \qquad (4.51)$$

which represents the equation for a generic node P. The coefficients A_l are obtained by summing the convection and diffusion contributions, see Eqs. (4.47), (4.48) and (4.50):

$$A_l = A_l^c + A_l^d \qquad (4.52)$$

where l represents any of the indices P, E, W, N, S. That A_P is equal to the negative sum of all neighbor coefficients is a feature of all conservative schemes and ensures that a uniform field is a solution of the discretized equations.

The above expressions are valid at all internal CVs. For CVs next to boundary, the boundary conditions require that the equations be modified somewhat. At the north and west boundaries, where ϕ is prescribed, the gradient in the normal direction is approximated using one-sided differences, e.g., at the west boundary:

$$\left(\frac{\partial \phi}{\partial x} \right)_w \approx \frac{\phi_P - \phi_W}{x_P - x_W} \,, \qquad (4.53)$$

where W denotes the boundary node whose location coincides with the cell-face center 'w'. This approximation is of first-order accuracy, but it is applied on a half-width CV. The product of the coefficient and the boundary value is added to the source term. For example, along the west boundary (CVs with index $i = 2$), $A_W \phi_W$ is added to Q_P and the coefficient A_W is set to zero. The same applies to the coefficient A_N at the north boundary.

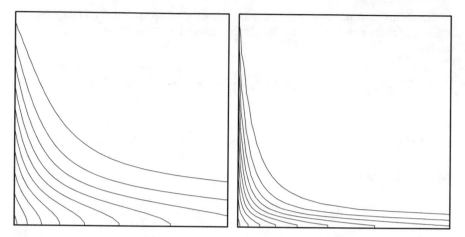

Fig. 4.9 Isolines of ϕ, from 0.05 to 0.95 with step 0.1 (top to bottom), for $\Gamma = 0.01$ (left) and $\Gamma = 0.001$ (right)

At the south boundary, the normal gradient of ϕ is zero which, when the above approximation is applied, means that the boundary values are equal to the values at CV centers. Thus, for cells with index $j = 2$, $\phi_S = \phi_P$ and the algebraic equation for those CVs is modified to:

$$(A_P + A_S)\,\phi_P + A_N\phi_N + A_W\phi_W + A_E\phi_E = Q_P\,, \qquad (4.54)$$

which requires adding A_S to A_P and then setting $A_S = 0$. The zero-gradient condition at the outlet (east) boundary is implemented in a similar way.

We now turn to the results. The isolines of ϕ calculated on a 40×40 CV uniform grid using CDS for the convection fluxes with two values of Γ: 0.001 and 0.01 ($\rho = 1.0$) are presented in Fig. 4.9. We see that transport by diffusion across the flow is much stronger for higher Γ, as expected.

In order to assess the accuracy of the prediction, we monitor the total flux of ϕ through the west (left) boundary, at which ϕ is prescribed. Because the convection flux is zero on this boundary, this quantity is obtained by summing diffusion fluxes over all CV faces along this boundary, which are approximated by Eqs. (4.49) and (4.53). Figure 4.10 shows the variation of the flux as the grid is refined for the UDS and CDS discretizations of the convection fluxes; the diffusion fluxes are always discretized using CDS. The grid was refined from 10×10 CV to 320×320 CV. On the coarsest grid, the CDS does not produce a meaningful solution for $\Gamma = 0.001$; convection dominates and, on such a coarse grid, the rapid change in ϕ over short distance near the west boundary (see Fig. 4.9) results in oscillations so strong that most iterative solvers fail to converge (the local cell Peclet numbers, $\mathrm{Pe}_\Delta = \rho u_x \Delta x / \Gamma$, range between 10 and 100 on this grid). (A converged solution could probably be obtained with the aid of deferred correction but it would be very inaccurate.) As the grid is refined, the CDS results converge monotonically towards a grid-independent solution. On

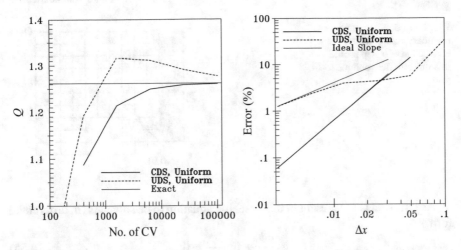

Fig. 4.10 Convergence of total flux of ϕ through the west wall (left) and the error in computed flux as a function of grid spacing, for $\Gamma = 0.001$

the 40×40 CV grid the local Peclet numbers range from 2.5 to 25, but there are no oscillations in the solution, as can be seen in Fig. 4.9.

The UDS solution does not oscillate on any grid, as expected. The convergence is, however, not monotonic: the flux on the two coarsest grids lies below the converged value; it is too high on the next grid and then approaches the correct result monotonically. By assuming second-order convergence of the CDS scheme, we estimated the grid-independent solution via Richardson extrapolation (see Sect. 3.9 for details) and were able to determine the error in each solution. The errors are plotted vs. normalized grid size ($\Delta x = 1$ for the coarsest grid) in Fig. 4.10 for both UDS and CDS. The expected slopes for first- and second-order schemes are also shown. The CDS error curve has the slope expected of a second-order scheme. The UDS error shows irregular behavior on the first three grids. From the fourth grid onwards the error curve approaches the expected slope. The UDS solution on the grid with 320×320 CV is still in error by over 1%; CDS produces a more accurate result on the 80×80 grid!

4.7.3 Testing the Numerical Diffusion

Another popular test case is the convection of a step profile in a uniform flow oblique to grid lines; see Fig. 4.11. It can be solved using the method described above by adjusting the boundary conditions (prescribed values of ϕ at west and south boundaries, outflow conditions at north and east boundaries). We show below the results obtained using the UDS and CDS discretizations.

Fig. 4.11 Geometry and boundary conditions for convection of a step profile in a uniform flow oblique to grid lines

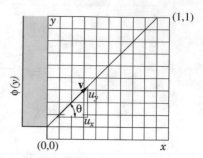

As diffusion is not present in this case, the equation to be solved is (in differential form):

$$u_x \frac{\partial \phi}{\partial x} + u_y \frac{\partial \phi}{\partial y} = 0 \ . \tag{4.55}$$

For this case, the UDS on a uniform grid in both directions gives the very simple equation:

$$u_x \frac{\phi_P - \phi_W}{\Delta x} + u_y \frac{\phi_P - \phi_S}{\Delta y} = 0 \ , \tag{4.56}$$

which is readily solved in a sequential manner without iteration. On the other hand, CDS gives a zero value for the coefficient on the main diagonal, A_P, making solution difficult. Most iterative solvers would fail to converge for this problem; however, by using the deferred-correction approach mentioned above and described in Sect. 5.6, it is possible to obtain the solution.

If the flow is parallel to the x-coordinate, both schemes give the correct result: the profile is simply convected downstream. When the flow is oblique to grid lines, UDS produces a smeared step profile at any downstream cross-section, while CDS produces oscillations. In Fig. 4.12 we show the profile of ϕ at $x = 0.5$ for the case when the flow is at 45° to the grid ($u_x = u_y = 1$; $\rho = 1$; at the west boundary $\phi = 0$ for $y < 0.1$ and $\phi = 1$ for $0.1 < y < 1$; at the south boundary, $\phi = 0$), obtained on a sequence of three uniform grids with 20×20, 40×40 and 80×80 CVs using three different discretizations: UDS, CDS, and a blend of 95% CDS and 5% UDS. The effect of numerical diffusion is clearly seen in the UDS solution; the step is highly smeared and the difference between solutions on consecutive grids is almost the same, indicated that the asymptotic convergence of first order is not yet reached—one would need to refine the grid many more times before this difference is halved when the mesh spacing is halved. On the other hand, CDS produces a profile with the proper steepness at the step, but it oscillates on both sides of the step and produces over- and undershoots. The amplitude of wiggles does not decrease with grid refinement—only the wavelength reduces. When 5% of UDS is blended with 95% of CDS, the amplitude of wiggles is dramatically reduced and it becomes significantly smaller as the grid is refined. In this example, physical diffusion is absent; in real flows, viscosity and diffusivity are always present and at a sufficiently fine grid,

Fig. 4.12 Profile of ϕ at $x = 0.5$ calculated on three grids using UDS (top), a blend of 95% CDS and 5% UDS (middle) and CDS (bottom)

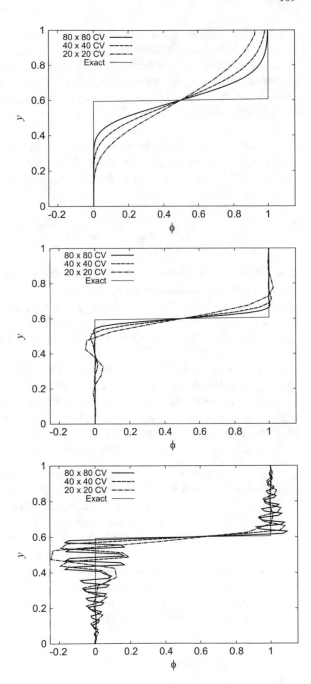

wiggles would disappear in CDS-solution, as demonstrated in Sect. 3.10. Local grid refinement would help localize and, perhaps even remove, the oscillations as will be discussed in Chap. 12. The oscillations could also be removed by locally introducing numerical diffusion (e.g., by blending CDS with UDS only where necessary rather than uniformly in whe whole solution domain, as was done above). This is sometimes done in compressible flows near shocks (see Sect. 4.4.6 for a systematic approach to such blending).

By using the *modified equation method* described in Sect. 4.4.6, one can show (using a Taylor-series expansion of the difference equation about the cell center) that the UDS method embodied in Eq. (4.56) is more nearly solving the convection-diffusion problem:

$$u_x \frac{\partial \phi}{\partial x} + u_y \frac{\partial \phi}{\partial y} = u_x \, \Delta x \, \frac{\partial^2 \phi}{\partial x^2} + u_y \, \Delta y \, \frac{\partial^2 \phi}{\partial y^2} \qquad (4.57)$$

than the original Eq. (4.55). Equation (4.57) is the *modified equation* for this problem. By transforming this equation into coordinates parallel and perpendicular to the flow, one can show that the effective diffusivity in the normal direction is:

$$\Gamma_{\text{eff}} = U \, \sin\theta \, \cos\theta (\Delta x \, \cos\theta + \Delta y \, \sin\theta) \,, \qquad (4.58)$$

where U is the magnitude of the velocity and θ is the angle of the flow with respect to the x-direction. A similar and widely quoted result was derived by de Vahl Davis (1972). As in Sect. 4.4.6, we see that, as a consequence of the truncation errors exposed by the modified equation, the FV discrete equation does not solve exactly the differential equation whose solution we sought.

To sum up our findings, we have shown:

- High-order schemes oscillate on coarse grids but converge to an accurate solution more rapidly than low-order schemes as the grid is refined.
- First-order UDS is inaccurate and should not be used. This scheme is mentioned because it is still used in some commercial codes. Users should be aware that high accuracy cannot be obtained on affordable grids with this method, especially in 3D. It introduces a large diffusive error in both the streamwise and normal directions.[3]
- CDS is the simplest scheme of second-order accuracy and offers a good compromise among accuracy, simplicity and efficiency. However, care is needed in convection-dominated problems and some kind of TVD-approach may be needed to avoid wiggles if the grid is not fine enough.

---

[3]The Spalding hybrid scheme (Sect. 4.4.5) can have this effect as well, as seen in Freitas et al. (1985), where the replacement of that scheme by QUICK (Sect. 4.4.3) in a simulation with a three-dimensional unsteady code revealed vortexes and other three-dimensional effects that had been hidden by the spurious numerical diffusion of the hybrid scheme.

Chapter 5
Solution of Linear Equation Systems

5.1 Introduction

In the previous two chapters we showed how the convection-diffusion equation may be discretized using FD and FV methods. In either case, the result of the discretization process is a system of algebraic equations, which are linear or non-linear according to the nature of the partial differential equation(s) from which they are derived. In the non-linear case, the discretized equations must be solved by an iterative technique that involves guessing a solution, linearizing the equations about that solution, and improving the solution; the process is repeated until a converged result is obtained. So, whether the equations are linear or not, efficient methods for solving linear systems of algebraic equations are needed.

The matrices derived from partial differential equations are always sparse, i.e., most of their elements are zero. Some methods for solving the equations that arise when structured grids are used will be described below; all of the non-zero elements of the matrices then lie on a small number of well-defined diagonals; we may take advantage of this structure. Some of the methods are applicable to matrices arising from unstructured grids as well.

The structure of the coefficient matrix for a 2D problem discretized with a five-point approximation (upwind or central difference) is shown in Fig. 3.6. The algebraic equation for one CV or grid node is given by Eq. (3.43), and the matrix version of the complete problem is given by Eq. (3.44), see Sect. 3.8, which is repeated here:

$$A\phi = Q . \tag{5.1}$$

In addition to describing some of the better solution methods for linear algebraic systems representing discretized partial differential equations, we shall discuss the solution of non-linear systems of equations in this chapter. However, we begin with linear equations. It is assumed that the reader has had some contact with methods for solving linear systems so the descriptions are brief.

© Springer Nature Switzerland AG 2020
J. H. Ferziger et al., *Computational Methods for Fluid Dynamics*,
https://doi.org/10.1007/978-3-319-99693-6_5

5.2 Direct Methods

The matrix A is assumed to be very sparse. In fact, the most complicated matrix we shall encounter is a banded matrix of block type; this greatly simplifies the task of solution but we shall briefly review methods for general matrices as methods for sparse matrices are closely related to them. For the description of methods designed to deal with full matrices, use of full-matrix notation (as opposed to the diagonal notation introduced earlier) is more sensible and will be adopted.

5.2.1 Gauss Elimination

The basic method for solving linear systems of algebraic equations is Gauss elimination. Its basis is the systematic reduction of large systems of equations to smaller ones. In this procedure, the elements of the matrix are modified but, as the dependent variable names do not change, it is convenient to describe the method in terms of the matrix alone:

$$
A = \begin{pmatrix}
A_{11} & A_{12} & A_{13} & \ldots & A_{1n} \\
A_{21} & A_{22} & A_{23} & \ldots & A_{2n} \\
\vdots & \vdots & \vdots & \ddots & \vdots \\
A_{n1} & A_{n2} & A_{n3} & \ldots & A_{nn}
\end{pmatrix}. \tag{5.2}
$$

The heart of the algorithm is the technique for eliminating A_{21}, i.e., replacing it with a zero. This is accomplished by multiplying the first equation (first row of the matrix) by A_{21}/A_{11} and subtracting it from the second row or equation; in the process, all of the elements in the second row of the matrix are modified as is the second element of the forcing vector on the right-hand side of the equation. The other elements of the first column of the matrix, $A_{31}, A_{41}, \ldots, A_{n1}$ are treated similarly; for example, to eliminate A_{i1}, the first row of the matrix is multiplied by A_{i1}/A_{11} and subtracted from the ith row. By systematically proceeding down the first column of the matrix, all of the elements below A_{11} are eliminated. When this process is complete, none of the equations 2, 3, $\ldots, n$ contain the variable ϕ_1; they are a set of $n - 1$ equations for the variables $\phi_2, \phi_3, \ldots, \phi_n$. The same procedure is then applied to this smaller set of equations—all of the elements below A_{22} in the second column are eliminated.

This procedure is carried out for columns 1, 2, 3, $\ldots, n - 1$. After this process is complete, the original matrix has been replaced by an upper triangular one:

$$
U = \begin{pmatrix}
A_{11} & A_{12} & A_{13} & \ldots & A_{1n} \\
0 & A_{22} & A_{23} & \ldots & A_{2n} \\
\vdots & \vdots & \vdots & \ddots & \vdots \\
0 & 0 & 0 & \ldots & A_{nn}
\end{pmatrix}. \tag{5.3}
$$

All of the elements except those in the first row differ from those in the original matrix A. As the elements of the original matrix will never be needed again, it is efficient to store the modified elements in place of the original ones. (In the rare case requiring that the original matrix be saved, a copy can be created prior to starting the elimination procedure.)

This portion of the algorithm just described is called *forward elimination*. The elements on the right-hand side of the equation, Q_i, are also modified in this procedure.

The upper triangular system of equations resulting from forward elimination is easily solved. The last equation contains only one variable, ϕ_n and is readily solved:

$$\phi_n = \frac{Q_n}{A_{nn}} . \tag{5.4}$$

The next to last equation contains only ϕ_{n-1} and ϕ_n and, once ϕ_n is known, it can be solved for ϕ_{n-1}. Proceeding upward in this manner, each equation is solved in turn; the ith equation yields ϕ_i:

$$\phi_i = \frac{Q_i - \sum_{k=i+1}^{n} A_{ik}\phi_k}{A_{ii}} . \tag{5.5}$$

The right-hand side is calculable because all of the ϕ_k appearing in the sum have already been evaluated. In this way, all of the variables may be computed. The part of the Gauss elimination algorithm which starts with the triangular matrix and computes the unknowns is called *back substitution*.

It is not difficult to show that, for large n, the number of operations required to solve a linear system of n equations by Gauss elimination is proportional to $n^3/3$. The bulk of this effort is in the forward elimination phase; the back substitution requires only $n^2/2$ arithmetic operations and is much less costly than the forward elimination. Gauss elimination is thus expensive but, for full matrices, it is as good as any method available. The high cost of Gauss elimination provides incentive to search for more efficient special solvers for matrices such as the sparse ones arising from the discretization of differential equations.

For large systems that are not sparse, Gauss elimination is susceptible to accumulation of errors (see Golub and van Loan 1996; Watkins 2010) which makes it unreliable if not modified. The addition of *pivoting* or interchange of rows in order to make the pivot elements (the diagonal elements that appear in the denominators) as large as possible, keeps the error growth in check. Fortunately, for sparse matrices, error accumulation is rarely a problem so this issue is not important here.

Gauss elimination does not vectorize or parallelize well and is rarely used without modification in CFD problems.

5.2.2 LU Decomposition

A number of variations on Gauss elimination have been proposed. Most are of little interest here. One variant of value to CFD is LU decomposition. It is presented without derivation.

We have seen that, in Gauss elimination, forward elimination reduces a full matrix to an upper triangular one. This process can be carried out in a more formal manner by multiplying the original matrix A by a lower triangular matrix. By itself, this is of little interest but, as the inverse of a lower triangular matrix is also lower triangular, this result shows that any matrix A, subject to some limitations that can be ignored here, can be factored into the product of lower (L) and upper (U) triangular matrices:

$$A = LU .\tag{5.6}$$

To make the factorization unique, we require that the diagonal elements of L, L_{ii}, all be unity; alternatively, one could require the diagonal elements of U to be unity.

What makes this factorization useful is that it is easily constructed. The upper triangular matrix U is precisely the one produced by the forward phase of Gauss elimination. Furthermore, the elements of L are the multiplicative factors (e.g., A_{ji}/A_{ii}) used in the elimination process. This allows the factorization to be constructed by a minor modification of Gauss elimination. Furthermore, the elements of L and U can be stored where the elements of A were.

The existence of this factorization allows the solution of the system of equations (5.1) in two stages. With the definition:

$$U\boldsymbol{\phi} = \mathbf{Y} ,\tag{5.7}$$

the system of equations (5.1) becomes:

$$L\mathbf{Y} = \mathbf{Q} .\tag{5.8}$$

The latter set of equations can be solved by a variation of the method used in the backward substitution phase of Gauss elimination in which one starts from the top rather than the bottom of the system. Once Eq. (5.8) has been solved for $\mathbf{Y}$, Eq. (5.7), which is identical to the triangular system solved in the back substitution phase of Gauss elimination, can be solved for $\boldsymbol{\phi}$.

The advantage of LU factorization over Gauss elimination is that the factorization can be performed without knowing the vector $\mathbf{Q}$. As a result, if many systems involving the same matrix are to be solved, considerable savings can be obtained by performing the factorization first; the systems can then be solved as required. As we shall see below, variations on LU factorization are the basis of some of the better iterative methods of solving systems of linear equations; this is the principal reason for introducing it here.

5.2.3 Tridiagonal Systems

When ordinary differential equations (1D problems) are finite differenced, for example, with the CDS approximation, the resulting algebraic equations have an especially simple structure. Each equation contains only the variables at its own node and its immediate left and right neighbors:

$$A_W^i \phi_{i-1} + A_P^i \phi_i + A_E^i \phi_{i+1} = Q_i \; . \tag{5.9}$$

The corresponding matrix A has non-zero terms only on its main diagonal (represented by A_P) and the diagonals immediately above and below it (represented by A_E and A_W, respectively). Such a matrix is called *tridiagonal*; systems containing tridiagonal matrices are especially easy to solve. The matrix elements are best stored as three $n \times 1$ arrays.

Gauss elimination is especially easy for tridiagonal systems: only one element needs to be eliminated from each row during the forward elimination process. When the algorithm has reached the ith row, only A_P^i needs to be modified; the new value is:

$$A_P^i = A_P^i - \frac{A_W^i A_E^{i-1}}{A_P^{i-1}} \; , \tag{5.10}$$

where this equation is to be understood in the programmer's sense that the result is stored in place of the original A_P^i. The forcing term is also modified:

$$Q_i^* = Q_i - \frac{A_W^i Q_{i-1}^*}{A_P^{i-1}} \; . \tag{5.11}$$

The back substitution part of the method is also simple. The ith variable is computed from:

$$\phi_i = \frac{Q_i^* - A_E^i \phi_{i+1}}{A_P^i} \; . \tag{5.12}$$

This tridiagonal solution method is sometimes called the *Thomas Algorithm* or the Tridiagonal Matrix Algorithm (TDMA). It is easily programmed (a FORTRAN code requires only eight executable lines) and, more importantly, the number of operations is proportional to n, the number of unknowns, rather than the n^3 of full matrix Gauss elimination. In other words, the cost per unknown is independent of the number of unknowns, which is almost as good a scaling as one could desire. The low cost suggests that this algorithm be employed whenever possible. Many solution methods take advantage of the low cost of this method by reducing the problem to one involving tridiagonal matrices.

5.2.4 Cyclic Reduction

There are cases even more special that allow still greater cost reduction than that offered by TDMA. An interesting example is provided by systems in which the matrix is not only tridiagonal but all of the elements on each diagonal are identical. The *cyclic reduction* method can be used to solve such systems with a cost per variable that actually decreases as the system becomes larger. Let us see how that is possible.

Suppose that, in the system (5.9), the coefficients A_W^i, A_P^i and A_E^i are independent of the index i; we may then drop the index. Then, for even values of i, we multiply row $i - 1$ by A_W/A_P and subtract it from row i. Then we multiply row $i + 1$ by A_E/A_P and subtract it from row i. This eliminates the elements to the immediate left and right of the main diagonal in the even-numbered rows but replaces the zero element two columns to the left of the main diagonal by $-A_W^2/A_P$ and the zero element two columns to the right of the main diagonal by $-A_E^2/A_P$; the diagonal element becomes $A_P - 2A_W A_E/A_P$. Because the elements in every even row are the same, the calculation of the new elements needs to be done only once; this is where the savings come from.

At the completion of these operations the even numbered equations contain only even indexed variables and constitute a set of $n/2$ equations for these variables; considered as a separate system, these equations are tridiagonal and the elements on each diagonal of the reduced matrix are again equal. In other words, the reduced set of equations has the same form as the original one but is half the size. It can be further reduced in the same way. If the number of equations in the original set is a power of two (or certain other convenient numbers), the method can be continued until only one equation remains; the latter is solved directly. The remaining variables can then be found by a variant of back substitution.

One can show that the cost of this method is proportional to $\log_2 n$, so that the cost per variable decreases with the number of variables. Although the method may seem rather specialized, there are CFD applications in which it plays a role. These are flows in very regular geometries such as the rectangular boxes which are used, for example, in direct or large-eddy simulations of turbulence and in some meteorological applications.

In these applications, cyclic reduction and related methods provide the basis for methods of solving elliptic equations such as Laplace and Poisson equations directly, that is, non-iteratively. Because the solutions are also exact in the sense that they contain no iteration error, this method is invaluable whenever it can be used.[1]

Cyclic reduction is closely related to the fast Fourier transform which is also used to solve elliptic equations in simple geometries. Fourier methods may also be used for evaluating derivatives, as was shown in Sect. 3.11.

---

[1]Bini et al. (2009) review the history, extensions, and new proofs and formulas for cyclic reduction. It is being used as a smoother for multigrid applications on highly parallel, graphics processors (GPUs) now being used for fluid flow calculations (Göddeke and Strzodka 2011).

5.3 Iterative Methods

5.3.1 Basic Concept

Any system of equations can be solved by Gauss elimination or LU decomposition. Unfortunately, the triangular factors of sparse matrices are not sparse, so the cost of these methods is quite high. Furthermore, the discretization error is usually much larger than the accuracy of the computer arithmetic so there is no reason to solve the system that accurately. Solution to somewhat more accuracy than that of the discretization scheme suffices.

This leaves an opening for iterative methods. They are used out of necessity for non-linear problems, but they are just as valuable for sparse linear systems. In an iterative method, one guesses a solution, and uses the equation to systematically improve it. If each iteration is cheap and the number of iterations is small, an iterative solver may cost less than a direct method. In CFD problems this is usually the case.

Consider the matrix problem represented by Eq. (5.1) which might result from FD or FV approximation of a flow problem. After n iterations we have an approximate solution ϕ^n which does not satisfy these equations exactly. Instead, there is a non-zero residual ρ^n:

$$A\phi^n = Q - \rho^n . \tag{5.13}$$

By subtracting this equation from Eq. (5.1), we obtain a relation between the iteration error defined by:

$$\epsilon^n = \phi - \phi^n , \tag{5.14}$$

where ϕ is the converged solution, and the residual:

$$A\epsilon^n = \rho^n . \tag{5.15}$$

The purpose of the iteration procedure is to drive the residual to zero; in the process, ϵ also becomes zero. To see how this can be done, consider an iterative scheme for a linear system; such a scheme can be written:

$$M\phi^{n+1} = N\phi^n + \mathbf{B} . \tag{5.16}$$

An obvious property that must be demanded of an iterative method is that the converged result satisfies Eq. (5.1). As, by definition, at convergence, $\phi^{n+1} = \phi^n = \phi$, we must have:

$$A = M - N \quad \text{and} \quad \mathbf{B} = Q , \tag{5.17}$$

or, we can write, more generally,

$$PA = M - N \quad \text{and} \quad \mathbf{B} = PQ , \tag{5.18}$$

where P is a non-singular, so-called *preconditioning matrix*. Preconditioning matrices, which can materially speed convergence of the iterations, are discussed in Sect. 5.3.6.1 below.

An alternative version of this iterative method may be obtained by subtracting $M\phi^n$ from each side of Eq. (5.16). We obtain:

$$M(\phi^{n+1} - \phi^n) = \mathbf{B} - (M - N)\phi^n \quad \text{or} \quad M\delta^n = \rho^n , \tag{5.19}$$

where $\delta^n = \phi^{n+1} - \phi^n$ is called the *correction* or update and is an approximation to the iteration error.

For an iterative method to be effective, solving the system (5.16) must be cheap and the method must converge rapidly. Inexpensive iteration requires that computation of $N\phi^n$ and solution of the system must both be easy to perform. The first requirement is easily met; because A is sparse, N is also sparse, and computation of $N\phi^n$ is simple. The second requirement means that the iteration matrix M must be easily inverted; from a practical point of view, M should be diagonal, tridiagonal, triangular, or, perhaps, block tridiagonal or triangular; another possibility is described below. For rapid convergence, M should be a good approximation to A, making $N\phi$ small in some sense. This is discussed in the next section.

5.3.2 Convergence

As we have noted, rapid convergence of an iterative method is key to its effectiveness. Here we give a simple analysis that is useful in understanding what determines the convergence rate and provides insight into how to improve it.

To begin, we derive the equation that determines the behavior of the iteration error. To find it, we recall that, at convergence, $\phi^{n+1} = \phi^n = \phi$, so that the converged solution obeys the equation:

$$M\phi = N\phi + \mathbf{B} . \tag{5.20}$$

Subtracting this equation from Eq. (5.16) and using the definition (5.14) of the iteration error, we find:

$$M\epsilon^{n+1} = N\epsilon^n \tag{5.21}$$

or

$$\epsilon^{n+1} = M^{-1}N\epsilon^n . \tag{5.22}$$

The iterative method converges if $\lim_{n \to \infty} \epsilon^n = 0$. The critical role is played by the eigenvalues λ_k and eigenvectors ψ^k of the iteration matrix $M^{-1}N$ which are defined by:

$$M^{-1}N\psi^k = \lambda_k\psi^k , \quad k = 1, \ldots, K , \tag{5.23}$$

where K is the number of equations (grid points). We assume that the eigenvectors form a complete set, i.e., a basis for $\mathbf{R}^n$, the vector space of all n-component vectors. If that is so, the initial error may be expressed in terms of them:

$$\epsilon^0 = \sum_{k=1}^{K} a_k \boldsymbol{\psi}^k , \qquad (5.24)$$

where the a_k are constants. Then the iterative procedure (5.22) yields:

$$\epsilon^1 = M^{-1} N \epsilon^0 = M^{-1} N \sum_{k=1}^{K} a_k \boldsymbol{\psi}^k = \sum_{k=1}^{K} a_k \lambda_k \boldsymbol{\psi}^k \qquad (5.25)$$

and, by induction, it is not difficult to show that

$$\epsilon^n = \sum_{k=1}^{K} a_k (\lambda_k)^n \boldsymbol{\psi}^k . \qquad (5.26)$$

It is clear that, if ϵ^n is to become zero when n is large, the necessary and sufficient condition is that all of the eigenvalues must be less than unity in magnitude. In particular, this must be true of the largest eigenvalue, whose magnitude is called the *spectral radius* of the matrix $M^{-1}N$. In fact, after a number of iterations, the terms in Eq. (5.26) that contain eigenvalues of small magnitude become very small and only the term containing the largest eigenvalue (which we can take to be λ_1 and assume to be unique) remains:

$$\epsilon^n \sim a_1 (\lambda_1)^n \boldsymbol{\psi}^1 . \qquad (5.27)$$

If convergence is defined as the reduction of the iteration error below some tolerance δ, we require:

$$a_1 (\lambda_1)^n \approx \delta . \qquad (5.28)$$

Taking the logarithm of both sides of this equation, we find an expression for the required number of iterations:

$$n \approx \frac{\ln\left(\dfrac{\delta}{a_1}\right)}{\ln \lambda_1} . \qquad (5.29)$$

We see that, if the spectral radius is very close to unity, the iterative procedure will converge very slowly.

As a simple (trivial might be more descriptive) example consider the case of a single equation (for which one would never dream of using an iterative method). Suppose we want to solve:

$$ax = b \qquad (5.30)$$

and we use the iterative method (note that $m = a + n$ and p is the iteration counter):

$$mx^{p+1} = nx^p + b .$$
(5.31)

Then the error obeys the scalar equivalent of Eq. (5.22):

$$\epsilon^{p+1} = \frac{n}{m}\epsilon^p .$$
(5.32)

We see that the error is reduced quickly if n/m is small, i.e., if n is small, which means that $m \approx a$. In constructing iterative methods for systems, we shall find that an analogous result holds: the more closely M approximates A, the more rapid the convergence.

In an iterative method it is important to be able to estimate the iteration error in order to decide when to stop iterating. Calculation of the eigenvalues of the iteration matrix is difficult (it is often not explicitly known), so approximations have to be used. We shall describe some methods of estimating the iteration error and criteria for stopping iterations later in this chapter.

5.3.3   Some Basic Methods

In the simplest method, the Jacobi method, M is a diagonal matrix whose elements are the diagonal elements of A. For the five-point discretization of Laplace equation, if each iteration is begun at the lower left (southwest) corner of the domain and we use the geographic notation introduced above, the method is:

$$\phi_P^{n+1} = \frac{Q_P - A_S\phi_S^n - A_W\phi_W^n - A_N\phi_N^n - A_E\phi_E^n}{A_P} .$$
(5.33)

It may be shown that, for convergence, this method requires a number of iterations proportional to the square of the number of grid points in one direction. This means that it is more expensive than a direct solver so there is little reason to use it.

In the Gauss–Seidel method, M is the lower triangular portion of A. As it is a special case of the SOR method given below, we shall not give the equations separately. It converges twice as fast as the Jacobi method but this is not enough of an improvement to be useful.

One of the better methods is an accelerated version of the Gauss–Seidel method called *successive over-relaxation* or SOR, which we shall describe below. For an introduction and analysis of the Jacobi and Gauss–Seidel methods, see an introductory text on numerical methods such as Ferziger (1998) or Press et al. (2007)

If each iteration is begun at the lower left (southwest) corner of the domain and we again use the geographic notation, the SOR method can be written:

$$\phi_P^{n+1} = \omega \frac{Q_P - A_S\phi_S^{n+1} - A_W\phi_W^{n+1} - A_N\phi_N^n - A_E\phi_E^n}{A_P} + (1 - \omega)\phi_P^n , \qquad (5.34)$$

where ω is the over-relaxation factor, which must be greater than 1 for acceleration, and n is the iteration counter. There is theory to guide the selection of the optimum over-relaxation factor for simple problems such as Laplace equation in a rectangular domain but it is hard to apply that theory to more complex problems; fortunately, the behavior of the method is usually similar to that found in the simple case. Generally, the larger the number of grid points, the larger the optimum over-relaxation factor (see Sect. 5.7). Typically, $1.6 \leq \omega \leq 1.9$ will work well; for $\omega = 2.0$ the scheme diverges. For values of ω less than the optimum, the convergence is monotonic and the rate of convergence increases as ω increases. When the optimum ω is exceeded, the convergence rate deteriorates and the convergence is oscillatory. This knowledge can be used to search for the optimum over-relaxation factor. When the optimum over-relaxation factor is used, the number of iterations is proportional to the number of grid points in one direction, a substantial improvement over the methods mentioned above. For $\omega = 1$, SOR reduces to the Gauss–Seidel method.

5.3.4 Incomplete LU Decomposition: Stone's Method

We make two observations. LU decomposition is an excellent general-purpose linear systems solver but it cannot take advantage of the sparseness of a matrix. In an iterative method, if M is a good approximation to A, rapid convergence results. These observations lead to the idea of using an approximate LU factorization of A as the iteration matrix, M i.e.:

$$M = LU = A + N , \qquad (5.35)$$

where L and U are both sparse and N is small.

A version of this method for symmetric matrices, known as *incomplete Cholesky* factorization, is often used in conjunction with conjugate gradient methods. Because the matrices that arise from discretizing convection-diffusion problems or the Navier–Stokes equations are not symmetric, this method cannot be applied to them. An asymmetric version of this method, called *incomplete* LU factorization or ILU, is possible but has not found widespread use. In the ILU method one proceeds as in LU decomposition but, for every element of the original matrix A that is zero, the corresponding element of L or U is set to zero. This factorization is not exact, but the product of these factors can be used as the matrix M of the iterative method. This method converges rather slowly.

Another incomplete lower-upper decomposition method, which has found use in CFD, was proposed by Stone (1968). This method, also called the *strongly implicit procedure* (SIP), is specifically designed for algebraic equations that are discretizations of partial differential equations and does not apply to generic systems of equations.

We shall describe the SIP method for the five-point computational molecule, i.e., a matrix with the structure shown in Fig. 3.6. The same principles can be used to construct solvers for 7-point (in 3D) and 9-point (for 2D non-orthogonal grids) computational molecules.

As in ILU, the L and U matrices have non-zero elements only on diagonals on which A has non-zero elements. The product of lower and upper triangular matrices with these structures has more non-zero diagonals than A. For the standard five-point molecule there are two more diagonals (corresponding to nodes NW and SE or NE and SW, depending on the ordering of the nodes in the vector), and for seven-point molecules in 3D, there are six more diagonals. For the ordering of nodes used in this book for 2D problems, the extra two diagonals correspond to the nodes NW and SE (see Table 3.2 for the correspondence of the grid indices (i, j) and the one-dimensional storage location index l).

To make these matrices unique, every element on the main diagonal of U is set to unity. Thus five sets of elements (three in L, two in U) need to be determined. For matrices of the form shown in Fig. 5.1, the rules of matrix multiplication give the elements of the product of L and U, $M = LU$:

$$M_W^l = L_W^l$$
$$M_{NW}^l = L_W^l U_N^{l-N_j}$$
$$M_S^l = L_S^l$$
$$M_P^l = L_W^l U_E^{l-N_j} + L_S^l U_N^{l-1} + L_P^l \qquad (5.36)$$
$$M_N^l = U_N^l L_P^l$$
$$M_{SE}^l = L_S^l U_E^{l-1}$$
$$M_E^l = U_E^l L_P^l$$

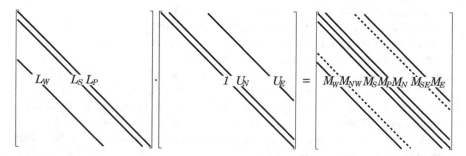

Fig. 5.1 Schematic presentation of the matrices L and U and the product matrix M; diagonals of M not found in A are shown by dashed lines

We wish to select L and U, such that M is as good an approximation to A as possible. At minimum, N contains the two non-zero diagonals of M that correspond to zero diagonals of A; see Eq. (5.36). An obvious choice is to let N have non-zero elements on only these two diagonals, and force the other diagonals of M to equal the corresponding diagonals of A. This is possible; in fact, this is the standard ILU method mentioned earlier. Unfortunately, this method converges slowly.

Stone (1968) recognized that convergence can be improved by allowing N to have non-zero elements on the diagonals corresponding to all seven non-zero diagonals of LU. The method is most easily derived by considering the vector $M\boldsymbol{\phi}$:

$$(M\boldsymbol{\phi})_P = M_P\phi_P + M_S\phi_S + M_N\phi_N + M_E\phi_E + M_W\phi_W +$$
$$M_{NW}\phi_{NW} + M_{SE}\phi_{SE} \, . \tag{5.37}$$

The last two terms are the 'extra' ones. Each term in this equation corresponds to a diagonal of $M = LU$.

The matrix N must contain the two 'extra' diagonals of M, and we want to choose the elements on the remaining diagonals so that $N\boldsymbol{\phi} \approx \mathbf{0}$ or, in other words,

$$N_P\phi_P + N_N\phi_N + N_S\phi_S + N_E\phi_E + N_W\phi_W + M_{NW}\phi_{NW} + M_{SE}\phi_{SE} \approx 0 \, . \tag{5.38}$$

This requires that the contribution of the two 'extra' terms in the above equation be nearly canceled by the contribution of other diagonals. Accordingly, Eq. (5.38) should reduce to the following expression:

$$M_{NW}(\phi_{NW} - \phi_{NW}^*) + M_{SE}(\phi_{SE} - \phi_{SE}^*) \approx 0 \, , \tag{5.39}$$

where ϕ_{NW}^* and ϕ_{SE}^* are approximations to ϕ_{NW} and ϕ_{SE}.

Stone's key idea is that, because the equations approximate an elliptic partial differential equation, the solution can be expected to be smooth. This being so, ϕ_{NW}^* and ϕ_{SE}^* can be approximated in terms of the values of ϕ at nodes corresponding to the diagonals of A. Stone proposed the following approximation (other approximations are possible; see Schneider and Zedan (1981), for an example):

$$\phi_{NW}^* \approx \alpha(\phi_W + \phi_N - \phi_P) \, ,$$
$$\phi_{SE}^* \approx \alpha(\phi_S + \phi_E - \phi_P) \, . \tag{5.40}$$

If $\alpha = 1$, these are second-order accurate interpolations but Stone found that stability requires $\alpha < 1$. These approximations are based on the connection to partial differential equations and make little sense for generic algebraic equations.

If these approximations are substituted into Eq. (5.39) and the result is equated to Eq. (5.38), we obtain all elements of N as linear combinations of M_{NW} and M_{SE}. The elements of M, Eq. (5.36), can now be set equal to the sum of elements of A and N. The resulting equations are not only sufficient to determine all of the elements of L and U, but they can be solved in sequential order beginning at the southwest corner of the grid:

$$L_{\mathrm{W}}^l = A_{\mathrm{W}}^l / \left(1 + \alpha U_{\mathrm{N}}^{l-N_j}\right),$$

$$L_{\mathrm{S}}^l = A_{\mathrm{S}}^l / \left(1 + \alpha U_{\mathrm{E}}^{l-1}\right),$$

$$L_{\mathrm{P}}^l = A_{\mathrm{P}}^l + \alpha\left(L_{\mathrm{W}}^l U_{\mathrm{N}}^{l-N_j} + L_{\mathrm{S}}^l U_{\mathrm{E}}^{l-1}\right) - L_{\mathrm{W}}^l U_{\mathrm{E}}^{l-N_j} - L_{\mathrm{S}}^l U_{\mathrm{N}}^{l-1}, \tag{5.41}$$

$$U_{\mathrm{N}}^l = \left(A_{\mathrm{N}}^l - \alpha L_{\mathrm{W}}^l U_{\mathrm{N}}^{l-N_j}\right)/L_{\mathrm{P}}^l,$$

$$U_{\mathrm{E}}^l = \left(A_{\mathrm{E}}^l - \alpha L_{\mathrm{S}}^l U_{\mathrm{E}}^{l-1}\right)/L_{\mathrm{P}}^l.$$

The coefficients must be calculated in this order. For nodes next to boundaries, any matrix element that carries the index of a boundary node is understood to be zero. Thus, along the west boundary ($i = 2$), elements with index $l - N_j$ are zero; along the south boundary ($j = 2$), elements with index $l - 1$ are zero; along the north boundary ($j = N_j - 1$), elements with index $l + 1$ are zero; finally, along the east boundary ($i = N_i - 1$), elements with index $l + N_j$ are zero.

We now turn to solving the system of equations with the aid of this approximate factorization. The equation relating the update to the residual is (see Eq. (5.19)):

$$LU\delta^{n+1} = \rho^n. \tag{5.42}$$

The equations are solved as in in generic LU decomposition. Multiplication of the above equation by L^{-1} leads to:

$$U\delta^{n+1} = L^{-1}\rho^n = \mathbf{R}^n. \tag{5.43}$$

$\mathbf{R}^n$ is easily computed:

$$R^l = \left(\rho^l - L_{\mathrm{S}}^l R^{l-1} - L_{\mathrm{W}}^l R^{l-N_j}\right)/L_{\mathrm{P}}^l. \tag{5.44}$$

This equation is to be solved by marching in the order of increasing l. When the computation of $\mathbf{R}$ is complete, we need to solve Eq. (5.43):

$$\delta^l = R^l - U_{\mathrm{N}}^l\delta^{l+1} - U_{\mathrm{E}}^l\delta^{l+N_j} \tag{5.45}$$

in the order of decreasing index l.

In the SIP method, the elements of the matrices L and U need be calculated only once, prior to the first iteration. On subsequent iterations, we need calculate only the residual, then $\mathbf{R}$ and finally δ, by solving the two triangular systems.

Stone's method usually converges in a small number of iterations. The rate of convergence can be improved by varying α from iteration to iteration (and point to point). These methods converge in fewer iterations but they require the factorization to be redone each time α is changed. Because computing L and U is as expensive as an iteration with a given decomposition, it is usually more efficient overall to keep α fixed.

Stone's method can be generalized to yield an efficient solver for the nine-diagonal matrices that arise when compact difference approximations are applied in two dimensions and for the seven-diagonal matrices that arise when central differences are used in three dimensions. A 3D (7-point) vectorized version is given by Leister and Perić (1994)); two 9-point versions for 2D problems are described by Schneider and Zedan (1981) and Perić (1987). Computer codes for five-diagonal (2D) and seven-diagonal (3D) matrices are available via the Internet; see the appendix for details. The performance of SIP for a model problem will be presented in Sect. 5.8.

Unlike other methods, Stone's method is both a good iterative technique in its own right and a good basis for conjugate gradient methods (where it is called a preconditioner) and multigrid methods (where it is used as a smoother). These methods will be described below.

5.3.5   ADI and Other Splitting Methods

A common method of solving elliptic problems is to add a term containing the first time derivative to the equation and solve the resulting parabolic problem until a steady state is reached. At that point, the time derivative is zero and the solution satisfies the original elliptic equation. Many iterative methods of solving elliptic equations, including most of those already described, can be interpreted in this way. In this section we present a method whose connection to parabolic equations is so close that it might not have been discovered if one were thinking only of elliptic equations.

Considerations of stability require methods for parabolic equations to be implicit in time. In two or three dimensions, this requires solution of a 2D or 3D elliptic problem at each time step; the cost can be enormous but it can be reduced considerably by using the *alternating direction implicit* or ADI method. We give only the simplest such method in 2D and a variant. ADI is the basis for many other methods; for more details of some of these methods see Hageman and Young (2004).

Suppose we want to solve 2D Laplace equation. Adding a time derivative to it converts it to the heat equation in two dimensions:

$$\frac{\partial \phi}{\partial t} = \Gamma \left(\frac{\partial^2 \phi}{\partial x^2} + \frac{\partial^2 \phi}{\partial y^2} \right) . \tag{5.46}$$

If this equation is discretized using the trapezoid rule in time (called the Crank-Nicolson method when applied to partial differential equations; see the next chapter), and central differences are used to approximate the spatial derivatives on a uniform grid, we obtain:

$$\frac{\phi^{n+1} - \phi^n}{\Delta t} = \frac{\Gamma}{2} \left[\left(\frac{\delta^2 \phi^n}{\delta x^2} + \frac{\delta^2 \phi^n}{\delta y^2} \right) + \left(\frac{\delta^2 \phi^{n+1}}{\delta x^2} + \frac{\delta^2 \phi^{n+1}}{\delta y^2} \right) \right] , \tag{5.47}$$

where we have used the shorthand notation:

$$\left(\frac{\delta^2\phi}{\delta x^2}\right)_{i,j} = \frac{\phi_{i+1,j} - 2\phi_{i,j} + \phi_{i-1,j}}{(\Delta x)^2}$$

$$\left(\frac{\delta^2\phi}{\delta y^2}\right)_{i,j} = \frac{\phi_{i,j+1} - 2\phi_{i,j} + \phi_{i,j-1}}{(\Delta y)^2}$$

for the spatial finite differences. Rearranging Eq. (5.47), we find that, at time step $n+1$, we have to solve the system of equations:

$$\left(1 - \frac{\Gamma\Delta t}{2}\frac{\delta^2}{\delta x^2}\right)\left(1 - \frac{\Gamma\Delta t}{2}\frac{\delta^2}{\delta y^2}\right)\phi^{n+1} =$$

$$\left(1 + \frac{\Gamma\Delta t}{2}\frac{\delta^2}{\delta x^2}\right)\left(1 + \frac{\Gamma\Delta t}{2}\frac{\delta^2}{\delta y^2}\right)\phi^n - \tag{5.48}$$

$$\frac{(\Gamma\Delta t)^2}{4}\frac{\delta^2}{\delta x^2}\left[\frac{\delta^2(\phi^{n+1} - \phi^n)}{\delta y^2}\right].$$

As $\phi^{n+1} - \phi^n \approx \Delta t\,\partial\phi/\partial t$, the last term is proportional to $(\Delta t)^3$ for small Δt. Because the FD approximation is of second order, for small Δt, the last term is small compared to the discretization error and may be neglected. The remaining equation can be factored into two simpler equations:

$$\left(1 - \frac{\Gamma\Delta t}{2}\frac{\delta^2}{\delta x^2}\right)\phi^* = \left(1 + \frac{\Gamma\Delta t}{2}\frac{\delta^2}{\delta y^2}\right)\phi^n , \tag{5.49}$$

$$\left(1 - \frac{\Gamma\Delta t}{2}\frac{\delta^2}{\delta y^2}\right)\phi^{n+1} = \left(1 + \frac{\Gamma\Delta t}{2}\frac{\delta^2}{\delta x^2}\right)\phi^* . \tag{5.50}$$

Each of these systems of equations is a set of tridiagonal equations that can be solved with the efficient TDMA method; this requires no iteration and is much cheaper than solving Eq. (5.47). Either Eq. (5.49) or (5.50), as a method in its own right, is only first-order accurate in time and conditionally stable but the combined method is second-order accurate and unconditionally stable! The family of methods based on these ideas are known as *splitting* or *approximate factorization* methods; a wide variety of them has been developed.

Neglect of the third-order term, which is essential to the factorization, is justified only when the time step is small. So, although the method is unconditionally stable, it may not be accurate in time if the time step is large. For elliptic equations, the objective is to obtain the steady-state solution as quickly as possible; this is best accomplished with the largest possible time step. However, the factorization error becomes large when the time step is large so the method loses some of its effectiveness. In fact, there is an optimum time step which gives the most rapid convergence. When this time step is used, the ADI method is very efficient—it converges in a number of iterations proportional to the number of points in one direction.

A better strategy uses different time steps for several iterations in a cyclic fashion. This approach can make the number of iterations for convergence proportional to the square root of the number of grid points in one direction, making ADI an excellent method.

Equations which involve the convection and source terms require some generalization of this method. In CFD, the pressure or pressure-correction equation is of the above type and the variants of the method just described are often used to solve it. ADI methods are very commonly used when solving compressible flow problems. They are also well adapted to parallel computation.

The method described in this section takes advantage of the structure of the matrix which is, in turn, due to the use of a structured grid. However, closer inspection of the development shows that the basis of the method is an *additive decomposition* of the matrix:

$$A = H + V , \tag{5.51}$$

where H is the matrix representing the terms contributed by the second derivative with respect to x and V, the terms coming from the second derivative in the y-direction.

There is no reason why other additive decompositions cannot be used. One useful suggestion is to consider the additive LU decomposition:

$$A = L + U . \tag{5.52}$$

This is different from the multiplicative LU decomposition of Sect. 5.2.2. With this decomposition, Eqs. (5.49) and (5.50) are replaced by:

$$\begin{aligned}
(I - L\,\Delta t)\phi^* &= (I + U\,\Delta t)\phi^n , \\
(I - U\,\Delta t)\phi^{n+1} &= (I + L\,\Delta t)\phi^* .
\end{aligned} \tag{5.53}$$

Each of these steps is essentially a Gauss–Seidel iteration. The rate of convergence of this method is similar to that of the ADI method given above. It also has the very important advantage that it does not rely on the structure of the grid or of the matrix and may therefore be applied to problems on unstructured grids as well as structured ones. However, it does not parallelize as well as the HV version of ADI.

5.3.6 Krylov Methods

In this section, we present methods based on techniques for solving non-linear equations. Non-linear solvers can be grouped into two broad categories: Newton-like methods (Sect. 5.5.1) and global methods. The former converge very quickly if an accurate estimate of the solution is available but may fail catastrophically if the initial guess is far from the exact solution. 'Far' is a relative term; it is different for each equation. One cannot determine whether an estimate is 'close enough' except by trial and error. Global methods are guaranteed to find the solution (if one exists) but are not very fast. Combinations of the two types of methods are often used; global methods are used initially and followed by Newton-like methods as convergence is approached.

The object of the methods in this section is essentially to create schemes for reducing large linear systems of equations to smaller problems by creating approximations for the large matrix and then using those in the iteration process; this is called a *projection method* because we project the problem from, say, an $N \times N$ to one of much smaller dimension. van der Vorst (2002) provides a context for this approach as follows. Given our equation set (5.1) with the matrix A, an iteration scheme at the kth step of the form

$$\phi_k = (I - A)\phi_{k-1} + \mathbf{Q}$$

yields[2]

$$\phi_k = \phi_0 + K^k(A; \rho^0) = \phi_0 + \{\rho^0, A\rho^0, \cdots, A^{k-1}\rho^0\} \qquad (5.54)$$

where the initial residual is $\rho^0 = \mathbf{Q} - A\phi_0$ if ϕ_0 is the initial guessed solution. The iteration process can be seen to generate approximate solutions in what are called shifted *Krylov* subspaces $K^k(A; \rho^0)$. This particular scheme (Richardson iteration), while straightforward, is neither efficient nor optimal. The Krylov subspace projection approach attempts then to improve upon it by constructing better approximate solutions.

van der Vorst (2002) succinctly describes three classes of methods for doing this:

1. The *Ritz–Galerkin approach* (R–G): Construct the ϕ_k for which the residual is orthogonal to the current subspace $\mathbf{Q} - A\phi_k \perp K^k(A; \rho_0)$.
2. The *minimum residual approach*: Identify the ϕ_k for which the Euclidean norm $\parallel \mathbf{Q} - A\phi_k \parallel_2$ is minimal over $K^k(A; \rho^0)$.
3. The *Petrov–Galerkin approach* (P–G): Find a ϕ_k so that the residual $\mathbf{Q} - A\phi_k$ is orthogonal to an other suitable k-dimensional space.

The methods presented below are from these classes; for reference, see also Saad (2003). The conjugate gradient method (CG) is an R-G approach, Biconjugate gradients and CGSTAB are P-G approaches, and GMRES is a minimum residual approach.

---

[2]For example, $\phi_3 = \phi_0 + \rho^0 + (I - A)\rho^0 + (I - A)(I - A)\rho^0$.

5.3.6.1 Conjugate Gradient Methods

Many global methods are descent methods.[3] These methods begin by converting the original system of equations into a minimization problem. Suppose that the set of equations to be solved is given by Eq. (5.1) and that the matrix A is symmetric and its eigenvalues are positive; such a matrix is called *positive definite*. (Most matrices associated with problems in fluid mechanics are not symmetric or positive definite so we will need to generalize this method later.) For positive definite matrices, solving the system of equations (5.1) is equivalent to the problem of finding the minimum of

$$F = \frac{1}{2}\phi^T A\phi - \phi^T Q = \frac{1}{2}\sum_{j=1}^{n}\sum_{i=1}^{n}A_{ij}\phi_i\phi_j - \sum_{i=1}^{n}\phi_i Q_i \qquad (5.55)$$

with respect to all the ϕ_i; this may be verified by taking the derivative of F with respect to each variable and setting it equal to zero. A way to convert the original system into a minimization problem that does not require positive definiteness is to take the sum of the squares of all of the equations but this introduces additional difficulties.

The oldest and best known method for seeking the minimum of a function is *steepest descent*. The function F may be thought of as a surface in (hyper-)space. Suppose we have some starting guess which may be represented as a point in that (hyper-)space. At that point, we find the steepest downward path from the surface; it lies in the direction opposite to the gradient of the function. We then search for the lowest point on that line. By construction, it has a lower value of F than the starting point; in this sense, the new estimate is closer to the solution. The new value is then used as the starting point for the next iteration and the process is continued until it converges. Unfortunately, while it is guaranteed to converge, the steepest-descent method often converges very slowly.

If the contour plot of the magnitude of the function F has a narrow valley, the method tends to oscillate back and forth across that valley and many steps may be required to find the solution. In other words, the method tends to use the same search directions over and over again.

Many improvements have been suggested. The easiest ones require the new search directions to be as different from the old ones as possible. Among these is the *conjugate gradient* method. We shall give only the general idea and a description of the algorithm here; more complete presentations can be found in Shewchuk (1994), Watkins (2010), or Golub and van Loan (1996).

The conjugate gradient method is based on a remarkable discovery: it is possible to minimize a function with respect to several directions simultaneously while searching in one direction at a time. This is made possible by a clever choice of directions. We shall describe this for the case of two directions; suppose we wish to find values of α_1 and α_2 in

---

[3]The discussion in this book covers linear systems. Chapter 14 of Shewchuk (1994) describes the nonlinear conjugate gradient method and preconditioning for it.

$$\phi = \phi^0 + \alpha_1\,\mathbf{p}^1 + \alpha_2\,\mathbf{p}^2\,, \tag{5.56}$$

which minimize F; that is, we try to minimize F in the $\mathbf{p}^1 - \mathbf{p}^2$ plane. This problem can be reduced to the problem of minimizing with respect to $\mathbf{p}^1$ and $\mathbf{p}^2$ individually provided that the two directions are conjugate in the following sense:

$$\mathbf{p}^1 \cdot A\,\mathbf{p}^2 = 0\,. \tag{5.57}$$

This property is akin to orthogonality; the vectors $\mathbf{p}^1$ and $\mathbf{p}^2$ are said to be conjugate with respect to the matrix A, which gives the method its name. A detailed proof of this statement and others cited below can be found in the book of Golub and van Loan (1996).

This property can be extended to any number of directions. In the conjugate gradient method, each new search direction is required to be conjugate to all the preceding ones. If the matrix is non-singular, as is the case in nearly all engineering problems, the directions are guaranteed to be linearly independent. Consequently, if exact (no round-off error) arithmetic were employed, the conjugate gradient method would converge exactly when the number of iterations is equal to the size of the matrix. This number can be quite large and, in practice, exact convergence is not achieved due to arithmetic errors. It is therefore wiser to regard the conjugate gradient method as an iterative method.

While the conjugate gradient method guarantees that the error is reduced on each iteration, the size of the reduction depends on the search direction. It is not unusual for this method to reduce the error only slightly for a number of iterations and then find a direction that reduces the error by an order of magnitude or more in one iteration. In fact, the rate of convergence of this method (and, indeed, of iterative methods in general) depends on the coefficient matrix A (Eq. (5.1)). The change in the unknowns ϕ relative to perturbation of the right-hand side of the equation, $\mathbf{Q}$, can be shown (Watkins 2010) to be related to

$$\kappa = \parallel A \parallel \parallel A^{-1} \parallel\,, \tag{5.58}$$

where $\parallel A \parallel$ represents the norm of A. Saad (2003) shows that this analysis can be extended to iterative procedures as follows. First define the residual norm,

$$\parallel \rho^k \parallel = \parallel \mathbf{Q} - A\phi^k \parallel\,,$$

where ϕ^k is an estimate of the solution after k iterations. Then, it will follow [after some work!] from Eq. (5.58) that the ratio of the norm of iteration error $\epsilon^k = \phi - \phi^k$ and the norm of solution vector ϕ is related to the ratio of the norm of residual to the norm of $\mathbf{Q}$ as follows:

$$\frac{\parallel \epsilon^k \parallel}{\parallel \phi \parallel} \leq \kappa \frac{\parallel \rho^k \parallel}{\parallel \mathbf{Q} \parallel}\,, \tag{5.59}$$

which gives an upper bound on the iteration error. Clearly, the magnitude of κ has a significant impact on the iterations. Now, from basic linear algebra (Golub and van Loan 1996), we recall that the eigenvalues of the matrix A are the values that allow non-zero solutions to

$$A\mathbf{x} = \lambda \mathbf{x} ,$$

where $\mathbf{x}$ are eigenvectors of A. From this result and Eq. (5.58) one can show that

$$\kappa = \frac{\lambda_{\max}}{\lambda_{\min}} , \tag{5.60}$$

when $\lambda_{\max}$ and $\lambda_{\min}$ are the largest and smallest eigenvalues of the matrix A. Thus, the rate of convergence depends on this so-called *condition number* κ which is a property of the coefficient matrix only. A useful way to explore condition numbers is to apply the "cond" function in MATLAB™ to various representative matrices.

The condition numbers of matrices that arise in CFD problems are usually approximately the square of the maximum number of grid points in any direction. With 100 grid points in each direction, the condition number should be about 10^4 and the standard conjugate gradient method would converge slowly. Although the conjugate gradient method is significantly faster than steepest descent for a given condition number, this basic method is not very useful.

This method can be improved by replacing the problem whose solution we seek by another one with the same solution but a smaller condition number. This is accomplished by *preconditioning*. One way to precondition the problem is to pre-multiply the equation by another (carefully chosen) matrix. As this would destroy the symmetry of the matrix, the preconditioning must take the following form:

$$C^{-1}AC^{-1}C\boldsymbol{\phi} = C^{-1}\mathbf{Q} . \tag{5.61}$$

The conjugate gradient method is applied to the matrix $C^{-1}AC^{-1}$, i.e., to the modified problem (5.61). If this is done and the residual form of the iterative method is used, the following algorithm results (for a detailed derivation, see Golub and van Loan 1996, or Shewchuk 1994). In this description, ρ^k is the residual at the kth iteration, $\mathbf{p}^k$ is the kth search direction, $\mathbf{z}^k$ is an auxiliary vector and α^k and β^k are parameters used in constructing the new solution, residual, and search direction. The algorithm can be summarized as follows:

- Initialize by setting: $k = 0$, $\boldsymbol{\phi}^0 = \boldsymbol{\phi}_{\text{in}}$, $\rho^0 = \mathbf{Q} - A\boldsymbol{\phi}_{\text{in}}$, $\mathbf{p}^0 = \mathbf{0}$, $s^0 = 10^{30}$
- Advance the counter: $k = k + 1$
- Solve the system: $M\mathbf{z}^k = \rho^{k-1}$
- Calculate: $s^k = \rho^{k-1} \cdot \mathbf{z}^k$
  $\beta^k = s^k/s^{k-1}$
  $\mathbf{p}^k = \mathbf{z}^k + \beta^k \mathbf{p}^{k-1}$
  $\alpha^k = s^k/(\mathbf{p}^k \cdot A\mathbf{p}^k)$

$$\boldsymbol{\phi}^k = \boldsymbol{\phi}^{k-1} + \alpha^k \mathbf{p}^k$$
$$\boldsymbol{\rho}^k = \boldsymbol{\rho}^{k-1} - \alpha^k A\mathbf{p}^k$$

- Repeat until convergence.

This algorithm involves solving a system of linear equations at the first step. The matrix involved is $M = C^{-1}$ where C is the preconditioning matrix, which is in fact never actually constructed. For the method to be efficient, M must be easy to invert. The choice of M used most often is incomplete Cholesky factorization of A, but in tests it was found that if $M = LU$, where L and U are the factors used in Stone's SIP method, faster convergence is obtained. Examples will be presented below. Saad (2003) gives an extensive discussion of preconditioners for both serial and parallel computations.

5.3.6.2 Biconjugate Gradients and CGSTAB

The conjugate gradient method given above is applicable only to symmetric systems; the matrices obtained by discretizing the Poisson equation are often symmetric (examples are the heat conduction equation and pressure or pressure-correction equations to be introduced in Chap. 7). To apply the method to systems of equations that are not necessarily symmetric (for example, any convection-diffusion equation), we need to convert an asymmetric problem to a symmetric one. There are a couple of ways of doing this of which the following is perhaps the simplest. Consider the system:

$$\begin{pmatrix} 0 & A \\ A^T & 0 \end{pmatrix} \cdot \begin{pmatrix} \boldsymbol{\psi} \\ \boldsymbol{\phi} \end{pmatrix} = \begin{pmatrix} \mathbf{Q} \\ \mathbf{0} \end{pmatrix}. \tag{5.62}$$

This system can be decomposed into two subsystems. The first is the original system; the second involves the transpose matrix and is irrelevant. (If there were a need to do so, one could solve a system of equations involving the transpose matrix at little extra cost.) When the preconditioned conjugate gradient method is applied to this system, the following method, called *biconjugate gradients*, results:

- Initialize by setting: $k = 0$, $\boldsymbol{\phi}^0 = \boldsymbol{\phi}_{\text{in}}$, $\boldsymbol{\rho}^0 = \mathbf{Q} - A\boldsymbol{\phi}_{\text{in}}$, $\overline{\boldsymbol{\rho}}^0 = \mathbf{Q} - A^T\boldsymbol{\phi}_{\text{in}}$, $\mathbf{p}^0 = \overline{\mathbf{p}}^0 = \mathbf{0}$, $s^0 = 10^{30}$
- Advance the counter: $k = k + 1$
- Solve the systems: $M\mathbf{z}^k = \boldsymbol{\rho}^{k-1}$, $M^T\overline{\mathbf{z}}^k = \overline{\boldsymbol{\rho}}^{k-1}$
- Calculate: $s^k = \mathbf{z}^k \cdot \overline{\boldsymbol{\rho}}^{k-1}$
  $\beta^k = s^k/s^{k-1}$
  $\mathbf{p}^k = \mathbf{z}^k + \beta^k \mathbf{p}^{k-1}$
  $\overline{\mathbf{p}}^k = \overline{\mathbf{z}}^k + \beta^k \overline{\mathbf{p}}^{k-1}$
  $\alpha^k = s^k/(\overline{\mathbf{p}}^k A\mathbf{p}^k)$
  $\boldsymbol{\phi}^k = \boldsymbol{\phi}^{k-1} + \alpha^k \mathbf{p}^k$
  $\boldsymbol{\rho}^k = \boldsymbol{\rho}^{k-1} - \alpha^k A\mathbf{p}^k$
  $\overline{\boldsymbol{\rho}}^k = \overline{\boldsymbol{\rho}}^{k-1} - \alpha^k A^T \overline{\mathbf{p}}^k$
- Repeat until convergence.

The above algorithm was published by Fletcher (1976). It requires almost exactly twice as much effort per iteration as the standard conjugate gradient method but converges in about the same number of iterations. It has not been widely used in CFD applications but it appears to be very robust (meaning that it handles a wide range of problems without difficulty).

Other variants of the biconjugate gradient method type which are more stable and robust have been developed. We mention here the CGS (conjugate gradient squared) algorithm, proposed by Sonneveld (1989); CGSTAB (CGS stabilized), proposed by van der Vorst and Sonneveld (1990) and another version by van der Vorst (1992); and GMRES (see Sect. 5.3.6.3 below). All of these can be applied to non-symmetric matrices and to both structured and unstructured grids. We give below the CGSTAB algorithm without formal derivation:

- Initialize by setting: $k = 0$, $\boldsymbol{\phi}^0 = \boldsymbol{\phi}_{in}$, $\boldsymbol{\rho}^0 = \mathbf{Q} - A\boldsymbol{\phi}_{in}$, $\mathbf{u}^0 = \mathbf{p}^0 = \mathbf{0}$
- Advance the counter $k = k + 1$ and calculate:
$$\beta^k = \boldsymbol{\rho}^0 \cdot \boldsymbol{\rho}^{k-1}$$
$$\omega^k = (\beta^k \gamma^{k-1})/(\alpha^{k-1}\beta^{k-1})$$
$$\mathbf{p}^k = \boldsymbol{\rho}^{k-1} + \omega^k(\mathbf{p}^{k-1} - \alpha^{k-1}\mathbf{u}^{k-1})$$
- Solve the system: $M\mathbf{z} = \mathbf{p}^k$
- Calculate: $\mathbf{u}^k = A\mathbf{z}$
$$\gamma^k = \beta^k/(\mathbf{u}^k \cdot \boldsymbol{\rho}^0)$$
$$\mathbf{w} = \boldsymbol{\rho}^{k-1} - \gamma^k\mathbf{u}^k$$

- Solve the system: $M\mathbf{y} = \mathbf{w}$
- Calculate: $\mathbf{v} = A\mathbf{y}$
$$\alpha^k = (\mathbf{v} \cdot \boldsymbol{\rho}^k)/(\mathbf{v} \cdot \mathbf{v})$$
$$\boldsymbol{\phi}^k = \boldsymbol{\phi}^{k-1} + \gamma^k\mathbf{z} + \alpha^k\mathbf{y}$$
$$\boldsymbol{\rho}^k = \mathbf{w} - \alpha^k\mathbf{v}$$
- Repeat until convergence.

Note that $\mathbf{u}$, $\mathbf{v}$, $\mathbf{w}$, $\mathbf{y}$ and $\mathbf{z}$ are auxiliary vectors and have nothing to do with the velocity vector or the coordinates y and z. The algorithm can be programmed as given above; computer codes for the conjugate gradient method with incomplete Cholesky preconditioning (ICCG, for symmetric matrices, both 2D and 3D versions) and the 3D CGSTAB solver are available via the Internet; see the Appendix for details.

5.3.6.3 Generalized Minimum Residual Method

The Generalized Minimum Residual (GMRES) method proposed by Saad and Schultz (1986) handles non-symmetric A. Saad (2003) describes the method in detail. GMRES has shortcomings, but has become popular because it is robust. Two issues are:

1. The iteration uses all previous search direction vectors to compute the next search direction and so the storage space and operation count grows linearly, which can be a significant problem if A is very large.

2. The iteration can stall when the matrix is not positive definite.

Amelioration is achieved by (1) restarting the iteration after a predetermined number of steps and (2) preconditioning the matrix (see Sect. 5.3.6.1) to reduce the number of iterations needed for convergence. In a series of papers (see, e.g., Armfield and Street 2004), we had success with the restarted, preconditioned GMRES method, finding it the most efficient for solving the Poisson and pressure-correction equations (see Sect. 7.1.5) compared to other solvers (including CG).

A basic GMRES algorithm to solve $A\boldsymbol{\phi} = \mathbf{Q}$ is (adapted from Golub and van Loan 1996, and Saad 2003):

- START: Set an error tolerance and choose m to limit number of iterations.
  Initialize by setting: $k = 0$, $\boldsymbol{\phi}^0 = \boldsymbol{\phi}_{in}$, $\rho^0 = \mathbf{Q} - A\boldsymbol{\phi}_{in}$, $h_{10} = \parallel \rho^0 \parallel_2$
- If $h_{k+1,k} > 0$, calculate
  $$\beta^{k+1} = \rho^k / k_{k+1,k}$$
  $$k = k + 1$$
  $$\rho^k = A\beta^k$$
  For $i = 1 : k$
  $$h_{ik} = \beta_i^T \rho^k$$
  $$\rho^k \leftarrow \rho^k - h_{ik}\beta^i$$
  Endfor
  $$h_{k+1,k} = \parallel \rho^k \parallel_2$$
  $$\boldsymbol{\phi}^k = \boldsymbol{\phi}^0 + Q_k y_k \text{ where } y_k \text{ is the solution}$$
  of the $(k + 1) \times k$ least squares problem:
  $$\parallel h_{10}e_1 - \tilde{H}_k y_k \parallel_2 = min$$
  If the tolerance is met, set $\boldsymbol{\phi} = \boldsymbol{\phi}^k$ and STOP
  Else, if $k \geq m$, set $\boldsymbol{\phi}^0 = \boldsymbol{\phi}^k$ and return to: START
  Return to: If $h_{k+1,k} > 0$

In the above algorithm there are three auxiliary matrices:

1. The columns of Q_k are the orthonormal Arnoldi vectors.
2. $\tilde{H}_k$ is an upper Hessenberg matrix containing the h_{ij}.
3. e_1 is the first column of the identity matrix I_n, so

$$e_1 = (1, 0, 0, \cdots, 0)^T$$

Finally, we note that $\parallel \cdot \parallel_2$ is the matrix norm.

5.3.7 Multigrid Methods

The final method for solving linear systems to be discussed here is the multigrid method; we will apply the method for flow calculation in Sect. 12.4. The basis for the multigrid concept is an observation about iterative methods. Their rate of convergence depends on the eigenvalues of the iteration matrix associated with the method. In particular, the eigenvalue(s) with largest magnitude (the *spectral radius* of the matrix) determines how rapidly the solution is reached; see Sect. 5.3.2. The eigenvector(s) associated with this eigenvalue(s) determines the spatial distribution of the iteration error and varies considerably from method to method. Let us briefly review the behavior of these entities for some of the methods presented above. The properties are given for the Laplace equation; most of them generalize to other elliptic partial differential equations.

For the Laplace equation, the two largest eigenvalues of the Jacobi method are real and of opposite sign. One eigenvector represents a smooth function of the spatial coordinates, the other, a rapidly oscillating function. The iteration error for the Jacobi method is thus a mixture of very smooth and very rough components; this makes acceleration difficult. On the other hand, the Gauss–Seidel method has a single real positive largest eigenvalue with an eigenvector that makes the iteration error a smooth function of the spatial coordinates.

The largest eigenvalues of the SOR method with optimum over-relaxation factor lie on a circle in the complex plane and there are a number of them; consequently, the error behaves in a very complicated manner. In ADI, the nature of the error depends on the parameter but tends to be rather complicated. Finally, SIP has relatively smooth iteration errors.

Some of these methods produce errors that are smooth functions of the spatial coordinates. Let us consider one of these methods. The iteration error ϵ^n and residual ρ^n after the nth iteration are related by Eq. (5.15). In the Gauss–Seidel and SIP methods, after a few iterations, the rapidly varying components of the error have been removed and the error becomes a smooth function of the spatial coordinates. If the error is smooth, the update (an approximation to the iteration error) can be computed on a coarser grid. On a grid twice as coarse as the original one in two dimensions, iterations cost $1/4$ as much; in three dimensions, the cost is $1/8$ the fine grid cost. Furthermore, iterative methods converge much faster on coarser grids. Gauss–Seidel converges four times as fast on a grid twice as coarse; for SIP the ratio is less favorable but still substantial.

This suggests that much of the work can be done on a coarser grid. To do this, we need to define: the relationship between the two grids, the finite-difference operator on the coarse grid, a method of smoothing (*restricting*) the residual from the fine grid to the coarse one and a method of interpolating (*prolonging*) the update or correction from the coarse grid to the fine one; the words in parentheses are special terms that are in common use in the multigrid literature. Many choices are available for each item; they affect the behavior of the method but, within the range of good choices, the differences are not great. We shall thus present just one good choice for each item.

In a finite-difference scheme, the coarse grid normally consists of every second line of the fine grid. In a finite-volume method, one usually takes the coarse grid CVs to be composed of 2 (in two dimensions, 4, and in three dimensions, 8) fine grid CVs; the coarse grid nodes then lie between the fine grid nodes.

Although there is no reason to use the multigrid method in one dimension (because the TDMA algorithm is very effective), it is easy to illustrate the principles of the multigrid method and to derive some of the procedures used in the general case. Thus consider the problem:

$$\frac{d^2\phi}{dx^2} = f(x) \tag{5.63}$$

for which the standard FD approximation on a uniform grid is:

$$\frac{1}{(\Delta x)^2}\left(\phi_{i-1} - 2\phi_i + \phi_{i+1}\right) = f_i . \tag{5.64}$$

After performing n iterations on the grid with Δx spacing, we obtain an approximate solution ϕ^n, and the above equation is satisfied to within the residual ρ^n:

$$\frac{1}{(\Delta x)^2}\left(\phi_{i-1}^n - 2\phi_i^n + \phi_{i+1}^n\right) = f_i - \rho_i^n . \tag{5.65}$$

Subtracting this equation from Eq. (5.64) gives

$$\frac{1}{(\Delta x)^2}\left(\epsilon_{i-1}^n - 2\epsilon_i^n + \epsilon_{i+1}^n\right) = \rho_i^n , \tag{5.66}$$

which is Eq. (5.15) for node i. This is the equation we want to iterate on the coarse grid.

To derive the discretized equations on the coarser grid, we note that control volume around node I of the coarse grid consists of the whole control volume around node i plus half of control volumes $i - 1$ and $i + 1$ of the fine grid (see Fig. 5.2). This suggests that we add one half of equation (5.66) with indices $i - 1$ and $i + 1$ to the full equation with index i, which leads to (superscript n being omitted):

$$\frac{1}{4(\Delta x)^2}\left(\epsilon_{i-2} - 2\epsilon_i + \epsilon_{i+2}\right) = \frac{1}{4}\left(\rho_{i-1} + 2\rho_i + \rho_{i+1}\right) . \tag{5.67}$$

Using the relationship between the two grids ($\Delta X = 2\,\Delta x$, see Fig. 5.2), this is equivalent to the following equation on the coarse grid:

$$\frac{1}{(\Delta X)^2}\left(\epsilon_{I-1} - 2\epsilon_I + \epsilon_{I+1}\right) = \overline{\rho}_I , \tag{5.68}$$

Fig. 5.2 The grids used in the multigrid technique in one dimension

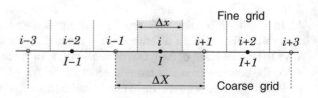

which also serves to define $\overline{\rho}_I$. The left-hand side of this equation is the standard approximation to the second derivative on the coarse grid, indicating that the obvious discretization on the coarse grid is a reasonable one. The right-hand side is a smoothing or filtering of the fine-grid forcing term and provides the natural definition of the smoothing or restriction operation.

The simplest prolongation or interpolation of a quantity from the coarse grid to the fine grid is linear interpolation. At coincident points of the two grids, the value at the coarse grid point is simply *injected* onto the corresponding fine grid point. At fine grid points that lie between the coarse grid points, the injected value is the average of the neighboring coarse grid values.

A two-grid iterative method is thus:

- On the fine grid, perform iterations with a method that gives a smooth error (restriction);
- Compute the residual on the fine grid;
- Restrict (smooth) the residual to the coarse grid;
- Perform iterations of the correction equation on the coarse grid;
- Prolong (interpolate) the correction to the fine grid;
- Update the solution on the fine grid;
- Repeat the entire procedure until the residual is reduced to the desired level.

It is natural to ask: why not use still coarser grids to further improve the rate of convergence? This is a good idea. In fact, one should continue the procedure until it becomes impossible to define a still coarser grid; on the coarsest grid, the number of unknowns is so small that the equations can be solved exactly at a negligible cost.

Multigrid is more a strategy than a particular method. Within the framework just described there are many parameters that can be selected more or less arbitrarily: the coarse grid structure, the restricter (smoother), the number of iterations on each grid, the order in which the various grids are visited, and the restriction and prolongation (interpolation) schemes are the most important of these. The rate of convergence does, of course, depend on the choices made but the range of performance between the worst and the best methods is probably less than a factor of two.

The most important property of the multigrid method is that the number of iterations on the finest grid required to reach a given level of convergence is roughly independent of the number of grid nodes. This is as good as one can expect to do—the computational cost is proportional to the number of grid nodes. In 2D and 3D problems with about 100 nodes in each direction, the multigrid method may converge in one-tenth to one-hundredth of the time required by the basic method. An example will be presented in Sect. 5.8.

The iterative method on which the multigrid method is based must be a good smoother; its convergence properties as a stand-alone method are less important. Gauss–Seidel and SIP are two good choices but there are other possibilities.

In two dimensions, there are many possibilities for the restriction (smoothing) operator. If the method described above is used in each direction, the result would be a nine-point scheme. A simpler, but nearly as effective, restriction is the five-point scheme:

$$\overline{\rho}_{I,J} = \frac{1}{8}\left(\rho_{i+1,j} + \rho_{i-1,j} + \rho_{i,j+1} + \rho_{i,j-1} + 4\,\rho_{i,j}\right). \tag{5.69}$$

Similarly, an effective prolongator is bilinear interpolation. In 2D, there are three kinds of points on the fine grid. Those which correspond to coarse grid points are given the value at the corresponding point. Ones which lie on lines connecting two coarse grid points receive the average of the two coarse grid values. Finally, the points at the centers of coarse grid volumes take the average of the four neighbor values. Similar schemes can be derived for FV methods and 3D problems.

The initial guess in an iterative solution method is usually far from the converged solution (a zero field is often used). It therefore makes sense to solve the equation first on a very coarse grid (which is cheap) and use that solution to provide a better guess for the initial field on the next finer grid. By the time we reach the finest grid, we already have a fairly good starting solution. Multigrid methods of this type are called *full-multigrid* (FMG) methods. The cost of obtaining the initial solution for the finest grid is usually more than compensated by the savings on fine grid iterations.

Finally, we remark that it is possible to construct a method in which one solves equations for approximations to the solution rather than for corrections at each grid. This is called the *full-approximation scheme* (FAS) and is often used for solving non-linear problems. It is important to note that the solution obtained on each grid in FAS is *not* the solution that would be obtained if that grid were used by itself but a smoothed version of the fine grid solution; this is achieved by passing a correction from each grid to the next coarser grid. One variant of the FAS scheme for the Navier–Stokes equations will be presented in Chap. 12.

For a detailed analysis of multigrid methods, see books by Briggs et al. (2000), Hackbusch (2003) and Brandt (1984). A 2D multigrid solver that uses the Gauss–Seidel, SIP, or ICCG methods as the smoother is available via the Internet; see the Appendix for details.

5.3.8   Other Iterative Solvers

There are many other iterative solvers that cannot be described in detail here. We mention the 'red–black' variation of the Gauss–Seidel solver which is often used in conjunction with multigrid methods. On a structured grid, the nodes are imagined to be 'colored' in the same way as a checkerboard. The method consists of two Jacobi steps: black nodes are updated first, then the red nodes. When the values at black nodes are updated, only the 'old' red values are used, see Eq. (5.33). On the next step, red values are recalculated using the updated black values. This alternate application of the Jacobi method to the two sets of nodes gives an overall method with the same convergence properties as the Gauss–Seidel method. The nice feature of the red-black Gauss–Seidel solver is that it both vectorizes and parallelizes well, because there are no data dependencies in either step.

Another practice often applied to multi-dimensional problems is the use of iteration matrices which correspond to lower-dimensional problems. One version of this is the ADI method described above which reduces a 2D problem to a sequence of 1D problems. The resulting tridiagonal problems are solved line-by-line. The direction of solution is changed from iteration to iteration to improve the rate of convergence. This method is usually used in the Gauss–Seidel fashion, i.e., new variable values from lines already visited are used.

A counterpart of the red–black Gauss–Seidel method is the 'zebra' line–by–line solver: first the solution is found on the even-numbered lines, then the odd-numbered lines are treated. This gives better parallelization and vectorization possibilities with no sacrifice in convergence properties.

It is also possible to use the 2D SIP method to solve 3D problems, applying it plane–by–plane and relegating the contributions from neighboring planes to the right-hand side of the equations. However, this method is neither cheaper nor faster than the 3D version of SIP, so it is not often used.

5.4 Coupled Equations and Their Solution

Most problems in fluid dynamics and heat transfer require solution of coupled systems of equations, i.e., the dominant variable of each equation occurs in some of the other equations. There are two types of approaches to such problems. In the first, all variables are solved for simultaneously. In the other, each equation is solved for its dominant variable, treating the other variables as known, and one iterates through the equations until the solution of the coupled system is obtained. The two approaches may also be mixed. We call these simultaneous and sequential methods, respectively, and they are described in more detail below.

5.4.1 Simultaneous Solution

In simultaneous methods, all the equations are considered part of a single system. The discretized equations of fluid mechanics have, after linearization, a block-banded structure. Direct solution of these equations would be very expensive, especially when the equations are non-linear and the problem is three-dimensional. Iterative solution techniques for coupled systems are generalizations of methods for single equations. The methods described above were chosen for their applicability to coupled systems. Simultaneous solution methods based on iterative solvers have been developed by several authors; see, e.g., papers by Galpin and Raithby (1986), Deng et al. (1994), and Weiss et al. (1999).

5.4.2 Sequential Solution

When the equations are linear and tightly coupled, the simultaneous approach is best. However, the equations may be so complex and non-linear that coupled methods are difficult and expensive to use. It may then be preferable to treat each equation as if it has only a single unknown, temporarily treating the other variables as known, using the best currently available values for them. The equations are then solved in turn, repeating the cycle until all equations are satisfied. When using this type of method, two points need to be borne in mind:

- Because some terms, e.g., the coefficients and source terms that depend on the other variables, change as the computation proceeds, it is inefficient to solve the equations accurately at each iteration. That being the case, direct solvers are unnecessary and iterative solvers are preferred. Iterations performed on each equation are called *inner* iterations.
- In order to obtain a solution which satisfies all of the equations, the coefficient matrices and source vectors must be updated after each cycle and the process repeated. The cycles are called *outer* iterations.

Optimization of this type of solution method requires careful choice of the number of inner iterations per outer iteration. It is also necessary to limit the change in each variable from one outer iteration to the next (under-relaxation), because a change in one variable changes the coefficients in the other equations, which may slow or prevent convergence. Unfortunately, it is hard to analyze the convergence of these methods so the selection of under-relaxation factors is largely empirical.

The multigrid method, which was described above as a convergence accelerator for inner iterations (linear problems), may be applied to coupled problems. It may also be used to accelerate the outer iterations as will be described in Chap. 12.

5.4.3 Under-Relaxation

We shall present one under-relaxation technique that is widely used. On the nth outer iteration, the algebraic equation for a generic variable, ϕ, at a typical point P may be written:

$$A_P \phi_P^n + \sum_l A_l \phi_l^n = Q_P , \tag{5.70}$$

where Q contains all the terms that do not depend explicitly on ϕ^n; the coefficients A_l and the source Q may involve ϕ^{n-1}. The discretization scheme is unimportant here. This equation is linear and the system of equations for the whole solution domain is solved usually iteratively (inner iterations).

In the early outer iterations, allowing ϕ to change by as much as Eq. (5.70) requires could cause instability, so we allow ϕ^n to change only a fraction α_ϕ of the would-be difference:

$$\phi^n = \phi^{n-1} + \alpha_\phi(\phi^{new} - \phi^{n-1}) , \qquad (5.71)$$

where ϕ^{new} is the result of Eq. (5.70) and the under-relaxation factor satisfies $0 < \alpha_\phi < 1$.

Because the old iterate is usually no longer required after the coefficient matrix and source vector are updated, the new solution can be written over it. Replacing ϕ^{new} in Eq. (5.71) by

$$\phi_P^{new} = \frac{Q_P - \sum_l A_l \phi_l^n}{A_P} , \qquad (5.72)$$

which follows from Eq. (5.70), leads to a modified equation at node P:

$$\underbrace{\frac{A_P}{\alpha_\phi}}_{A_P^*} \phi_P^n + \sum_l A_l \phi_l^n = \underbrace{Q_P + \frac{1-\alpha_\phi}{\alpha_\phi} A_P \phi_P^{n-1}}_{Q_P^*} , \qquad (5.73)$$

where A_P^* and Q_P^* are modified main diagonal matrix elements and source vector components. This modified equation is solved within inner iterations. When the outer iterations converge, the terms involving α_ϕ cancel out and we obtain the solution of the original problem.

This kind of under-relaxation was proposed by Patankar (1980). It has a positive effect on many iterative solution methods because the diagonal dominance of the matrix A is increased (the element A_P^* is larger than A_P, while A_l remain the same). It is more efficient than explicit application of the expression (5.71).

Optimum under-relaxation factors are problem dependent. A good strategy is to use a small under-relaxation factor in the early iterations and increase it towards unity as convergence is approached. Some guidance for selecting the under-relaxation factors for solving the Navier–Stokes equations will be given in Chaps. 7, 8, 9 and 12. Under-relaxation may be applied not only to the dependent variables but also to individual terms in the equations. It is often necessary to do so when the fluid properties (viscosity, density, Prandtl number etc.) depend on the solution and need be updated.

We mentioned above that iterative solution methods can often be regarded as solving an unsteady problem until a steady state is reached. Control of the time step is then important in controlling the evolution of the solution. In the next chapter we shall show that time step may be interpreted as an under-relaxation factor. The under-relaxation scheme described above may be interpreted as using different time steps at different nodes.

5.5 Non-linear Equations and Their Solution

As mentioned above, there are two types of techniques for solving non-linear equations: Newton-like and global. The former are much faster when a good estimate of the solution is available but the latter are guaranteed not to diverge; there is a trade-off between speed and security. Combinations of the two methods are often used. There is a vast literature devoted to methods for solving non-linear equations, and state-of-the-art is still evolving. We cannot cover even a substantial fraction of the methods here and give a short overview of some methods.

5.5.1 Newton-Like Techniques

The master method for solving non-linear equations is Newton's method. Suppose that one needs to find the root of a single algebraic equation $f(x) = 0$. Newton's method linearizes the function about an estimated value of x using the first two terms of the Taylor series:

$$f(x) \approx f(x_0) + f'(x_0)(x - x_0) \ . \tag{5.74}$$

Setting the linearized function equal to zero provides a new estimate of the root:

$$x_1 = x_0 - \frac{f(x_0)}{f'(x_0)} \quad \text{or, in general,} \quad x_k = x_{k-1} - \frac{f(x_{k-1})}{f'(x_{k-1})} \tag{5.75}$$

and we continue until the change in the root $x_k - x_{k-1}$ is as small as desired. The method is equivalent to approximating the curve representing the function by its tangent at x_k. When the estimate is close enough to the root, this method converges quadratically, i.e., the error at iteration $k + 1$ is proportional to the square of the error at iteration k. This means that only a few iterations are needed once the solution estimate is close to the root. For that reason, it is employed whenever it is feasible to do so.

Newton's method is easily generalized to systems of equations. A generic system of non-linear equations can be written:

$$f_i(x_1, x_2, \ldots, x_n) = 0 \ , \quad i = 1, 2, \ldots, n \ . \tag{5.76}$$

This system can be linearized in exactly the same way as the single equation. The only difference is that now we need to use multi-variable Taylor series:

$$f_i(x_1, x_2, \ldots, x_n) = f_i(x_1^k, x_2^k, \ldots, x_n^k) + \sum_{j=1}^{n}(x_j^{k+1} - x_j^k)\frac{\partial f_i(x_1^k, x_2^k, \ldots, x_n^k)}{\partial x_j} \ , \tag{5.77}$$

for $i = 1, 2, \ldots, n$. When this is set to zero, we have a system of linear algebraic equations that can be solved by Gauss elimination or some other technique. The matrix of the system is the set of partial derivatives:

$$a_{ij} = \frac{\partial f_i(x_1^k, x_2^k, \ldots, x_n^k)}{\partial x_j}, \quad i = 1, 2, \ldots, n, \quad j = 1, 2, \ldots, n, \quad (5.78)$$

which is called the *Jacobian* of the system. The system of equations is:

$$\sum_{j=1}^{n} a_{ij}(x_j^{k+1} - x_j^k) = -f_i(x_1^k, x_2^k, \ldots, x_n^k), \quad i = 1, 2, \ldots, n. \quad (5.79)$$

For an estimate that is close to the correct root, Newton's method for systems converges as rapidly as the method for a single equation. However, for large systems, the rapid convergence is more than offset by its principal disadvantage. For the method to be effective, the Jacobian has to be evaluated at each iteration. This presents two difficulties. The first is that, in the general case, there are n^2 elements of the Jacobian and their evaluation becomes the most expensive part of the method. The second is that a direct method of evaluating the Jacobian may not exist; many systems are such that the equations are implicit or they may be so complicated that differentiation is all but impossible.

For generic systems of non-linear equations, secant methods are much more effective. For a single equation, the secant method approximates the derivative of the function by the secant drawn between two points on the curve (see Ferziger 1998, or Moin 2010). This method converges more slowly than Newton's method, but as it does not require evaluation of the derivative, it may find the solution at lower overall cost and can be applied to problems in which direct evaluation of the derivative is not possible. There are a number of generalizations of the secant method to systems, most of which are quite effective but, as they have not been applied in CFD, we shall not review them here.

5.5.2 Other Techniques

The usual approach to the solution of coupled non-linear equations is the sequential decoupled method described in the previous section. The non-linear terms (convection flux, source term) are usually linearized using a *Picard iteration* approach. For the convection terms, this means that the mass flux is treated as known, so the non-linear convection term in the equation for the u_i momentum component is approximated by:

$$\rho u_j u_i \approx (\rho u_j)^o u_i, \quad (5.80)$$

where index o denotes that the values are taken from the result of the previous outer iteration. Similarly, the source term is decomposed into two parts:

$$q_\phi = b_0 + b_1 \phi . \tag{5.81}$$

The portion b_0 is absorbed into the right-hand side of the algebraic equation, while b_1 contributes to the coefficient matrix A. A similar approach can be used for the non-linear terms that involve more than one variable.

This kind of linearization requires many more iterations than a coupled technique using Newton-like linearization. However, the computing effort per iteration is much smaller and the number of outer iterations can be reduced using multigrid techniques, which makes this approach attractive.

Newton's method is sometimes used to linearize the non-linear terms; for example, the convection term in the equation for the u_i momentum component can be expressed as:

$$\rho u_j u_i \approx \rho u_j^o u_i + \rho u_i^o u_j - \rho u_j^o u_i^o . \tag{5.82}$$

Non-linear source terms can be treated in the same way. This leads to a coupled linear system of equations which is difficult to solve, and the convergence is not quadratic unless the full Newton technique is used. However, special coupled iterative techniques which benefit from this linearization technique can be developed, as shown by Galpin and Raithby (1986).

5.6 Deferred-Correction Approaches

If all terms containing the nodal values of the unknown variable are kept on the left-hand side of Eq. (3.43), the computational molecule may become very large. Because the size of the computational molecule affects both the storage requirements and the effort needed to solve the linear equation system, we would like to keep it as small as possible; usually, only the nearest neighbors of node P are kept on the left-hand sides of the equations. However, approximations which produce such simple computational molecules are usually not accurate enough, so we are often forced to use approximations that refer to more nodes than just the nearest neighbors.

One way around this problem would be to leave only the terms containing the nearest neighbors on the left-hand side of Eq. (3.43) and bring all other terms to the right-hand side. This requires that these terms be evaluated using values from the previous iteration. However, this is not a good practice and may lead to the divergence of the iterations because the terms treated explicitly may be substantial. To prevent divergence, strong under-relaxation of the changes from one iteration to the next would be required (see Sect. 5.4.3), resulting in slow convergence.

A better approach is to compute the terms that are approximated with a higher-order approximation explicitly and put them on the right-hand side of the equation. Then one takes a simpler approximation to these terms (one that gives a small computational molecule) and puts it both on the left-hand side of the equation (with unknown variable values) and on the right-hand side (computing it explicitly using existing values). The right-hand side is now the difference between two approximations of the same term, and is likely to be small. Thus it should cause no problems in the iterative solution procedure. Once the iterations converge, the low-order approximation terms drop out and the obtained solution corresponds to the higher-order approximation.

Because iterative methods are usually necessary due to the non-linearity of the equations to be solved, adding a small term to the part treated explicitly increases the computing effort by only a small amount. On the other hand, both the memory and computing time required are greatly reduced when the size of the computational molecule in the part of the equation treated implicitly is small.

We shall refer to this technique very often. It is used when treating higher-order approximations, grid non-orthogonality, and corrections needed to avoid undesired effects like oscillations in the solution. Because the right-hand side of the equation can be regarded as a "correction" this method is called deferred correction. Here its use will be demonstrated in conjunction with Padé schemes in FD (see Sect. 3.3.3) and with higher-order interpolations in FV-methods (see Sect. 4.4.4).

If Padé schemes are to be used in implicit FD methods, deferred correction must be employed because approximation of the derivative at one node involves derivatives at neighboring nodes. One approach is to use the "old values" of the derivatives at neighboring nodes and variable values at distant nodes. These are usually taken from the result of the preceding iteration and we have:

$$\left(\frac{\partial \phi}{\partial x}\right)_i = \beta \frac{\phi_{i+1} - \phi_{i-1}}{2 \Delta x} + \gamma \left(\frac{\phi_{i+2} - \phi_{i-2}}{4 \Delta x}\right)^{\text{old}} - \alpha \left(\frac{\partial \phi}{\partial x}\right)_{i+1}^{\text{old}} - \alpha \left(\frac{\partial \phi}{\partial x}\right)_{i-1}^{\text{old}}. \tag{5.83}$$

In this case, only the first term on the right-hand side of this equation will be moved to the left-hand side of the equation to be solved at the new outer iteration.

However, this approach may affect the convergence rate adversely because the implicitly treated part is not an approximation to the derivative but, rather, some multiple of it. The following version of deferred correction is more efficient:

$$\left(\frac{\partial \phi}{\partial x}\right)_i = \frac{\phi_{i+1} - \phi_{i-1}}{2 \Delta x} + \left[\left(\frac{\partial \phi}{\partial x}\right)_i^{\text{Padé}} - \frac{\phi_{i+1} - \phi_{i-1}}{2 \Delta x}\right]^{\text{old}}. \tag{5.84}$$

Here, the complete second-order CDS approximation is used on the left-hand side. On the right-hand side we have the difference between the explicitly computed Padé scheme derivative and the explicitly computed CDS approximation. This gives a more balanced expression because, where the second-order CDS is accurate enough, the term in square brackets is negligible. Instead of CDS, we could use UDS for the implicit part; in that case, the UDS approximation should be used on both sides of the equation.

Deferred correction is also useful in FV-methods when higher-order schemes are used (see Sect. 4.4.4). Higher-order flux approximations are computed *explicitly* and this approximation is then combined with an implicit lower-order approximation (which uses only variable values at nearest neighbors) in the following way (first suggested by Khosla and Rubin 1974):

$$F_e = F_e^L + \left(F_e^H - F_e^L \right)^{\text{old}} .$$

(5.85)

F_e^L stands for the approximation by some lower-order scheme (UDS is often used for convection and CDS for diffusion fluxes) and F_e^H is the higher-order approximation. The term in brackets is evaluated using values from the previous iteration, as indicated by the superscript 'old'. It is normally small compared to the implicit part, so its explicit treatment does not affect the convergence significantly.

The same approach can be applied to all high-order approximations including spectral methods. Although deferred correction increases the computation time per iteration relative to that for a pure low-order scheme, the additional effort is much smaller than that needed to treat the entire higher-order approximation implicitly. One can also multiply the "old" term with a blending factor between zero and unity to produce a mixture of the pure low-order and pure high-order schemes. This is sometimes used to avoid oscillations which result when higher-order schemes are used on grids that are not sufficiently fine. For example, when flow around a body is computed, one would like to use a fine grid near the body and a coarser grid far from it. A high-order scheme may produce oscillations in the coarse-grid region, thus spoiling the whole solution. As the variables vary slowly in the coarse-grid region, we may reduce the order of approximation there without affecting the solution in the fine-grid region; this can be achieved by using a blending factor in the outer region only.

More details on other uses of deferred-correction approach will be given in subsequent chapters.

5.7 Convergence Criteria and Iteration Errors

When using iterative solvers, it is important to know when to quit. The most common procedure is based on the difference between two successive iterates; the procedure is stopped when this difference, measured by some norm, is less than a pre-selected value. Unfortunately, this difference may be small when the error is not small and a proper normalization is essential.

From the the analysis presented in Sect. 5.3.2, we find (see Eqs. (5.14) and (5.27)):

$$\delta^n = \phi^{n+1} - \phi^n \approx (\lambda_1 - 1)(\lambda_1)^n a_1 \psi_1 ,$$

(5.86)

where δ^n is the difference between the solution at iterations $n + 1$ and n, and λ_1 is the largest eigenvalue or *spectral radius* of the iteration matrix. It can be estimated from (Ferziger 1998):

$$\lambda_1 \approx \frac{\|\delta^n\|}{\|\delta^{n-1}\|} , \qquad (5.87)$$

for sufficiently large n and where $\|\mathbf{a}\|$ represents the norm (e.g., root mean square or L_2 norm) of $\mathbf{a}$.

Once an estimate of the eigenvalue is available, it is not difficult to estimate the iteration error. In fact, by rearranging Eq. (5.86), we find:

$$\epsilon^n = \phi - \phi^n \approx \frac{\delta^n}{\lambda_1 - 1} . \qquad (5.88)$$

A good estimate of the iteration error is therefore:

$$\|\epsilon^n\| \approx \frac{\|\delta^n\|}{\lambda_1 - 1} \qquad (5.89)$$

This error estimate can be computed from the two successive iterates of the solution. Although the method is designed for linear systems, all systems are essentially linear near convergence; as this is the time when error estimates are most needed, the method can be applied to non-linear systems as well.

Unfortunately, iterative methods often have complex eigenvalues. When this is the case, the error reduction is not exponential and may not be monotonic. Because the equations are real, complex eigenvalues must occur as conjugate pairs. Their estimation requires an extension of the above procedure. In particular, data from more iterates are required. Some of the ideas used below are found in Golub and van Loan (1996).

If the eigenvalues of largest magnitude are complex, there are at least two of them and Eq. (5.27) must be replaced by

$$\epsilon^n \approx a_1 (\lambda_1)^n \boldsymbol{\psi}_1 + a_1^* (\lambda_1^*)^n \boldsymbol{\psi}_1^* , \qquad (5.90)$$

where * indicates the conjugate of a complex quantity. As before, we subtract two successive iterates to obtain δ^n; see Eq. (5.86). If we further let:

$$\boldsymbol{\omega} = (\lambda_1 - 1) a_1 \boldsymbol{\psi}_1 , \qquad (5.91)$$

then the following expression is obtained:

$$\delta^n \approx (\lambda_1)^n \boldsymbol{\omega} + (\lambda_1^*)^n \boldsymbol{\omega}^* . \qquad (5.92)$$

Because the magnitude of the eigenvalue λ_1 is the quantity of greatest interest, we write:

$$\lambda_1 = \ell\, e^{i\vartheta} . \tag{5.93}$$

A straightforward calculation then shows that:

$$z^n = \delta^{n-2} \cdot \delta^n - \delta^{n-1} \cdot \delta^{n-1} = 2\ell^{2n-2}|\omega|^2[\cos(2\vartheta) - 1] , \tag{5.94}$$

from which it is easy to show that:

$$\ell = \sqrt{\frac{z^n}{z^{n-1}}} \tag{5.95}$$

is an estimate of the magnitude of the eigenvalue.

Estimation of the error requires further approximations. The complex eigenvalues cause the errors to oscillate and the shape of the error is not independent of the iteration number, even for large n. To estimate the error, we compute δ^n and ℓ from expressions given above. Due to the complex eigenvalues and eigenvectors, the result contains terms proportional to the cosine of the phase angle. As we are interested only in magnitudes, we assume that these terms are zero in an average sense and drop them. This allows us to find a simple relationship between the two quantities:

$$\epsilon^n \approx \frac{\delta^n}{\sqrt{\ell^2 + 1}} . \tag{5.96}$$

This is the desired estimate of the error. Due to the oscillations in the solution, the estimate may not be accurate on any particular iteration, but, as we shall show below, it is quite good on the average.

In order to remove some of the effects of the oscillation, the eigenvalue estimates should be averaged over a range of iterations. Depending on the problem and the number of anticipated iterations, the averaging range may vary from 1 to 50 (typically 1% of the expected number of iterations).

Finally, we want a method that can treat both real and complex eigenvalues. The error estimator for the complex case (5.96) gives low estimates if the principal eigenvalue (λ_1) is real. Also, the contribution of λ_1 to z^n drops out in this case so the eigenvalue estimate is quite poor. However, this fact can be used to determine whether λ_1 is real or complex. If the ratio

$$r = \frac{z^n}{|\delta^n|^2} \tag{5.97}$$

is small, the eigenvalue is probably real; if r is large, the eigenvalue is probably complex. For real eigenvalues, r tends to be smaller than 10^{-2} and, for complex eigenvalues, $r \approx 1$. One can therefore adopt a value of $r = 0.1$ as an indicator of type of eigenvalue and use the appropriate expression for the error estimator.

A compromise is to use the reduction of the residual as a stopping criterion. Iteration is stopped when the residual norm has been reduced to some fraction of its original size (usually by three or four orders of magnitude). As we have shown, the iteration error is related to the residual via Eq. (5.15) so reduction of the residual is accompanied by reduction of the iteration error. If the iteration is started from zero initial values, then the initial error is equal to the solution itself. When the residual level has fallen say three to four orders of magnitude below the initial level, the error is likely to have fallen by a comparable amount, i.e., it is of the order of 0.1% of the solution. The residual and the error usually do not fall in the same way at the beginning of iteration process; caution is also needed because, if the matrix is poorly conditioned, the error may be large even when the residual is small.

Many iterative solvers require calculation of the residual. The above approach is then attractive as it requires no additional computation. The norm of residual prior to the first inner iteration provides a reference for checking the convergence of inner iterations. At the same time it provides a measure of the convergence of the outer iterations. Experience shows that inner iterations can be stopped when the residual has fallen by one to two orders of magnitude. Outer iterations should not be stopped before the residual has been reduced by three to five orders of magnitude, depending on the desired accuracy. The sum of absolute residuals (the L_1 norm) can be used instead of the r.m.s. (L_2) norm. The convergence criterion should be more stringent on refined grids, because the discretization errors are smaller on them than on coarse grids.

If the order of the initial error is known, it is possible to monitor the norm of the difference between two iterates and compare it with the same quantity at the beginning of the iteration process. When the difference norm has fallen three to four orders of magnitude, the error has usually fallen by a comparable amount.

Both of these methods are only approximate; however, they are better than the criterion based on the non-normalized difference between two successive iterates.

In order to test the method of estimating iteration errors, we first present the solution of a linear 2D problem using the SOR solver. The linear problem is Laplace equation in the square domain $\{0 < x < 1; \ 0 < y < 1\}$ with Dirichlet boundary conditions chosen to correspond to the solution $\phi(x, y) = 100\,xy$. The advantage of this choice is that the second-order central-difference approximation to the converged solution is exact on any grid so the actual difference between the present iterate and the converged solution is easily computed. The initial guess of the solution is zero everywhere within the domain. We chose the SOR method as the iterative technique because, if the relaxation parameter is greater than the optimum value, the eigenvalues are complex.

Results are shown for uniform grids with 20×20 and 80×80 CV in Figs. 5.3 and 5.4. In each case, the norms of the exact iteration error, the error estimate using the above described technique, the difference between two iterates and the residual are shown. For both cases the results of calculation for two values of the relaxation parameter are shown: one below the optimum value, which has real eigenvalues,

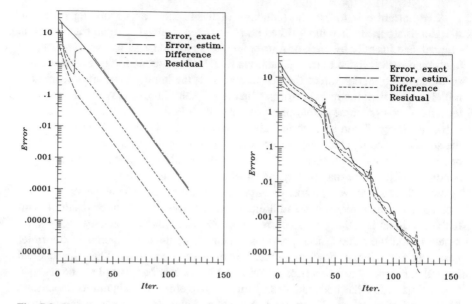

Fig. 5.3 Variation of the norm of the exact iteration error, error estimate, residual and difference between two iterations for the Laplace problem with the SOR solver on a 20 × 20 CV grid: relaxation parameter smaller (left) and larger (right) than the optimum, which is about 1.73

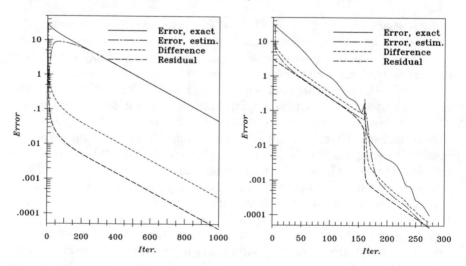

Fig. 5.4 Variation of the norm of the exact iteration error, error estimate, residual and difference between two iterations for the Laplace problem with the SOR solver on a 80 × 80 CV grid: relaxation parameter smaller (left) and larger (right) than the optimum, which is about 1.92

and one above the optimum, leading to complex eigenvalues.[4] For the case of real eigenvalues, smooth exponential convergence results. The error estimate is almost exact in this case (except in the initial period). However, the norms of the residual and difference between two iterates initially fall too rapidly and do not follow the fall of the iteration error. This effect is more pronounced as the grid is refined. On the 80 × 80 CV grid, the residual norm is quickly reduced by two orders of magnitude while the error is only slightly reduced. Once the asymptotic reduction rate is achieved, the slopes of all four curves are the same.

In the case for which the eigenvalues of the iteration matrix are complex, the convergence is not monotonic—there are oscillations in the error. The comparison of predicted and exact errors is in this case also quite satisfactory. All of the above-mentioned convergence criteria seem to be equally good in this case.

Further examples of the estimation of iteration errors, especially for the outer iterations in case of solving coupled flow problems, will be presented later.

5.8 Examples

In the previous chapter we presented solutions of some 2D problems without discussing the methods of solution. We shall now show the performance of various solvers for the case of scalar transport in a stagnation point flow; see Sect. 4.7 for the description of problem and discretization techniques used to derive the linear algebraic equations.

We consider the case with $\Gamma = 0.01$ and uniform grids with 20 × 20, 40 × 40 and 80 × 80 CVs. The equation matrix A is not symmetric and, in case of CDS discretization, it is not diagonally dominant. In a diagonally-dominant matrix, the element on the main diagonal satisfies the following condition:

$$A_P \geq \sum_l |A_l| . \tag{5.98}$$

It can be shown that a sufficient condition for convergence of iterative solution methods is that the above relation be satisfied, and that inequality must apply at least at one node. This condition is satisfied only by UDS discretization of convection terms. While simple solvers like Jacobi and Gauss–Seidel usually diverge when the above condition is violated, ILU, SIP and conjugate gradient solvers are less sensitive to diagonal dominance of the matrix.

We considered five solvers:

- Gauss–Seidel, denoted by GS;
- Line Gauss–Seidel using TDMA on lines $x = $ const., denoted by LGS-X;
- Line Gauss–Seidel using TDMA on lines $y = $ const., denoted by LGS-Y;

---

[4]For a simple rectangular geometry and Dirichlet boundary conditions (Brazier 1974): $\omega = 2/(1 + \sin(\pi/N_{CV}))$.

Table 5.1 Numbers of iterations required by various solvers to reduce the L_1 residual norm by four orders of magnitude when solving the 2D scalar transport problem in stagnation point flow

Scheme	Grid	GS	LGS-X	LGS-Y	LGS-ADI	SIP
UDS	20×20	68	40	35	18	14
	40×40	211	114	110	52	21
	80×80	720	381	384	175	44
CDS	20×20	–	–	–	12	19
	40×40	163	95	77	39	19
	80×80	633	349	320	153	40

- Line Gauss–Seidel using TDMA alternately on lines $x =$ const. and $y =$ const., denoted by LGS-ADI;
- Stone's ILU method, denoted by SIP.

Table 5.1 shows numbers of iterations required by the above solvers to reduce the absolute residual sum by four orders of magnitude.

From the table we see that LGS-X and LGS-Y solvers are about twice as fast as GS; LGS-ADI is about twice as fast as LGS-X, and on the finer grids, SIP is about four times as fast as LGS-ADI. For the GS and LGS solvers, the number of iterations increases by about a factor of four each time the grid is refined; the factor is smaller in case of SIP and LGS-ADI, but, as we shall see in the next example, the factor increases and asymptotically approaches four in the limit of very fine grids.

Another interesting observation is that the GS and LGS solvers do not converge on a 20×20 CV grid with CDS discretization. This is due to the fact that the matrix is not diagonally dominant in this case. Even on the 40×40 CV grid, the matrix is not completely diagonally dominant, but the violation is in the region of uniform distribution of the variable (low gradients) so the effect on the solver is not as severe. The LGS-ADI and SIP solvers are not affected.

We turn now to a test case for which an analytical solution exists and the CDS approximation produces an exact solution on any grid. This helps in the evaluation of the iteration error but the behavior of the solver does not benefit so this case is quite suitable for assessing solver performance. We are solving the Laplace equation with Dirichlet boundary conditions, for which the exact solution is $\phi = xy$. The solution domain is a rectangle, the solution is prescribed at all boundaries and the initial values in the interior are all zero. The initial error is thus equal to the solution and is a smooth function of spatial coordinates. Discretization is performed using FV method described in the previous chapter and CDS scheme. Because the convection is absent, the problem is fully elliptic.

The solvers considered are:

- Gauss–Seidel solver, denoted by GS;
- Line Gauss–Seidel solver using TDMA alternately along lines $x =$ const. and $y =$ const., denoted by LGS-ADI;
- ADI solver described in Sect. 5.3.5;

Table 5.2 Numbers of iterations required by various solvers to reduce the normalized L_1 error norm below 10^{-5} for the 2D Laplace equation with Dirichlet boundary conditions on a square domain $X \times Y = 1 \times 1$ with a uniform grid in both directions

Grid	GS	LGS-ADI	ADI	SIP	ICCG	MG-GS	MG-SIP
8×8	74	22	16	8	7	12	7
16×16	292	77	31	20	13	10	6
32×32	1160	294	64	67	23	10	6
64×64	4622	1160	132	254	46	10	6
128×128	–	–	274	1001	91	10	6
256×256	–	–	–	–	181	10	6

Table 5.3 Numbers of iterations required by ADI solver as a function of the time step size (uniform grid in both directions, 64×64 CV)

$1/\Delta t$	80	68	64	60	32	16	8
No. Iter.	152	134	132	134	234	468	936

- Stone's ILU method, denoted by SIP;
- Conjugate gradient method, preconditioned using incomplete Cholesky decomposition, denoted ICCG;
- Multigrid method using GS as a smoother, denoted MG-GS;
- Multigrid method using SIP as a smoother, denoted MG-SIP.

Table 5.2 shows results obtained on uniform grids on a square solution domain. LGS-ADI is again about four times as fast as GS, and the SIP is about four times as fast as LGS-ADI. ADI is less efficient than SIP on coarse grids but, when the optimum time step is chosen, the number of iterations only doubles when the number of grid points in one direction is doubled, so for fine grids ADI is quite effective. When the time step is varied in a cyclic fashion, the solver becomes even more efficient. This is also true of SIP, but cyclic variation of the parameter increases the cost per iteration. The number of iterations required by ADI to reach convergence for various time steps is given in Table 5.3. The optimum time step is reduced by a factor of two when the grid is refined.

ICCG is substantially faster than SIP; the number of iterations doubles when the grid is refined, so its advantage is bigger on fine grids. Multigrid solvers are very efficient; with SIP as the smoother, only 6 iterations on the finest grid are required. In the MG solvers, the coarsest level was 2×2 CV, so there were three levels on the 8×8 CV grid and eight levels on the 256×256 CV grid. One iteration was performed on the finest grid and on all grids after prolongation, while 4 iterations were performed during the restriction phase. In SIP, the parameter α was set to 0.92. No attempt was made to find optimum values of the parameters; the results obtained are representative enough to show the trends and the relative performance of the various solvers. Note also that the computing effort per iteration is different for each

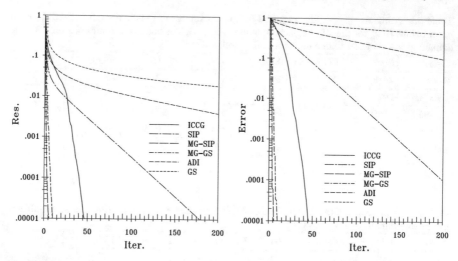

Fig. 5.5 Variation of the L_1 norm of residual (left) and iteration error (right) as a function of the number of performed iterations for various solvers and a 64×64 CV grid

solver. Using the cost of a GS iteration as a reference, we find the following relative costs: LGS-ADI—2.5, ADI—3.0, SIP—4.0 for the first iteration and 2.0 afterwards, ICCG—4.5 for the first iteration and 3.0 afterwards. For MG methods, one needs to multiply the number of iterations on the finest grid by roughly 1.5 to account for the cost of iterations on coarse grids. MG-GS is therefore computationally the most efficient solver in this case.

Because the rate of convergence is different for each solver, the relative cost depends on how accurately we want to solve the equations. To analyze this issue we have plotted the variation of the sum of absolute residuals, and the variation of iteration error with iterations in Fig. 5.5. Two observations can be made:

- The fall of the residual sum is irregular initially, but after a certain number of iterations, the convergence rate becomes constant. An exception is the ICCG solver, which becomes faster as iterations go on. When a very accurate solution is needed, MG solvers and ICCG are the best choice. If—as is the case when solving non-linear problems—moderate accuracy is needed, SIP becomes competitive, and even ADI may be good enough in this case.
- The initial reduction of residual norm is not accompanied by an equal reduction of iteration error for the GS, SIP, ICCG and ADI solvers. Only MG solvers reduce the error and the residual norm at the same pace.

These conclusions are quite general, although there are problem-dependent features. We shall show similar results for the Navier–Stokes equations in Chap. 8.

Because the SIP solver is used in CFD on structured grids, we show the dependence of the number of iterations required to reach convergence on the parameter α in Fig. 5.6. For $\alpha = 0$ the SIP reduces to the standard ILU solver. With the optimum value of α, SIP is about six times as fast as ILU. The problem with SIP is that the optimum

Fig. 5.6 Number of
iterations required to reduce
the L_1 residual norm below
10^{-4} in the above 2D
Laplace problem using SIP
solver, as a function of the
parameter α

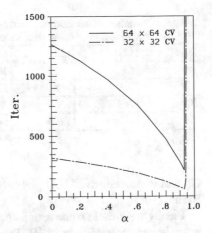

Table 5.4 Numbers of iterations required by various solvers to reduce the normalized L_1 residual norm below 10^{-5} for the 2D Laplace equation with Dirichlet boundary conditions on a rectangular domain $X \times Y = 10 \times 1$ with uniform grid in both directions

Grid	GS	LGS-ADI	SIP	ICCG	MG-GS	MG-SIP
8×8	74	5	4	4	54	3
16×16	293	8	6	6	140	4
32×32	1164	18	13	11	242	5
64×64	4639	53	38	21	288	6
128×128	–	189	139	41	283	6
256×256	–	–	–	82	270	6

value of α lies at the end of the range of usable values: for α slightly larger than the optimum value, the method does not converge. The optimum value usually lies between 0.92 and 0.96. It is safe to use $\alpha = 0.92$, which is usually not optimum, but it provides about five times the speed of the standard ILU method.

Some solvers are affected by cell aspect ratio, because the magnitudes of coefficients become non-uniform. Using a grid with $\Delta x = 10 \, \Delta y$ makes A_N and A_S 100 times larger than A_W and A_E (see example section of the previous chapter). To investigate this effect we solved the Laplace equation problem described above on a rectangular region $X \times Y = 10 \times 1$ using the same number of grid nodes in each direction. Table 5.4 shows numbers of iterations required to reduce the normalized L_1 residual norm below 10^{-5} for various solvers. The GS solver is not affected, but it is no longer a suitable smoother for the MG method. LGS-ADI and SIP solvers become substantially faster compared to the square grid problem. ICCG also performs slightly better. MG-SIP is not affected but MG-GS deteriorates considerably.

This behavior is typical and is also found in convection-diffusion problems (although the effect is less pronounced) and on non-uniform grids which have both small and large cell aspect ratios. A mathematical explanation for the worsening of

Table 5.5 Numbers of iterations required by various solvers to reduce the L_1 residual norm below 10^{-4} for the 3D Poisson equation with Neumann boundary conditions

Grid	GS	SIP	ICCG	CGSTAB	FMG-GS	FMG-SIP
8^3	66	27	10	7	10	6
16^3	230	81	19	12	10	6
32^3	882	316	34	21	9	6
64^3	–	1288	54	41	7	6

GS and improvement of ILU performance with increasing aspect ratio is given by Brandt (1984).

Finally we present some results for the solution of a Poisson equation with Neumann boundary conditions in 3D. The pressure and pressure-correction equations in CFD are of this type. The equation solved is:

$$\frac{\partial^2 \phi}{\partial x^2} + \frac{\partial^2 \phi}{\partial y^2} + \frac{\partial^2 \phi}{\partial z^2} = \sin(x^*\pi)\,\sin(y^*\pi)\,\sin(z^*\pi)\,, \tag{5.99}$$

where $x^* = x/X$, $y^* = y/Y$, $z^* = z/Z$, and X, Y, Z are the dimensions of the solution domain. The equation is discretized using the FV method. The sum of the source terms over the domain is zero, and Neumann boundary conditions (zero gradient normal to boundary) were specified at all boundaries. In addition to GS, SIP and ICCG solvers introduced above, we used also the CGSTAB method with incomplete Cholesky preconditioning. The initial solution is zero. The numbers of iterations required to reduce the normalized sum of absolute residuals four orders of magnitude are presented in Table 5.5.

The conclusions reached from this exercise are similar to those drawn from 2D problems with Dirichlet boundary conditions. When an accurate solution is required, GS and SIP become inefficient on fine grids; conjugate gradient solvers are a better choice, and multigrid methods are best. The FMG strategy, in which the solution on a coarse grid provides the initial solution for the next finer grid, is better than straight multigrid. FMG with ICCG or CGSTAB as a smoother requires even fewer iterations (three to four on the finest grid), but the computing time is higher than for MG-SIP. The FMG principle can be applied to other solvers as well.

Chapter 6
Methods for Unsteady Problems

6.1 Introduction

When computing unsteady flows, we have a fourth coordinate direction to consider: *time*. Just as with the space coordinates, time must be discretized. We can consider the time "grid" in either the finite difference spirit, as discrete points in time, or in a finite volume view as "time volumes". The major difference between the space and time coordinates lies in the direction of influence: whereas a force at any space location may (in elliptic problems) influence the flow anywhere else, forcing at a given instant will affect the flow only in the future—there is no backward influence. Unsteady flows are, therefore, parabolic-like in time. This means that no conditions can be imposed on the solution (except at the boundaries) at any time after the initiation of the calculation, which has a strong influence on the choice of solution strategy. To be faithful to the nature of time, essentially all solution methods advance in time in a step-by-step or "marching" manner. These methods are very similar to ones applied to initial value problems for ordinary differential equations (ODEs) so we shall give a brief review of such methods in the next section.

6.2 Methods for Initial Value Problems in ODEs

6.2.1 Two-Level Methods

For initial value problems, it is sufficient to consider the first-order ordinary differential equation with an initial condition:

$$\frac{d\phi(t)}{dt} = f(t, \phi(t)) ; \quad \phi(t_0) = \phi^0 . \tag{6.1}$$

© Springer Nature Switzerland AG 2020
J. H. Ferziger et al., *Computational Methods for Fluid Dynamics*,
https://doi.org/10.1007/978-3-319-99693-6_6

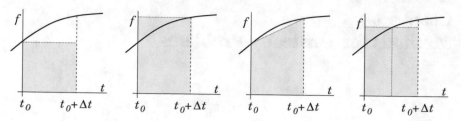

Fig. 6.1 Approximation of the time integral of $f(t)$ over an interval Δt (from left to right: explicit Euler, implicit Euler, trapezoid rule and midpoint rule, respectively)

The basic problem is to find the solution ϕ a short time Δt after the initial point. The solution at $t_1 = t_0 + \Delta t$, ϕ^1, can be regarded as a new initial condition and the solution can be advanced to $t_2 = t_1 + \Delta t$, $t_3 = t_2 + \Delta t$, ... etc.

The simplest methods can be constructed by integrating Eq. (6.1) from t_n to $t_{n+1} = t_n + \Delta t$:

$$\int_{t_n}^{t_{n+1}} \frac{d\phi}{dt}\, dt = \phi^{n+1} - \phi^n = \int_{t_n}^{t_{n+1}} f(t, \phi(t))\, dt\, , \tag{6.2}$$

where we use the shorthand notation $\phi^{n+1} = \phi(t_{n+1})$. This equation is exact. However, the right-hand side cannot be evaluated without knowing the solution; so some approximation is necessary. The mean-value theorem of calculus guarantees that if the integrand is evaluated at the proper point $t = \tau$ between t_n and t_{n+1}, the integral is equal to $f(\tau, \phi(\tau))\, \Delta t$ but this is of little use because τ is unknown. We therefore use some approximate numerical quadrature to evaluate the integral.

Four relatively simple procedures are given below; a geometric picture is provided in Fig. 6.1.

If the integral on the right-hand side of Eq. (6.2) is estimated using the value of the integrand at the initial point, we have:

$$\phi^{n+1} = \phi^n + f(t_n, \phi^n)\, \Delta t\, , \tag{6.3}$$

which is known as the *explicit* or *forward Euler* method.

If, instead, we use the final point in estimating the integral, we obtain the *implicit* or *backward Euler* method:

$$\phi^{n+1} = \phi^n + f(t_{n+1}, \phi^{n+1})\, \Delta t\, . \tag{6.4}$$

Still another method can be obtained by using the midpoint of the interval:

$$\phi^{n+1} = \phi^n + f\left(t_{n+\frac{1}{2}}, \phi^{n+\frac{1}{2}}\right) \Delta t\, , \tag{6.5}$$

which is known as the *midpoint rule* and may be regarded as the basis of a widely-used method for solving partial differential equations—the leapfrog method.

Finally, one can use straight line interpolation between the initial and final points to construct the approximation:

$$\phi^{n+1} = \phi^n + \frac{1}{2}\left[f(t_n, \phi^n)) + f(t_{n+1}, \phi^{n+1})\right]\Delta t , \qquad (6.6)$$

which is called the *trapezoid rule* and is the basis for a popular method of solving partial differential equations—the Crank–Nicolson method.

Collectively, these methods are called two-level methods because they involve the values of the unknown at only two times (the midpoint rule may or may not be regarded as a two-level method, depending on what further approximations are employed). Analysis of these methods can be found in texts on the numerical solution of ordinary differential equations (see, for example, Ferziger 1998, or Moin 2010) and will not be repeated here.

We shall simply review some of their most important properties. First we note that all of the methods but the first require the value of $\phi(t)$ at some point other than $t = t_n$ (which is the initial point of the integration interval at which the solution is known). Therefore, for these methods, the right-hand side cannot be calculated without further approximation or iteration. Thus, the first method belongs to the class called *explicit* methods while all of the others are *implicit*.

All methods produce good solutions if Δt is small. However, the behavior of methods for large step size is important because, in problems with widely varying time scales (including many problems in fluid mechanics), the goal is often to compute the slow, long term behavior of the solution and the short time scales are merely a nuisance. Problems with a wide range of time scales are called *stiff* and are the greatest difficulty one faces in the solution of ordinary differential equations. It is therefore important to inquire about the behavior of methods when the step size is large. This raises the issue of *stability*.

There are a number of definitions of stability in the literature. We shall use a rough definition that calls a method stable if it produces a bounded solution when the solution of the underlying differential equation is also bounded. For the explicit Euler method, stability requires:

$$\left|1 + \Delta t \frac{\partial f(t, \phi)}{\partial \phi}\right| < 1 , \qquad (6.7)$$

which, if $f(t, \phi)$ is allowed to have complex values, requires that $\Delta t\, \partial f(t, \phi)/\partial \phi$ be restricted to the unit circle with center at -1. (Complex values must be considered because higher-order systems may have complex eigenvalues. Only values with zero or negative real part are of interest because they lead to bounded solutions.) A method with this property is called *conditionally stable*; for real values of f, Eq. (6.7) reduces to (see Eq. (6.1)):

$$\left| \Delta t \frac{\partial f(t, \phi)}{\partial \phi} \right| < 2 \, . \tag{6.8}$$

All of the other methods defined above are *unconditionally stable*, i.e., they produce bounded solutions for any time step if $\partial f(t, \phi)/\partial \phi < 0$. However, the implicit Euler method tends to produce smooth solutions even when Δt is very large while the trapezoid rule frequently yields solutions which oscillate with little damping. Consequently, the implicit Euler method tends to behave well, even for non-linear equations, while the trapezoid rule may be unstable for non-linear problems.

Finally, the question of accuracy needs to be considered. On a global scale, it is difficult to say much about this issue due to the incredible variety of equations that one might need to deal with. For a single small step it is possible to use Taylor series to show that the explicit Euler method, starting with the known solution at t_n, yields the solution at time $t_n + \Delta t$ with an error proportional to $(\Delta t)^2$. However, as the number of steps required to compute to some finite final time $t = t_0 + T$ is inversely proportional to Δt, and an error is incurred on each step, the error will at the end be proportional to Δt itself. Therefore, explicit Euler method is a first-order method. The implicit Euler method is also a first-order method, while the trapezoid and midpoint rule methods have errors proportional to $(\Delta t)^2$ and are therefore second-order methods. It can be shown that second order is the highest order achievable by a two-level scheme.

It is important to note that the order of a method is not the sole indicator of its accuracy. While it is true that, for small enough step size, a high-order method will have a smaller error than a low-order one, it is also true that two methods of the same order may have errors that differ by as much as an order of magnitude. The order determines only the *rate* at which the error goes to zero as the step size goes to zero, and this only after the step size has become small enough. 'Small enough' is both method and problem dependent and cannot be determined a priori.

When the step size is small enough, one can estimate the discretization error in the solution by comparing solutions obtained using two different step sizes. This method, known as *Richardson extrapolation*, has been described in Chap. 3, and applies to both spatial and temporal discretization errors. The error can also be estimated by analyzing the difference in solutions produced by two schemes of different order; this will be discussed in Chap. 12.

6.2.2   Predictor-Corrector and Multipoint Methods

The properties that we have found for two-level methods are quite general. Explicit methods are very easy to program and use little computer memory and computation time per step but are unstable if the time step is large. On the other hand, implicit methods require iterative solution to obtain the values at the new time step. This makes them harder to program and they use more computer memory and time per time step, but they are much more stable. (The implicit methods described above are unconditionally stable; this is not true of all implicit methods but they are generally more stable than their explicit counterparts.) One might ask whether it is possible to combine the best of the two methods. Predictor-corrector methods are an attempt to do this.

A wide variety of predictor-corrector methods has been developed; for the present, we shall give just one, which is so well-known that it is often called *the* predictor-corrector method. In this method, the solution at the new time step is *predicted* using the explicit Euler method:

$$\phi_{n+1}^* = \phi^n + f(t_n, \phi^n) \Delta t , \tag{6.9}$$

where the $*$ indicates that this is not the final value of the solution at t_{n+1}. Rather, the solution is *corrected* by applying the trapezoid rule using ϕ_{n+1}^* to compute the derivative:

$$\phi^{n+1} = \phi^n + \frac{1}{2}\left[f(t_n, \phi^n) + f(t_{n+1}, \phi_{n+1}^*)\right] \Delta t . \tag{6.10}$$

This method can be shown to be second-order accurate (the accuracy of the trapezoid rule) but has roughly the stability of the explicit Euler method. One might think that by iterating the corrector, the stability could be improved but this turns out not to be the case because this iteration procedure converges to the trapezoid rule solution only if Δt is small enough.

This predictor-corrector method belongs to the two-level family, for which the highest accuracy possible is second order. For higher-order approximations one must use information at more points. The additional points may be ones at which data has already been computed or points between t_n and t_{n+1}, which are used strictly for computational convenience; the former are called multipoint methods, the latter, Runge–Kutta methods. Here, we shall present multipoint methods; Runge–Kutta methods are presented in the next section.

The best known multipoint methods, the Adams methods, are derived by fitting a polynomial to the derivatives at a number of points in time. If a Lagrange polynomial (Ferziger 1998, or Moin 2010) is fit to $f(t_{n-m}, \phi^{n-m})$, $f(t_{n-m+1}, \phi^{n-m+1})$, ..., $f(t_n, \phi^n)$, and the result is used to compute the integral in Eq. (6.2), we obtain an explicit method of order $m + 1$; methods of this type are called *Adams–Bashforth* methods. For the solution of partial differential equations, only the lower-order methods are used. The first-order method is explicit Euler while the second and third-order methods are:

$$\phi^{n+1} = \phi^n + \frac{\Delta t}{2}\left[3 f(t_n, \phi^n) - f(t_{n-1}, \phi^{n-1})\right] \tag{6.11}$$

and

$$\phi^{n+1} = \phi^n + \frac{\Delta t}{12}\left[23 f(t_n, \phi^n) - 16 f(t_{n-1}, \phi^{n-1}) + 5 f(t_{n-2}, \phi^{n-2})\right]. \tag{6.12}$$

If data at t_{n+1} is included in the interpolation polynomial, implicit methods, known as *Adams–Moulton* methods, are obtained. The first-order method is implicit Euler, the second-order one is trapezoid rule and the third-order method is:

$$\phi^{n+1} = \phi^n + \frac{\Delta t}{12}\left[5 f(t_{n+1}, \phi^{n+1}) + 8 f(t_n, \phi^n) - f(t_{n-1}, \phi^{n-1})\right]. \qquad (6.13)$$

A common method uses the $(m - 1)$st-order Adams–Bashforth method as a predictor and the mth-order Adams–Moulton method as a corrector. Thus, predictor-corrector methods of any order can be obtained.

The multipoint approach has the advantage that it is relatively easy to construct methods of any order. These methods are also easy to use and program. A final advantage is that they require only one evaluation of the derivative per time step (which may be very complicated, especially in applications involving partial differential equations), making them relatively cheap. (The values of $f(t, \phi(t))$ are required several times, but they can be stored once calculated, so that only one evaluation per time step is required.) Their principal disadvantage is that, because they require data from many points prior to the current one, they cannot be started using only data at the initial time point. One has to use other methods to get calculation started. One approach is to use a small step size and a lower-order method (so that the desired accuracy is achieved) and slowly increase the order as more data become available.

These methods are the basis for many accurate ordinary differential equation solvers. In these solvers, error estimators are used to determine the accuracy of the solution at every step. If the solution is not accurate enough, the order of the method is increased up to the maximum order allowed by the program. On the other hand, if the solution is much more accurate than necessary, the order of the method might be reduced to save computing time. Because the step size is difficult to change in multipoint methods, this is done only when the maximum order of the method has already been reached.[1]

Because multipoint methods use data from several time steps, they may produce non-physical solutions. Space does not permit inclusion of the analysis here but it is worth noting that the instabilities of multipoint methods are most often due to the non-physical solutions. These may be suppressed, but not entirely, by a careful choice of the starting method. It is common for these methods to give an accurate solution for some time and then begin to behave badly as the non-physical component of the solution grows. A common remedy for this problem is to restart the method every so often, a trick that is effective but may reduce the accuracy and/or the efficiency of the scheme.

A special multi-point (three-level) scheme which deserves mentioning is the *leapfrog method*. It is essentially the application of the midpoint-rule integration to a time interval of size $2\Delta t$:

---

[1] The quadrature approximations used above assume a uniform time step; if the time step is allowed to vary, the coefficients multiplying the function values at different time levels become complicated functions of step sizes, as we saw in Chap. 3 for finite differences in space.

$$\phi_i^{n+1} = \phi_i^{n-1} + f(t_n, \phi^n) \, 2\Delta t \ . \tag{6.14}$$

Although one integrates over the interval $2\Delta t$ wide, the marching step is Δt, so the integration intervals overlap. This scheme has been widely used in the past and some of its features will be described in Sect. 6.3.1.2.

6.2.3 Runge–Kutta Methods

The difficulties in starting multipoint methods can be overcome by using points between t_n and t_{n+1} rather than earlier points. Methods of this kind are called Runge–Kutta methods, which are very popular in fluid mechanics applications. These methods can be derived for any order of accuracy in a systematic way; we show three levels here. In each case, the level of the scheme is also reflective of its accuracy.

The first-order method is actually just the explicit Euler method described above, but the second-order Runge–Kutta method (RK2) consists of two steps. The first may be regarded as a half-step predictor based on the explicit Euler method; it is followed by a midpoint-rule corrector which makes the method second order:

$$\phi_{n+\frac{1}{2}}^* = \phi^n + \frac{\Delta t}{2} \, f(t_n, \phi^n) \ , \tag{6.15}$$

$$\phi^{n+1} = \phi^n + \Delta t \, f\left(t_{n+\frac{1}{2}}, \phi_{n+\frac{1}{2}}^*\right) \ . \tag{6.16}$$

This method is easy to use and is self-starting, i.e., it requires no data other than the initial condition required by the differential equation itself. In fact, it is very similar in many ways to the predictor-corrector method described above. However, Durran (2010) shows that the RK2-schemes have an amplitude factor greater than unity and so will amplify errors.

6.2.3.1 Third-Order Runge–Kutta Methods

The third-order Runge–Kutta methods (RK3) are finding use in meteorological simulation codes, where they are replacing the leapfrog-method described in the previous section. As with other higher-order Runge–Kutta schemes, the third-order scheme is not unique. One of the best known versions is *Heun's method*. It consists of three steps; the first is the explicit Euler predictor over 1/3rd of the time step:

$$\phi_{n+\frac{1}{3}}^* = \phi^n + \frac{\Delta t}{3} \, f(t_n, \phi^n) \ . \tag{6.17}$$

It is followed by a midpoint-rule approximation applied to 2/3rds of the time step (i.e., centered around $t + \frac{\Delta t}{3}$):

$$\phi^*_{n+\frac{2}{3}} = \phi^n + \frac{2\Delta t}{3} f\left(t_{n+\frac{1}{3}}, \phi^*_{n+\frac{1}{3}}\right) . \tag{6.18}$$

The final step can be viewed as a trapezoid-rule approximation over the whole time step, whereby the value of the integrand at the end of the time step is obtained by linear extrapolation through values at t_n and $t_{n+\frac{2}{3}}$,

$$f(t_{n+1}, \phi^{n+1}) \approx \frac{3}{2} f\left(t_{n+\frac{2}{3}}, \phi^*_{n+\frac{2}{3}}\right) - \frac{1}{2} f(t_n, \phi^n) ,$$

leading to:

$$\phi^{n+1} = \phi^n + \frac{\Delta t}{4}\left[f(t_n, \phi^n) + 3 f\left(t_{n+\frac{2}{3}}, \phi^*_{n+\frac{2}{3}}\right)\right] . \tag{6.19}$$

As noted above third-order Runge–Kutta schemes are not unique. Another version that is used in geophysical applications is described next.

Wicker and Skamarock (2002) proposed a novel scheme for use in codes that use time-splitting.[2] Their scheme has been implemented in the Advanced Research (ARF) version of the Weather Research and Forecasting (WRF) Model (Skamarock and Klemp 2008). For the "slow or low frequency" (the meteorologically significant) modes of motion, it has been shown to be easily adaptable for stable time-splitting, with excellent stability properties for both oscillatory and damped modes of motion. Other schemes are used for acoustic and gravity wave motions.

The Wicker and Skamarock integration takes three steps to advance from t_n to $t_{n+1} = t_n + \Delta t$:

$$\phi^*_{n+\frac{1}{3}} = \phi_n + \frac{\Delta t}{3} f(t_n, \phi^n) ,$$

$$\phi^{**}_{n+\frac{1}{2}} = \phi_n + \frac{\Delta t}{2} f\left(t_{n+\frac{1}{3}}, \phi^*_{n+\frac{1}{3}}\right) , \tag{6.20}$$

$$\phi_{n+1} = \phi_n + \Delta t\, f\left(t_{n+\frac{1}{2}}, \phi^{**}_{n+\frac{1}{2}}\right) .$$

Purser (2007) showed that this is not a true Runge–Kutta scheme and, while it is third-order accurate for linear equations, it is only second-order accurate for non-linear equations (albeit with a considerably smaller error in tests when compared to an RK2 scheme).

---

[2]In many geophysical applications, certain motions, e.g., acoustic waves, move very rapidly relative to others, e.g., winds or ocean currents. It is useful then to actually split the computation and carry out separate calculations for the rapid and slow parts of the system (Klemp et al. 2007, or Blumberg and Mellor 1987).

6.2.3.2   Fourth-Order Runge–Kutta Methods

Historically, the most popular Runge–Kutta method is the fourth-order version (RK4). The first two steps of this method use an explicit-Euler predictor and an implicit-Euler corrector at $t_{n+\frac{1}{2}}$. This is followed by a midpoint-rule predictor for the full step and a Simpson's rule final corrector that gives the method its fourth order. The method is:

$$\phi^*_{n+\frac{1}{2}} = \phi^n + \frac{\Delta t}{2} \, f(t_n, \phi^n) \,, \tag{6.21}$$

$$\phi^{**}_{n+\frac{1}{2}} = \phi^n + \frac{\Delta t}{2} \, f\left(t_{n+\frac{1}{2}}, \phi^*_{n+\frac{1}{2}}\right) \,, \tag{6.22}$$

$$\phi^*_{n+1} = \phi^n + \Delta t \, f\left(t_{n+\frac{1}{2}}, \phi^{**}_{n+\frac{1}{2}}\right) \,, \tag{6.23}$$

$$\phi^{n+1} = \phi^n + \frac{\Delta t}{6}\left[f(t_n, \phi^n) + 2 f\left(t_{n+\frac{1}{2}}, \phi^*_{n+\frac{1}{2}}\right) + \right. \\ \left. 2 f\left(t_{n+\frac{1}{2}}, \phi^{**}_{n+\frac{1}{2}}\right) + f(t_{n+1}, \phi^*_{n+1})\right]. \tag{6.24}$$

A number of variations on this method have been developed. In particular, there are several methods which add a fifth step of either fourth or fifth order to allow estimation of the error and, thereby, the possibility of automatic error control.

As is readily seen from the methods given above, an nth-order Runge–Kutta method requires that the derivative be evaluated n times per time step, making these methods more expensive than multipoint methods of comparable order. In partial compensation, the Runge–Kutta methods of a given order are more accurate (i.e., the coefficient of the error term is smaller) and more stable than the multipoint methods of the same order. Purser (2007) reviews the systematic strategy for generating RK-methods which includes Butcher's array for display of the coefficients and the necessary constraints needed to make each method consistent (see also http://en. wikipedia.org/wiki/Butcher_tableau or Butcher 2008).

6.2.4   Other Methods

There are many other possibilities to advance the solution in time. We describe only one multi-point fully implicit scheme which is widely used in engineering CFD.

A fully implicit three-level second-order scheme can be constructed by integrating over a time interval Δt centered around t_{n+1} (i.e., from $t_{n+1} - \Delta t/2$ to $t_{n+1} + \Delta t/2$) and applying the midpoint rule to both the left and right-hand sides of the Eq. (6.2). The time derivative at t_{n+1} is approximated by differentiating a parabola forced through solutions at three time levels, $t_{n-1}, t_n,$ and t_{n+1}, leading to:

$$\left(\frac{d\phi}{dt}\right)_{n+1} \approx \frac{3\,\phi^{n+1} - 4\,\phi^n + \phi^{n-1}}{2\,\Delta t} \,. \tag{6.25}$$

The right-hand side is evaluated only at t_{n+1}. Because both sides of the equation are second-order approximations of the integrand at the center of the integration interval, application of the midpoint rule to approximate the integral means simply multiplying both sides by Δt—a step that can be omitted. This leads to the following time-advancing method:

$$\phi^{n+1} = \frac{4}{3}\phi^n - \frac{1}{3}\phi^{n-1} + \frac{2}{3}\,f(t_{n+1}, \phi^{n+1})\,\Delta t \,. \tag{6.26}$$

The scheme is fully implicit, because f is evaluated only at the new time level. It is of second order and very easy to implement, but as it is implicit, it requires solving an algebraic equation system at each time step.

This scheme is widely used in engineering applications of CFD; all major commercial CFD-packages as well as public codes offer this scheme for time marching. We shall discuss the reasons in more detail in Sect. 6.3.2.4, but because of scheme's popularity, we give here also its version for variable time steps. There is a conceptual difference relative to other methods, which we need to highlight.

Following the notation from Fig. 6.2 for non-uniform time steps, by fitting a parabola through values of ϕ at three time levels the following expression for the first derivative at t_{n+1} is obtained:

$$\left(\frac{\partial\phi}{\partial t}\right)_{n+1} \approx \frac{\phi_{n+1}\left[(1+\epsilon)^2 - 1\right]\quad \phi_n(1+\epsilon)^2 + \phi_{n-1}}{(t_{n+1} - t_n)\epsilon(1+\epsilon)} \,, \tag{6.27}$$

where the following shorthand notation has been used:

Fig. 6.2 Approximation of time derivative of ϕ at t_{n+1} and the time integral of $f(t)$ using midpoint rule and non-uniform time steps

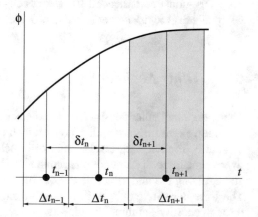

$$\epsilon = \frac{t_n - t_{n-1}}{t_{n+1} - t_n} = \frac{\delta t_n}{\delta t_{n+1}} . \tag{6.28}$$

It is important to note that the location of time levels at which the solution is computed (t_{n-1}, t_n and t_{n+1}) is now different: it is not at the end of user-specified integration intervals (as is the case with all other schemes), but at their center. Thus,

$$t_n = \sum_{i=1}^{n-1} \Delta t_i + \frac{\Delta t_n}{2} = t_{n-1} + \frac{1}{2}(\Delta t_{n-1} + \Delta t_n) .$$

Note also that this scheme, being a multi-level type, cannot be used in the first time step at the start of a computation; one has to use a different scheme to get started. It is common to start with implicit Euler scheme and a smaller time step.

6.3 Application to the Generic Transport Equation

We next consider application of some of the methods given above to the generic transport equation (1.28). In Chaps. 3 and 4, discretization of the convection and diffusion fluxes and source terms for steady problems was discussed. These terms can be treated in the same way for unsteady flows; however, the question of the time at which the fluxes and sources are to be evaluated must be answered.

If the conservation equation is rewritten in a form which resembles the ordinary differential equation (6.1), e.g.:

$$\frac{\partial(\rho\phi)}{\partial t} = -\nabla \cdot (\rho\phi\mathbf{v}) + \nabla \cdot (\Gamma \nabla\phi) + q_\phi = f(t, \phi(t)) , \tag{6.29}$$

any method of time integration can be used. The function $f(t, \phi)$ represents the sum of convection, diffusion and source terms, all of which now appear on the right-hand side of the equation. Because these terms are not known, we must use one of the quadrature approximations introduced above. The convection, diffusion and source terms are discretized, using one of the methods presented in Chaps. 3 or 4, at one or more time levels. If an explicit method is used for time integration, these terms have to be evaluated only at times for which the solution is already known, so they can be calculated. For an implicit method, the discretized right-hand side of the above equation is required at the new time level, for which the solution is not known yet. Therefore, an algebraic system of equations, which differs from the one obtained for steady problems, must be solved. We shall analyze properties of some of the common schemes when applied to a 1D problem below; solutions of a 1D and a 2D problem will be discussed in the examples section.

6.3.1 Explicit Methods

6.3.1.1 Explicit Euler Method

The simplest method is explicit Euler in which all fluxes and sources are evaluated using known values at t_n. In the equation for a CV or grid point, the only unknown at the new time level is the value at that node; the neighbor values are all evaluated at earlier time levels. Thus one can explicitly calculate the new value of the unknown at each node.

In order to study properties of the explicit Euler and other simple schemes, we consider the 1D version of Eq. (6.29) with constant velocity, constant fluid properties, and no source terms:

$$\frac{\partial \phi}{\partial t} = -u \frac{\partial \phi}{\partial x} + \frac{\Gamma}{\rho} \frac{\partial^2 \phi}{\partial x^2} \ .$$

(6.30)

This equation is often used in the literature as a model equation for the Navier–Stokes equations. It is the time-dependent version of the Eq. (3.55) used to illustrate methods for steady problems. Like that equation, it assumes that the important balance is between advection and streamwise diffusion, a balance that rarely occurs in real flows. For this reason, one must be careful about extending what is learned from this equation to the Navier–Stokes equations. Despite this important shortcoming, we can learn something by considering Eq. (6.30).

We first assume that the spatial derivatives are approximated using CDS and that the grid is uniform in x-direction. In this case the same algebraic equation results from both FD and FV discretizations. The new variable value, ϕ_i^{n+1}, is:

$$\phi_i^{n+1} = \phi_i^n + \left[-u \frac{\phi_{i+1}^n - \phi_{i-1}^n}{2\,\Delta x} + \frac{\Gamma}{\rho} \frac{\phi_{i+1}^n + \phi_{i-1}^n - 2\,\phi_i^n}{(\Delta x)^2} \right] \Delta t \ ,$$

(6.31)

which can be rewritten:

$$\phi_i^{n+1} = (1 - 2d)\,\phi_i^n + \left(d - \frac{c}{2}\right) \phi_{i+1}^n + \left(d + \frac{c}{2}\right) \phi_{i-1}^n \ ,$$

(6.32)

where we introduced the dimensionless parameters:

$$d = \frac{\Gamma\,\Delta t}{\rho(\Delta x)^2} \quad \text{and} \quad c = \frac{u\,\Delta t}{\Delta x} \ .$$

(6.33)

The parameter d is the ratio of time step Δt to the characteristic diffusion time $\rho(\Delta x)^2/\Gamma$, which is roughly the time required for a disturbance to be transmitted by diffusion over a distance Δx. The second quantity (c) is the ratio of time step Δt to the characteristic convection time, $\Delta x/u$, the time required for a disturbance to be convected over a distance Δx. This ratio is called the *Courant number* and is one of the key parameters in computational fluid dynamics.[3]

If ϕ were the temperature (a possibility), Eq. (6.32) would need to satisfy several conditions. By virtue of diffusion, an increase in temperature at any of the three points x_{i-1}, x_i and x_{i+1} at the old time level should increase the temperature at point x_i at the new time level. The same can be said for the points x_{i-1} and x_i with respect to convection, assuming $u > 0$. If ϕ represents the concentration of a substance, it should not be negative.

The possibility that some of the coefficients of ϕ_i^{n-1} and ϕ_{i+1}^{n-1} in Eq. (6.32) can become negative should alert us to possible trouble and demands a more detailed analysis. To the extent possible, we want this analysis to mimic the one used for ordinary differential equations. A simple way to do this was invented by *von Neumann* for whom the method is named. He argued that the boundary conditions are rarely the cause of problems (there are exceptions that are not relevant here) so why not ignore them altogether? If that is done, the analysis is simplified. Because this method of analysis can be applied to essentially all of the methods discussed in this chapter, we shall describe it in a little detail; for further details, see Moin (2010) or Fletcher (1991).

In essence the idea can be arrived at as follows. The set of Eq. (6.32) can be written in matrix form:

$$\boldsymbol{\phi}^{n+1} = A\boldsymbol{\phi}^n , \tag{6.34}$$

where the elements of the tridiagonal matrix A can be derived by inspection of Eq. (6.32). This equation gives the solution at the new step in terms of the solution at the previous step. The solution at t_{n+1} can thus be obtained by repetitive multiplication of the initial solution $\boldsymbol{\phi}^0$ by the matrix A. The question is: do the differences between solutions at successive time steps (for non-varying boundary conditions), measured in any convenient way, increase, decrease, or stay the same as n is increased? For example, one measure is the norm:

$$\epsilon = ||\boldsymbol{\phi}^n - \boldsymbol{\phi}^{n-1}|| = \sqrt{\sum_i (\phi_i^n - \phi_i^{n-1})^2} . \tag{6.35}$$

The differential equation requires this quantity to decrease with time through the action of diffusion. Eventually, a steady-state solution will be obtained if the boundary conditions do not vary. Naturally, we would like the numerical method to preserve this property of the exact equations.

This issue is closely connected with the eigenvalues of the matrix A. If some of them are greater than 1, it is not difficult to show that ϵ will grow with n while, if all of them are smaller than 1, ϵ will decay. Normally, the eigenvalues of a matrix are difficult to estimate and, for a problem more complicated than this, we would be in serious trouble. The saving feature of this problem is that, because each diagonal of

---

[3]This number is often called the CFL number, where CFL stands for the initials of R. Courant, K. Friedrichs and H. Lewy, who first defined it in their paper from 1928.

the matrix is constant, the eigenvectors are easily found. They can be represented in terms of sines and cosines but it is simpler to use the complex exponential form:

$$\phi_j^n = \sigma^n e^{i\alpha j} , \tag{6.36}$$

where $i = \sqrt{-1}$ and α is a wavenumber that can be chosen arbitrarily; the choice will be discussed below. If Eq. (6.36) is substituted into Eq. (6.32), the complex exponential term $e^{i\alpha j}$ is common to every term and can be removed and we obtain an explicit expression for the eigenvalue σ:

$$\sigma = 1 + 2d(\cos\alpha - 1) + i c \sin\alpha . \tag{6.37}$$

The magnitude of this quantity is what is important. Because the magnitude of a complex quantity is the sum of the squares of the real and imaginary parts, we have:

$$\sigma^2 = [1 + 2d(\cos\alpha - 1)]^2 + c^2 \sin^2\alpha . \tag{6.38}$$

We now investigate the conditions for σ^2 to be smaller than unity.

Because there are two independent parameters in the expression for σ, it is simplest to consider special cases first. When there is no diffusion (so the diffusion parameter $d = 0$), $\sigma > 1$ for any α and this method is unstable for any value of c, i.e., the method is *unconditionally unstable*, rendering it useless. On the other hand, when there is no convection (so the Courant number $c = 0$), we find that σ is maximum when $\cos\alpha = -1$ so the method is stable provided $d < \frac{1}{2}$, i.e., it is *conditionally stable*.

Direct appeal to our intuition above for Eq. (6.32) that the coefficients of all old nodal values should be positive leads to similar conclusions: $d < 0.5$ and $c < 2d$. The first condition leads to the limit on Δt:

$$\Delta t < \frac{\rho(\Delta x)^2}{2\Gamma} . \tag{6.39}$$

The second requirement imposes no limit on the time step, but gives a relation between convection and diffusion coefficients:

$$\frac{\rho u \Delta x}{\Gamma} < 2 \quad \text{or} \quad \text{Pe}_\Delta < 2 , \tag{6.40}$$

i.e., the cell Peclet number should be smaller than two. This has already been mentioned as a sufficient (but not necessary) condition for boundedness of solutions obtained using CDS for convection fluxes.

Because the method is based on a combination of the explicit Euler method for ordinary differential equations and the central-difference approximation for the spatial derivatives, it inherits the accuracy of each. The method is therefore first order in time and second order in space. The requirement that $d < 0.5$ means that, each time the spatial mesh is halved, the time step has to be reduced by a factor of four. This makes the scheme unsuitable for problems which do not require high temporal resolution (slowly varying solutions, solutions approaching steady state); these are normally the cases in which one would like to use first-order methods in time. An example which illustrates the stability problem will be shown below.

The problem of instability in connection with the condition of Eq. (6.40) was recognized in the 1920s by Courant and Friedrichs and they suggested a cure which is still used today. They noted that, for problems dominated by convection, it is possible for the coefficient of ϕ_{i+1}^{n-1} in Eq. (6.32) to be negative and suggested that the problem could be cured by using *upwind differences*. Instead of evaluating the convection term by a CDS-approximation as we have done above, we use UDS (see Chap. 3). We then get, in place of Eq. (6.31):

$$\phi_i^{n+1} = \phi_i^n + \left[-u \frac{\phi_i^n - \phi_{i-1}^n}{\Delta x} + \frac{\Gamma}{\rho} \frac{\phi_{i+1}^n + \phi_{i-1}^n - 2\phi_i^n}{(\Delta x)^2} \right] \Delta t . \tag{6.41}$$

which yields, in place of Eq. (6.32):

$$\phi_i^{n+1} = (1 - 2d - c)\phi_i^n + d\phi_{i+1}^n + (d + c)\phi_{i-1}^n . \tag{6.42}$$

Because the coefficients of the neighbor values are always positive, they cannot contribute to unphysical behavior of instability. However, the coefficient ϕ_i^n can be negative, thus creating a potential problem. For this coefficient to be positive, the time step should satisfy the following condition:

$$\Delta t < \frac{1}{\frac{2\Gamma}{\rho(\Delta x)^2} + \frac{u}{\Delta x}} . \tag{6.43}$$

When convection is negligible, the restriction on the time step required for stability is the same as Eq. (6.39). For negligible diffusion, the criterion to be satisfied is:

$$c < 1 \quad \text{or} \quad \Delta t < \frac{\Delta x}{u} , \tag{6.44}$$

i.e., the Courant number should be smaller than unity.

One can also apply the von Neumann stability analysis to this problem; the analysis is in accord with the conclusions just reached. Thus, unlike the central-difference method, the upwind approximation provides some stability which, combined with its ease of use, made this type of method very popular for many years. It continues to be used today. When both convection and diffusion are present, the stability criterion is more complicated. Rather than dealing with this complexity, most people require that each individual criterion be satisfied, a condition that may be a little more restrictive than necessary but is safe.

This method has first-order truncation errors in both space and time and requires very small step sizes in both variables if errors are to be kept small. For this reason it is seldom used.

The restriction on the Courant number also has the interpretation that a fluid particle should not move more than one grid length in a single time step. This restriction on the rate of information propagation appears very reasonable but can limit the rate of convergence when methods of this kind are employed for the solution of steady-state problems or when using locally highly refined grids (e.g., near solid walls).

Other explicit schemes may be based on other methods for ordinary differential equations. Indeed, all of the methods described earlier have been used at one time or another in CFD. A central-difference-based three-time-level method, known as the *leapfrog method*, will be described next.

6.3.1.2   Leapfrog Method

When the leapfrog method (see Sect. 6.2.2 for its definition) is applied to the generic transport equation, Eq. (6.30), in which CDS is used for spatial discretization, we obtain:

$$\phi_i^{n+1} = \phi_i^{n-1} + \left[-u \frac{\phi_{i+1}^n - \phi_{i-1}^n}{2\,\Delta x} + \frac{\Gamma}{\rho} \frac{\phi_{i+1}^n + \phi_{i-1}^n - 2\,\phi_i^n}{(\Delta x)^2} \right] 2\Delta t \ . \qquad (6.45)$$

In terms of the dimensionless coefficients d and c, the above equation reads:

$$\phi_i^{n+1} = \phi_i^{n-1} - 4d\phi_i^n + (2d - c)\phi_{i+1}^n + (2d + c)\phi_{i-1}^n \ . \qquad (6.46)$$

In this method, the heuristic requirements for a physically realistic simulation of heat conduction are never satisfied, because the coefficient of ϕ_i^n is unconditionally negative! Indeed, this scheme is unconditionally unstable, and it appears not to be useful for the numerical solution of unsteady problems. However, we need to take a closer look at the role of diffusion in the transport equation. Mesinger and Arakawa (1976) give a clear description.

Unfortunately, due to the construction of the leapfrog method, it has both a "correct" physical mode and an erroneous "computational" mode (all three-level explicit methods exhibit these modes). A von Neumann stability analysis of Eq. (6.46) shows that the method is always unstable, but when $d = 0$, that is, the diffusion is absent, the stability analysis reveals that the leapfrog scheme for the resulting linear convection equation is stable for Courant number $c \leq 1$ and has some numerical dispersion, but no numerical diffusion. The two modes are uncoupled and the computational mode grows over time. On the other hand, the instability is very weak if the time step is small and wave-like solutions are very weakly damped compared to other methods. For these reasons, this method is actually used (with some tricks to stabilize it) in a number of applications, especially in meteorology and oceanography, where convection dominates and diffusion is small.

Williams (2009) lists a large number of numerical codes in geophysical fluid mechanics that use the leapfrog method and notes that dominantly they use the Robert–Asselin (RA) time filter (Asselin 1972) to suppress the method's computational mode which infects the solution. The RA-scheme modifies ϕ_i^n after calculation of ϕ_i^{n+1} by

$$\phi_i^n \leftarrow \phi_i^n + \frac{\nu}{2} \left(\phi_i^{n-1} - 2\phi_i^n + \phi_i^{n+1} \right) = \phi_i^n + d_f . \tag{6.47}$$

This correction d_f is a central second-derivative time filter. The filter parameter ν is made as small as possible (0.01–0.2) consistent with removal of the spurious computational mode. This filtering reduces the leapfrog scheme accuracy from second to first order and damps the solution's amplitude. Williams (2009) modification of the scheme uses

$$\phi_i^n \leftarrow \phi_i^n + \alpha d_f , \tag{6.48}$$
$$\phi_i^{n+1} \leftarrow \phi_i^{n+1} + (1 - \alpha) d_f . \tag{6.49}$$

This reduces the undesired numerical damping and increases the solution accuracy (Amezcua et al. 2011). Williams shows that $\alpha \geq 0.5$ is required for stability and that $\alpha = 0.53$ and $\nu = 0.2$ offer significant improvement over the RA-scheme; Amezcua et al. use $\nu = 0.1$. The Williams' RAW-scheme (so named by Amezcua et al. 2011) is being adopted in meteorological codes which choose to keep the leapfrog time advancement. Some code managers are choosing to implement alternative schemes such as the third-order Runge–Kutta schemes described in Sect. 6.2.3.

6.3.2 Implicit Methods

6.3.2.1 Implicit Euler Method

If stability is a prime requirement, the analysis of methods for ordinary differential equations suggests use of the backward or implicit Euler method. Applied to the generic transport equation (6.30), with the CDS approximation to the spatial derivatives, it gives:

$$\phi_i^{n+1} = \phi_i^n + \left[-u \frac{\phi_{i+1}^{n+1} - \phi_{i-1}^{n+1}}{2\,\Delta x} + \frac{\Gamma}{\rho} \frac{\phi_{i+1}^{n+1} + \phi_{i-1}^{n+1} - 2\phi_i^{n+1}}{(\Delta x)^2} \right] \Delta t , \tag{6.50}$$

or, rearranged and in terms of the dimensionless coefficients c and d introduced in Eq. (6.33):

$$(1 + 2d)\,\phi_i^{n+1} + \left(\frac{c}{2} - d \right) \phi_{i+1}^{n+1} + \left(-\frac{c}{2} - d \right) \phi_{i-1}^{n+1} = \phi_i^n . \tag{6.51}$$

The above equation may be written:

$$A_P \phi_i^{n+1} + A_E \phi_{i+1}^{n+1} + A_W \phi_{i-1}^{n+1} = Q_P, \qquad (6.52)$$

where the matrix coefficients are (see Eq. (6.33)):

$$A_E = \frac{c}{2} - d; \quad A_W = -\frac{c}{2} - d;$$
$$A_P = 1 + 2d = 1 - (A_E + A_W); \quad Q_P = \phi_i^n \qquad (6.53)$$

In this method, all of the fluxes and source terms are evaluated in terms of the unknown variable values at the new time level. The result is a system of algebraic equations very similar to the one obtained for steady problems; actually, the only difference lies in an additional contribution to the coefficient A_P and to the source term Q_P, which stem from the unsteady term.

As was the case for ordinary differential equations, use of the implicit Euler method allows arbitrarily large time steps to be taken; this property is useful in studying flows with slow transients or steady flows. Problems may arise when CDS is used on coarse grids (if the Peclet number is too large in regions of strong change in variable gradient); oscillatory solutions are produced but the scheme remains stable.

The shortcomings of this method are its first-order truncation error in time and the need to solve a large coupled set of equations at each time step. It also requires much more storage than the explicit scheme, because the entire coefficient matrix A and the source vector have to be stored. The advantage is the possibility of using a large time step, which may result in a more efficient procedure despite the shortcomings, especially when marching towards a steady-state solution.

6.3.2.2 Crank–Nicolson Method

The second-order accuracy of the trapezoid-rule method and its relative simplicity suggest its application to partial differential equations when time accuracy is of importance. It is then known as the Crank–Nicolson method. In particular, when applied to the 1D generic transport equation with CDS discretization of spatial derivatives one has:

$$\phi_i^{n+1} = \phi_i^n + \frac{\Delta t}{2} \left[-u \frac{\phi_{i+1}^{n+1} - \phi_{i-1}^{n+1}}{2\Delta x} + \frac{\Gamma}{\rho} \frac{\phi_{i+1}^{n+1} + \phi_{i-1}^{n+1} - 2\phi_i^{n+1}}{(\Delta x)^2} \right] +$$
$$\frac{\Delta t}{2} \left[-u \frac{\phi_{i+1}^n - \phi_{i-1}^n}{2\Delta x} + \frac{\Gamma}{\rho} \frac{\phi_{i+1}^n + \phi_{i-1}^n - 2\phi_i^n}{(\Delta x)^2} \right]. \qquad (6.54)$$

The scheme is implicit; the contributions from fluxes and sources at the new time level give rise to a coupled set of equations similar to those of implicit Euler scheme. The above equation can be re-written as:

$$A_P \phi_i^{n+1} + A_E \phi_{i+1}^{n+1} + A_W \phi_{i-1}^{n+1} = Q_i^t , \qquad (6.55)$$

where (in terms of dimensionless coefficients c and d, introduced in Eq. (6.33)):

$$
\begin{aligned}
A_E &= \frac{c}{4} - \frac{d}{2} ; \quad A_W = -\frac{c}{4} - \frac{d}{2} , \\
A_P &= 1 + d = 1 - (A_E + A_W) , \\
Q_i^t &= (1 + A_E + A_W)\phi_i^n - A_E \phi_{i+1}^n - A_W \phi_{i-1}^n .
\end{aligned}
\qquad (6.56)
$$

The term Q_i^t represents an "additional" source term, which contains the contribution from the previous time level; it remains constant during iterations at the new time level. The equation may also contain a source term dependent on the new solution, so the above term needs to be stored separately.

This scheme requires very little more computational effort per step than the first-order implicit Euler scheme. Von Neumann stability analysis shows that the scheme is unconditionally stable, but oscillatory solutions (and even instability) are possible for large time steps. This may be attributed to the possibility of the coefficient of ϕ_i^n becoming negative at large Δt, but it is guaranteed to be positive if $1 - d > 0$ or $\Delta t < \rho(\Delta x)^2/\Gamma$, which is twice the maximum step size allowed by the explicit Euler method. In practice, much larger time steps can be used without producing oscillations; the limit is problem-dependent.

If the modified equation method[4] (Sect. 4.4.6) is applied to Eq. (6.54), the result is (when retaining only the leading-order truncation terms; cf., Fletcher 1991, Table 9.3)

$$\frac{\partial \phi}{\partial t} + u \frac{\partial \phi}{\partial x} - \frac{\Gamma}{\rho} \frac{\partial^2 \phi}{\partial x^2} = -u \frac{\Delta x^2}{6}\left(1 + \frac{c^2}{2}\right)\frac{\partial^3 \phi}{\partial x^3} + \frac{\Gamma}{\rho}\frac{\Delta x^2}{12}(1 + 3c^2)\frac{\partial^4 \phi}{\partial x^4} . \tag{6.57}$$

Two things are apparent. First, the remaining third-order derivative in the truncation error signals that the scheme is dispersive, meaning that a well-formed initial condition will be distorted over time because its components move at different speeds in the solution. Fletcher (1991) shows that this effect can be significant! Second, the scheme has a fourth-order derivative with a positive coefficient, which will work to detract slightly from the physical diffusion.[5] Warming and Hyett (1974) and Donea

---

[4]Recall that the procedure is (1) expand the terms in the difference scheme in a two-dimensional (x, t) Taylor series about a single point (in this case, x_i, t_n) to obtain an expanded partial differential equation, (2) use this partial differential equation itself (and derivatives of it) to replace all of the higher-order and mixed time terms with spatial derivative terms in the expanded PDE, and (3) re-arrange to have the original PDE on the left-hand side and the remaining terms on the right-hand side. Warming and Hyett (1974)'s Procedure Table is useful when doing the method by hand. The remaining terms are the truncation error, i.e., the difference between what the difference equation solves and the original equation. Keep the lowest-order ones to see the most important physical effects.

[5]For diffusion (dissipation) to occur, the coefficients of the even-order derivatives in the modified equation must have alternating signs, the one for the second-order term being positive and the one for the fourth-order term being negative, and so on. Note that second-order physical diffusion is present in the original equation here.

et al. (1987) show that a necessary condition for stability of a scheme is for the leading even-order term in the modified equation to be diffusive; however, application of the von Neumann method may be necessary for a complete stability analysis.

The Crank–Nicolson scheme can be regarded as an equal blend of first-order explicit and implicit Euler schemes. Only for equal blending is second-order accuracy obtained; for other blending factors, which may vary in space and time, the method remains first-order accurate. The stability is increased if the implicit contribution is increased, but the accuracy is reduced as described in the next section.

6.3.2.3 θ–Method

The explicit and implicit Euler schemes and the Crank–Nicolson scheme can be cast together into a more general (weighted) scheme that allows a choice between stability and accuracy, which can be important in nonlinear problems. For the generic transport equation, to advance from t_n to $t_{n+1} = t_n + \Delta t$, we write, in lieu of Eq. (6.54):

$$
\phi_i^{n+1} = \phi_i^n + \theta \Delta t \left[-u \frac{\phi_{i+1}^{n+1} - \phi_{i-1}^{n+1}}{2\Delta x} + \frac{\Gamma}{\rho} \frac{\phi_{i+1}^{n+1} + \phi_{i-1}^{n+1} - 2\phi_i^{n+1}}{(\Delta x)^2} \right] +
$$
$$
(1-\theta)\Delta t \left[-u \frac{\phi_{i+1}^n - \phi_{i-1}^n}{2\Delta x} + \frac{\Gamma}{\rho} \frac{\phi_{i+1}^n + \phi_{i-1}^n - 2\phi_i^n}{(\Delta x)^2} \right] . \tag{6.58}
$$

The scheme is implicit unless $\theta = 0$, in which case it reverts to the forward Euler scheme. For $\theta = 1$, one retrieves the backward Euler scheme, while for $\theta = 0.50$, the Crank–Nicolson scheme from just above emerges. It follows that the solution method described there in Eqs. (6.55) and (6.56) can be applied. When $\theta < \frac{1}{2}$, the method is unstable according to the von Neumann method. The method is stable for $\frac{1}{2} \le \theta \le 1$, and, as expected, is most accurate for $\theta = \frac{1}{2}$, i.e., second order. For $\theta > \frac{1}{2}$, the method is first-order accurate and dissipative. While codes using this method can be run for tests with $\theta = \frac{1}{2}$, 0.5 is at the limit of stability and so codes run for real world simulations typically employ $0.52 \le \theta \le 0.60$.

This θ–method was described in Casulli and Cattani (1994) and subsequently used by Casulli in an impressive set of papers related to free-surface and non-hydrostatic flows. Fringer et al. (2006) employed the method in a code for non-hydrostatic coastal ocean simulation.

6.3.2.4   Three-Time-Level Method

A fully implicit scheme of second-order accuracy can be obtained by using a quadratic backward approximation in time, as described in Sect. 6.2.4. For the 1D generic transport equation and CDS discretization in space we obtain:

$$\rho \frac{3\phi_i^{n+1} - 4\phi_i^n + \phi_i^{n-1}}{2\Delta t} \Delta t = \\ \left[-\rho u \frac{\phi_{i+1}^{n+1} - \phi_{i-1}^{n+1}}{2\Delta x} + \Gamma \frac{\phi_{i+1}^{n+1} + \phi_{i-1}^{n+1} - 2\phi_i^{n+1}}{(\Delta x)^2} \right] \Delta t . \tag{6.59}$$

The resulting algebraic equation can be written:

$$A_{\mathrm{P}}\phi_i^{n+1} + A_{\mathrm{E}}\phi_{i+1}^{n+1} + A_{\mathrm{W}}\phi_{i-1}^{n+1} = 2\phi_i^n - \frac{1}{2}\phi_i^{n-1} . \tag{6.60}$$

The coefficients A_{E} and A_{W} are the same as those for the implicit Euler scheme, see Eq. 6.53. The central coefficient has now a stronger contribution from the time derivative:

$$A_{\mathrm{P}} = -(A_{\mathrm{E}} + A_{\mathrm{W}}) + \frac{3}{2} = \frac{3}{2} + 2d , \tag{6.61}$$

and the source term contains contribution from time t_{n-1}; see Eq. 6.60.

This three-time-level (TTL) scheme is easier to implement than the Crank–Nicolson (CN) scheme; it is also less prone to producing oscillatory solutions, although this may happen with large values of Δt. One has to store the variable values from three time levels, but the memory requirements are the same as for the Crank–Nicolson scheme. The scheme is second-order accurate in time and one can show that this scheme is unconditionally stable. We also see from Eq. (6.60) that the coefficient of the old value at node i is always positive; however, the coefficient of the value at t_{n-1} is always negative, which is why the scheme may produce oscillatory solutions if the steps are large.

If the modified equation method (Sect. 4.4.6) is applied to Eq. (6.59), the result is (retaining the leading-order truncation terms; cf., Fletcher 1991, Table 9.3)

$$\frac{\partial \phi}{\partial t} + u \frac{\partial \phi}{\partial x} - \frac{\Gamma}{\rho} \frac{\partial^2 \phi}{\partial x^2} = -u \frac{\Delta x^2}{6}(1 + 2c^2)\frac{\partial^3 \phi}{\partial x^3} + \frac{\Gamma}{\rho} \frac{\Delta x^2}{12}(1 + 12c^2)\frac{\partial^4 \phi}{\partial x^4} . \tag{6.62}$$

From Eq. (6.62), we conclude that, as for the CN method, the truncation error is $O(\Delta x^2)$. Interestingly, in the limit of small c^2, the CN and TTL errors are the same, but for $c^2 = u\frac{\Delta t}{\Delta x} = O(1)$, the TTL error is 2 times larger for dispersion and 3.25 times larger for "anti" diffusion/dissipation.

This scheme can be blended with the first-order implicit Euler scheme. Only the contributions to the central coefficient and source term need be modified in the manner of the deferred-correction approach described in Sect. 5.6. This is useful when starting the calculation, because only one old-level solution is available. Also, if one is after a steady-state solution, switching to the implicit Euler scheme ensures stability and allows large time steps to be used. Blending in a small amount of the first-order scheme helps prevent oscillations which contributes to the esthetics of the solution (the accuracy is no better without oscillations, but it looks nicer graphically).

If oscillations do occur, one has to reduce the time step as the oscillations are an indication of large temporal discretization errors. This comment does not apply to schemes which are only conditionally stable.

This scheme is available in most commercial and public CFD-codes. One reason for its popularity is the fact that it is fully implicit, i.e., the convection, diffusion and source terms are computed only at the new time level, as in the first-order implicit Euler scheme. This means that one can change the grid between two time steps; the only information required from previous time levels are the variable values at grid points, so old solutions need to be interpolated to the new grid. However, one does not need fluxes or source terms from previous time levels, as is the case with Crank–Nicolson scheme.

6.3.3 Other Methods

The schemes described above are the ones most often used in general-purpose CFD codes. For special purposes, for example, in large eddy and direct simulation of turbulence, one often uses higher-order schemes, such as third or fourth-order Runge–Kutta or Adams methods. Usually, higher-order temporal discretization is used when the spatial discretization is also of higher order, which is the case when the solution domain is of regular shape so that higher-order methods in space are easy to apply. Application of higher-order methods for ordinary differential equations to CFD-problems is straightforward.

One can blend any two schemes in a similar way as explicit and implicit Euler schemes are blended in the θ-method. Usually, one scheme is of higher order but prone to stability problems under certain conditions; the other is usually unconditionally stable but less accurate (typically the implicit Euler scheme). This blending can be applied locally, when and where the higher-order scheme encounters problems. However, identifying such conditions and determining the optimal blending factor to maximize both the accuracy and stability is not so trivial to achieve. Although we have not provided an example of such a method, we wanted to point out this possibility; similar approaches to spatial discretization are explained in Sect. 4.4.6.

6.4 Examples

In order to demonstrate the performance of some of the methods described above, we look first at the unsteady version of the example problem of Chap. 3. The problem to be solved is given by Eq. (6.30), with the following initial and boundary conditions: at $t = 0$, $\phi_0 = 0$; for all subsequent times, $\phi = 0$ at $x = 0$, $\phi = 1$ at $x = L = 1$, $\rho = 1$, $u = 1$ and $\Gamma = 0.1$. Because the boundary conditions do not change in time, the solution develops from the initial zero field to the steady state solution given in Sect. 3.10. For spatial discretization, second-order CDS is used. For temporal

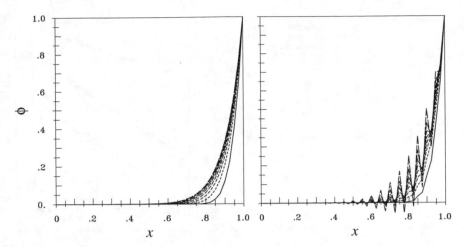

Fig. 6.3 Time evolution of the solution by explicit Euler method using time steps $\Delta t = 0.003$ (left; $d < 0.5$) and $\Delta t = 0.00325$ (right, $d > 0.5$)

discretization, we use both explicit and implicit first-order Euler methods, the Crank–Nicolson method and the fully-implicit method with three time levels.

We first demonstrate what happens when the explicit Euler scheme (which is only conditionally stable) is used with time steps which violate the stability condition. Figure 6.3 shows the evolution of the solution over a small time period calculated using a time step slightly below and another slightly above the critical value given by Eq. (6.39). When the time step is larger than the critical value, oscillations are generated which grow unboundedly with time. A few time steps later than the last solution shown in Fig. 6.3, the numbers become too large to be handled by the computer. With implicit schemes, no problems occurred even when very large time steps were used.

In order to investigate the accuracy of temporal discretization, we look at the solution at the node at $x = 0.95$ of a uniform grid with 41 nodes ($\Delta x = 0.025$) at time $t = 0.01$. We performed calculations up to that time using 5, 10, 20 and 40 time steps with the four above-mentioned schemes. The convergence of ϕ at $x = 0.95$ for $t = 0.01$ as the time step size is reduced is shown in Fig. 6.4. The implicit Euler and the three-level method under-predict, while the explicit Euler and the Crank–Nicolson method over-predict the correct value. All schemes show monotonic convergence towards the time-step independent solution.

Because we do not have an exact solution to compare with, we obtained an accurate reference solution at time $t = 0.01$ using the Crank–Nicolson scheme (the most accurate method) with $\Delta t = 0.0001$ (100 time steps). This solution is much more accurate than any of the above solutions so it can be treated as an exact solution for error estimation purposes. By subtracting solutions mentioned above from this reference solution, we obtained estimates of the temporal discretization error for each scheme and time step size. The spatial discretization error is the same in all

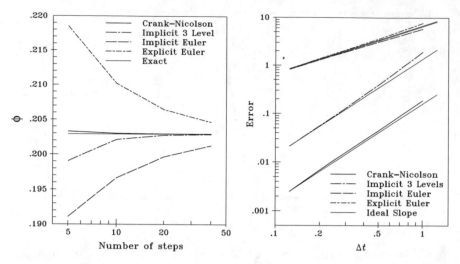

Fig. 6.4 Convergence of ϕ at $x = 0.95$ and $t = 0.01$ as the time step size is reduced (left) and temporal discretization errors (right) for various time integration schemes

cases and plays no role here. The errors thus obtained are plotted against time step size in Fig. 6.4.

The two Euler methods show the expected first-order behavior: the error is reduced by one order of magnitude when the time step is reduced by the same amount. The two second order schemes show also the expected error reduction rate, which closely follow the ideal slope. However, the Crank–Nicolson method gives a more accurate solution because its initial error is much smaller. The three-level scheme is started by the implicit Euler method, which resulted in a large initial error. Because, in this problem, the temporal variation is monotonic from the initial towards steady state, the initial error remains important throughout the solution. The error reduction rate is the same in both methods, but the error level is in this case determined by its initial value.

We next examine the 2D test case of Chap. 4, which involves heat transfer from a wall with prescribed temperature, exposed to a stagnation point flow; see Sect. 4.7. The initial solution is again $\phi_0 = 0$; the boundary conditions do not change with time and are the same as in the steady state problem investigated in Sect. 4.7, with $\rho = 1.2$ and $\Gamma = 0.1$. Spatial discretization is by CDS, and a uniform grid with 20×20 CV is used. Linear equation systems in case of implicit schemes are solved using the SIP-solver (described in Chap. 5), and the iteration error was reduced below 10^{-5}. We compute the time evolution of solution towards the steady state. Figure 6.5 shows isotherms at four time instants.

In order to investigate the accuracy of the various schemes in this case, we look at the heat flux through the isothermal wall at time $t = 0.12$. The variation of the heat flux, Q, as a function of the time step size is shown in Fig. 6.6. As in the previous test case, the results obtained using second-order schemes change little as the number

Fig. 6.5 Isotherms in the unsteady 2D problem at times $t = 0.2$ (upper left), $t = 0.5$ (upper right), $t = 1.0$ (lower left) and $t = 2.0$ (lower right), calculated on a uniform 20×20 CV grid using the CDS for spatial and the Crank–Nicolson method for temporal discretization

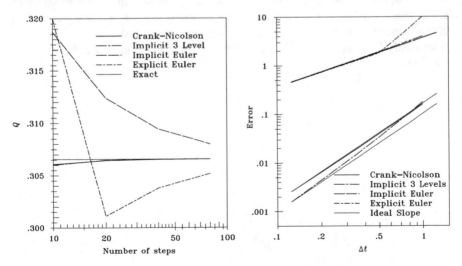

Fig. 6.6 Heat flux through the isothermal wall at $t = 0.12$ (left) and temporal discretization errors in calculated wall heat flux (right) as a function of the time step size for various schemes (spatial discretization by CDS, 20×20 CV uniform grid)

of time steps is increased, while the first-order schemes are far less accurate. The explicit Euler scheme does not converge monotonically; the solution obtained with the largest time step lies on the opposite side of the time-step independent value from the values obtained using smaller time steps.

For the sake of error estimation, we obtained an accurate reference solution by using very small time step ($\Delta t = 0.0003$, 400 steps to $t = 0.12$) with the Crank–Nicolson scheme. The spatial discretization was the same in all cases, so again the spatial discretization errors cancel out. By subtracting the heat flux value calculated using different schemes and time steps from the reference solution, we obtain the estimates of the temporal discretization error. These are plotted against normalized time step (normalized with the largest time step) in Fig. 6.6.

Again, the expected asymptotic convergence rates for first- and second-order schemes are obtained. However, the lowest error is now obtained with the second-order scheme with three time levels. This is, as in the previous example, due to the dominant effect of the initial error, which turns out to be smaller when the three-level scheme is started with implicit Euler method than in the Crank–Nicolson scheme. With both second-order schemes the errors are much smaller than with first-order schemes: the result is more accurate with second-order schemes using the largest time step than with first-order schemes and an eight times smaller time step!

Although we were dealing with simple unsteady problems which have a smooth transition from the initial to the steady state, we see that first-order Euler schemes are very inaccurate. In both test cases the second-order schemes had errors more than two orders of magnitude lower than the Euler schemes (whose errors were of the order of 1% with the smallest time step). One can expect that in transient flows even larger discrepancies can occur. Of the first-order schemes only the implicit Euler method can be used when steady-state solutions are sought; for transient flow problems, second (or higher) order schemes are recommended.

Unsteady flow problems will be discussed in Sects. 8.4.2 and 9.12.

Chapter 7
Solution of the Navier–Stokes Equations: Part 1

The solution of the Navier–Stokes equations is a principal topic of this book; thus, we have given the material significant attention and organized it in two main parts, each constituting one chapter. In this chapter we cover basic issues and features of the equations and the methods of solution; implementation details for some commonly used methods and examples of their application are presented in the next chapter. This layout aims to provide a convenient structure for neophytes and/or professionals, allowing either systematic study from the beginning or direct access to particular issues, methods or tools for those who are, for example, (1) using commercial codes and want to understand available methods in those codes or (2) code developers who need to choose among alternatives.

7.1 Basics

In Chaps. 3, 4, and 6 we dealt with the discretization of a generic conservation equation. The discretization principles described there apply to the momentum and continuity equations (which we shall collectively call the Navier–Stokes equations). We reprise here an integral form of those equations:

$$\frac{\partial}{\partial t} \int_V \rho u_i \, dV + \int_S \rho u_i \mathbf{v} \cdot \mathbf{n} \, dS = \int_S \mathbf{t}_i \cdot \mathbf{n} \, dS + \int_V \rho b_i \, dV , \qquad (7.1)$$

$$\frac{\partial}{\partial t} \int_V \rho \, dV + \int_S \rho \mathbf{v} \cdot \mathbf{n} \, dS = 0 , \qquad (7.2)$$

and, recalling from Eq. (1.18) that $\mathbf{t}_i = \tau_{ij}\mathbf{i}_j - p\mathbf{i}_i$, we give a differential form:

$$\frac{\partial(\rho u_i)}{\partial t} + \frac{\partial(\rho u_j u_i)}{\partial x_j} = \frac{\partial \tau_{ij}}{\partial x_j} - \frac{\partial p}{\partial x_i} + \rho b_i , \qquad (7.3)$$

© Springer Nature Switzerland AG 2020
J. H. Ferziger et al., *Computational Methods for Fluid Dynamics*,
https://doi.org/10.1007/978-3-319-99693-6_7

$$\frac{\partial \rho}{\partial t} + \nabla \cdot (\rho \mathbf{v}) = 0 \,, \tag{7.4}$$

from Eqs. (1.17), (1.21), (1.5), and (1.6); see Sect. 1.4 for the definition of the stress and forcing terms. Next, we shall describe how the terms in the momentum equations which differ from those in the generic conservation equation are treated.

The unsteady and advection terms in the momentum equations have the same form as in the generic conservation equation. The diffusion (viscous) terms are similar to their counterparts in the generic equation but, because the momentum equations are vector equations, these contributions become a bit more complex and their treatment needs to be considered in more detail. The momentum equations also contain a contribution from the pressure, which has no analog in the generic equation. It may be regarded either as a source term (treating the pressure gradient as body force—non-conservatively) or as a surface force (conservative treatment) but, due to the close connection of the pressure and the continuity equation, it requires special attention. Finally, the fact that the principal variable is a vector allows more freedom in the choice of a grid.

7.1.1 Discretization of Convection and Viscous Terms

The convection term in the momentum equation is non-linear; its differential and integral forms read (see Eqs. (7.1) and (7.3) above):

$$\frac{\partial(\rho u_i u_j)}{\partial x_j} \quad \text{and} \quad \int_S \rho u_i \mathbf{v} \cdot \mathbf{n} \, dS \,. \tag{7.5}$$

The treatment of the convection term in the momentum equations follows that of the convection term in the generic equation; any of the methods described in Chaps. 3 and 4 can be used.

The viscous terms in the momentum equations correspond to the diffusion term in the generic equation; their differential and integral forms are:

$$\frac{\partial \tau_{ij}}{\partial x_j} \quad \text{and} \quad \int_S (\tau_{ij} \mathbf{i}_j) \cdot \mathbf{n} \, dS \,, \tag{7.6}$$

where, for a Newtonian fluid and incompressible flow:

$$\tau_{ij} = \mu \left(\frac{\partial u_i}{\partial x_j} + \frac{\partial u_j}{\partial x_i} \right) \,. \tag{7.7}$$

Because the momentum equations are vector equations, the viscous term is more complicated than the generic diffusion term. The part of the viscous term in the momentum equations which corresponds to the diffusion term in the generic conservation equation is

$$\frac{\partial}{\partial x_j}\left(\mu\frac{\partial u_i}{\partial x_j}\right) \quad \text{and} \quad \int_S \mu\,\nabla u_i\cdot\mathbf{n}\,dS\,. \tag{7.8}$$

This term can be discretized using any of the approaches described for the corresponding terms of the generic equation in Chaps. 3 and 4 but it is only one contribution of viscous effects to the ith component of momentum.

Equations (1.28), (1.26), (1.21) and (1.17) allow us to identify the other viscous effects as the contributions of the bulk viscosity (which is non-zero only in compressible flows) and a further contribution due to the spatial variability of the viscosity. For incompressible flow with constant fluid properties, these contributions disappear (thanks to the continuity equation).

The extra terms which are non-zero when the viscosity is spatially variable in an incompressible flow may be treated in the same manner as the terms (7.8):

$$\frac{\partial}{\partial x_j}\left(\mu\frac{\partial u_j}{\partial x_i}\right) \quad \text{and} \quad \int_S\left(\mu\frac{\partial u_j}{\partial x_i}\mathbf{i}_j\right)\cdot\mathbf{n}\,dS\,, \tag{7.9}$$

where $\mathbf{n}$ is the unit outward normal to the surface of the control volume and summation on j applies. As noted above, for constant μ, this term vanishes. For this reason, this term is often treated explicitly even when implicit solution methods are used. It is argued that even when the viscosity varies this term is small compared to the term (7.8) so its treatment has only a slight impact on the rate of convergence. However, this argument applies strictly only in an integral sense; the extra term may be quite large on any one CV face.

7.1.2 Discretization of Pressure Terms and Body Forces

As noted in Chap. 1, we usually deal with the "pressure" in terms of the combination $p - \rho_0\mathbf{g}\cdot\mathbf{r} + \mu\frac{2}{3}\nabla\cdot\mathbf{v}$. In incompressible flows, the last term is zero. One form of the momentum equations (see Eq. (7.3)) contains the gradient of this quantity which may be approximated by the FD methods described in Chap. 3. However, as the pressure and velocity nodes on the grid may not coincide, the approximations used for their derivatives may differ.

In FV methods, the pressure term is usually treated as a surface force (conservative approach), i.e., in the equation for u_i the integral

$$-\int_S p\,\mathbf{i}_i\cdot\mathbf{n}\,dS \tag{7.10}$$

is required. Methods described in Chap. 4 for the approximation of surface integrals can then be used. As we shall show below, the treatment of this term and the arrangement of variables on the grid play an important role in assuring the computational efficiency and accuracy of the numerical solution method.

Alternatively, the pressure can be treated non-conservatively, by retaining the above integral in its volumetric form:

$$-\int_V \nabla p \cdot \mathbf{i}_i \, dV \ . \tag{7.11}$$

In this case, the derivative (or, for non-orthogonal grids, all three derivatives) needs to be approximated at one or more locations within the CV. The non-conservative approach introduces a global non-conservative error; although this error tends to zero as the grid size goes to zero, it may be significant for finite grid size.

The difference between the two approaches is significant only in the FV methods. In FD methods, there is no distinction between the two versions, although one can produce both conservative and non-conservative approximations.

Other body forces, like the non-conservative ones arising when covariant or contravariant velocities are used in non-Cartesian coordinate systems are easy to treat in finite difference schemes: they are usually simple functions of one or more variables and can be evaluated using techniques described in Chap. 3. If these terms involve the unknowns, as for example, the component of the viscous term in cylindrical coordinates:

$$-2\mu \frac{v_r}{r^2} \ ,$$

they may be treated implicitly. This is usually done when the contribution of this term to the central coefficient A_P in the discretized equation is positive, in order to avoid destabilization of the iterative solution scheme by reducing the diagonal dominance of the matrix. Otherwise, the extra term is treated explicitly.

In FV-methods, these terms are integrated over the CV volume. Usually, the midpoint-rule approximation is used, so that the value at CV-center is multiplied by cell volume. More elaborate schemes are possible but rarely used.

In some cases the non-conservative terms, considered as body forces, dominate the transport equation (e.g., when swirling flows are computed in polar coordinates or when flows are treated in a rotating coordinate frame, for example, in turbomachinery flows). The treatment of the non-linear source terms and the variable coupling may then become very important.

7.1.3 Conservation Properties

The Navier–Stokes equations have the property that the momentum in any control volume (microscopic or macroscopic) is changed only by flow through the surface, forces acting on the surface, and volumetric body forces. This important property is inherited by the discretized equations if the FV-approach is used and the surface fluxes for adjacent control volumes are identical. If this is done, then the integral over the entire domain, being the sum of the integrals over the microscopic control volumes, reduces to a sum over the surface of the domain. Overall mass conservation follows in the same way from the continuity equation.

Energy conservation is a more complex issue. In incompressible isothermal flows, the only energy of significance is kinetic energy.[1] When heat transfer is important, the kinetic energy is generally small compared to the thermal energy so the equation introduced to account for energy transport is a conservation equation for thermal energy. So long as the temperature-dependence of the fluid properties is not significant, the thermal energy equation can be solved after solution of the momentum equations is complete. The coupling is then entirely one-way and the energy equation becomes an equation for the transport of a passive scalar, the case treated in Chaps. 3–6.

An equation for the kinetic energy can be derived by taking the scalar product of the momentum equation with the velocity, a procedure which mimics the derivation of the energy equation in classical mechanics. Note that, in contrast to compressible flow, for which there is a separate conservation equation for the total energy, in incompressible isothermal flows both momentum and energy conservation are consequences of the same equation; this poses the problems that are the subject of this section.

We shall be interested principally in the kinetic energy conservation equation for a macroscopic control volume, which may be either the entire considered domain or one of the small CVs used in a finite volume method. If the local kinetic energy equation obtained in the manner just described is integrated over a control volume, we obtain, after using Gauss' theorem:

$$
\frac{\partial}{\partial t} \int_V \rho \frac{v^2}{2} \, dV = - \int_S \rho \frac{v^2}{2} \mathbf{v} \cdot \mathbf{n} \, dS - \int_S p \mathbf{v} \cdot \mathbf{n} \, dS + \int_S (\mathbf{S} \cdot \mathbf{v}) \cdot \mathbf{n} \, dS -
$$
$$
\int_V (\mathbf{S} : \nabla \mathbf{v} - p \nabla \cdot \mathbf{v} + \rho \mathbf{b} \cdot \mathbf{v}) \, dV . \tag{7.12}
$$

Here $\mathbf{S}$ stands for the viscous part of the stress tensor whose components are τ_{ij} defined in Eq. (1.13), i.e., $\mathbf{S} = \mathbf{T} + p\mathbf{I}$. The first term in the volume integral on the right-hand side disappears if the flow is inviscid; the second is zero if the flow is incompressible; the third is zero in the absence of body forces.

Several points relating to this equation are worth mentioning.

- The first three terms on its right side are integrals over the surface of the control volume. This means that the kinetic energy in the control volume is not changed by the action of convection and/or pressure within the control volume. In the absence of viscosity, only flow of energy through the surface or work done by the forces acting at the surface of the control volume can affect the kinetic energy within it;

---

[1]In flows where the density of the fluid varies with height in a gravitational field, the flow is said to be *stratified*, and the flow may carry heavy fluid up or light fluid down so that it now has a density different from its surroundings; the fluid then has not only kinetic energy, but also energy as a result of its position, called *potential energy*. The result is buoyant forces that are touched on later in this chapter and are very important in both meteorology and oceanography.

kinetic energy is then globally conserved in this sense. This is a property that we would like to preserve in a numerical method.

- Guaranteeing global energy conservation in a numerical method is a worthwhile goal, but not an easily attained one. Because the kinetic energy equation is a consequence of the momentum equation and not a distinct conservation law, it cannot be enforced separately.

- If a numerical method is energy-conservative and the net energy flux through the surface is zero, then the total kinetic energy in the domain does not grow with time. If such a method is used, the velocity at every point in the domain must remain bounded, providing an important kind of numerical stability. Indeed, energy methods (which sometimes have no connection to physics) are often used to prove stability of numerical methods. Energy conservation does not say anything about the convergence or accuracy of a method. Accurate solutions may be obtained with methods that are not conservative of kinetic energy. However, kinetic energy conservation is especially important in computing unsteady flows.

- Because the kinetic energy equation is a consequence of the momentum equations and not independently enforceable in a numerical method, global kinetic energy conservation must be a consequence of the *discretized* momentum equations. It is thus a property of the discretization method, but not an obvious one. To see how it might arise, we form the kinetic energy equation corresponding to the discretized momentum equations by taking the scalar product of the latter with the velocity and summing over all control volumes. We shall consider the result term-by-term.

- The pressure gradient terms are especially important, so let us look into them further. To get the pressure gradient term into the form displayed in Eq. (7.12), we used the following equality:

$$\mathbf{v} \cdot \nabla p = \nabla \cdot (p\mathbf{v}) - p \nabla \cdot \mathbf{v} . \tag{7.13}$$

For incompressible flows, $p \nabla \cdot \mathbf{v} = 0$ so only the first term on the right-hand side remains. As it is a divergence, its volume integral can be converted to a surface integral. As already noted, this means that the pressure influences the overall kinetic energy budget only by its action at the surface. We would like the discretization to retain this property. Let us see how this might happen.

If $G_i p$ represents the numerical approximation to the ith component of the pressure gradient, then, when the discretized u_i-momentum equation is multiplied by u_i, the pressure gradient term gives a contribution $\sum u_i G_i p \Delta V$. Energy conservation requires that this contribution be equal to (cf. Eq. (7.13)):

$$\sum_{i=1}^{N} u_i G_i p \Delta V = \sum_{S_b} p v_n \Delta S - \sum_{N} p D_i u_i \Delta V , \tag{7.14}$$

where the superscript N indicates that the sum is over all CVs (grid nodes), S_b is the boundary of the solution domain, v_n is the velocity component normal to the boundary and $D_i u_i$ is the discretized velocity divergence used in the continuity equation. If this is so, $D_i u_i = 0$ at each node, so the second term in the above equation is zero. The equality of the left and right-hand sides can then be ensured only if G_i and D_i are compatible in the following sense:

$$\sum_{i=1}^{N} (u_i G_i p + p D_i u_i) \Delta V = \text{surface terms} . \tag{7.15}$$

This states that the approximation of the pressure gradient and the divergence of the velocity must be compatible if kinetic energy conservation is to hold. Once either approximation is chosen, the freedom to choose the other is lost.

To make this more concrete, assume that the pressure gradient is approximated with backward differences and the divergence operator with forward differences (the usual choice on a staggered grid). The one-dimensional version of Eq. (7.15) on a uniform grid then reads:

$$\sum_{i=1}^{N} [(p_i - p_{i-1})u_i + (u_{i+1} - u_i)p_i] = u_{N+1} p_N - u_1 p_0 . \tag{7.16}$$

The only two terms that remain when the sum is taken are the "surface terms" on the right-hand side. The two operators are therefore compatible in the above sense. Conversely, if forward differences were used for the pressure gradient, the continuity equation would need to use backward differences. If central differences are used for one, they are required for the other.

The requirement that only boundary terms remain when the sum over all CVs (grid nodes) is taken applies to the other two conservative terms, the convection and viscous stress terms. Satisfaction of this requirement is not easy in any case and is especially difficult for arbitrary and unstructured meshes (see, e.g., Mahesh et al. 2004, for staggered and colocated unstructured meshes). If a method is not energy conservative on uniform regular grids, it will certainly not be so on more complicated ones. On the other hand, a method which is conservative on uniform grids might be nearly so on complex grids.

- A Poisson equation is often used to compute the pressure. As we shall see, it is derived by taking the divergence of the momentum equation. The Laplacian operator in the Poisson equation is thus the product of the divergence operator in the continuity equation and the gradient operator in the momentum equation, i.e., $L = D(G())$. The approximation of the Poisson equation cannot be selected independently; it must be consistent with the divergence and gradient operators if mass conservation is to obtain. Energy conservation adds the further requirement that the divergence and gradient approximations be consistent in the sense defined above.

- For an incompressible flow without body forces, the only remaining volume integral is the viscous term. For a Newtonian fluid, this term becomes:

$$-\int_V \tau_{ij} \frac{\partial u_j}{\partial x_i}\, dV\ .$$

Inspection reveals that the integrand is a sum of squares (see, e.g., the definition of τ_{ij} in Eq. (7.7)) so this term is always negative (or zero). It represents the irreversible (in the thermodynamic sense) conversion of kinetic energy of the flow into internal energy of the fluid and is called *viscous dissipation*. As incompressible flows are usually low speed flows, the addition to the internal energy is rarely significant but the loss of kinetic energy is often quite important to the flow. In compressible flows, the energy transfer is often important to both sides.

• The time-differencing method can destroy the energy conservation property. In addition to the requirements on the spatial discretization mentioned above, the approximation of the time derivatives should be properly chosen.

The Crank–Nicolson scheme is a particularly good choice. In it, the time derivatives in the momentum equations are approximated by:

$$\frac{\rho\,\Delta V}{\Delta t}\left(u_i^{n+1} - u_i^n\right)\ .$$

If we take the scalar product of this term with $u_i^{n+1/2}$, which in the Crank–Nicolson scheme is approximated by $(u_i^{n+1} + u_i^n)/2$, the result is the change in the kinetic energy:

$$\frac{\rho\,\Delta V}{\Delta t}\left[\left(\frac{v^2}{2}\right)^{n+1} - \left(\frac{v^2}{2}\right)^n\right]\ ,$$

where $v^2 = u_i u_i$ (summation implied). With proper choices of the approximations to the other terms, the Crank–Nicolson scheme is energy conservative.

The fact that momentum and energy conservation are both governed by the same equation makes construction of numerical approximations that conserve both properties difficult. As already noted, kinetic energy conservation cannot be enforced independently. If the momentum equations are written in strong conservation form and a finite volume method is used then global momentum conservation is usually assured. The construction of energy conservative methods is a hit or miss affair. One selects a method and determines whether it is conservative or not; if not, adjustments are made until conservation is achieved.

An alternative method of guaranteeing kinetic energy conservation is to use a different form of the momentum equations. For example, one could use the following equation for incompressible flows:

$$\frac{\partial u_i}{\partial t} + \epsilon_{ijk} u_j \omega_k = \frac{\partial\left(\dfrac{p}{\rho} + \dfrac{1}{2} u_j u_j\right)}{\partial x_i} + \nu \frac{\partial^2 u_i}{\partial x_j \partial x_j}\ , \qquad (7.17)$$

where ϵ_{ijk} is the Levi-Civita symbol (it is $+1$ if $\{ijk\} = \{123\}$ or an even permutation of it, it is -1 if $\{ijk\}$ is an odd permutation of $\{123\}$ such as $\{321\}$ and zero otherwise). ω is the vorticity defined by Eq. (7.105). Energy conservation follows from this form of the momentum equation by symmetry; when the equation is multiplied by u_i, the second term on the left-hand side is identically zero as a consequence of the antisymmetry property of ϵ_{ijk}. However, because this is not a conservative form of the momentum equation, construction of a momentum conserving method requires care.

Kinetic energy conservation is of particular importance in computing complex unsteady flows. Examples include the simulation of global weather patterns and simulations of turbulent flows. Lack of guaranteed energy conservation in these simulations often leads to growth of the kinetic energy and instability. For steady flows, energy conservation is less important but it does prevent certain types of misbehavior by the iterative solution method.

Kinetic energy is not the only quantity whose conservation is desirable but cannot be independently enforced: angular momentum is another such quantity. Flows in rotating machinery, internal combustion engines and many other devices exhibit pronounced rotation or swirl. If the numerical scheme does not conserve global angular momentum, the calculation is likely to get into trouble. Central difference schemes are generally much better than upwind schemes with respect to angular momentum conservation.

7.1.4 Choice of Variable Arrangement on the Grid

Now let us turn to the discretizations. The first issue is to select the points in the domain at which the values of the unknown dependent variables are to be computed. There is more to this than one might think. Basic features of numerical grids were outlined in Chap. 2. There are, however, many variants of the distribution of computational points within the solution domain. The basic arrangements associated with the FD and FV discretization methods were shown in Figs. 3.1 and 4.1. These arrangements may become more complicated when coupled equations for vector fields (like the Navier–Stokes equations) are being solved. These issues are discussed below.

7.1.4.1 Colocated Arrangement

The obvious choice is to store all the variables at the same set of grid points and to use the same control volumes for all variables; such an arrangement is called colocated, see Fig. 7.1. Because many of the terms in each of the equations are essentially identical, the number of coefficients that must be computed and stored is minimized and the programming is simplified by this choice. Furthermore, when multigrid procedures are used, the same restriction and prolongation operators for transfer of information between the various grids can be used for all variables.

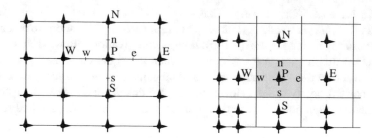

Fig. 7.1 Colocated arrangement of velocity components and pressure on a FD (left) and FV (right) grid

The colocated arrangement also has significant advantages in complicated solution domains, especially when the boundaries have slope discontinuities or the boundary conditions are discontinuous. A set of control volumes can be designed to fit the boundary including the discontinuity. Other arrangements of the variables lead to some of the variables being located at singularities of the grid, which may lead to singularities in the discretized equations.

The colocated arrangement was out of favor for a long time for incompressible flow computation due to the difficulties with pressure-velocity coupling and the occurrence of oscillations in the pressure. From the time the staggered grid was introduced in the mid-1960s until the early 1980s, the colocated arrangement was hardly used. Then, use of non-orthogonal grids became more commonplace as problems in complex geometries began to be tackled. The staggered approach can be used in generalized coordinates only when contravariant (or other grid-oriented) components of vectors and tensors are the working variables. This complicates the equations by introducing curvature terms that are difficult to treat numerically, and may create non-conservative errors when the grid is not smooth as will be shown in Chap. 9. When improved pressure-velocity coupling algorithms were developed in the 1980s, the popularity of the colocated arrangement began to rise. Nowadays all major commercial and public-domain CFD codes use the colocated arrangement of variables. The advantages will be demonstrated below.

7.1.4.2   Staggered Arrangements

There is no need for all variables to share the same grid; a different arrangement may turn out to be advantageous. In Cartesian coordinates, the staggered arrangement introduced by Harlow and Welsh (1965) offers several advantages over the colocated arrangement. This arrangement is shown in Fig. 7.2. Several terms that require interpolation with the colocated arrangement, can be calculated (to a second-order approximation) without interpolation. This can be seen from the x-momentum CV shown in Fig. 8.1. Both the pressure and diffusion terms are very naturally approximated by central-difference approximations without interpolation, because the pres-

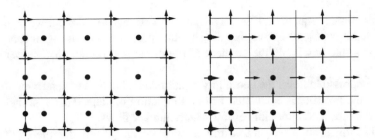

Fig. 7.2 Fully staggered arrangement of velocity components and pressure (right) and a partially staggered arrangement (left) on a FV-grid

sure nodes lie at CV-face centers and the velocity derivatives needed for the diffusion terms are readily computed at the CV-faces. Also, evaluation of mass fluxes in the continuity equation on the faces of a pressure CV is straightforward. Details are presented below.

Perhaps the biggest advantage of the staggered arrangement is the strong coupling between the velocities and the pressure. This helps to avoid some types of convergence problems and oscillations in pressure and velocity fields, an issue that will be further discussed later.

The numerical approximation on a staggered grid is also conservative of kinetic energy, which has advantages that were discussed earlier. The proof of this statement is straightforward but lengthy and will not be given here.

Other staggering methods have been suggested, for example, the partially staggered ALE (Arbitrary Lagrangian–Eulerian) method (Hirt et al. 1997; Donea et al. 2004), in which both velocity components are stored at the corners of the pressure CVs, see Fig. 7.2. This variant has some advantages when the grid is non-orthogonal, an important one being that the pressure at the boundary need not be specified. However, it also has drawbacks, notably the possibility of producing oscillatory pressure or velocity fields.

Other arrangements have not gained wide popularity and will not be further discussed here.

7.1.5 Calculation of the Pressure

Solution of the Navier–Stokes equations is complicated by the lack of an independent equation for the pressure, whose gradient contributes to each of the three momentum equations. Furthermore, while the mass conservation equation for compressible flow has density as the dominant variable, the continuity equation does not have a dominant variable in incompressible flows, mass conservation becoming then a kinematic constraint on the velocity field rather than a dynamic equation. One way out of this difficulty is to construct the pressure field so as to guarantee satisfaction of the continuity equation. This may seem a bit strange at first, but we shall show below that it is possible. Note that the absolute pressure is of no significance in an incompressible flow; only the gradient of the pressure (pressure difference) affects the flow.

In compressible flows the continuity equation can be used to determine the density and the pressure is calculated from an equation of state. This approach is not appropriate for incompressible or low Mach number flows.

Within this section we present the basic philosophy behind some of the most popular methods of pressure-velocity coupling. Chapter 8 presents sets of discretized equations and other guidance which form the basis for writing computer codes.

7.1.5.1   The Pressure Equation and Its Solution

The momentum equations clearly determine the respective velocity components so their roles are well defined. This leaves the continuity equation, which does not contain the pressure, to determine the pressure. How can this be done? The most common method is based on combining the two equations.

The form of the continuity equation suggests that we take the divergence of the momentum Eq. (1.15). The continuity equation can be used to simplify the resulting equation, leaving a Poisson-equation for the pressure:

$$\nabla \cdot (\nabla p) = -\nabla \cdot \left[\nabla \cdot (\rho \mathbf{v} \mathbf{v} - \mathbf{S}) - \rho \mathbf{b} + \frac{\partial (\rho \mathbf{v})}{\partial t} \right] . \qquad (7.18)$$

In Cartesian coordinates this equation reads:

$$\frac{\partial}{\partial x_i} \left(\frac{\partial p}{\partial x_i} \right) = -\frac{\partial}{\partial x_i} \left[\frac{\partial}{\partial x_j} \left(\rho u_i u_j - \tau_{ij} \right) \right] + \frac{\partial (\rho b_i)}{\partial x_i} + \frac{\partial^2 \rho}{\partial t^2} . \qquad (7.19)$$

For the case of constant density, viscosity, and body force, this equation simplifies further; the viscous and unsteady terms disappear by virtue of the continuity equation leaving:

$$\frac{\partial}{\partial x_i} \left(\frac{\partial p}{\partial x_i} \right) = -\frac{\partial}{\partial x_i} \left[\frac{\partial (\rho u_i u_j)}{\partial x_j} \right] . \qquad (7.20)$$

The pressure equation can be solved by one of the numerical methods for elliptic equations described in Chaps. 3 and 4. It is important to note that the right-hand side of the pressure equation is a sum of derivatives of terms in the momentum equations; these must be approximated in a manner consistent with their treatment in the equations from which they are derived. In what follows there are two approaches for obtaining the pressure. In the first case, we seek to find the pressure itself at the next iteration or next time step in each scheme. Alternatively, we can define the new pressure in terms of an older value (at previous iteration level or time step) plus a pressure correction, i.e., $p^{\text{new}} = p^{\text{old}} + p'$. We then seek to find the much smaller pressure correction p'. As might be expected there are usually advantages to this approach.

The Laplacian operator in the pressure equation is the product of the divergence operator originating from the continuity equation and the gradient operator that comes from the momentum equations. In a numerical approximation, it is important that the consistency of these operators be maintained, i.e., the approximation of the Poisson equations is defined as the product of the divergence and gradient approximations used in the basic equations. To emphasize the importance of this issue, the two derivatives of the pressure in the above equations were separated: the outer derivative stems from the continuity equation while the inner derivative arises from the momentum equations. The outer and inner derivatives may be discretized using different schemes—they have to be those used in the momentum and continuity equations. Violation of this constraint leads to lack of satisfaction of the continuity equation (which, unfortunately, we will see later (Sect. 8.2) is an issue for colocated-grid methods).

A pressure equation of this kind is used to calculate the pressure in both explicit and implicit solution methods. *To maintain consistency among the approximations used, it is best to derive the equation for the pressure from the discretized momentum and continuity equations rather than by approximating the above Poisson equation.* The pressure equation can also be used to calculate the pressure from a velocity field obtained by solving the vorticity-streamfunction equations, see Sect. 7.2.4.

7.1.5.2 Note on Pressure and Incompressibility

Suppose that we have a velocity field $\mathbf{v}^*$, which does not satisfy the continuity condition; for example, $\mathbf{v}^*$ may have been obtained by time-advancing the Navier–Stokes equations without invoking continuity. We wish to create a new velocity field $\mathbf{v}$ which:

- satisfies continuity,
- is as close as possible to the original field $\mathbf{v}^*$.

Mathematically we can pose this problem as one of minimizing:

$$\tilde{R} = \frac{1}{2} \int_V [\mathbf{v}(\mathbf{r}) - \mathbf{v}^*(\mathbf{r})]^2 \, dV , \qquad (7.21)$$

where $\mathbf{r}$ is the position vector and V is the domain over which the velocity field is defined, subject to the continuity constraint

$$\nabla \cdot \mathbf{v}(\mathbf{r}) = 0 \qquad (7.22)$$

being satisfied everywhere in the field. The question of boundary conditions will be dealt with below.

This is a standard type of problem of the calculus of variations. A useful way of dealing with it is to introduce a Lagrange multiplier. The original problem (7.21) is replaced by the problem of minimizing:

$$R = \frac{1}{2} \int_V [\mathbf{v}(\mathbf{r}) - \mathbf{v}^*(\mathbf{r})]^2 \, dV - \int_V \lambda(\mathbf{r}) \, \nabla \cdot \mathbf{v}(\mathbf{r}) \, dV \,, \tag{7.23}$$

where λ is the Lagrange multiplier. The inclusion of the Lagrange multiplier term does not affect the minimum value because the constraint (7.22) requires that term to be zero.

Suppose that the function that minimizes the functional R is $\mathbf{v}^+$; of course, $\mathbf{v}^+$ also satisfies (7.22). Thus:

$$R_{\min} = \frac{1}{2} \int_V [\mathbf{v}^+(\mathbf{r}) - \mathbf{v}^*(\mathbf{r})]^2 \, dV \,. \tag{7.24}$$

If $R_{\min}$ is a true minimum, then any deviation from $\mathbf{v}^+$ must produce a second-order change in R. Thus suppose that:

$$\mathbf{v} = \mathbf{v}^+ + \delta \mathbf{v} \,, \tag{7.25}$$

where $\delta \mathbf{v}$ is small but arbitrary. When $\mathbf{v}$ is substituted into the expression (7.23), the result is $R_{\min} + \delta R$ where:

$$\delta R = \int_V \delta \mathbf{v}(\mathbf{r}) \cdot [\mathbf{v}^+(\mathbf{r}) - \mathbf{v}^*(\mathbf{r})] \, dV - \int_V \lambda(\mathbf{r}) \nabla \cdot \delta \mathbf{v}(\mathbf{r}) \, dV \,. \tag{7.26}$$

We have dropped the term proportional to $(\delta \mathbf{v})^2$ as it is of second order. Now, integrating the last term by parts and applying Gauss's theorem, we obtain:

$$\delta R = \int_V \delta \mathbf{v}(\mathbf{r}) \cdot [\mathbf{v}^+(\mathbf{r}) - \mathbf{v}^*(\mathbf{r}) + \nabla \lambda(\mathbf{r})] \, dV + \int_S \lambda(\mathbf{r}) \, \delta \mathbf{v}(\mathbf{r}) \cdot \mathbf{n} \, dS \,. \tag{7.27}$$

On the parts of the domain surface on which a boundary condition on $\mathbf{v}$ is given (walls, inflow), it is presumed that both $\mathbf{v}$ and $\mathbf{v}^+$ satisfy the given condition so $\delta \mathbf{v}$ is zero there. These portions of the boundaries make no contribution to the surface integral in Eq. (7.27) so no condition on λ is required on them; however, a condition will be developed below. On those parts of the boundary where other types of boundary conditions are given (symmetry planes, outflows) $\delta \mathbf{v}$ is not necessarily zero; to make the surface integral vanish, we need to require that $\lambda = 0$ on these portions of the boundary.

If δR is to vanish for arbitrary $\delta \mathbf{v}$, we must require that the volume integral in Eq. (7.27) also vanishes, i.e.:

$$\mathbf{v}^+(\mathbf{r}) - \mathbf{v}^*(\mathbf{r}) + \nabla \lambda(\mathbf{r}) = 0 \,. \tag{7.28}$$

Finally, we recall that $\mathbf{v}^+(\mathbf{r})$ must satisfy the continuity equation (7.22). Taking the divergence of Eq. (7.28) and applying this condition, we find:

$$\nabla^2 \lambda(\mathbf{r}) = \nabla \cdot \mathbf{v}^*(\mathbf{r}) \,, \tag{7.29}$$

which is a Poisson equation for $\lambda(\mathbf{r})$. On those portions of the boundary on which boundary conditions are given on $\mathbf{v}$, $\mathbf{v}^+ = \mathbf{v}^*$. Equation (7.28) shows that in this case $\nabla \lambda(\mathbf{r}) = 0$ and we have a boundary condition on λ.

If Eq. (7.29) and the boundary conditions are satisfied, the velocity field will be divergence-free. It is also useful to note that this entire exercise can be repeated with the continuous operators replaced by discrete ones.

Once the Poisson equation is solved, the corrected velocity field is obtained from Eq. (7.28) written in the form:

$$\mathbf{v}^+(\mathbf{r}) = \mathbf{v}^*(\mathbf{r}) - \nabla \lambda(\mathbf{r}) \,. \tag{7.30}$$

This shows that the Lagrange multiplier $\lambda(\mathbf{r})$ essentially plays the role of the pressure and again shows that, in incompressible flows, the function of the pressure is to allow continuity to be satisfied.

7.1.6 Initial and Boundary Conditions for the Navier–Stokes Equations

Values of all variables need to be initialized before the iterative solution procedure is started. In the case of steady-state flows, the initial values are of no importance as far as the final solution is concerned; however, they influence the convergence rate and the overall computing effort. It is therefore desirable to initialize the solution such that the difference from final solution is as small as possible. However, in many practical situations it is difficult to provide initial fields which are a good estimate of the final solution; in most cases, we start with some trivial initialization (e.g., zero or constant non-zero velocity, constant pressure and temperature).

When unsteady flows are computed, the requirements on initial conditions are more stringent. The velocity and pressure fields which we specify at $t = 0$ must satisfy the Navier–Stokes equations. Because the solution at $t = \Delta t$ depends strongly on the solution at $t = 0$, any errors in the initial solution will be carried over to future time steps. Only when periodic flows (or flows with a stochastic nature, e.g., direct or large-eddy simulations of turbulent flows, cf. Chap. 10) are computed is the effect of initial conditions lost after a while. In any case, inappropriate initial conditions can cause considerable trouble, so care is needed.

Boundary conditions need to be applied at every time step; they may be constant or vary in time. Everything that has been said about boundary conditions in Chaps. 3

Fig. 7.3 On the boundary
conditions for velocities at a
wall and a symmetry plane

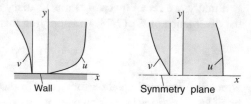

Wall   Symmetry plane

and 4 for the generic conservation equation applies to the momentum equations.
Some special features will be addressed in this section.

At a wall the no-slip boundary condition applies, i.e., the velocity of the fluid is
equal to the wall velocity, a Dirichlet boundary condition. However, there is another
condition that can be directly imposed in a FV method; the normal viscous stress is
zero at a wall. This follows from the continuity equation, e.g., for a wall at $y = 0$
(see Fig. 7.3):

$$\left(\frac{\partial u}{\partial x}\right)_{\text{wall}} = 0 \Rightarrow \left(\frac{\partial v}{\partial y}\right)_{\text{wall}} = 0 \Rightarrow \tau_{yy} = 2\mu\left(\frac{\partial v}{\partial y}\right)_{\text{wall}} = 0 . \qquad (7.31)$$

Therefore, the diffusion flux in the v equation at the south boundary is:

$$F_s^d = \int_{S_s} \tau_{yy}\,dS = 0 . \qquad (7.32)$$

This should be implemented directly, rather than using only the condition that $v = 0$
at the wall. Because at cell center $v_P \neq 0$, we would obtain a non-zero derivative in
the discretized flux expression if this were not done; $v = 0$ is used as a boundary
condition in the continuity equation. The shear stress can be calculated by using a
one-sided approximation of the derivative $\partial u/\partial y$; one possible approximation is (for
the u equation and the situation from Fig. 7.4):

$$F_s^d = \int_{S_s} \tau_{xy}\,dS = \int_{S_s} \mu\frac{\partial u}{\partial y}\,dS \approx \mu_s\left(\frac{\partial u}{\partial y}\right)_s S_s . \qquad (7.33)$$

The derivative of u with respect to y at the wall can be approximated in several
ways. One approach is to set the location of node S to be identical with the face
centroid "s", see the left-hand side of Fig. 7.4. The other approach is to set the node
S outside the computational domain, as if the near-boundary cell was mirrored at the
boundary, see the right-hand side of Fig. 7.4.

In the first-named case, the following simple approximation which assumes a
linear variation of u with y is:

$$\left(\frac{\partial u}{\partial y}\right)_s \approx \frac{u_P - u_s}{y_P - y_s} . \qquad (7.34)$$

Fig. 7.4 On the implementation of boundary conditions for velocities on Cartesian grids

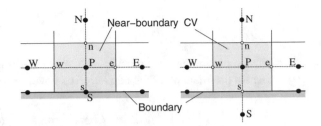

Because this is a one-sided approximation, it is only first-order accurate. However, if this approximation over the half-cell width is used at the boundary together with a central-difference approximation in the interior, the solution will converge with second order. The accuracy can be increased by using a second-order one-sided approximation, based on the assumption of a quadratic variation of u with y (a parabola is fitted to values at boundary location "s" and cell centers P and N, see the left-hand side of Fig. 7.4), which for a uniform grid reads:

$$\left(\frac{\partial u}{\partial y}\right)_s \approx \frac{9u_P - 8u_s - u_N}{6(y_P - y_s)} = \frac{9u_P - 8u_s - u_N}{6(\Delta y/2)} . \tag{7.35}$$

If "boundary" nodes are placed outside of the solution domain, then the same central-difference approximation for the derivative of u with respect to y can be used at both the boundary face "s" and the internal face "n", i.e.:

$$\left(\frac{\partial u}{\partial y}\right)_s \approx \frac{u_P - u_S}{y_P - y_S} = \frac{u_P - u_S}{\Delta y} . \tag{7.36}$$

However, the boundary condition specifies $u_s = u_{\text{wall}}$ and thus the value of u at the auxiliary node S needs to be obtained by extrapolation, using the specified boundary value and one or more values at internal cell centers. The simplest approximation based on linear extrapolation leads to:

$$u_S \approx 2u_s - u_P . \tag{7.37}$$

A more accurate approximation is obtained by using quadratic extrapolation (the same that was used to obtain the approximation of the derivative in Eq. (7.35)); for a uniform grid one obtains:

$$u_S \approx \frac{8}{3}u_s - 2u_P + \frac{1}{3}u_N . \tag{7.38}$$

In corresponding expressions for non-uniform grids, which are easily derived, the multipliers of nodal variable values become functions of grid spacings.

At a symmetry plane we have the opposite situation: the shear stress is zero, but the normal stress is not, because (for the situation from Fig. 7.3):

$$\left(\frac{\partial u}{\partial y}\right)_{\text{sym}} = 0 ; \quad \left(\frac{\partial v}{\partial y}\right)_{\text{sym}} \neq 0 . \tag{7.39}$$

The diffusion flux in the u equation is zero, and the diffusion flux in the v equation requires an approximation of the derivative of v with respect to y:

$$F_s^d = \int_{S_s} \tau_{yy} \, dS = \int_{S_s} 2\mu \frac{\partial v}{\partial y} \, dS \approx \mu_s \left(\frac{\partial v}{\partial y}\right)_s . \tag{7.40}$$

The boundary condition specifies $v_s = 0$ (no flow through symmetry plane). The same approximations that were introduced above for $(\partial u/\partial y)_s$ at a wall boundary can be applied to v at the symmetry boundary.

In a FV method using a staggered grid, the pressure is not required at boundaries (except if the pressure at a boundary is specified; this is handled in Chap. 11). This is due to the fact that the nearest CV for the velocity component normal to the boundary extends only up to the center of the scalar CV, where the pressure is calculated. When a colocated arrangement is used, all CVs extend to the boundary and we need the boundary pressure to calculate the pressure forces in the momentum equations. We have to use extrapolation from the interior to obtain the pressure at the boundaries. In most cases, linear extrapolation is sufficiently accurate for a second-order method, but quadratic extrapolation is even better. However, there are cases in which a large pressure gradient near a wall is needed in the equation for the normal velocity component to balance a body force (buoyancy, centrifugal force etc.). If the pressure extrapolation is not accurate, this condition may not be satisfied and large normal velocities near the boundary may result. This can be avoided by calculating the normal velocity component for the first CV from the continuity equation, by adjusting the pressure extrapolation, or by local grid refinement.

The boundary conditions for the pressure-correction equation (defined in Sect. 7.2.2) also deserve some attention. When the mass flux through a boundary is prescribed, the mass flux correction in the pressure-correction equation is zero there. This condition should be directly implemented in the continuity equation when deriving the pressure-correction equation. It is equivalent to specifying a Neumann boundary condition (zero gradient) for the pressure correction. A discussion of the boundary conditions for the pressure-correction in fractional-step methods is included in Sect. 8.3.4.

At the outlet, if the inlet mass fluxes are given, extrapolation of the velocity to the boundary (zero gradient, e.g. $u_E = u_P$) can usually be used for steady flows when the outflow boundary is far from the region of interest and the Reynolds number is large. The extrapolated velocity is then corrected to give exactly the same total mass flux as at inlet (this cannot be guaranteed by any extrapolation). The corrected velocities are then considered as prescribed for the following outer iteration and the mass flux correction at the outflow boundary is set to zero in the continuity equation. This leads to the pressure-correction equation having Neumann conditions on all boundaries and makes it *mathematically singular*. To make the solution unique, one may take the pressure at one point to be fixed, so the pressure correction calculated at that point is subtracted from all the corrected pressures. Another choice is to set the mean pressure to some value, say zero. As a practical matter, it is not necessary to take these steps because most of the methods begin each step with a starting pressure and the actual pressure itself is not important in general in incompressible flows because only the gradient matters. In case where the actual pressure is known or needed, it is most often defined by the physics of the situation, e.g., a free-surface flow with the pressure known on the surface. Common sense suggests care in any case. If the mean pressure becomes very large compared to differences in pressure between grid points in the gradient, numerical precision may drop and assigning a fixed pressure at a point will prevent this.

Another case is obtained when the pressure difference between the inlet and outlet boundaries is specified. Then the velocities at these boundaries cannot be specified—they have to be computed so that the pressure loss is the specified value. This can be implemented in several ways. In any case the boundary velocity has to be extrapolated from the inner nodes (in a manner similar to the interpolation for cell faces in a colocated arrangement) and then corrected. An example of how specified static pressure can be handled is given in Chap. 11.

7.1.7 Illustrative Simple Schemes

To begin our actual solution of the Navier–Stokes equations, we describe two simple schemes. The first is an explicit time-stepping method for unsteady flows. The second scheme introduces us to the additional features of an implicit scheme. Detailed examination of common schemes begins then in Sect. 7.2.

7.1.7.1 A Simple Explicit Time-Advancing Method

Let us look at a method for the unsteady equations that illustrates how the numerical Poisson equation for the pressure is constructed and the role it plays in enforcing continuity. The choice of the approximations to the spatial derivatives is not important here so the semi-discretized (discrete in space but not time) momentum equations are written symbolically as:

$$\frac{\partial (\rho u_i)}{\partial t} = -\frac{\delta (\rho u_i u_j)}{\delta x_j} - \frac{\delta p}{\delta x_i} + \frac{\delta \tau_{ij}}{\delta x_j} = H_i - \frac{\delta p}{\delta x_i} , \qquad (7.41)$$

where $\delta/\delta x$ represents a discretized spatial derivative (which could represent a different approximation in each term) and H_i is shorthand notation for the advection and viscous terms whose treatment is of no importance here.

For simplicity, assume that we wish to solve Eq. (7.41) with the explicit Euler method for time advancement. We then have:

$$(\rho u_i)^{n+1} - (\rho u_i)^n = \Delta t \left(H_i^n - \frac{\delta p^n}{\delta x_i} \right).$$ (7.42)

To apply this method, the velocity at time step n is used to compute H_i^n and, if the pressure is available, $\delta p^n / \delta x_i$ may also be computed. This gives an estimate of ρu_i at the new time step $n + 1$. In general, this velocity field does not satisfy the continuity equation which we want to enforce:

$$\frac{\delta (\rho u_i)^{n+1}}{\delta x_i} = 0.$$ (7.43)

We have stated an interest in incompressible flows, but these include flows with variable density; this is emphasized by including the density. To see how continuity may be enforced, let us take the numerical divergence (using the numerical operators used to approximate the continuity equation) of Eq. (7.42). The result is:

$$\frac{\delta (\rho u_i)^{n+1}}{\delta x_i} - \frac{\delta (\rho u_i)^n}{\delta x_i} = \Delta t \left[\frac{\delta}{\delta x_i} \left(H_i^n - \frac{\delta p^n}{\delta x_i} \right) \right].$$ (7.44)

The first term is the divergence of the new velocity field, which we want to be zero. *The second term is zero if continuity was enforced at time step n; we shall assume that this is the case but, if it is not, this term should be left in the equation. Retaining this term is necessary when an iterative method is used to solve the Poisson equation for the pressure and the iterative process is not converged completely.* Similarly, the divergence of the viscous component of H_i should be zero for constant ρ, but a non-zero value is easily accounted for. With all of this taken into account, the result is the discrete Poisson equation for the pressure p^n:

$$\frac{\delta}{\delta x_i} \left(\frac{\delta p^n}{\delta x_i} \right) = \frac{\delta H_i^n}{\delta x_i}.$$ (7.45)

Note that the operator $\delta / \delta x_i$ outside the parentheses is the divergence operator inherited from the continuity equation, while $\delta p / \delta x_i$ is the pressure gradient from the momentum equations. If the pressure p^n satisfies this discrete Poisson equation, the velocity field at time step $n + 1$ will be divergence-free (in terms of the discrete divergence operator). Note that the time step to which this pressure belongs is arbitrary. If the pressure gradient term had been treated implicitly, we would have p^{n+1} in place of p^n but everything else would remain unchanged.

This provides the following algorithm for time-advancing the Navier–Stokes equations:

- Start with a velocity field u_i^n at time t_n which is assumed divergence-free. (As noted, if it is not divergence-free, this can be corrected.)
- Compute the combination, H_i^n, of the advection and viscous terms and its divergence (both need to be retained for later use).

- Solve the Poisson equation for the pressure p^n.
- Compute the velocity field at the new time step. It will be divergence-free.
- The stage is now set for the next time step.

Methods similar to this are commonly used to solve the Navier–Stokes equations when an accurate time history of the flow is required. The principal differences in practice are that time-advancement methods more accurate than the first-order Euler method are used and that some of the terms may be treated implicitly. Some of these methods will be described later.

We have shown how solving the Poisson equation for the pressure can assure that the velocity field satisfies the continuity equation, i.e., that it is divergence-free. This idea runs through many of the methods used to solve both the steady and unsteady Navier–Stokes equations.

7.1.7.2 A Simple Implicit Time-Advancing Method

To see what additional difficulties arise when an implicit method is used to solve the Navier–Stokes equations, let us construct such a method. Because we are interested in illuminating certain issues, let us use a scheme based on the simplest implicit method, the backward or implicit Euler method. If we apply this method to Eq. (7.41), we have:

$$(\rho u_i)^{n+1} - (\rho u_i)^n = \Delta t \left(-\frac{\delta(\rho u_i u_j)}{\delta x_j} - \frac{\delta p}{\delta x_i} + \frac{\delta \tau_{ij}}{\delta x_j} \right)^{n+1}. \qquad (7.46)$$

We see immediately that there are difficulties that were not present in the explicit method described in the preceding section. Let us consider these one at a time.

First, there is a problem with the pressure. The divergence of the velocity field at the new time step must be zero. This can be accomplished in much the same way as in the explicit method. We take the divergence of Eq. (7.46), assume that the velocity field at time step n is divergence-free (this can be corrected for if necessary) and demand that the divergence at the new time step $n + 1$ also be zero. This leads to the Poisson equation for the pressure:

$$\frac{\delta}{\delta x_i} \left(\frac{\delta p}{\delta x_i} \right)^{n+1} = \frac{\delta}{\delta x_i} \left(\frac{-\delta(\rho u_i u_j)}{\delta x_j} \right)^{n+1}. \qquad (7.47)$$

The problem is that the term on the right-hand side cannot be computed until the computation of the velocity field at time $n + 1$ is completed and vice versa. As a result, the Poisson equation and the momentum equations have to be solved simultaneously. That can only be done with some type of iterative procedure.

Next, even if the pressure were known, Eq. (7.46) are a large system of non-linear equations which must be solved for the velocity field. The structure of this system of equations is similar to the structure of the matrix of the finite-differenced Laplace operator from the pressure equation. However, because momentum equations contain contributions from advection terms and Dirichlet boundary conditions are applied at some boundaries (e.g., inlet, walls, symmetry planes), their solution is usually somewhat easier than the solution of the pressure or pressure-correction equation. This will be demonstrated in examples at the end of this chapter.

If one wishes to solve Eq. (7.46) accurately (as is the case in non-iterative fractional-step methods to be presented in the next section), the best procedure is to adopt the converged results from the preceding time step as the initial guess for the new velocity field and then converge to the solution at the new time step using the Newton iteration method (Sect. 5.5.1) or a secant method designed for systems (Ferziger 1998; Moin 2010).

Having seen how both explicit and implicit schemes may be constructed to solve the Navier–Stokes equations, we shall now study some of the more commonly used methods for solving them. Our coverage is not exhaustive and other methods can be found in the literature, e.g., the exact projection method of Chang et al. (2002).

7.2 Calculation Strategies for Steady and Unsteady Flows

In this section, a set of commonly used solution methods is described. Methods are given that apply to both steady and unsteady flows and use iterative and non-iterative approaches. All have elements similar to those explicated in the previous Sect. 7.1.7. In particular, for incompressible flow, both steady and unsteady problems are solved as marching schemes (either in time or through a set of iterations) leading to the need to solve a Poisson equation (for pressure, a pressure correction, or the streamfunction). Similar methods are grouped together below.

7.2.1 Fractional-Step Methods

The idea of integrating the unsteady Navier–Stokes equations forward in time in a segregated manner was chronicled in the seminal papers of Harlow and Welsh (1965) and Chorin (1968). Many of the methods described herein are derived from or built upon these early works. Indeed, Patankar and Spalding (1972) note the influence of these papers on the development of the SIMPLE algorithm (described in the next section). While SIMPLE is an iterative method, such an approach is not required and indeed one can even ignore the pressure in the predictor step. However, all of these methods and generally speaking those that follow use the pressure in the incompressible flow to enforce continuity. Armfield (1991, 1994) and Armfield and Street (2002) provide background and context for many fractional-step methods.

The basic theme of the fractional-step method is to (i) find an accurate estimate of the velocity field at the next time step using available pressure information (if desired) and the current velocity field, (ii) solve a Poisson equation for the new pressure or an equation for a correction to the old pressure, and (iii) use the new pressure or the corrected pressure to update the estimated velocities to new, mass-conserving velocities at the next time step. No iteration is used, but it can be performed, and we will discuss that later.

Kim and Moin (1985) created a scheme for incompressible flow on a staggered grid using a fractional step in which the pressure was not used in the predictor, but was calculated and then used in the corrector step. Zang et al. (1994) extended that method to colocated grids and curvilinear coordinates. Both Kim and Moin and Zang et al. used an approximate factorization ADI scheme (see, e.g., Beam and Warming 1976, or Sect. 5.3.5) in the predictor step to allow the velocity field to be estimated by solution of tridiagonal matrices in each coordinate direction; the factorization approach itself has an error of $O((\Delta t)^3)$, i.e., one order smaller than the hoped for $O((\Delta t)^2)$ of the overall scheme. Important in these formulations is the implicit treatment of the viscous (or turbulent) terms; this avoids time-step limitations on diffusion-like terms which are very stringent for explicit schemes (see, Ferziger 1998). These methods have proved popular and have been used widely, and the approximate factorization ADI scheme is common in other fractional-step methods. In addition, Zang et al. (1994) used a compact pressure stencil for the FV formulation and interpolation of the CV-centered velocities on the colocated grid to the volume faces to aid in application of the continuity constraint. Ye et al. (1999) discuss the value of this approach and Kim and Choi (2000) extended it to unstructured grids.

Gresho (1990) coined names for the fractional-step schemes and they are useful as a shorthand. The P1 method sets the pressure field to zero in the momentum equation used to estimate the new velocity field and the pressure Poisson equation is then solved for the new pressure. The Kim and Moin (1985) and Zang et al. (1994) schemes are P1 and solve for a pseudo-pressure that is related to the actual pressure which is rarely needed or calculated. The P2 method sets the pressure in the momentum equation equal to the value from the previous time step, and the pressure Poisson equation is then solved to find a pressure correction. In a P3 method the pressure used in the momentum equation is extrapolated with a second-order accurate scheme from the two previous pressure fields. We will comment on this last scheme later. In what follows, we will, in this section, give a general description of a P2 scheme; in Sect. 8.3 we give some details on the implementation of schemes for both staggered and colocated grids; and finally in Sect. 8.4 we examine results obtained for both steady and unsteady flows in lid and buoyancy-driven flows in rectangular enclosures, including a direct comparison with alternative solution methods.

Here, we describe fractional-step methods following closely the papers of Armfield and Street (2000, 2003, 2004). An important result (Fig. 7.5) from Armfield and Street (2003) is that in P2 schemes, the pressure is second-order accurate in time.[2] In addition, we write the equations in vector and operator form because it is particularly easy to understand the set up and flow of the method when it is written this way.

The governing equations are the three-dimensional Navier–Stokes equations for an incompressible fluid (see Sects. 1.3 and 1.4 and cf. Eqs. (7.3) and (7.4)):

---

[2]In this case the second-order extrapolation $p^{n+1} = (3/2)p^{n+1/2} - (1/2)p^{n-1/2}$ is used.

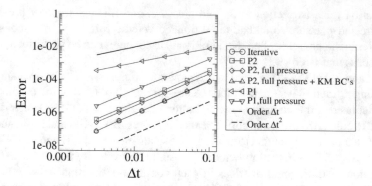

Fig. 7.5 Pressure accuracy for various methods for natural convection in a cavity (from Armfield and Street 2003; reprinted with permission)

$$\frac{\partial (\rho \mathbf{v})}{\partial t} + \nabla \cdot (\rho \mathbf{vv}) = \nabla \cdot \mathbf{S} - \nabla p \,, \tag{7.48}$$

$$\nabla \cdot (\rho \mathbf{v}) = 0 \,, \tag{7.49}$$

where $\mathbf{v} = (u_i)$ is the velocity, p is the pressure, and, as before, $\mathbf{S} = \mathbf{T} + p\mathbf{I}$ represents the viscous part of the stress tensor (see Eqs. (1.9) and (1.13)). Note that we have neglected body forces here as they have no impact on the methodology.

The goal of the method is second-order accuracy in space and time. Toward this, Eqs. (7.48) and (7.49) are discretized in time by use of the second-order Adams–Bashforth method (Sect. 6.2.2) for the convection terms and Crank–Nicolson (Sect. 6.3.2.2) for the viscous terms. Thus, the method is semi-implicit. The discretized equations can be written as:

$$\frac{(\rho \mathbf{v})^{n+1} - (\rho \mathbf{v})^{n}}{\Delta t} + C(\mathbf{v}^{n+1/2}) = -G(p^{n+1/2}) + \frac{L(\mathbf{v}^{n+1}) + L(\mathbf{v}^{n})}{2} \,, \tag{7.50}$$

$$D(\rho \mathbf{v})^{n+1} = 0 \,, \tag{7.51}$$

where the discrete terms include the velocity $\mathbf{v}$, pressure p, convection operator $C(\mathbf{v}) = \nabla \cdot (\rho \mathbf{vv})$, gradient operator $G() = G_i() = \nabla()$ (Eq. (1.15)), the divergence operator $D() = D_i() = \nabla \cdot ()$ (Eq. (1.6)), and the linear Laplace-like operator for viscous terms $L(\mathbf{v}) = \nabla \cdot \mathbf{S}$. If standard second-order, central differences (CDS) are used, the scheme is then second order in time and space when centered at the $n + 1/2$ time step. Note that the Adams–Bashforth representation yields

$$C(\mathbf{v}^{n+1/2}) \sim \frac{3}{2} C(\mathbf{v}^{n}) - \frac{1}{2} C(\mathbf{v}^{n-1}) + O((\Delta t)^{2}) \,. \tag{7.52}$$

Our fractional-step method now solves Eqs. (7.50) and (7.51) by

1. finding an estimate of the new velocity $\mathbf{v}^*$ using the old value of the pressure $p^{n-1/2}$ by solving the following equation:

$$\frac{(\rho\mathbf{v})^* - (\rho\mathbf{v})^n}{\Delta t} + C(\mathbf{v}^{n+1/2}) = -G(p^{n-1/2}) + \frac{L(\mathbf{v}^*) + L(\mathbf{v}^n)}{2} , \qquad (7.53)$$

2. defining a pressure correction p' according to

$$p^{n+1/2} = p^{n-1/2} + p' , \qquad (7.54)$$

3. next defining a correction to $\mathbf{v}^*$ by requiring that the corrected velocity satisfies the following approximate version of Eq. (7.50) (in which $\mathbf{v}^{n+1}$ in the L-operator is replaced by $\mathbf{v}^*$, for reasons which will become obvious shortly):

$$\frac{(\rho\mathbf{v})^{n+1} - (\rho\mathbf{v})^n}{\Delta t} + C(\mathbf{v}^{n+1/2}) = -G(p^{n+1/2}) + \frac{L(\mathbf{v}^*) + L(\mathbf{v}^n)}{2} . \qquad (7.55)$$

By subtracting Eq. (7.53) from Eq. (7.55), the following expression for the corrected velocity is obtained:

$$(\rho\mathbf{v})^{n+1} = (\rho\mathbf{v})^* - \Delta t\, G(p') , \qquad (7.56)$$

and finally

4. finding p' by substituting Eq. (7.56) into the continuity equation (7.51) to obtain

$$D(G(p')) = \frac{D(\rho\mathbf{v})^*}{\Delta t} . \qquad (7.57)$$

Given knowledge of p', the final velocity $\mathbf{v}^{n+1}$ and pressure $p^{n+1/2}$ are given by Eqs. (7.56) and (7.54), respectively. In a typical case, the convection term is discretized in space using the QUICK scheme (Sect. 4.4.3), other terms are discretized with CDS, Eq. (7.53) is solved by using an ADI scheme (Sects. 5.3.5 or 8.3.2), and Eq. (7.57) is solved by a Poisson solver, e.g., a preconditioned restarted GMRES (Sect. 5.3.6.3). Armfield and Street (1999) report that four sweeps of the ADI scheme were sufficient to obtain accurate solutions for the cases that they tested.

Once the solution is advanced by a time step it is appropriate to ascertain if any errors were made. We first rewrite Eq. (7.56) in the form:

$$(\rho\mathbf{v})^* = (\rho\mathbf{v})^{n+1} + \Delta t\, G(p') , \qquad (7.58)$$

and now insert this in the difference Eq. (7.55) which the corrected velocity and pressure satisfy (using the definition from Eq. (7.54)) to obtain

$$\frac{(\rho \mathbf{v})^{n+1} - (\rho \mathbf{v})^n}{\Delta t} + C(\mathbf{v}^{n+1/2}) = -G(p^{n+1/2}) + \frac{L(\mathbf{v}^{n+1}) + L(\mathbf{v}^n)}{2}$$

$$+ \frac{1}{2}\Delta t\, L\left(\frac{1}{\rho}G(p')\right). \quad (7.59)$$

Comparing this equation (which we have actually solved) to Eq. (7.50) (which we wanted to solve) shows that the equation solved contains an extra term—the last term in the above equation—which represents the error introduced by not correcting $\mathbf{v}^*$ in the L-operator; recall that, essentially, $L = D(G())$ in the last term above. By recognizing that the pressure correction can be expressed as

$$p' = p^{n+1/2} - p^{n-1/2} \approx \frac{\partial p}{\partial t}\Delta t. \quad (7.60)$$

we see that this additional term is proportional to $(\Delta t)^2$ and so consistent with the basic discretization (Armfield and Street 2002). In sum, independent of whether the grid is staggered or non-staggered, the P2 method described here does not provide an exact solution of the discrete equations. While the velocity field will be divergence-free (to the accuracy of the pressure Poisson solution), this error means that the updated velocity field plus the updated pressure does not satisfy the discrete momentum equations exactly. While the error is second order in time (meaning that it reduces by a factor of 4 when the time step is halved), it may still be significant if the selected time step is not small enough. Unfortunately, what is small enough is problem-dependent.

The method described above represents just one of numerous possible choices. Obviously, one can easily implement a different explicit time-advancing scheme for the convection term, without altering the pressure-correction equation. Note that in the algorithm described above, pressure is computed at $t_{n+1/2}$ while velocities are computed at t_{n+1}; if both quantities are needed at the same time level, one has to be interpolated. However, one can apply the Crank–Nicolson time-advancing scheme to the pressure term as was done to viscous terms; the only difference is that now half of the pressure term (coming from time level t_n) will be fixed and only the implicit half (coming from time level t_{n+1}) will be corrected.

One can also advance viscous terms in time using an explicit scheme; this may be acceptable if there are no extremely small cells in the solution domain, so that the stability limit on time-step size is not too stringent. In that case, the momentum Eq. (7.50) can be solved explicitly without the need to solve an equation system.

Things become a bit more complicated if one chooses implicit time-advancing for both convection and diffusion terms. Now the non-linear convection terms need to be linearized. A Picard-iteration scheme is the usual choice, but there are still many possibilities for approximating the part that is assumed known. This largely depends on whether one wants to iterate within a time step, or whether the momentum equation is to be solved only once. In the former case the choice is obvious: one uses the value from the previous iteration:

$$(\rho \mathbf{v v})^{n+1} \approx (\rho \mathbf{v})^{m-1} \mathbf{v}^m . \tag{7.61}$$

Here m is the iteration counter. The question is only what to do for the first iteration? If more iterations are to follow, the choice is not critical—as the iterations converge, the values at two subsequent iterations will be practically the same. If one wants a non-iterative scheme, then the choice for the explicit part becomes critical. It should be a second-order approximation of the solution at the new time level if second-order accuracy of the whole scheme is to be retained. For this purpose one can use any of the Adams–Bashforth schemes of second or higher order, such as the one used for the complete convection in the algorithm presented above. Actually, in an implicit method all variable values should be initialized at the new time level using explicit time-advancing schemes of second order, especially when the grid is non-orthogonal and the discretization involves deferred corrections which are computed using prevailing variable values. Otherwise, the overall scheme will be first order if simply the values from the previous time step are used.

With implicit iterative methods, there are even more choices to be made: one can iterate first on the momentum equation alone, in order to update non-linearities and deferred corrections, and then proceed to a single pressure-correction step, in which the equation for pressure correction has to be solved to a relatively tight tolerance in order to ensure that the continuity requirement is sufficiently enforced. The other option is to extend the iteration loop to solve both the momentum and the pressure-correction equation, in order to update both the non-linearities and the pressure-velocity coupling. Here is an example algorithm for such an iterative scheme:

1. At the mth iteration within the new time step, solve the momentum equation for the estimate of the new time level solution of the following form, using the fully-implicit three-time-level scheme of second order for time integration (see Sect. 6.3.2.4):

$$\frac{3(\rho \mathbf{v})^* - 4(\rho \mathbf{v})^n + (\rho \mathbf{v})^{n-1}}{2\Delta t} + C(\mathbf{v}^*) = L(\mathbf{v}^*) - G(p^{m-1}) , \tag{7.62}$$

   where $\mathbf{v}^*$ is the predictor value for $\mathbf{v}^m$; it needs to be corrected to enforce continuity.
2. Require that the corrected velocity and pressure satisfy this form of the momentum equation:

$$\frac{3(\rho \mathbf{v})^m - 4(\rho \mathbf{v})^n + (\rho \mathbf{v}^{n-1}}{2\Delta t} + C(\mathbf{v}^*) = L(\mathbf{v}^*) - G(p^m) . \tag{7.63}$$

By subtracting Eq. (7.62) from Eq. (7.63) we obtain the following relation between velocity and pressure correction:

$$\frac{3}{2\Delta t}\left[(\rho \mathbf{v})^m - (\rho \mathbf{v})^*\right] = -G(p') \quad \Rightarrow \quad (\rho \mathbf{v})' = -\frac{2\Delta t}{3}G(p') . \tag{7.64}$$

3. Impose the continuity requirement on velocity $\mathbf{v}^m$,

$$D(\rho\mathbf{v})^m = 0 \quad \Rightarrow \quad D(G(p')) = \frac{3}{2\Delta t} D(\rho\mathbf{v})^* , \tag{7.65}$$

and solve the resulting pressure-correction equation.
4. Increase the iteration counter and repeat steps 1–3 until residuals become sufficiently small. Then set $\mathbf{v}^{n+1} = \mathbf{v}^m$, $p^{n+1} = p^m$ and proceed to the next time level.

Note that the above pressure-correction equation looks exactly the same as that from the previous version, Eq. (7.57), except for the 3/2 multiplier; however, the right-hand side is different, because $\mathbf{v}^*$ came from a different form of the momentum equation. In this case one does not have to solve a linear momentum equation or pressure-correction equation to a very tight tolerance, because the solution will be continued in the next iteration; usually, reducing residuals one order of magnitude in each iteration is sufficient if three or more iterations per time step are performed.

A non-iterative version of the above algorithm is easily derived; it uses an explicit estimate of convection fluxes at t_{n+1} using Adams–Bashforth scheme similar to that of Eq. (7.52), only that we now need the estimate at t_{n+1} rather than at $t_{n+1/2}$:

$$C(\mathbf{v}^{n+1}) \approx 2C(\mathbf{v}^n) - C(\mathbf{v}^{n-1}) . \tag{7.66}$$

The outer iteration loop is dropped, momentum equations and the pressure-correction equation are solved only once per time step (but now to a tighter tolerance).

The above iterative implicit fractional-step method (which we will often refer to using the acronym IFSM) is very similar to the SIMPLE algorithm described in the next section. The subtle difference will be discussed at the end of the next section. Computer codes that incorporate both the iterative and the non-iterative algorithm are available (see the Appendix) and some results from their application will be shown in the examples-section at the end of the next chapter.

7.2.2   SIMPLE, SIMPLER, SIMPLEC and PISO

As noted in Chap. 6, many methods for steady problems can be regarded as solving an unsteady problem until a steady state is reached. The principal difference is that, when solving an unsteady problem, the time step is chosen so that an accurate history is obtained while, when a steady solution is sought, large time steps are used to try to reach the steady state quickly. Implicit methods are preferred for steady and slow-transient flows, because they have less stringent time step restrictions than explicit schemes (indeed, they may not have any!). They are also often used for time-accurate solution of transient problems (especially when using commercial CFD-codes, which often do not offer explicit versions), in particular when the grid is locally refined such that explicit methods would require smaller time steps for stability reasons than would be required for a sufficient solution accuracy.

Many solution methods developed for steady incompressible flows are implicit; some of the most popular ones can be regarded as variations on the method of the preceding section. They use a pressure (or pressure-correction) equation to enforce mass conservation at each step or, in the language preferred for steady solvers, each outer iteration. We now look at some of these methods. In this section, we give a general description of SIMPLE and related schemes; in Sects. 8.1 and 8.2 we detail implementation of SIMPLE and IFSM schemes for both staggered and colocated grids; and finally in Sect. 8.4.1 we examine results of SIMPLE and IFSM simulations on both grid types.

We use a similar notation to that of the preceding section, in order to demonstrate both similarities and differences between fractional-step and SIMPLE-type methods.

The starting point is Eqs. (7.48) and (7.49) given in the preceding section. The major difference between SIMPLE-type methods (which include SIMPLEC and PISO, all found in most commercial CFD-codes) and fractional-step methods is that the former are usually fully implicit (but the Crank–Nicolson scheme can also be used). This means that all fluxes and source terms are computed at the new time level; values from previous time levels appear only in the discretized time derivative. Most codes offer the option between an implicit-Euler scheme (1st order, not suitable for time-accurate simulations) and a quadratic backward scheme (also called a three-time-level scheme; 2nd order in time and suitable for time-accurate simulations). If we write the discretized momentum and continuity equations using the same operator notation from the preceding section, the two versions of discretized momentum equation are as follows:

$$\frac{(\rho\mathbf{v})^{n+1} - (\rho\mathbf{v})^n}{\Delta t} + C(\mathbf{v}^{n+1}) = L(\mathbf{v}^{n+1}) - G(p^{n+1}) \,, \tag{7.67}$$

$$\frac{3(\rho\mathbf{v})^{n+1} - 4(\rho\mathbf{v})^n + (\rho\mathbf{v})^{n-1}}{2\Delta t} + C(\mathbf{v}^{n+1}) = L(\mathbf{v}^{n+1}) - G(p^{n+1}) \,. \tag{7.68}$$

Because the first equation uses the implicit Euler scheme, it is first order in time (the approximation of the time derivative is the first-order backward scheme with respect to the time level at which all other terms are evaluated). The second equation uses a time-derivative approximation which is second-order accurate at the time level at which all other terms are evaluated; it is obtained by differentiating a quadratic interpolation in time that involves the new time level, as explained in Sect. 6.3.2.4. Because convection, diffusion and source terms are always evaluated at t_{n+1}, it is easy to switch between the two schemes and even to blend them.

We shall assume here that the flow is incompressible and that density is constant; we show in Chap. 11 how the method can be extended to compressible flow. Under this assumption there is no time derivative in the continuity equation, so for both time-marching schemes, it is the same as in the preceding section, see Eq. (7.51).

Because all terms are discretized using a fully-implicit scheme, we have to linearize the non-linear terms and solve equations iteratively for $\mathbf{v}^{n+1}$ and p^{n+1}. If we are computing an unsteady flow and time accuracy is required, iteration must be continued within each time step until the entire system of non-linear equations is satisfied to within a narrow tolerance. For steady flows, the tolerance can be much more generous; one can then either take an infinite time step and iterate until the steady non-linear equations are satisfied, or march in time without requiring full satisfaction of the non-linear equations at each time step (in that case, one usually performs only one iteration per time step).

The iterations within one time step, in which the non-linear and coupling terms are updated, are called *outer iterations* to distinguish them from the *inner iterations* performed on linear systems with fixed coefficients.

We now drop the superscript $(n + 1)$ and use an outer iteration counter m to denote the current estimate of the new solution; as these iterations converge, we obtain $\mathbf{v}^{n+1} \approx \mathbf{v}^m$. We assume that linearization is performed using Picard-iteration, as shown in Eq. (7.61) in the previous section. For steady-state problems and iterative time-marching schemes (SIMPLE, SIMPLEC or PISO), this is almost exclusively used. Only the first iteration on the new time level is critical if time-accurate solutions are to be obtained efficiently. If one simply takes the value from the preceding time step to represent the $(m = 0)$-value, the number of required iterations within time step will be higher than if a better estimate (e.g., using second-order explicit time-marching schemes) is used. The issue is especially critical for the PISO-algorithm, which solves the momentum equations only once per time step; thus, there is no way to improve the initial error by iteration. This issue will be addressed again at the end of this section. Source terms and variable fluid properties are treated in a similar way, i.e., parts of those terms are evaluated at the previous iteration.

Linearized momentum equations are solved sequentially.[3] When all implicitly discretized terms are grouped together, we obtain for each velocity component a matrix equation of the form:

$$A^{m-1} u_i^m = Q_i^{m-1} - G_i(p^m) \,, \tag{7.69}$$

where G_i is a shorthand notation for the i-component of the gradient operator. The source term Q contains all of the terms that may be explicitly computed in terms of u_i^{m-1} as well as any body force or other linearized or deferred-correction terms that may depend on the u_i^{n+1} or other variables at the new time level (e.g., temperature). It also contains parts of the unsteady term that refer to solution at previous time steps, see Eqs. (7.67) and (7.68). Note that the matrix A may not be the same for all velocity components, but we shall ignore this for the time being. We assume here that the FD-method is used to discretize equations, but the same procedure applies to the FV-method as well—the above equation needs only to be multiplied by cell volume.

---

[3]Coupled (or monolithic) solvers are also available and most commercial packages now offer them; discussion of solver alternatives can be found in the code documentation or in the literature, e.g., Heil et al. (2008) or Malinen (2012). See also Chap. 11 for a brief description of one such method.

For the sake of clarity, we shall drop the superscript $(m-1)$ for the matrix A and source term Q, now that we have recognized that these do depend on the solution but are computed using values from the previous outer iteration.

A single row of the above equation reads:

$$A_P u_{i,P}^m + \sum_k A_k u_{i,k}^m = Q_P - \left(\frac{\delta p^m}{\delta x_i}\right)_P. \qquad (7.70)$$

The pressure term is written in symbolic difference form to emphasize the independence of the solution method from the discretization approximation for the spatial derivatives. The discretizations of the spatial derivatives may be of any order or any type described in Chaps. 3 and 4. Note that coefficients A_k contain contributions from discretized convection and diffusion terms, while the diagonal coefficient A_P contains, in addition, contributions from the unsteady term (see Eqs. (7.67) and (7.68)).

We now return to the simpler matrix notation of Eq. (7.69) but split the matrix A into a diagonal part A_D and off-diagonal part A_{OD}, for reasons that will become obvious shortly. We also replace superscript m denoting values from the current outer iteration by one or more asterisks, depending at which level of approximation we are.

At outer iteration m, we solve this equation first, using the pressure from the previous iteration:

$$(A_D + A_{OD})u_i^* = Q - G_i(p^{m-1}). \qquad (7.71)$$

The velocity field obtained by solving this equation, $\mathbf{v}^{m*}$, will in general not satisfy the continuity equation. This is what we want to enforce by correcting both pressure and velocities,

$$p^* = p^{m-1} + p', \quad u_i^{**} = u_i^* + u_i'. \qquad (7.72)$$

The relation between velocity and pressure correction is obtained by requiring that the corrected velocity and pressure satisfy the following simplified version of Eq. (7.69):

$$A_D u_i^{**} + A_{OD} u_i^* = Q - G_i(p^*). \qquad (7.73)$$

Now by subtracting Eq. (7.71) from Eq. (7.73), we obtain the following relation between velocity and pressure corrections:

$$A_D u_i' = -G_i(p') \quad \Rightarrow \quad u_i' = -(A_D)^{-1} G_i(p'). \qquad (7.74)$$

This relation is simple because the diagonal matrix can be easily inverted; if u_i^{**} was applied to the off diagonal matrix in Eq. (7.73) as well, the relation would be too complicated for the purpose of deriving a pressure-correction equation. The simplification is justified by the fact that, as outer iterations converge, all corrections tend to zero, so the final solution is not affected. However, this simplification does affect the convergence rate and we shall describe later how this can be improved by properly selecting the under-relaxation factors.

We now require that corrected velocities u_i^{**} satisfy the discretized continuity equation, thus:

$$D(\rho\mathbf{v})^* + D(\rho\mathbf{v}') = 0 . \tag{7.75}$$

By expressing u_i' via p' with the help of expression (7.74), we obtain the pressure-correction equation:

$$D(\rho(A_D)^{-1}G(p')) = D(\rho\mathbf{v})^* . \tag{7.76}$$

Note that this equation has the same form as Eq. (7.57) from the previous section; we shall discuss the similarity and differences further below.

This method is essentially a variation on the one presented in the preceding section. Methods of this kind, which first construct a velocity field that does not satisfy the continuity equation and then correct it by subtracting something (usually a pressure gradient) are known as *projection methods*. From the perspective of vector mathematics, the pressure acts through the continuity restraint as an operator which projects the divergent velocity vector field into a non-divergent vector field (see Kim and Moin 1985).

Once the pressure correction has been solved for, the velocities and pressure are updated using Eqs. (7.74) and (7.72). These are taken to represent the solution at outer iteration m and the new iteration can start. This is known as the SIMPLE algorithm (Caretto et al. 1972); SIMPLE is an acronym for "Semi-Implicit Method for Pressure-Linked Equations". We shall discuss its properties below.

Rather than neglecting the velocity correction in off-diagonal terms, one can approximate its effect. This can be achieved by approximating the velocity correction u_i' at any node by a weighted mean of the neighbor values, for example,

$$u_{i,P}' \approx \frac{\sum_k A_k u_{i,k}'}{\sum_k A_k} \quad \Rightarrow \quad \sum_k A_k u_{i,k}' \approx u_{i,P}' \sum_k A_k . \tag{7.77}$$

This allows us to approximate $A_{OD}u_i'$ which appears in the relation between velocity and pressure corrections when Eq. (7.71) is subtracted from Eq. (7.69) rather than (7.73):

$$A_D u_i' + A_{OD}u_i' = -G_i(p') . \tag{7.78}$$

By using expression (7.77), we can write a simplified version of the above equation as follows (the simplification is much less crude than neglecting the second term on the left-hand side, as is done in SIMPLE):

$$u_i' = -(A_D + \tilde{A}_D)^{-1}G_i(p') . \tag{7.79}$$

where $\tilde{A}_D$ represents the sum of off-diagonal matrix elements, see Eq. (7.77).

It is worth noting that, when FV-methods are used, the contribution from convection and diffusion fluxes to the diagonal matrix element is equal to the negative sum of contributions to off-diagonal elements; see Sect. 8.1 for further details. In the absence of other contributions, this would make $A_D + \tilde{A}_D = 0$ in the above equation. Fortunately, either a contribution from the unsteady term occurs in A_D when transient problems are solved, or it is divided by an under-relaxation factor $\alpha_u < 1$ when steady-state problems are solved; see Sect. 5.4.3 for details on this kind of under-relaxation. Thus, A_D is positive and always larger in magnitude than the sum of off-diagonal elements in $\tilde{A}_D$ (which are usually negative). However, the sum $A_D + \tilde{A}_D$ appearing in Eq. (7.79) is much smaller than A_D appearing in the corresponding expression from SIMPLE, see Eq. (7.74). Because the gradient of pressure correction is multiplied by the reciprocal value of this term, for the same velocity correction in both methods, the pressure correction from this corrected approach (called SIMPLEC; see below) must be much smaller than the one from SIMPLE. This is why SIMPLE needs an under-relaxation of the pressure correction (which is over-predicted due to the simplifications in its derivation), as will be explained below.

The next step is to require that corrected velocities satisfy the continuity equation, which leads to a pressure-correction equation of the same form as Eq. (7.76), except that A_D is replaced by the sum $A_D + \tilde{A}_D$. This is known as the SIMPLEC algorithm (SIMPLE-Corrected) (Van Doormal and Raithby 1984).

Still another method of this general type is derived by considering the SIMPLE step to be a predictor followed by a series of corrector steps. We simply continue the process of correcting velocities and pressure by adding one more asterisk; in the first correction following SIMPLE, we want velocities and pressure to satisfy this form of the momentum equation:

$$A_D u_i^{***} + A_{OD} u_i^{**} = Q - G_i(p^{**}) . \tag{7.80}$$

Now by subtracting Eq. (7.73) from Eq. (7.80), we obtain the following relation between the second velocity and pressure corrections:

$$A_D u_i'' + A_{OD} u_i' = -G_i(p'') \quad \Rightarrow \quad u_i'' = -(A_D)^{-1}(A_{OD} u_i' + G_i(p'')) . \tag{7.81}$$

Here u_i' is known from the previous step. We now require that u_i^{***} also satisfies the continuity equation:

$$D(\rho \mathbf{v})^{**} + D(\rho \mathbf{v}'') = 0 . \tag{7.82}$$

By expressing u_i'' via p'' with the help of expression (7.81) and recognizing that u_i^{**} already satisfies the continuity equation, we obtain the second pressure-correction equation:

$$D(\rho(A_D)^{-1} G(p'')) = D(\rho(A_D)^{-1} A_{OD} u_i') . \tag{7.83}$$

This process can be continued by adding one more asterisk to each term in Eq. (7.80).

Note that the pressure-correction equations for every step have the same coefficient matrix, which can be exploited in some solvers (a factorization of the matrix may be stored and reused). This procedure is known as the PISO-algorithm (Pressure Implicit with Splitting of Operators) (Issa 1986). Some commercial and open CFD-packages offer this algorithm in addition to SIMPLE or SIMPLEC. Usually, 3–5 correction steps are performed.

Finally, another similar method was proposed by Patankar (1980) and was called SIMPLER (SIMPLE-Revised). In it, the pressure-correction Eq. (7.76) is solved first and velocities are corrected as in SIMPLE. The new pressure field is calculated from the pressure equation obtained by taking divergence of Eq. (7.69). This algorithm has not found a widespread use because it offers no advantages over the other alternatives discussed so far.

As already noted, due to the neglect of the effect of velocity corrections in off-diagonal terms, the SIMPLE algorithm does not converge rapidly; actually, if both pressure and velocities are corrected using the computed pressure correction and equations given above, it may not converge at all unless the time step is very small. Especially for steady flows computed using an infinite time step, its performance depends greatly on the value of the under-relaxation parameter used in the momentum equations. It has been first found by trial and error that convergence can be improved if one adds only a portion of p' to p^{m-1}, i.e., if one updates the pressure, contrary to the statement in expression (7.72), after the pressure-correction equation is solved as follows:

$$p^m = p^{m-1} + \alpha_p p' , \tag{7.84}$$

where $0 \le \alpha_p \le 1$. SIMPLEC, SIMPLER and PISO do not need under-relaxation of the pressure correction on Cartesian grids. We shall show in the next chapter that discretization and solution of the pressure-correction equation on non-orthogonal grids may require under-relaxation due to the deferred-correction approach applied to (or the neglect of) terms related to grid non-orthogonality.

One can derive an optimum relation between the under-relaxation factors for velocities and pressure by requiring that SIMPLE and SIMPLEC produce the same velocity correction, because in the latter the effect of neighbor velocity corrections was approximated rather than neglected (see Eqs. (7.79) and (7.74)):

$$-\frac{1}{A_P} \left(\frac{\delta p'}{\delta x_i} \right)_P^{\text{SIMPLE}} = -\frac{1}{A_P + \sum_k A_k} \left(\frac{\delta p'}{\delta x_i} \right)_P^{\text{SIMPLEC}} . \tag{7.85}$$

This expression can be re-written as follows:

$$\left(\frac{\delta p'}{\delta x_i} \right)_P^{\text{SIMPLEC}} = \frac{A_P + \sum_k A_k}{A_P} \left(\frac{\delta p'}{\delta x_i} \right)_P^{\text{SIMPLE}} . \tag{7.86}$$

Optimal under-relaxation is crucial for steady-state problems; in that case (without a contribution from the unsteady term) and in the absence of contributions from source terms, we can write by looking at Eq. 7.86 (cf. Sect. 5.4.3):

$$A_P = \frac{-\sum_k A_k}{\alpha_u} . \tag{7.87}$$

When this relation is inserted into Eq. (7.86), we obtain:

$$\left(\frac{\delta p'}{\delta x_i}\right)_P^{\text{SIMPLEC}} = (1 - \alpha_u) \left(\frac{\delta p'}{\delta x_i}\right)_P^{\text{SIMPLE}} , \tag{7.88}$$

which suggests that pressure correction resulting from SIMPLE should be multiplied by

$$\alpha_p = 1 - \alpha_u \tag{7.89}$$

in order to obtain the same velocity correction that would result from SIMPLEC.[4] We shall show in Sect. 8.4 how under-relaxation factors for velocity and pressure affect the efficiency of SIMPLE in an example.

The solution algorithm for this class of methods can be summarized as follows (see Fig. 7.6):

1. Start calculation of the fields at the new time t_{n+1} using the latest solution u_i^n and p^n as starting estimates for u_i^{n+1} and p^{n+1}.
2. Start the outer iteration loop with iteration counter m.
3. Assemble and solve the linearized algebraic equation systems for the velocity components (momentum equations) to obtain u_i^*.
4. Assemble and solve the pressure-correction equation to obtain p'.
5. Correct the velocities and pressure to obtain the velocity field u_i^{**}, which satisfies the continuity equation, and the new pressure p^*. For SIMPLE and SIMPLEC, these are the final values for the iteration m; for SIMPLER, only the velocity is final.
   For the PISO-algorithm, solve the second pressure-correction equation and correct both velocities and pressure again. Repeat this until corrections become small enough and proceed to the next time step.
   For SIMPLER, solve the pressure equation for p^m after u_i^m is obtained above.
6. If additional transport equations need to be solved (e.g., for temperature, species, turbulence quantities etc.), this is done here; corrected mass fluxes, velocities and pressure are used where needed.
7. If fluid properties are variable, they can now be re-computed using updated variable values at cell centers.
8. Return to step 2 and repeat, using u_i^m and p^m as improved estimates for u_i^{n+1} and p^{n+1}, until all corrections are negligibly small.
9. Advance to the next time step.

Methods of this kind are fairly efficient for solving steady state problems; their convergence can be further improved by the multigrid strategy, as will be demon-

---

[4]This relation was first derived by Raithby and Schneider (1979) and later re-discovered by Perić (1985) using different arguments.

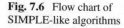

Fig. 7.6 Flow chart of SIMPLE-like algorithms

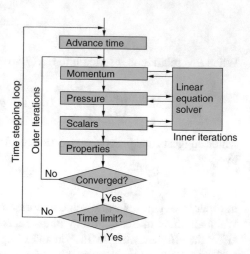

strated in Chap. 12. There are many derivatives of the above methods which are named differently, but they all have roots in the ideas described above and will not be listed here. We shall show below that the artificial compressibility method can also be interpreted in a similar way.

A nice feature of SIMPLE-like methods is that they are easily extended to solve additional transport equations, as outlined in step 6. Also, variable fluid properties are easily handled: they are simply assumed constant during one outer iteration and re-computed at the end of the loop, when all variables are updated. This is shown in Fig. 7.6; the outer-iterations loop is easily extended to update any non-linearity or deferred correction, or to solve additional equations. The same is true for the implicit iterative fractional-step method described in the previous section, which essentially follows the same flow chart.

By comparing the above class of methods with fractional-step methods presented in the previous section, one can see that they are very similar to the implicit version of the latter. The only difference is that SIMPLE-like methods, when deriving the pressure-correction equation, update the new velocity not only in the unsteady term (as fractional-step methods do), but also in the diagonal part of discretized convection and diffusion fluxes. This difference is not so important when unsteady flows are computed using small time steps, but it does become significant if large time steps are used to march towards a steady-state solution. We look into this issue in more detail in the following section.

7.2.2.1 SIMPLE Versus Iterative Implicit Fractional-Step Method

A closer look at Eqs. (7.57) and (7.65) reveals that, as the time step tends to infinity, the source term in the Poisson equation for pressure correction tends to zero. This clearly shows that fractional-step method in the given form cannot be used to solve steady-state problems, i.e., when the unsteady term is missing or time steps are too large.

If we examine the pressure-correction equation from SIMPLE-like methods, we see that they do not suffer from the same problem, cf. Eq. (7.76). This can be better analyzed if we compare the pressure-correction equation for a single grid point for both approaches. For SIMPLE, we get from Eq. (7.76):

$$\frac{\delta}{\delta x_i}\left(\frac{\rho}{A_P}\frac{\delta p'}{\delta x_i}\right) = \frac{\delta(\rho u_i^*)}{\delta x_i} . \tag{7.90}$$

If we assume that ρ/A_P does not vary spatially in the vicinity of the grid point (which is generally not true but also not an unreasonable assumption for the present purpose), we can re-write the above equation as follows:

$$\frac{\delta}{\delta x_i}\left(\frac{\delta p'}{\delta x_i}\right) = \frac{A_P}{\rho}\frac{\delta(\rho u_i^*)}{\delta x_i} . \tag{7.91}$$

For the fractional-step method, Eq. (7.65) delivers:

$$\frac{\delta}{\delta x_i}\left(\frac{\delta p'}{\delta x_i}\right) = \frac{3}{2\Delta t}\frac{\delta(\rho u_i^*)}{\delta x_i} . \tag{7.92}$$

Now we see that, if $3/(2\Delta t) = A_P/\rho$, the two equations would essentially be the same.

Let us now see what hides behind A_P. From Eq. (7.68) we see that it certainly contains $3\rho/(2\Delta t)$, plus contributions from the discretized convection and viscous term, which are of the order of $\rho u_i/\Delta x$ and $\mu/(\Delta x)^2$, respectively (depending on which approximations were used to discretize these terms). One can show that, for incompressible flows and all conservative schemes, this contribution equals $-\sum_k A_k$, where A_k are off-diagonal elements from the matrix of the discretized and linearized momentum equation (see Eq. (8.20)). Thus, for the three-time-level scheme:

$$A_P = \frac{3\rho}{2\Delta t} - \sum_k A_k , \tag{7.93}$$

where the summation goes over all neighbor grid points contained in the computational molecule for node P. Obviously, if the contribution from convection and diffusion is omitted in SIMPLE, it becomes the implicit fractional-step method described in the previous section, because the velocity corrections would also be the same (see Eqs. (7.56) and (7.81)):

$$u_i' = -\frac{2\Delta t}{3\rho}\frac{\delta p'}{\delta x_i} \quad \text{(IFSM) ,} \tag{7.94}$$

$$u_i' = -\frac{1}{A_P}\frac{\delta p'}{\delta x_i} \quad \text{(SIMPLE) .} \tag{7.95}$$

When time steps are very small, the contribution to A_P from neighbor coefficients becomes small compared to $\rho/\Delta t$, so one can expect that transient SIMPLE and the iterative implicit fractional-step method would behave similarly. However, for larger time steps, there are differences resulting from both under-relaxation and contributions from convection and diffusion fluxes to the diagonal element of the pressure-correction equation matrix in SIMPLE which are absent in the fractional-step method. We compare the performance of the two methods in Sect. 8.4 for both steady and unsteady flows.

7.2.2.2   Under-Relaxation and Time-Marching in SIMPLE

When SIMPLE-like methods are used to compute steady-state flows, they usually drop the unsteady term, thus assuming an infinitely large time step. However, the method would not work without invoking under-relaxation, as described in Sect. 5.4.3. There is a strong similarity between the algebraic equations resulting from the use of under-relaxation when solving steady problems and those resulting from applying the implicit Euler scheme to unsteady equations. Both under-relaxation and implicit time discretization result in an additional source term and a contribution to the central coefficient A_P.

If we require that the diagonal coefficient from time-marching scheme given in Eq. (7.93) is equal to the one used in steady-state computation with under-relaxation, we obtain the following relation (note that off-diagonal coefficients are typically all negative):

$$\frac{\rho}{\Delta t} - \sum_k A_k = -\frac{\sum_k A_k}{\alpha_u} \, , \tag{7.96}$$

where k runs over neighbor cell centers from the computational molecule. From this equation we can derive expressions for an equivalent time step expressed as a function of under-relaxation factor α_u and vice-versa, which would lead to an identical behavior of the two computation approaches:

$$\Delta t = \frac{\rho \alpha_u}{-(1 - \alpha_u) \sum_k A_k} \quad \text{or} \quad \alpha_u = \frac{-\Delta t \sum_k A_k}{-\Delta t \sum_k A_k + \rho} \, . \tag{7.97}$$

These expressions were derived assuming that the FD-method is used; for the FV-methods, ρ just needs to be replaced by ρV, where V is cell volume.

In the iteration at the new time step, the usual initial guess is the solution from the preceding step. If the final steady state is the only result of interest and the details of the development from the initial guess to the final stage are not of importance, it is sufficient to perform only one iteration per time step. Then one does not have to store the old solution—it is needed only to assemble the matrix and source terms.

The major difference between using time-marching and under-relaxation is that using the same time step for all grid points in time marching is equivalent to using a variable under-relaxation factor; conversely, use of a constant under-relaxation factor is equivalent to applying a different time step to each grid point.

It is important to note that, if only one iteration is performed at each time step, the scheme does not retain all of the stability of the implicit Euler method. There is then a limitation on the time step that can be employed in such time-marching schemes. On the other hand, when under-relaxation is used with outer iterations in steady-state computations, the choice of the parameter α_u is also limited and certainly has to be smaller than unity. Typical values range between 0.7 and 0.9, depending on the problem and grid quality (lower for poor quality grid and stiff problems).

We compared here SIMPLE algorithm operating in time-marching and under-relaxation modes. The same analysis applies to the comparison of the iterative implicit fractional-step method presented in Sect. 7.2.1 and SIMPLE in under-relaxation mode. With a suitable choice of time step in one and under-relaxation parameter in the other, they can be made to perform more or less in exactly the same way.

Note that under-relaxation is usually retained in SIMPLE-like methods even when they are used to compute unsteady flows, i.e., the diagonal matrix element always has the form:

$$A_P = \frac{\dfrac{\rho}{\Delta t} - \sum_k A_k}{\alpha_u}. \tag{7.98}$$

Without under-relaxation, the method would suffer from a limit to the size of time step that could be used when the non-linearity and coupling effects are strong (e.g., turbulent flows computed using the Reynolds-averaged Navier–Stokes equations, coupled with a turbulence model). Because outer iterations are performed within a time step anyway, it is safer to allow a small amount of under-relaxation in any case; only for very small time steps, e.g., in direct or large-eddy simulation of turbulence, can the under-relaxation factor be set to unity.

7.2.3 Artificial Compressibility Methods

Compressible flow is an area of fluid mechanics of great importance. Its applications, especially in aerodynamics and turbine engine design, have caused a great deal of attention to be focused on development of methods for the numerical solution of the compressible flow equations. Many such methods have been developed. An obvious question is whether they can be adapted to the solution of the incompressible flows. We show here how these methods may be adapted to incompressible flow and present a description of a few of the key properties of artificial compressibility methods. See Kwak and Kiris (2011, Chap. 4) for an in-depth discussion of this method and Louda et al. (2008) for a recent application to turbulent flows using the Reynolds-averaged Navier–Stokes equations.

The major difference between the equations of compressible flow and those of incompressible flow is their mathematical character. The compressible flow equations are hyperbolic which means that they have real characteristics on which signals travel at finite propagation speeds; this reflects the ability of compressible fluids to support sound waves. By contrast, we have seen that the incompressible equations have a mixed parabolic-elliptic character. If methods for compressible flow are to be used to compute incompressible flow, the character of the equations will need to be modified.

The difference in character can be traced to the lack of a time-derivative term in the incompressible continuity equation. The compressible version contains the time derivative of the density. So, the most straightforward means of giving the incompressible equations hyperbolic character is to append a time derivative to the continuity equation. Because density is constant, adding $\partial \rho / \partial t$, i.e., using the compressible equation, is not possible. Time derivatives of velocity components appear in the momentum equations so they are not logical choices. That leaves the time derivative of the pressure as the clear choice.

Addition of a time derivative of the pressure to the continuity equation means that we are no longer solving the equations for a truly incompressible flow. As a result, the time history generated cannot be accurate and the applicability of artificial compressibility methods to unsteady incompressible flows is a questionable enterprise although it has been attempted. On the other hand, at convergence, the time derivative is zero and the solution satisfies the incompressible equations. This approach was first proposed by Chorin (1967) and a number of versions differing mainly in the underlying compressible flow method used, have been presented in the literature. As noted above, the essential idea is to add a time derivative of pressure to the continuity equation:

$$\frac{1}{\beta} \frac{\partial p}{\partial t} + \nabla \cdot (\rho \mathbf{v}) = 0 \,, \tag{7.99}$$

where β is an artificial compressibility parameter whose value is key to the performance of this method. Clearly, the larger the value of β, the more "incompressible" the equations; however, large β makes the equations very stiff numerically. We shall consider only the case of constant density, but the method can be applied to variable density flows.

For solving these equations, many methods are available. In fact, because each equation now contains a time derivative, methods employed to solve them can be modeled after ones used to solve ordinary differential equations presented in Chap. 6. Because the artificial compressibility method is principally intended for steady flows, implicit methods should be favored. Another important point is that the principal difficulty faced in compressible flow, namely the possibility of transition from subsonic to supersonic flow and, especially, the possible existence of shock waves can be avoided. The best choice for a solution method for two or three-dimensional problems is an implicit method that does not require the solution of a full two or three-dimensional problem at each time step, which means that an alternating-direction-implicit or approximate factorization method is the best choice. An example of a scheme for deriving a pressure equation using artificial compressibility is presented below.

The simplest scheme uses a first-order explicit discretization in time; it enables pointwise calculation of pressure, but places a severe restriction on the size of time step. Because the time development of pressure is not important and we are interested in obtaining the steady-state solution as quickly as possible, the fully implicit Euler scheme is a better choice. To connect this method with the ones described in previous sections, we note that the intermediate velocity field $\mathbf{v}^*$, obtained by solving the momentum equation using old pressure, does not satisfy the incompressible continuity equation, so it needs to be corrected. The velocity correction needs to be linked to the pressure correction. The corrections are defined as:

$$\mathbf{v}^{n+1} = \mathbf{v}^* + \mathbf{v}' \quad \text{and} \quad p^{n+1} = p^n + p' \, . \tag{7.100}$$

It is obvious from momentum equation that the velocity correction must be proportional to the gradient of pressure; suitable relations have been derived in both fractional-step and SIMPLE-like methods, let us adopt the definition from SIMPLE method, see Eq. (7.74):

$$\mathbf{v}' = -(A_{\mathrm{D}})^{-1} G(p') \, , \tag{7.101}$$

where A_{D} is the diagonal part of the coefficient matrix from the discretized momentum equation and G is the discrete gradient operator. With the definitions introduced in the above two equations, the discretized version of the modified continuity equation (7.99) can be written as:

$$\frac{p'}{\beta \, \Delta t} - D\left[\rho (A_{\mathrm{D}})^{-1} G(p')\right] = -D(\rho \mathbf{v}^*) \, , \tag{7.102}$$

where D represents the discrete divergence operator.

This equation is the same as the pressure-correction equation of the SIMPLE method, Eq. (7.76), except for the first term on the left-hand side. It introduces an additional contribution to the diagonal matrix element of the pressure-correction equation and thus acts as under-relaxation (see Sect. 5.4.3 on more details about this kind of under-relaxation). Because of this additional term, corrected velocity at the new time level, $\mathbf{v}^{n+1}$, will also not satisfy the incompressible continuity equation; however, as we approach the steady state, all corrections tend to zero and thus the correct equation will be satisfied.

It thus appears that all the pressure calculation methods presented so far, although arrived at via slightly different routes, reduce to the same basic method. Again, it is important that the pressure derivatives inside brackets are approximated in the same way as in the momentum equations, while the outer derivative is the one from the continuity equation.

The crucial factor for convergence of a method based on artificial compressibility is the choice of parameter β. The optimum value is problem dependent, although some authors have suggested an automatic procedure for choosing it. A very large value would require the corrected velocity field to satisfy the incompressible continuity equation. In the above version of the method, this corresponds to the SIMPLE scheme without under-relaxation of pressure correction; the procedure would then converge only for small Δt. However, if only a portion of p' is added to the pressure as in SIMPLE, infinitely large β could be used. Indeed, SIMPLE could be regarded as a special version of an artificial-compressibility method using infinite β.

The lowest value of β allowed can be determined by looking at the propagation speed of pressure waves. The pseudo-sound speed is:

$$c = \sqrt{v^2 + \beta} \, .$$

By requiring pressure waves to propagate much faster than the vorticity spreads, the following criterion can be derived for a simple channel flow (see Kwak et al. 1986):

$$\beta \gg \left[1 + \frac{4}{\mathrm{Re}} \left(\frac{x_{\mathrm{ref}}}{x_\delta} \right)^2 \left(\frac{x_L}{x_{\mathrm{ref}}} \right) \right]^2 - 1 \, ,$$

where x_L is the distance between inlet and outlet, x_δ is half the distance between two walls and x_{ref} is the reference length. Typical values of β used in various methods based on artificial compressibility were in the range between 0.1 and 10.

Obviously, $1/(\beta \Delta t)$ should be small compared to the coefficients arising from the second term in Eq. (7.102) if the corrected velocity field is to closely satisfy the continuity equation. This is a necessity if rapid convergence is to be obtained. For some iterative solution methods (e.g. using domain decomposition technique in parallel processing or block-structured grids in complex geometries) it has been found useful to divide the A_P coefficient of the pressure-correction equation in SIMPLE by a factor smaller than unity (0.95 to 0.99). This is equivalent to the artificial compressibility method with $1/(\beta \Delta t) \approx (0.01 \text{ to } 0.05) A_P$.

7.2.4   Streamfunction-Vorticity Methods

For incompressible two-dimensional flows with constant fluid properties, the Navier–Stokes equations can be simplified by introducing the *streamfunction* ψ and *vorticity* ω as dependent variables. These two quantities are (in two dimensions) defined in terms of Cartesian velocity components by:

$$\frac{\partial \psi}{\partial y} = u_x \, , \qquad \frac{\partial \psi}{\partial x} = -u_y \, , \tag{7.103}$$

and

$$\omega = \frac{\partial u_y}{\partial x} - \frac{\partial u_x}{\partial y} \, . \tag{7.104}$$

Lines of constant ψ are streamlines (lines which are everywhere parallel to the flow), giving this variable its name. The vorticity is associated with rotational motion; Eq. (7.104) is a special case of the more general definition that applies in 3D as well:

$$\boldsymbol{\omega} = \nabla \times \mathbf{v} . \tag{7.105}$$

In two-dimensional flows, the vorticity vector is orthogonal to the plane of flow and Eq. (7.105) reduces to Eq. (7.104). The principal reason for introducing the stream-function is that, for flows in which ρ, μ and $\mathbf{g}$ are constant, the continuity equation is identically satisfied and need not be dealt with explicitly. Substitution of Eqs. (7.103) into the definition of the vorticity (7.104) leads to a kinematic equation connecting the streamfunction and the vorticity:

$$\frac{\partial^2 \psi}{\partial x^2} + \frac{\partial^2 \psi}{\partial y^2} = -\omega . \tag{7.106}$$

Finally, by differentiating the x and y momentum equations with respect to y and x, respectively, and subtracting the results from each other we obtain the dynamic equation for the vorticity:

$$\rho \frac{\partial \omega}{\partial t} + \rho u_x \frac{\partial \omega}{\partial x} + \rho u_y \frac{\partial \omega}{\partial y} = \mu \left(\frac{\partial^2 \omega}{\partial x^2} + \frac{\partial^2 \omega}{\partial y^2} \right) . \tag{7.107}$$

The pressure does not appear in either of these equations, i.e., it has been eliminated as a dependent variable. Thus the Navier–Stokes equations have been replaced by a set of just two partial differential equations, in place of the three for the velocity components and pressure. This reduction in the number of dependent variables and equations is what makes this approach attractive.

The two equations are coupled through the appearance of u_x and u_y (which are derivatives of ψ) in the vorticity equation and by the vorticity ω acting as the source term in the Poisson equation for ψ. The velocity components are obtained by differentiating the streamfunction. If it is needed, the pressure can be obtained by solving the Poisson equation as described in Sect. 7.1.5.1.

A solution method for these equations is the following. Given an initial velocity field, the vorticity is computed by differentiation. The dynamic vorticity equation is then used to compute the vorticity at the new time step; any standard time-advance method may be used for this purpose. Having the vorticity, it is possible to compute the streamfunction at the new time step by solving the Poisson equation; any iterative scheme for elliptic equations may be used. Finally, having the streamfunction, the velocity components are easily obtained by differentiation and we are ready to begin the calculation for the next time step.

A problem with this approach lies in the boundary conditions, especially in complex geometries. Because the flow is parallel to them, solid boundaries and symmetry planes are surfaces of constant streamfunction. However, the values of the stream-function at these boundaries can be calculated only if velocities are known. A more difficult problem is that neither vorticity nor its derivatives at the boundary are usually known in advance. For example, vorticity at a wall is equal to $\omega_{\text{wall}} = -\tau_{\text{wall}}/\mu$, where τ_{wall} is the wall shear stress, which is usually the quantity we seek to determine. Boundary values of the vorticity may be calculated from the streamfunction

Fig. 7.7 Finite difference
grid for the calculation of
flow over a rib, showing
protruding corners A and B
where special treatment is
necessary to determine
boundary values of vorticity

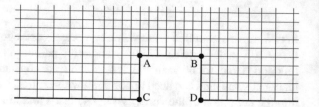

by differentiation using one-sided finite differences in the direction normal to the boundary, see Spotz and Carey (1995) or Spotz (1998). This approach usually slows down the convergence rate.

Calhoun (2002) presented a novel fractional-step method for problems with highly irregular boundaries. She employed an embedded (*aka* immersed) boundary method. While Calhoun treated general shapes of body within her domain, the vorticity is singular at sharp corners of a boundary and special care is required in treating them. For example, at the corners labeled A and B in Fig. 7.7 the derivatives of $\partial u_y / \partial x$ and $\partial u_x / \partial y$ are not continuous, which means that the vorticity ω is also not continuous there and cannot be computed using the approach described above. Some authors extrapolate the vorticity from the interior to the boundary but this does not provide a unique result at A and B either. It is possible to derive analytical behavior of the vorticity near a corner and use it to correct the solution but this is difficult as each special case must be treated separately. A simpler but efficient way of avoiding large errors (which may be convected downstream) is to locally refine the grid around singularities.

The vorticity-streamfunction approach saw considerable use for two-dimensional incompressible flows. It has become less popular in recent years because its extension to three-dimensional flows is difficult (but feasible; see the vorticity-vector-potential schemes on non-staggered grids by Weinan and Liu 1997; Wakashima and Saitoh 2004). Both the vorticity and streamfunction become three-component vectors in three dimensions so one has a system of six partial differential equations in place of the four that are necessary in a velocity-pressure formulation. The three-dimensional schemes also inherit the difficulties in dealing with variable fluid properties, compressibility, and boundary conditions that were described above for two-dimensional flows.

Chapter 8
Solution of the Navier–Stokes Equations: Part 2

We have described discretization methods for the various terms in the transport equations. The linkage of the pressure and the velocity components in incompressible flows was demonstrated and a few solution methods have been given. Many other methods of solving the Navier–Stokes equations can be devised. It is impossible to describe all of them here. However, many of them have elements in common with the methods already described. Familiarity with these methods should allow the reader to understand the others.

We describe below in detail some methods that are representative of a larger group of methods. First implicit methods (SIMPLE and iterative implicit fractional-step method) using the pressure-correction equation and staggered grids are described in enough detail to allow straightforward translation into a computer code. The corresponding codes with many comments pointing to corresponding equations in the book are available via Internet; see appendix for details. Working in a similar mode we then describe these methods for a colocated grid, and, finally, we discuss some additional issues for implementation of fractional-step methods on both staggered and colocated grids. The chapter closes with examples of both steady and unsteady flow simulations.

8.1 Implicit Iterative Methods on a Staggered Grid

In this section we present two implicit finite volume schemes that use the pressure-correction method on a staggered two-dimensional Cartesian grid. One is based on the SIMPLE-algorithm and the other on fractional-step method; we shall call the latter "Implicit Fractional-Step Method" (IFSM). They can be used to compute both steady-state and unsteady flows. Solution methods for complicated geometries are described in the next chapter.

© Springer Nature Switzerland AG 2020
J. H. Ferziger et al., *Computational Methods for Fluid Dynamics*,
https://doi.org/10.1007/978-3-319-99693-6_8

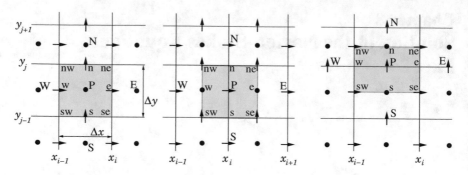

Fig. 8.1 Control volumes for a staggered grid: for mass conservation and scalar quantities (left), for x-momentum (center) and for y-momentum (right)

The Navier–Stokes equations in integral form read:

$$\int_S \rho \mathbf{v} \cdot \mathbf{n} \, dS = 0 \,, \tag{8.1}$$

$$\frac{\partial}{\partial t} \int_V \rho u_i \, dV + \int_S \rho u_i \mathbf{v} \cdot \mathbf{n} \, dS = \int_S \tau_{ij} \mathbf{i}_j \cdot \mathbf{n} \, dS - \int_S p \mathbf{i}_i \cdot \mathbf{n} \, dS + \int_V (\rho - \rho_0) g_i \, dV \,. \tag{8.2}$$

For convenience, it is assumed that the only body force is buoyancy. The macroscopic momentum flux vector $\mathbf{t}_i$, see Eq. (1.18), is split into a viscous contribution $\tau_{ij} \mathbf{i}_j$ and a pressure contribution $p \mathbf{i}_i$. We assume the density constant except in the buoyancy term, i.e., we use the Boussinesq approximation. The mean gravitational force is incorporated into the pressure term, as shown in Sect. 1.4.

Typical staggered control volumes are shown in Fig. 8.1. The control volumes for u_x and u_y are displaced with respect to the control volume for the continuity equation. For non-uniform grids, the velocity nodes are not at the centers of their control volumes. Cell faces 'e' and 'w' for u_x and 'n' and 's' for u_y lie midway between the nodes. For convenience, we shall sometimes use u instead of u_x and v instead of u_y.

Most of the discretization and solution algorithm is identical for both SIMPLE and IFSM; the differences are limited to the pressure-correction equation where they will be emphasized. We will use the second-order implicit three-time-level scheme described in Sect. 6.2.4 for integration in time. This leads to the following approximation of the unsteady term:

$$\left[\frac{\partial}{\partial t} \int_V \rho u_i \, dV \right]_P \approx \frac{\rho \, \Delta V}{2 \, \Delta t} \left(3u_i^{n+1} - 4u_i^n + u_i^{n-1} \right)_P = A_P^t u_{i,P}^{n+1} - Q_{u_i}^t \,,$$

where

$$A_{\mathrm{P}}^{\mathrm{t}} = \frac{3\rho\,\Delta V}{2\Delta t} \quad \text{and} \quad Q_{u_i}^{\mathrm{t}} = \frac{\rho\,\Delta V}{2\Delta t}\left(4u_i^n - u_i^{n-1}\right)_{\mathrm{P}}\,. \tag{8.3}$$

We shall drop from now on the superscript $n + 1$; all terms are evaluated at t_{n+1}, unless stated otherwise. Because the scheme is implicit, the equations require iterative solution. If the time steps are as small as those used in explicit schemes, one or two iterations per time step will suffice. For flows with slow transients, we may use larger time steps and more iterations per time step will be necessary. As noted earlier, these iterations are called outer iterations to distinguish them from the inner iterations used to solve linear equations such as the pressure-correction equation. We assume that one of the solvers described in Chap. 5 is used for the latter and shall concentrate on the outer iterations.

We now consider the approximation of the convection and diffusion fluxes and the source terms. The surface integral may be split into four CV-face integrals. Let us concentrate attention on CV-face 'e'; the other faces are treated in the same way, and the results can be obtained by index substitution. We adopt the second-order central-difference approximations presented in Chap. 4. Fluxes are approximated by assuming that the value of a quantity at a CV-face center represents the mean value over the face (midpoint rule approximation). On the mth outer iteration, all nonlinear terms are approximated by a product of an 'old' (from the preceding outer iteration) and a 'new' value. Thus, in discretizing the momentum equations, the mass flux through each CV-face is evaluated using the existing velocity field and is assumed known:

$$\dot{m}_{\mathrm{e}}^m = \int_{S_{\mathrm{e}}} \rho \mathbf{v} \cdot \mathbf{n}\,\mathrm{d}S \approx (\rho u)_{\mathrm{e}}^{m-1} S_{\mathrm{e}}\,. \tag{8.4}$$

This kind of linearization is essentially the first step of Picard iteration; other linearizations for implicit schemes were described in Chap. 5. Unless specifically stated otherwise, all variables in the remainder of this section belong to the mth outer iteration. The mass fluxes (8.4) satisfy the continuity equation on the 'scalar' CV, see Fig. 8.1. Mass fluxes at the faces of the momentum CVs must be obtained by interpolation; ideally, these fluxes would provide mass conservation for the momentum CV but this can be guaranteed only to the accuracy of the interpolation. Another possibility is to use the mass fluxes from the scalar CV faces. Because the east and west faces of a u-CV are halfway between scalar CV faces, the mass fluxes can be calculated as:

$$\dot{m}_{\mathrm{e}}^u = \frac{1}{2}(\dot{m}_{\mathrm{P}} + \dot{m}_{\mathrm{E}})^u\,; \quad \dot{m}_{\mathrm{w}}^u = \frac{1}{2}(\dot{m}_{\mathrm{W}} + \dot{m}_{\mathrm{P}})^u\,. \tag{8.5}$$

The mass fluxes through the north and south faces of the u-CV can be approximated as half the sum of the two scalar CV-face mass fluxes:

$$\dot{m}_{\mathrm{n}}^u = \frac{1}{2}(\dot{m}_{\mathrm{ne}} + \dot{m}_{\mathrm{nw}})^u\,; \quad \dot{m}_{\mathrm{s}}^u = \frac{1}{2}(\dot{m}_{\mathrm{se}} + \dot{m}_{\mathrm{sw}})^u\,. \tag{8.6}$$

The superscript u denotes that the indexes refer to the u-CV, see Fig. 8.1. The sum of the four mass fluxes for the u-CV is thus half the sum of mass fluxes into the two adjacent scalar CVs. They therefore satisfy the continuity equation for the double scalar CV so the mass fluxes through the u-CV faces also conserve mass. This result also holds for v-momentum CVs. It is necessary to ensure that the mass fluxes through the momentum CVs satisfy the continuity equation; otherwise, momentum will not be conserved.

The convection flux of u_i-momentum through the 'e'-face of a u_i-CV is then (see Sect. 4.2 and Eq. (8.4)):

$$F_{i,e}^c = \int_S \rho u_i \mathbf{v} \cdot \mathbf{n}\, dS \approx \dot{m}_e u_{i,e} \,. \tag{8.7}$$

The CV-face value of u_i used in this expression need not be the one used to calculate the mass flux, although an approximation of the same accuracy is desirable. Linear interpolation is the simplest second-order approximation; see Sect. 4.4.2 for details. We call this a *central-difference scheme* (CDS), although no differencing is involved. This is because, on uniform grids, it results in the same algebraic equations as the CDS finite-difference method.

Some iterative solvers fail to converge when applied to the algebraic equation systems derived from central-difference approximations of convection fluxes. This is because the matrices may not be diagonally dominant; these equations are best solved using the *deferred correction* approach described in Sect. 5.6. In this method, the flux is expressed as:

$$F_{i,e}^c = \dot{m}_e u_{i,e}^{\mathrm{UDS}} + \dot{m}_e (u_{i,e}^{\mathrm{CDS}} - u_{i,e}^{\mathrm{UDS}})^{m-1} \,, \tag{8.8}$$

where superscripts CDS and UDS denote approximation by central and upwind differences, respectively (see Sect. 4.4). The term in brackets is evaluated using values from the previous iteration while the matrix is computed using the UDS-approximation. At convergence, the UDS-contributions cancel out, leaving a CDS-solution. This procedure usually converges at approximately the rate obtained for a pure upwind approximation.

The two schemes may also be blended; this is achieved by multiplying the explicit part (the term in brackets in Eq. (8.8)) by a factor $0 \leq \gamma \leq 1$. This practice can remove the oscillations obtained with central differences on coarse grids. However, it improves the esthetics of the results at the cost of decreasing the accuracy. Blending may be used locally, e.g., to allow calculation of flows with shocks with CDS; this is preferable to applying it everywhere.

Calculation of the diffusion fluxes requires evaluation of the stresses τ_{xx} and τ_{yx} at the CV face 'e'. Because the outward unit normal vector at this CV-face is $\mathbf{i}$, we have:

$$F_{i,e}^d = \int_{S_e} \tau_{ix}\, dS \approx (\tau_{ix})_e S_e \,, \tag{8.9}$$

where $S_e = y_j - y_{j-1} = \Delta y$ for the u-CV and $S_e = \frac{1}{2}(y_{j+1} - y_{j-1})$ for the v-CV. The stresses at CV face require approximation of the derivatives; central difference approximations lead to:

$$(\tau_{xx})_e = 2\left(\mu \frac{\partial u}{\partial x}\right)_e \approx 2\mu \frac{u_E - u_P}{x_E - x_P} , \tag{8.10}$$

$$(\tau_{yx})_e = \mu\left(\frac{\partial v}{\partial x} + \frac{\partial u}{\partial y}\right)_e \approx \mu \frac{v_E - v_P}{x_E - x_P} + \mu \frac{u_{ne} - u_{se}}{y_{ne} - y_{se}} . \tag{8.11}$$

Note that τ_{xx} is evaluated at the 'e' face of the u-CV, and τ_{yx} at the 'e' face of the v-CV, so the indexes refer to locations on the appropriate CVs, see Fig. 8.1. Thus, u_{ne} and u_{se} on the v-CV are actually nodal values of the u-velocity and no interpolation is necessary.

At the other CV-faces we obtain similar expressions. For u-CVs, we need to approximate τ_{xx} at the faces 'e' and 'w', and τ_{xy} at the faces 'n' and 's'. For v-CVs, τ_{yx} is needed at the faces 'e' and 'w' and τ_{yy} at the faces 'n' and 's'.

The pressure terms are approximated by:

$$Q_u^p = -\int_S p\,\mathbf{i} \cdot \mathbf{n}\,dS \approx -(p_e S_e - p_w S_w)^{m-1} \tag{8.12}$$

for the u-equation and

$$Q_v^p = -\int_S p\,\mathbf{j} \cdot \mathbf{n}\,dS \approx -(p_n S_n - p_s S_s)^{m-1} \tag{8.13}$$

for the v-equation. There are no pressure force contributions from the 'n' and 's' CV faces to the u-equation or from the 'e' and 'w' faces to the v-equation on a Cartesian grid.

If buoyancy force is present, it is approximated by:

$$Q_{u_i}^b = \int_V (\rho - \rho_0)g_i\,dV \approx (\rho_P^{m-1} - \rho_0)g_i\,\Delta V , \tag{8.14}$$

where $\Delta V = (x_e - x_w)(y_n - y_s) = \frac{1}{2}(x_{i+1} - x_{i-1})(y_j - y_{j-1})$ for the u-CV and $\Delta V = \frac{1}{2}(x_i - x_{i-1})(y_{j+1} - y_{j-1})$ for the v-CV. Any other body force can be approximated in the same way.

The approximation to the complete u_i-momentum equation is:

$$A_P^t u_{i,P} + F_i^c = F_i^d + Q_i^p + Q_i^b + Q_i^t , \tag{8.15}$$

where

$$F^c = F_e^c + F_w^c + F_n^c + F_s^c \quad \text{and} \quad F^d = F_e^d + F_w^d + F_n^d + F_s^d . \tag{8.16}$$

If ρ and μ are constant, part of the diffusion flux term cancels out by virtue of the continuity equation, see Sect. 7.1. (It may not exactly cancel out in the numerical approximation but the equations may be simplified by deleting those terms prior to discretization). For example, in the u-equation, the τ_{xx} term on the 'e' and 'w' faces will be reduced by half, and in the τ_{yx} term at the 'n' and 's' faces, the $\partial v / \partial x$ contribution is removed. Even when ρ and μ are not constant, the sum of these terms contributes in only a minor way to F^d. This is why an explicit 'diffusion source term', e.g., for u:

$$Q_u^d = \left[\mu_e S_e \frac{u_E - u_P}{x_E - x_P} - \mu_w S_w \frac{u_P - u_W}{x_P - x_W} + \mu_n S_n \frac{v_{ne} - v_{nw}}{x_{ne} - x_{nw}} - \mu_s S_s \frac{v_{se} - v_{sw}}{x_{se} - x_{sw}} \right]^{m-1}$$

(8.17)

is usually calculated from the previous outer iteration $m - 1$ and treated explicitly. Only $F^d - Q_u^d$ is treated implicitly. A consequence of this approximation is that, on a colocated grid, the matrix implied by Eq. (8.15) is identical for all three velocity components.

When the approximations for all the fluxes and source terms are substituted into Eq. (8.15), we obtain an algebraic equation of the form:

$$A_P^u u_P + \sum_k A_k^u u_k = Q_P^u, \quad k = E, W, N, S .$$

(8.18)

The equation for v has the same form. The coefficients depend on the approximations used; for the UDS-approximations applied above, the coefficients of the u-equation are:

$$A_E^u = \min(\dot{m}_e^u, 0) - \frac{\mu_e S_e}{x_E - x_P}, \quad A_W^u = \min(\dot{m}_w^u, 0) - \frac{\mu_w S_w}{x_P - x_W},$$

$$A_N^u = \min(\dot{m}_n^u, 0) - \frac{\mu_n S_n}{y_N - y_P}, \quad A_S^u = \min(\dot{m}_s^u, 0) - \frac{\mu_s S_s}{y_P - y_S},$$

(8.19)

$$A_P^u = A_P^t - \sum_k A_k^u, \quad k = E, W, N, S .$$

Note that the contribution from convection and diffusion fluxes to the central coefficient A_P can be expressed as the negative sum of neighbor coefficients; actually, when this contribution is extracted from discretized flux approximations, one obtains:

$$A_P^u = A_P^t - \sum_k A_k^u + \sum_k \dot{m}_k .$$

(8.20)

The last term on the right-hand side represents the discretized continuity equation which, for incompressible flows, equals zero and is thus omitted in Eq. (8.19). This is true for all conservative schemes.

Note also that $\dot{m}_w$ for the CV centered around node P equals $-\dot{m}_e$ for the CV centered around node W. The source term Q_P^u contains not only the pressure and buoyancy terms but also the portion of convection and diffusion fluxes resulting from deferred correction and the contribution of the unsteady term, i.e.:

$$Q_P^u = Q_u^p + Q_u^b + Q_u^c + Q_u^d + Q_u^t ,\tag{8.21}$$

where

$$Q_u^c = \left[(F_u^c)^{\mathrm{UDS}} - (F_u^c)^{\mathrm{CDS}}\right]^{m-1} .\tag{8.22}$$

This 'convection source' is calculated using the velocities from the previous outer iteration $m - 1$.

The coefficients in the v equation are obtained in the same way and have the same form; however, the grid locations 'e', 'n' etc. have different coordinates, see Fig. 8.1.

The linearized momentum equations are solved with the sequential solution method (see Sect. 5.4), using the 'old' mass fluxes and the pressure from the previous outer iteration. This produces new velocities u^* and v^* which do not necessarily satisfy the continuity equation so:

$$\dot{m}_e^* + \dot{m}_w^* + \dot{m}_n^* + \dot{m}_s^* = \Delta\dot{m}_P^* ,\tag{8.23}$$

where the mass fluxes are calculated according to Eq. (8.4) using u^* and v^*. Because the arrangement of variables is staggered, the cell-face velocities on a mass CV are the nodal values. The indexes below refer to this CV, see Fig. 8.1, unless otherwise stated.

The derivation of a discrete pressure-correction equation follows slightly different routes in SIMPLE and IFSM; we proceed first with the SIMPLE algorithm.

8.1.1 SIMPLE for Staggered Grids

The velocity components u^* and v^* calculated from the momentum equations can be expressed as follows (by dividing Eq. (8.18) by A_P and writing the pressure term explicitly; note that index 'e' on a mass CV represents index P on a u-CV):

$$u_e^* = \tilde{u}_e^* - \frac{S_e}{A_P^u}(p_E - p_P)^{m-1} ,\tag{8.24}$$

where $\tilde{u}_e^*$ is shorthand notation for

$$\tilde{u}^* = \frac{Q_P^u - Q_u^P - \sum_k A_k^u u_k^*}{A_P} . \tag{8.25}$$

Analogously, v_n^* can be expressed as:

$$v_n^* = \tilde{v}_n^* - \frac{S_n}{A_P^v}(p_N - p_P)^{m-1} . \tag{8.26}$$

The velocities u^* and v^* need to be corrected to enforce mass conservation. This is done, as outlined in Sect. 7.2.2, by correcting the pressure, leaving $\tilde{u}$ and $\tilde{v}$ unchanged. The corrected velocities—the final values for the mth outer iteration $-u^m = u^* + u'$ and $v^m = v^* + v'$ are required to satisfy the linearized momentum equations, which is possible only if the pressure is corrected. So we write:

$$u_e^m = \tilde{u}_e^* - \frac{S_e}{A_P^u}(p_E - p_P)^m , \tag{8.27}$$

and

$$v_n^m = \tilde{v}_n^* - \frac{S_n}{A_P^v}(p_N - p_P)^m , \tag{8.28}$$

where $p^m = p^{m-1} + p'$ is the new pressure. The indexes refer to the mass CV. The relation between velocity and pressure corrections is obtained by subtracting Eq. (8.24) from Eq. (8.27):

$$u_e' = -\frac{S_e}{A_P^u}(p_E' - p_P') = -\left(\frac{\Delta V}{A_P^u}\frac{\delta p'}{\delta x}\right)_e . \tag{8.29}$$

Analogously, one obtains:

$$v_n' = -\frac{S_n}{A_P^v}(p_N' - p_P') = -\left(\frac{\Delta V}{A_P^v}\frac{\delta p'}{\delta y}\right)_n . \tag{8.30}$$

The corrected velocities are required to satisfy the continuity equation, so we substitute u^m and v^m into the expressions for mass fluxes, Eq. (8.4), and use Eq. (8.23):

$$(\rho S u')_e - (\rho S u')_w + (\rho S v')_n - (\rho S v')_s + \Delta \dot{m}_P^* = 0 . \tag{8.31}$$

Finally, substitution of the above expressions (8.29) and (8.30) for u' and v' into the continuity equation leads to the pressure-correction equation:

$$A_P^p p_P' + \sum_k A_k^p p_k' = -\Delta \dot{m}_P^* , \tag{8.32}$$

where the coefficients are:

$$A_E^p = -\left(\frac{\rho S^2}{A_P^u}\right)_e \; , \quad A_W^p = -\left(\frac{\rho S^2}{A_P^u}\right)_w \; ,$$

$$A_N^p = -\left(\frac{\rho S^2}{A_P^v}\right)_n \; , \quad A_S^p = -\left(\frac{\rho S^2}{A_P^v}\right)_s \; , \qquad (8.33)$$

$$A_P^p = -\sum_k A_k^p \; , \quad k = E, W, N, S \; .$$

After the pressure-correction equation has been solved, the velocities and pressure are corrected. As noted in Sect. 7.2.2, if one tries to calculate steady flows using very large time steps, the momentum equations must be under-relaxed as described in Sect. 5.4.2, so only part of the pressure correction p' is added to p^{m-1}. Under-relaxation may also be required in unsteady calculations with large time steps.

The corrected velocities satisfy the continuity equation to the accuracy with which the pressure-correction equation is solved. However, they do not satisfy the non-linear momentum equation because $\tilde{u}$ and $\tilde{v}$ were not corrected in Eqs. (8.27) and (8.28), so we have to begin another outer iteration. When both the continuity and momentum equations are satisfied to the desired tolerance, u'_i and p' will be negligible and we can proceed to the next time level. To begin the iterations at the new time step, the solution at the previous time step provides the initial guess. This may be improved by use of extrapolation; any of the explicit time-advancing schemes can serve as the predictor. For small time steps extrapolation is fairly accurate and saves a few iterations.

If a steady-state flow is computed, one can take a single infinite time step; all contributions resulting from the unsteady term drop out and one continues with outer iterations until all corrections become negligible. When computing unsteady flows, the number of outer iterations per time step is usually in the range between 3 (for small time steps and high under-relaxation factors) and 10 (for larger time steps and moderate under-relaxation factors). Some examples will be presented in Sect. 8.4. A computer code employing this algorithm is available via the Internet; see the appendix for details.

Note that the coefficients in the pressure-correction equation, (8.33), are proportional to the face area squared. Thus, the ratio A_E/A_N is proportional to the square of the aspect ratio $a_r = \Delta y/\Delta x$. If cells are highly stretched (e.g., near walls, when the viscous sublayer is to be resolved), this can make the pressure-correction equation stiff and more difficult to solve. Aspect ratios greater than 100 should be avoided, if possible, because the coefficients in one direction (e.g., normal to the wall in the above-mentioned example) are then more than 10^4 times larger than in the other directions.

The above algorithm is easily modified to give the SIMPLEC method described in Sect. 7.2.2. The pressure-correction equation has the form (8.32), but in expressions for coefficients, Eq. (8.33), A_P^u and A_P^v are replaced by $A_P^u + \sum_k A_k^u$ and $A_P^v + \sum_k A_k^v$, respectively. The extension to the PISO algorithm is also straightforward. The second pressure-correction equation has the same coefficient matrix as the first one, but the source term is now based on $\tilde{u}_i'$. This contribution was neglected in the first pressure-correction equation, but it can now be calculated using the first velocity correction u_i'.

Discretizations of higher order are easily incorporated into the above solution strategy. The implementation of boundary conditions was discussed in Sect. 7.1.6.

8.1.2 IFSM for Staggered Grids

The major difference between SIMPLE and IFSM is that the latter always uses finite time steps, even when a steady-state flow is computed. However, it does not require under-relaxation, so one only needs to choose an appropriate time step for each application.

The relation between velocity and pressure correction is derived following Eq. (7.64), i.e. velocity is only corrected in the unsteady term, leaving u_i^* in both convection and diffusion fluxes:

$$\frac{3\rho \Delta V}{2\Delta t}(u_e^m - u_e^*) = -S_e(p_E' - p_P') \Rightarrow u_e' = -\frac{2\Delta t S_e}{3\rho \Delta V}(p_E' - p_P') . \qquad (8.34)$$

Because $\Delta V = S_e(x_E - x_P) = S_n(y_N - y_P)$, we obtain:

$$u_e' = -\frac{2\Delta t}{3\rho(x_E - x_P)}(p_E' - p_P') = -\frac{2\Delta t}{3\rho}\left(\frac{\delta p'}{\delta x}\right)_e \qquad (8.35)$$

and analogously:

$$v_n' = -\frac{2\Delta t}{3\rho(y_N - y_P)}(p_N' - p_P') = -\frac{2\Delta t}{3\rho}\left(\frac{\delta p'}{\delta y}\right)_n . \qquad (8.36)$$

Substituting these expressions into continuity equation (8.31) leads to the same form of the pressure-correction equation as in the case of SIMPLE, Eq. (8.32), only the coefficients are different; instead of Eq. (8.33), we now have:

$$A_E^p = -\frac{2\Delta t S_e}{3(x_E - x_P)} , \quad A_W^p = -\frac{2\Delta t S_w}{3(x_P - x_W)} ,$$

$$A_N^p = -\frac{2\Delta t S_n}{3(y_N - y_P)} , \quad A_S^p = -\frac{2\Delta t S_s}{3(y_P - y_S)} , \qquad (8.37)$$

$$A_P^p = -\sum_k A_k^p , \quad k = E, W, N, S .$$

When computing unsteady flows, one chooses the time step so that flow variation is adequately resolved (e.g., order of 100 time steps per period if the flow is periodic). The number of required outer iterations per time step depends on the time-step size, as for the SIMPLE-method.

When marching towards a steady state, one still has to use a finite time step, but it can then be relatively large. It is often advantageous to use smaller time steps initially, when the changes in solution are relatively large (especially if the initialization is not a good approximation of the final solution); as the steady state is approached, the time step can be gradually increased. The maximum allowable time step size is problem-dependent, but it is usually related to the CFL-number:

$$\text{CFL}_x = \frac{u \Delta t}{\Delta x} \quad \text{and} \quad \text{CFL}_y = \frac{v \Delta t}{\Delta y} . \tag{8.38}$$

Values between 1 and 100 are usually appropriate; see Sect. 8.4 for some examples.

8.2 Implicit Iterative Methods for Colocated Grids

It was mentioned earlier that a colocated arrangement of variables on a numerical grid creates problems which caused it to be out of favor for some time. Here, we shall first show why the problems occur and then present a cure.

8.2.1 Treatment of Pressure for Colocated Variables

We start by looking at a finite-difference scheme and the simple time-advance method presented in Sect. 7.1.7.1. There we derived the discrete Poisson equation for the pressure, which can be written:

$$\frac{\delta}{\delta x_i} \left(\frac{\delta p^n}{\delta x_i} \right) = \frac{\delta H_i^n}{\delta x_i} , \tag{8.39}$$

where H_i^n is the shorthand notation for the sum of the advection and viscous terms:

$$H_i^n = -\frac{\delta (\rho u_i u_j)^n}{\delta x_j} + \frac{\delta \tau_{ij}^n}{\delta x_j} \tag{8.40}$$

(summation on j is implied). The discretization scheme used to approximate the derivatives is not important in Eq. (8.39); that is why symbolic notation is used. Also, the equation is not specific to any grid arrangement.

Let us now look at the colocated arrangement shown in Fig. 8.2 and various difference schemes for the pressure gradient terms in the momentum equations and for the

Fig. 8.2 Control volume in
a colocated grid and notation
used

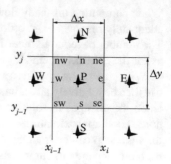

divergence in the continuity equation. We start by considering a forward-difference
scheme for pressure terms and a backward-difference scheme for the continuity equation. Section 7.1.3 shows that this combination is energy conserving. For simplicity
we assume that the grid is uniform with spacings Δx and Δy.

By approximating the outer difference operator $\delta/\delta x_i$ in the pressure equation
with the backward difference scheme, we obtain:

$$\frac{\left(\frac{\delta p^n}{\delta x}\right)_P - \left(\frac{\delta p^n}{\delta x}\right)_W}{\Delta x} + \frac{\left(\frac{\delta p^n}{\delta y}\right)_P - \left(\frac{\delta p^n}{\delta y}\right)_S}{\Delta y} = \frac{H^n_{x,P} - H^n_{x,W}}{\Delta x} + \frac{H^n_{y,P} - H^n_{y,S}}{\Delta y}.$$

(8.41)

Denoting the right-hand side as Q^H_P and using the forward-difference approximations
for the pressure derivatives, we arrive at:

$$\frac{\frac{p^n_E - p^n_P}{\Delta x} - \frac{p^n_P - p^n_W}{\Delta x}}{\Delta x} + \frac{\frac{p^n_N - p^n_P}{\Delta y} - \frac{p^n_P - p^n_S}{\Delta y}}{\Delta y} = Q^H_P .$$

(8.42)

The system of algebraic equations for the pressure then takes the form:

$$A^p_P p^n_P + \sum_k A^p_k p^n_k = -Q^H_P , \quad k = E, W, N, S ,$$

(8.43)

where the coefficients are:

$$A^p_E = A^p_W = -\frac{1}{(\Delta x)^2} , \quad A^p_N = A^p_S = -\frac{1}{(\Delta y)^2} , \quad A^p_P = -\sum_k A^p_k .$$

(8.44)

One can verify that the FV-approach would reproduce Eq. (8.42) if the CV shown
in Fig. 8.2 is used for both the momentum equations and the continuity equation,
and if the following approximations are used: $u_e = u_P$, $p_e = p_E$; $v_n = v_P$, $p_n = p_N$;
$u_w = u_W$, $p_w = p_P$; $v_s = v_S$, $p_s = p_P$.

The pressure or pressure-correction equation has the same form as the one obtained on a staggered grid with central-difference approximations; this is because approximation of a second derivative by a product of forward and backward-difference approximations for first derivatives gives the central-difference approximation. However, the momentum equations now suffer from use of a first-order approximation to the major driving force term—the pressure gradient. It is better to use higher-order approximations.

Now consider what happens if we choose central-difference approximations for both the pressure gradient in the momentum equations and the divergence in the continuity equation. Approximating the outer difference operator in Eq. (8.39) by central differences, we obtain:

$$
\frac{\left(\frac{\delta p^n}{\delta x}\right)_E - \left(\frac{\delta p^n}{\delta x}\right)_W}{2\Delta x} + \frac{\left(\frac{\delta p^n}{\delta y}\right)_N - \left(\frac{\delta p^n}{\delta y}\right)_S}{2\Delta y} = \frac{H^n_{x,E} - H^n_{x,W}}{2\Delta x} + \frac{H^n_{y,N} - H^n_{y,S}}{2\Delta y} .
$$
(8.45)

We again denote the right-hand side as Q^H_P; however, this quantity is not the one obtained previously. Inserting the central-difference approximations for pressure derivatives, we find:

$$
\frac{\frac{p^n_{EE} - p^n_P}{2\Delta x} - \frac{p^n_P - p^n_{WW}}{2\Delta x}}{2\Delta x} + \frac{\frac{p^n_{NN} - p^n_P}{2\Delta y} - \frac{p^n_P - p^n_{SS}}{2\Delta y}}{2\Delta y} = Q^H_P .
$$
(8.46)

The system of algebraic equations for the pressure has the form:

$$
A^P_P p^n_P + \sum_k A^P_k p^n_k = -Q^H_P , \quad k = \text{EE, WW, NN, SS}
$$
(8.47)

where the coefficients are:

$$
A^P_{EE} = A^P_{WW} = -\frac{1}{(2\Delta x)^2} , \quad A^P_{NN} = A^P_{SS} = -\frac{1}{(2\Delta y)^2} , \quad A^P_P = -\sum_k A^P_k .
$$
(8.48)

This equation has the same form as Eq. (8.43) but it involves nodes which are $2\Delta x$ or $2\Delta y$ apart! It is a discretized Poisson equation on a grid twice as coarse as the basic one but the equations split into four unconnected systems, one with i and j both even, one with i even and j odd, one with i odd and j even, and one with both odd. Each of these systems gives a different solution. For a flow with a uniform pressure field, the checkerboard pressure distribution shown in Fig. 8.3 satisfies these equations and could be produced. However, the pressure gradient is not affected and the velocity field may be smooth. There is also the possibility that one may not be able to obtain a converged steady-state solution.

A similar result is obtained with the finite volume approach if the CV-face values of the fluxes are calculated by linear interpolation of the two neighbor nodes.

Fig. 8.3 Checkerboard pressure field, made of four superimposed uniform fields on 2Δ-spacing, which is interpreted by CDS as a uniform field

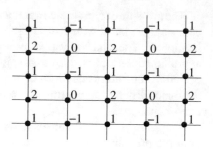

The source of the above problem may be traced to using $2\Delta x$ approximations to the first derivatives. Various cures have been proposed. In incompressible flows, the absolute pressure level is unimportant—only the differences matter. Unless the absolute value of the pressure is specified somewhere, the pressure equation is singular and has an infinite number of solutions, all differing by a constant. This makes a simple cure possible: filtering out the oscillations, as was done by van der Wijngaart (1990).

We shall present one approach to dealing with the pressure-velocity coupling on colocated grids that has found widespread use in complicated geometries and is simple and effective. It is used in almost all commercial and public CFD-codes.

On staggered grids, central-difference approximations are based on Δx-differences. Can we do the same with the colocated arrangement? A Δx-approximation of the outer first derivative in the pressure Eq. (8.39) has the form:

$$\frac{\left(\frac{\delta p^n}{\delta x}\right)_e - \left(\frac{\delta p^n}{\delta x}\right)_w}{\Delta x} + \frac{\left(\frac{\delta p^n}{\delta y}\right)_n - \left(\frac{\delta p^n}{\delta y}\right)_s}{\Delta y} = \frac{H_{x,e}^n - H_{x,w}^n}{\Delta x} + \frac{H_{y,n}^n - H_{y,s}^n}{\Delta y} .$$
(8.49)

The problem is that the values of pressure derivatives and quantities H are not available at cell-face locations, so we have to use interpolation. Let us choose linear interpolation, which has the same accuracy as the CDS approximation of the derivatives. Also let the inner derivatives of the pressure in Eq. (8.39) be approximated by central differences. Linear interpolation of cell-center derivatives leads to:

$$\left(\frac{\delta p^n}{\delta x}\right)_e \approx \frac{1}{2}\left(\frac{p_E - p_W}{2\,\Delta x} + \frac{p_{EE} - p_P}{2\,\Delta x}\right) .$$
(8.50)

With this interpolation the pressure Eq. (8.46) is recovered.

We could evaluate the pressure derivatives at cell faces using central differences and Δx spacing as follows:

$$\left(\frac{\delta p^n}{\delta x}\right)_e \approx \frac{p_E - p_P}{\Delta x} .$$
(8.51)

If this approximation is applied at all cell faces, we arrive at the following pressure equation (which is also valid on non-uniform grids):

$$\frac{\dfrac{p_E^n - p_P^n}{\Delta x} - \dfrac{p_P^n - p_W^n}{\Delta x}}{\Delta x} + \frac{\dfrac{p_N^n - p_P^n}{\Delta y} - \dfrac{p_P^n - p_S^n}{\Delta y}}{\Delta y} = Q_P^H, \qquad (8.52)$$

which is the same as Eq. (8.42), except that the right-hand side is now obtained by interpolation:

$$Q_P^H = \frac{\overline{(H_x^n)}_e - \overline{(H_x^n)}_w}{\Delta x} + \frac{\overline{(H_y^n)}_n - \overline{(H_y^n)}_s}{\Delta y}. \qquad (8.53)$$

Use of this approximation eliminates the oscillation in the pressure field but, in order to accomplish this, we have introduced an inconsistency in the treatment of the pressure gradient in the momentum and pressure equations. Let us compare the two approximations. It is easy to show that the left-hand sides of Eqs. (8.52) and (8.46) differ by:

$$R_P^p = \frac{4\,p_E + 4\,p_W - 6\,p_P - p_{EE} - p_{WW}}{4(\Delta x)^2} + \frac{4\,p_N + 4\,p_S - 6\,p_P - p_{NN} - p_{SS}}{4(\Delta y)^2}, \qquad (8.54)$$

which represents a central difference approximation to the fourth-order pressure derivatives:

$$R_P^p = -\frac{(\Delta x)^2}{4}\left(\frac{\partial^4 p}{\partial x^4}\right)_P - \frac{(\Delta y)^2}{4}\left(\frac{\partial^4 p}{\partial y^4}\right)_P. \qquad (8.55)$$

Expression (8.54) is easily obtained by applying the standard CDS-approximation of the second derivative twice, see Sect. 3.4.

This difference tends to zero as the grid is refined and the error introduced is of the same magnitude as the error in the basic discretization and so does not add significantly to the latter. However, the energy-conserving property of the scheme is destroyed in this process. Indeed, Ham and Iaccarino (2004) have shown (in the context of non-iterative (fractional-step) methods) that the net effect on the kinetic energy is dissipative.

The above result was derived for second-order CDS-discretization and linear interpolation. A similar derivation can be constructed for any discretization scheme and interpolation. Let us see how the above idea translates into implicit pressure-correction methods using FV discretization.

8.2.2 SIMPLE for Colocated Grids

Implicit solution of the momentum equations discretized with a colocated FV-method follows the lines of the previous section for the staggered arrangement. One has only to bear in mind that the CVs for all variables are now the same. The pressures at the cell-face centers, which are not nodal locations, have to be obtained by interpolation; linear interpolation is a suitable second-order approximation, but higher-order methods can be used. The gradients at the CV-center, which are needed for the calculation of cell-face velocities, can be obtained using Gauss' theorem. The pressure forces in the x and y direction are summed over all faces and divided by the cell volume to yield the corresponding mean pressure derivative, e.g.:

$$\left(\frac{\delta p}{\delta x}\right)_P = \frac{Q_u^p}{\Delta V} , \tag{8.56}$$

where Q_u^p stands for the sum of pressure forces in the x-direction over all CV faces, see Eq. (8.12). On Cartesian grids this reduces to the standard CDS-approximation.

Solution of the linearized momentum equations produces u^* and v^*. For the discretized continuity equation, we need the cell-face velocities which have to be calculated by interpolation; linear interpolation is the obvious choice. The pressure-correction equation of the SIMPLE algorithm can be derived following the lines of Sects. 7.2.2 and 8.1. The interpolated cell-face velocities needed in the continuity equation involve interpolated pressure gradients, so their correction is proportional to the interpolated pressure-correction gradient (see Eq. (8.29)):

$$u'_e = -\overline{\left(\frac{\Delta V}{A_P^u}\frac{\delta p'}{\delta x}\right)}_e . \tag{8.57}$$

On uniform grids, the pressure-correction equation derived using this expression for the cell-face velocity corrections corresponds to Eq. (8.46). On non-uniform grids, the computational molecule of the pressure-correction equation involves the nodes P, E, W, N, S, EE, WW, NN and SS. As shown in the preceding section, this equation may have oscillatory solutions. Although the oscillations can be filtered out (see van der Wijngaart 1990), the pressure-correction equation becomes complex on arbitrary grids and the convergence of the solution algorithm may be slow. A compact pressure-correction equation similar to the staggered grid equation can be obtained using the approach discussed in the preceding section. It is described below.

It was shown in the preceding section that the interpolated pressure gradients can be replaced by compact central-difference approximations at the cell faces. The interpolated cell-face velocity is thus modified by the difference between the interpolated pressure gradient and the gradient calculated at the cell face (see Eqs. (8.27) and (8.28)):

$$u_e^* = \overline{(u^*)}_e - \Delta V_e \overline{\left(\frac{1}{A_P^u}\right)}_e\left[\left(\frac{\delta p}{\delta x}\right) - \overline{\left(\frac{\delta p}{\delta x}\right)}\right]_e^{m-1} . \tag{8.58}$$

This correction to the interpolated cell-face velocity has become known as *Rhie–Chow correction* (Rhie and Chow 1983). An overbar denotes interpolation, and the volume centered around a cell face is defined by

$$\Delta V_e = (x_E - x_P)\,\Delta y$$

for Cartesian grids.

This procedure adds a correction to the interpolated velocity that is proportional to the third derivative of the pressure multiplied by $(\Delta x)^2/4$; the fourth derivative for cell center results from applying the divergence operator. In a second-order scheme, the pressure derivative at the cell face is calculated using CDS, see Eq. (8.51). If the CDS-approximation (8.51) is used on non-uniform grids, the cell-center pressure gradients should be interpolated with weights 1/2, because this approximation does not 'see' the grid non-uniformity.

The correction will be large if the pressure oscillates rapidly; the third derivative is then large and will activate the pressure-correction and smooth out the pressure.

The correction to the cell-face velocity in the SIMPLE method is now:

$$u'_e = -\Delta V_e \overline{\left(\frac{1}{A_P^u}\right)_e}\left(\frac{\delta p'}{\delta x}\right)_e = -S_e \overline{\left(\frac{1}{A_P^u}\right)_e}\,(p'_E - p'_P)\,, \qquad (8.59)$$

with corresponding expressions at other cell faces. When these are inserted into the discretized continuity equation, the result is again the pressure-correction equation (8.32). The only difference is that the coefficients $1/A_P^u$ and $1/A_P^v$ at the cell faces are not the nodal values, as in the staggered arrangement, but are interpolated cell-center values.

Because the correction term in Eq. (8.58) is multiplied by $1/A_P^u$, the value of the under-relaxation parameter contained in them may affect the converged cell-face velocity. However, there is little reason for concern, because the difference in the two solutions obtained using different under-relaxation parameters is much smaller than the discretization error, as will be shown in the examples below. We also show that the implicit algorithm using colocated grids has the same convergence rate, dependence on under-relaxation factors, and computing cost as the staggered grid algorithm. Furthermore, the difference between solutions obtained with different variable arrangements is also much smaller than the discretization error.

We have derived the pressure-correction equation on colocated grids for second-order approximations. The method can be adapted to approximations of higher order; it is important that the differentiation and interpolation be of the same order. For a description of a fourth-order method, see Lilek and Perić (1995).

8.2.3 IFSM for Colocated Grids

The IFSM method described earlier for staggered grids is easily extended to colocated grids. Actually, the pressure-correction equation is the same in both cases, see Eqs. (8.32) and (8.37). The only difference lies in the computation of cell-face velocities needed to compute the mass fluxes. In the case of staggered grids, velocities are stored at the faces of the continuity-CV, so no interpolation is needed. On colocated grids, face velocities have to be computed from nodal values using interpolation. When computing unsteady flows using IFSM, linear interpolation without Rhie–Chow correction usually works fine. However, in applications presented in Sect. 8.4 we applied the following correction to the linearly interpolated velocities at cell faces:

$$u_e^* = \overline{(u^*)}_e - \frac{\Delta t}{\rho} \left[\overline{\left(\frac{\delta p}{\delta x}\right)} - \overline{\left(\frac{\delta p}{\delta x}\right)} \right]_e^{m-1} . \tag{8.60}$$

This expression is the same as the one used in SIMPLE, except for a different multiplier of the term in square brackets, cf. Eqs. (8.58) and (8.60). The following relation between velocity and pressure correction is obtained for IFSM following the same approach as in SIMPLE:

$$u_e' = -\frac{\Delta t}{\rho} \left(\frac{\delta p'}{\delta x}\right)_e = -\frac{\Delta t}{\rho (x_E - x_P)} (p_E' - p_P') . \tag{8.61}$$

Analogous expressions follow for v_n^* and v_n'. A computer code that incorporates this algorithm is available in the Internet; see the appendix for details.

8.3   Non-iterative Implicit Methods for Unsteady Flows

Non-iterative implicit methods allow larger time steps to be used than with fully explicit methods. Because in the case of incompressible flow a Poisson-type equation for pressure or pressure correction has to be solved in any case, implicit methods are usually preferred. This is especially important when very fine grids are created locally (boundary layer near wall, around sharp edges etc.), because diffusion terms then induce a too stringent stability limit to the time-step size if the method is fully explicit.

The term "non-iterative" means here that the *outer iteration loop* is missing: the momentum and pressure-correction equation are solved only once per time step. Many versions of fractional-step method (FSM) fall into this category; PISO may or may not be considered non-iterative in this sense (it solves the momentum equation only once, but an outer iteration loop encompasses the pressure-correction equation and explicit velocity corrections).

The major differences between the IFSM presented earlier and the non-iterative versions of FSM are: (i) the latter use explicit time-advancing schemes for all or part of convection fluxes, and (ii) the linear equations are solved to a tighter tolerance than in iterative methods to ensure that iteration errors are small enough and both momentum and continuity equations are sufficiently satisfied. Non-iterative methods introduce a splitting error because they only correct velocities in the unsteady term; this error is usually proportional to $(\Delta t)^2$. Both SIMPLE and IFSM eliminate the splitting error through outer iterations. Usually, one time step in non-iterative FSM or

PISO costs (in terms of computing effort) as much as 2–3 outer iterations in SIMPLE or IFSM. We shall point to this fact again when presenting example computations in Sect. 8.4.

Many of the tools and algorithms described in the previous sections on SIMPLE and related methods can be used in fractional-step methods. Accordingly, this section is not as detailed as the previous ones, but rather focuses on some items typical of non-iterative fractional-step methods to highlight particular steps. In what follows we review the major steps in the method, beginning with the Adams–Bashforth discretization. Next, we tackle in order: the approximate factorization ADI, the pressure Poisson equation, and initial and boundary conditions. Finally, we comment on the differences in efficiency and accuracy between single-step and iterative fractional-step approaches.

Both colocated and staggered grids are considered, as in the previous sections. However, in the absence of a picture just at hand, it is easy to forget where the variables are placed and what the appropriate control volumes are for colocated and staggered grids. For example, given that the velocity components are on the cell faces in the staggered grid, one might naively wonder why it is necessary to use QUICK in the convection terms on a staggered grid. For this reason, we reproduce here in Fig. 8.4 the grids and relevant control volumes. For the colocated grid, we recall now

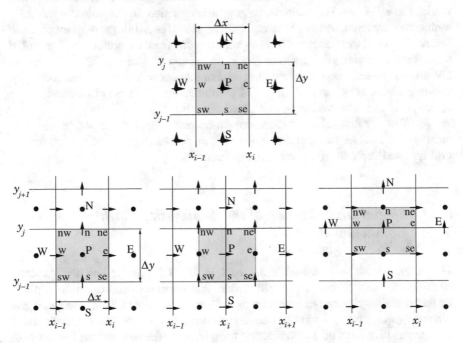

Fig. 8.4 Control volumes and notation: for all variables on colocated grids (upper), for scalar variables (lower left), for x-momentum (lower center) and for y-momentum (lower right) on staggered grids

that there is but one control volume for velocities and pressure and that all of the variables are defined at point P on that grid for the FV-method that we are using here. Thus, interpolation is needed to find variable values on the CV-faces because in the FV-method it is the *fluxes through the CV-faces* that are summed in each case. In the continuity equation, we need velocity components at CV-faces to compute mass fluxes; in all other transport equations, we need both the *convecting* mass flux and the *convected* variable (velocity components, temperature etc.) to compute the convection fluxes. Pressure also needs to be interpolated to obtain the pressure force at each cell face.

On the other hand, on staggered grids there are separate control volumes for each velocity component and for the continuity equation and scalar variables (e.g., temperature; pressure and fluid properties are also stored at nodes centered in this CV). In the lower part of Fig. 8.4, the velocities are known where the arrows are while the pressure is known where the dots are. For the continuity equation, no interpolation is needed: because velocities are stored at CV-faces, mass fluxes can be computed directly (to within a second-order approximation using midpoint rule). However, to compute convection fluxes in all other transport equations, interpolation is needed: for scalar variables (like temperature), we need to interpolate the variables stored at cell centers to obtain the convected value at the face centroid, while for velocity CVs, interpolation is needed to compute both the convecting (mass flux) and convected velocity at cell faces. For example, to get the convection flux at point 'e' in the x-momentum equation, we need to interpolate the velocities, while the relevant pressure forces at east and west faces can be computed directly from the pressure values stored there (on a Cartesian grid, pressure forces on 'n' and 's' faces of the x-momentum CV are not relevant because they have no component in x-direction; accordingly, pressure forces at 'e' and 'w' faces of y-momentum CV are not relevant, while forces at 'n' and 's' faces can be computed directly).

Note that, for integral approximations other than midpoint rule, interpolation of nodal variable values to integration points within cell faces would be needed for any grid type and variable arrangement in any case.

8.3.1   *Spatial Discretization of the Adams–Bashforth Convection Term*

The Adams–Bashforth discretization used in Eq. (7.50) is typical of fractional-step methods and crucial because it makes that portion of the equation explicit, while maintaining second-order accuracy in time. Both colocated and staggered grid schemes require interpolation to define the convection terms. For example, using the conservative form of the convection term requires the momentum flux on each face of the control volume. For a uniform Cartesian mesh then, the x-component is $(\rho uu)_e$ for the x-momentum CV in Fig. 8.4. Drawing from Eq. (4.24) and assuming that u_e is positive:

$$(\rho u u)_e = (\rho u)_e (u_e)_{\text{QUICK}} = (\rho u)_e \left(\frac{6}{8} u_P + \frac{3}{8} u_E - \frac{1}{8} u_W \right) =$$

$$\frac{(\rho u)_E + (\rho u)_P}{2} \left(\frac{6}{8} u_P + \frac{3}{8} u_E - \frac{1}{8} u_W \right) . \tag{8.62}$$

This result, *which is valid for either staggered or colocated grids*, can be compared to Eq. (8.7) in implicit iterative methods; note that in both cases, the QUICK-interpolation is applied to the *convected* quantity and not to the convecting velocity which defines the mass flux. The latter is usually interpolated linearly, when required (on colocated grids and for momentum-CVs on staggered grids).

Variable values required at other than nodal points are typically obtained by linear interpolation. However, other interpolations are sometimes useful, e.g., the face velocities in mass fluxes may be approximated by higher-order interpolations and/or pressure-weighted terms in the formation of the pressure Poisson equation (see, e.g., Rhie and Chow 1983; Armfield 1991; Zang et al. 1994; Ye et al. 1999). Likewise, the QUICK-term can be replaced by alternative discretizations such as those given in Sect. 4.4, e.g., UDS, CDS, CUI, or TVD.

8.3.2 An Alternating-Direction Implicit Scheme

The ADI-scheme, often called an approximate-factorization ADI-scheme, can be constructed by following the outline in Sect. 5.3.5. We transform Eq. (7.53) (for Cartesian coordinates) to

$$\left(\rho - \frac{\Delta t}{2} L \right) \mathbf{v}^* = \left(\rho + \frac{\Delta t}{2} L \right) \mathbf{v}^n + \Delta t \, [AP] , \tag{8.63}$$

where AP includes the known advection and pressure terms from previous time steps. Now, if we assume that density and viscosity are constant, we can re-arrange this equation to the form:

$$(1 - A_1 - A_2 - A_3) \mathbf{v}^* = (1 + A_1 + A_2 + A_3) \mathbf{v}^n + \frac{\Delta t}{\rho} [AP] , \tag{8.64}$$

where

$$A_1 = \frac{\mu \Delta t}{2\rho} \frac{\delta^2}{\delta x_1^2} , \quad A_2 = \frac{\mu \Delta t}{2\rho} \frac{\delta^2}{\delta x_2^2} , \quad A_3 = \frac{\mu \Delta t}{2\rho} \frac{\delta^2}{\delta x_3^2} .$$

Factoring Eq. (8.64) yields:

$$(1 - A_1)(1 - A_2)(1 - A_3) \mathbf{v}^* = (1 + A_1)(1 + A_2)(1 + A_3) \mathbf{v}^n +$$
$$\frac{\Delta t}{\rho} [AP] + O((\Delta t)^2)(\mathbf{v}^* - \mathbf{v}^n) . \tag{8.65}$$

However, $\mathbf{v}^* - \mathbf{v}^n \sim \Delta t \frac{\partial \mathbf{v}}{\partial t}$ so the last term is proportional to $(\Delta t)^3$ for small Δt and can be neglected. Equation (8.65) reduces the solution process to the sequential inversion of tridiagonal matrices in alternating directions. Thus, using the component notation for the velocities (cf. Eqs. (5.49) and (5.50)):

$$\left(1 - \frac{\mu \Delta t}{2\rho} \frac{\delta^2}{\delta x^2}\right) \overline{v_i}^x = \left(1 + \frac{\mu \Delta t}{2\rho} \frac{\delta^2}{\delta x^2}\right) v_i^n + \frac{\Delta t}{\rho} [AP] , \qquad (8.66)$$

$$\left(1 - \frac{\mu \Delta t}{2\rho} \frac{\delta^2}{\delta y^2}\right) \overline{v_i}^y = \left(1 + \frac{\mu \Delta t}{2\rho} \frac{\delta^2}{\delta y^2}\right) \overline{v_i}^x . \qquad (8.67)$$

$$\left(1 - \frac{\mu \Delta t}{2\rho} \frac{\delta^2}{\delta z^2}\right) v_i^* = \left(1 + \frac{\mu \Delta t}{2\rho} \frac{\delta^2}{\delta z^2}\right) \overline{v_i}^y . \qquad (8.68)$$

Each of these systems of equations has a tridiagonal coefficient matrix and thus can be solved with the efficient TDMA method; this requires no iteration. However, this solution method is only applicable to structured grids (both Cartesian and non-orthogonal ones); it cannot be applied with such simplicity and efficiency to unstructured grids.

8.3.3   The Poisson Equation for Pressure

The Poisson equation for pressure can be solved by any elliptic equation solver, e.g., multigrid, ADI, conjugate gradient, GMRES, etc. (see Chap. 5). Here, we address some issues related to the Poisson equation discretization.

1. For non-iterative fractional-step methods, Hirt and Harlow (1967) argued that, given that a Poisson equation is solved iteratively at every step to enforce continuity, one should take care to avoid accumulation of incompressibility errors. Their insight was that one can (1) carry the iterative solution of the equation to a high level of accuracy or (2) use some self-correcting procedure. Interestingly, for Eq. (7.44) it was noted that formulation of the pressure Poisson equation begins with inclusion of the divergence of the velocity at both the current (n) and future $(n + 1)$ time steps; we can show that the formulation laid out in Sect. 7.2.1 actually includes the ability to account for a divergent v^n. We begin by writing Eq. (7.53) without the viscous terms (they play no role in this illustration) as

$$(\rho \mathbf{v})^* = (\rho \mathbf{v})^n - \Delta t \left[\frac{3}{2} C(\mathbf{v}^n) - \frac{1}{2} C(\mathbf{v}^{n-1})\right] - \Delta t G(p^{n-1/2}) , \qquad (8.69)$$

and introducing it into the pressure Poisson equation (7.57) which was derived to force the divergence of $\mathbf{v}^{n+1}$ to be zero. The result is

$$D(G(p')) = \frac{D(\rho\mathbf{v})^*}{\Delta t} =$$

$$\frac{D(\rho\mathbf{v})^n}{\Delta t} - D\left(\left[\frac{3}{2}C(\mathbf{v}^n) - \frac{1}{2}C(\mathbf{v}^{n-1})\right]\right) - D(G(p^{n-1/2})) =$$

$$\frac{D(\rho\mathbf{v})^n}{\Delta t} - \cdots . \text{QED} \qquad (8.70)$$

Accordingly, any error in the divergence in the previous step is fed back as a correction in the next step. Hirt and Harlow (1967) show that this cuts down the accumulation of incompressibility errors in cases where, e.g., computing time is saved by terminating iteration with a smaller number of iterations and so less accuracy.

2. The set-up and strategy for staggered grids are essentially the same as those described in Sect. 8.1 so we shall not focus on staggered grids here; rather we spend our time on colocated grids. Issues and solutions related to the calculation of pressure on a colocated grid were discussed for SIMPLE and related schemes in Sect. 8.2.1. The checkerboard pressure pattern for the sparse Laplacian form was demonstrated and an alternative compact Laplacian was derived. However, there and in Sects. 7.1.3 and 7.1.5.1, the loss of energy conservation when the compact form is used was noted. We describe here two alternative formulations that do not experience an uncoupling of the pressure field; one uses the compact Laplacian with its accompanying continuity error and one uses a modified method with the sparse Laplacian.

3. On a colocated grid using the equations described in Sect. 7.2.1, the pressure Poisson equation (7.57) can be discretized in two ways.[1] We first discretize the equation in two dimensions by using CDS on the Laplacian directly; this yields (cf. Eq. (8.52)):

$$\left(\frac{p'_E - 2p'_P + p'_W}{(\Delta x)^2}\right) + \left(\frac{p'_N - 2p'_P + p'_S}{(\Delta y)^2}\right) =$$
$$\frac{1}{\Delta t}\left(\frac{(\rho u)_e - (\rho u)_w}{\Delta x} + \frac{(\rho v)_n - (\rho v)_s}{\Delta y}\right)^* . \qquad (8.71)$$

On the other hand, constructing the equation by using the discrete forms of the divergence and gradient operators as indicated in Sect. 7.1.5.1 yields (see Fig. 3.5 for $2\Delta x$, etc. grid points, namely, EE, WW, etc.)

$$\left(\frac{p'_{EE} - 2p'_P + p'_{WW}}{4(\Delta x)^2}\right) + \left(\frac{p'_{NN} - 2p'_P + p'_{SS}}{4(\Delta y)^2}\right) =$$
$$\frac{1}{\Delta t}\left(\frac{(\rho u)_e - (\rho u)_w}{\Delta x} + \frac{(\rho v)_n - (\rho v)_s}{\Delta y}\right)^* , \qquad (8.72)$$

---

[1]Both of the discretizations can be derived formally using FV methods, the differences being how the face fluxes on the control volume are defined. See, e.g., Ye et al. (1999) and Fletcher (1991), V.I, Sect. 5.2.

Using the compact Laplacian of Eq. (8.71) actually produces an error in continuity equal to

$$-\Delta t \left[(\Delta x)^2 \frac{\partial^4 p'}{\partial x^4} + (\Delta y)^2 \frac{\partial^4 p'}{\partial y^4} \right] =$$
$$-(\Delta t)^2 \left[(\Delta x)^2 \frac{\partial^5 p}{\partial t \partial x^4} + (\Delta y)^2 \frac{\partial^5 p}{\partial t \partial y^4} \right] \tag{8.73}$$

because $p' \sim \Delta t (\partial p / \partial t)$. The equivalent result was obtained in Sect. 8.2.1. In contrast, for staggered grids, the FV formulation using the discrete divergence and gradient forms gives both a compact Laplacian and no continuity error.

Equation (8.71) performed well in the *non-iterative* simulations of Armfield (2000); Armfield and Street (2005) and Armfield et al. (2010). There, at each time step, using the current p' values, the CV velocities are corrected such that, for example,

$$u_P^{n+1} = u_P^* - \frac{\Delta t}{2\rho \Delta x} (p_E' - p_W') \tag{8.74}$$

and each of the CV-face velocities is also corrected according to, for example,

$$(\rho u)_e^{n+1} = (\rho u)_e^* - \frac{\Delta t}{\Delta x} (p_E' - p_P') \tag{8.75}$$

which yields a divergence-free velocity field for these convecting velocities. It was observed that higher-order errors in the pressure appear to limit the growth of grid-scale error in the pressure field. Also, the feedback of continuity error into the next time step (Eq. (8.70)) reduces the accumulation of continuity error.

4. Use of the sparse form (8.72) is inefficient and leads to pressure oscillations in iterative colocated schemes (see Sect. 8.2.1). Armfield et al. (2010) made an improvement in the sparse scheme for the non-iterative case (cf. the staggered-grid fractional-step method of Choi and Moin 1994). The modified scheme is as follows (the original fractional-step procedure is described in Sect. 7.2.1): Solve Eqs. (7.50) and (7.51) by

a. finding an estimate of the new velocity $\mathbf{v}^*$ using the old value of the pressure $p^{n-1/2}$:

$$\frac{(\rho \mathbf{v})^* - (\rho \mathbf{v})^n}{\Delta t} + \left[\frac{3}{2} C(\mathbf{v}^n) - \frac{1}{2} C(\mathbf{v}^{n-1}) \right] =$$
$$-G(p^{n-1/2}) + \frac{L(\mathbf{v}^*) + L(\mathbf{v}^n)}{2}, \tag{8.76}$$

b. adding the old pressure gradient to the estimated velocity field

$$(\rho \hat{v})^* = (\rho \mathbf{v})^* + \Delta t \, G(p^{n-1/2}), \tag{8.77}$$

and so approximately canceling the pressure gradient in the previous equation. The cancellation would be exact in the absence of the implicit viscous term.

c. defining a correction to $\hat{v}^*$ of the form

$$(\rho v)^{n+1} = (\rho \hat{v})^* - \Delta t\, G(p^{n+1/2}) , \tag{8.78}$$

and finally

d. finding $p^{n+1/2}$ by substituting Eq. (8.78) into the continuity equation (7.51) to obtain

$$D(G(p^{n+1/2})) = \frac{D(\hat{v}^*)}{\Delta t} . \tag{8.79}$$

Using the sparse Laplacian (8.72) produces

$$\left(\frac{p_{EE} - 2p_P + p_{WW}}{4\Delta x^2}\right)^{n+1/2} + \left(\frac{p_{NN} - 2p_P + p_{SS}}{4\Delta y^2}\right)^{n+1/2} =$$
$$\frac{1}{\Delta t}\left(\frac{\hat{u}_e^* - \hat{u}_w^*}{\Delta x} + \frac{\hat{v}_n^* - \hat{v}_s^*}{\Delta y}\right) . \tag{8.80}$$

The initial pressure used at each time step in the Poisson equation solver should be the pressure from the previous time step.

In Sect. 7.2.1, we used the results of application of the fractional-step method to show in Eq. (7.59) that an $O(\Delta t)^2$ error $(1/2)\Delta t\, L(G(p'))$ was made, but it was consistent with the basic discretization. Following the same procedure here shows precisely the same error, i.e.,

$$\frac{1}{2}\Delta t\, L(G(p^{n+1/2} - p^{n-1/2})) = \frac{1}{2}\Delta t\, L(G(p')) . \tag{8.81}$$

This means that the extra pressure cancellation step in this procedure reduces the fractional-step error by an order of magnitude compared to the P1 method; again recall that $L = D(G(\ ))$. Results presented in Armfield et al. (2010) show that this sparse scheme has essentially no divergence error. However, their compact pressure-correction scheme (which was derived by directly differencing the Poison equation; see warnings in Sects. 7.1.3 or 7.1.5.1) has a divergence error that is approximately the same irrespective of the degree to which the pressure Poisson equation is converged.

Consistent with the result obtained for the methods' errors, both schemes have approximately the same accuracy. No grid scale oscillations were observed with either scheme, but solving for a new full pressure in the modified scheme prevents grid scale oscillations from accumulating in any case.

8.3.4 Initial and Boundary Conditions

The initial condition on the velocity should be divergence-free. Also, most of the fractional-step schemes employ the Adams–Bashforth scheme for convection. Because Adams–Bashforth schemes are multilevel methods, solutions cannot be started using only data at the initial time point. One has to use other methods to get the calculation started, e.g., the Crank–Nicolson method with iteration for the pressure; see Sect. 6.2.2. Often, the pressure-correction field computed at the first step can include a first-order-in-time error because the initial pressure is everywhere zero and prescribed at the wrong time, i.e., not at a $1/2$ time level; Fringer et al. (2003) give an example. However, this issue will not arise if a multilevel start is used because the full pressure is calculated at the right time, i.e., halfway through the time step, until enough temporal data is collected to use the regular method.

Boundary conditions are a more complex issue. For the original P1-schemes, special intermediate boundary conditions were needed after the first step of estimating the new velocity (Kim and Moin 1985; Zang et al. 1994). In general, however, the physical boundary conditions can be used for velocities and scalars, while for the pressure the conditions needed depend on the method as follows:

1. For staggered grids, no pressure boundary condition is required, but setting the gradient of the pressure correction (or pseudo-pressure in P1-schemes) normal to surfaces equal to zero is appropriate.[2]
2. For colocated schemes using auxiliary nodes outside boundaries, the normal momentum equation at the immediate interior node requires the pressure at the immediate exterior node, so a high-order extrapolation from the interior is used. Again, setting the gradient of the pressure correction (or pseudo-pressure in P1 schemes) normal to surfaces equal to zero is appropriate.[3]
3. For the tangential velocity at a wall, it is worth noting that, while a boundary condition can be set for the estimated velocity in the velocity-estimation step of the method, the projection step, which is essentially an irrotational step, cannot constrain the tangential velocity correction; thus, a small error is made (Armfield and Street 2002). It is less for schemes with a small fractional-step error (see Eq. (7.59) and (8.81)), i.e., when the correction by the divergence constraint is small because the original velocity estimate is better.

---

[2]This is true where there are Dirichlet *velocity* boundary conditions; at outflows where the velocity has a Neumann boundary condition, there are a number of options for pressure boundary conditions, depending on the flow. Professional code documentation usually describes these options. See also Sani et al. (2006).

[3]Ditto.

8.3.5   Iterative Versus Non-iterative Schemes

The question always arises as to whether or not there is a benefit in using iteration in the fractional-step method, i.e., returning after completion of the pressure correction and velocity update to use this information in a second (or third, ...) pass through the process. Two issues arise: accuracy and efficiency. Note that this discussion applies to classical fractional-step methods in which convection terms are approximated using explicit Adams–Bashforth schemes; iterative fully-implicit schemes were described in the previous section.

Iterative schemes were examined in the context of both staggered and colocated grids by Armfield and Street (2000, 2002, 2003, 2004). The results do not differ significantly with grid type, but do vary according to scheme. For all of the tests the iterative schemes are less efficient in that they require more CPU time to reach a prescribed level of accuracy, but because they remove the splitting error $(1/2)\Delta t\, L(G(p'))$ by driving the pressure correction or pressure difference in successive iterations to zero, the iterative methods are more accurate for a given grid and time-step.

Armfield and Street (2004) examined iterative, P2, and P3 schemes plus a new "pressure" scheme on staggered grids using the basic equation set (7.50) and (7.51).

P2 Method: The P2 method is that described by Eqs. (7.53–7.57) to obtain $\mathbf{v}^{n+1}$ and $p^{n+1/2}$ with the fixed projection error noted just above and given in Eq. (7.59).

Iterative Method: The iterative method simply cycles through the equations until the pressure correction is driven to zero and the solution satisfies both the momentum and continuity equations. Note that the stability limit of the iterative method is not significantly improved compared to the non-iterative version of the same algorithm, because the limit is imposed by the explicit Adams–Bashforth treatment of convection.

P3 Method: The P3 method is not iterative. A similar method was suggested by Gresho (1990), but apparently not implemented due to stability concerns. The basic idea is to get a better estimate for the pressure in the first step of the scheme by using a second-order extrapolation

$$\tilde{p}^{n+1/2} = 2p^{n-1/2} - p^{n-3/2} \tag{8.82}$$

so Eq. (7.53) becomes

$$\frac{(\rho\mathbf{v})^* - (\rho\mathbf{v})^n}{\Delta t} + \left[\frac{3}{2}C(\mathbf{v}^n) - \frac{1}{2}C(\mathbf{v}^{n-1})\right] = -G(\tilde{p}^{n+1/2}) + \frac{L(\mathbf{v}^*) + L(\mathbf{v}^n)}{2}, \tag{8.83}$$

and the pressure correction is given by

$$p^{n+1/2} = \tilde{p}^{n+1/2} + p'. \tag{8.84}$$

The pressure in the momentum equation is now approximated to second order in time and the fixed projection error is third order so the solution will be more accurate than that given by P2. Because of the second-order extrapolation for the pressure and for convection, the first two steps of the scheme need to use a different method; Kirkpatrick and Armfield (2008) use Crank–Nicolson for all terms in the momentum equation and iteration for the pressure.

Pressure Method: The idea here is to solve directly for the pressure correction first, given $p^{n-1/2}$, via the divergence of the momentum equations as

$$D(G(p')) = \frac{D(\rho \mathbf{v})^n}{\Delta t} - D(1.5\,C(\mathbf{v}^n) - 0.5\,C(\mathbf{v}^{n-1})) +$$
$$D(L(1.5\,\mathbf{v}^n - 0.5\,\mathbf{v}^{n-1})) - D(G(p^{n-1/2}))\,. \qquad (8.85)$$

The pressure $p^{n+1/2} = p^{n-1/2} + p'$ is then used in the momentum equations to find the new velocity. This completes the time step.

Each of the above schemes was solved using the tools described above, including QUICK, ADI (with four sweeps of the full system used), and GMRES. The results given here are from a test case of natural convection in a two-dimensional square cavity with Ra $= 6 \times 10^5$ and Pr $= 7.5$ (see Sect. 8.4); the flow development from the initial condition (fluid at rest with a constant temperature) to the state at $t = 2$ was simulated. The time step was varied to examine convergence, but the grid was kept fixed at a 50×50 uniform mesh. The code was run from a non-dimensionalized time $t = 0$ to 2 with steps varying in size from 0.003125 to 0.1. A benchmark with a time step $= 7.8125 \times 10^{-4}$ was run to allow assessment of error as the L_2 norm of the difference between a run and the benchmark. While the ADI-sweeps were limited, the number of GMRES-sweeps depended on the case; up to a hundred for the tightest convergence criteria on the non-iterative cases to only five sweeps per iterative step. More details can be found in the paper. The basic physics for natural convection is explained in the examples section to follow.

Figure 8.5 shows the error versus time step of the four schemes, where the error is the average of the pressure, velocity and temperature errors. It is clear that removing the projection error in the P3 and Pressure schemes is successful in bringing their error down to the level of that for the iteration scheme. On the other hand, Fig. 8.6

Fig. 8.5 Comparison of accuracy for four FV fractional-step schemes on a staggered grid (From Armfield and Street 2004; reprinted with permission)

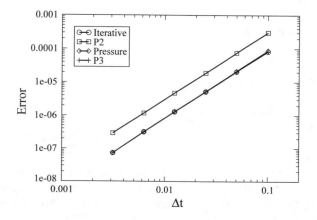

Fig. 8.6 Comparison of efficiency for four FV fractional-step schemes on a staggered grid (From Armfield and Street 2004; reprinted with permission)

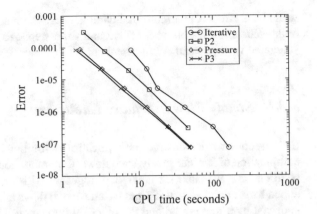

shows that the iterative scheme is the least efficient among the four, while the P3 and Pressure schemes are most efficient. In these last cases, while the schemes are not iterative, their set-up means that the effort to achieve a level of accuracy is significantly reduced, i.e., for a given level of accuracy they require only 50% of the CPU time of the iterative scheme and 60% of the time of the P2 scheme. Shen (1993) notes that a P3-like method could lead to solutions unbounded in time; that behavior was not observed in any of these tests or in the more extensive tests of Kirkpatrick and Armfield (2008).

The conclusion from this test case (unsteady flow developing from an initial condition towards a steady-state) may not be representative of all unsteady flows. However, it is fair to say that, for the same time step size, the non-iterative versions of the fractional-step method will typically need half the computing time of the iterative version. The fully-implicit iterative methods described in the previous section allow substantially larger time steps to be used than methods in which convection terms are treated explicitly (whether iterative or non-iterative); we present some results from both the fully-implicit iterative method and a version of P2 non-iterative method in the next section.

8.4 Examples

In this section, we give examples using three different solution techniques: SIMPLE and implicit iterative fractional-step method (IFSM) for steady-state flows and SIM-PLE, IFSM and a version of P2 non-iterative fractional-step method for unsteady flows. First, using staggered and colocated grids we treat steady flow in a square enclosure with two configurations. We begin with the lid-driven cavity and then examine a case where buoyancy is the driving force. Then, we use colocated grids to examine the features of unsteady, periodic flows driven by an oscillating lid or oscillating hot wall temperature. We assess both iteration and discretization errors, as well as the effects of various other parameters on both accuracy and efficiency (e.g., under-relaxation factors in SIMPLE and time-step size in fractional-step methods, staggered vs. colocated arrangement etc.).

8.4.1   Steady Flow in Square Enclosures

In this section we demonstrate the application of implicit iterative solution methods to computation of laminar steady-state flows. Computer codes used to obtain presented solutions are available for download, together with necessary input data; see the appendix for details. As test cases we chose two flows in square enclosures; one flow is driven by a moving lid and the other by buoyancy. The geometry and boundary conditions are shown schematically in Fig. 8.7. Both test cases have been used by many authors and accurate solutions are available in the literature; e.g., see Ghia et al. (1982) and Hortmann et al. (1990). We compare the performance of SIMPLE and the iterative fractional-step method (IFSM) and in particular demonstrate how to estimate iteration and discretization errors. Both staggered and colocated grids are used.

We first consider the lid-driven cavity flow, which is a popular benchmark problem. See Erturk (2009) for an insightful review. The moving lid creates a strong vortex and a sequence of weaker vortexes in the lower two corners (simulations at higher Reynolds numbers show that a third vortex forms in the upper left corner as well). A sample non-uniform grid and the streamlines for the Reynolds number, based on cavity height H and lid velocity U_L, $\mathrm{Re} = U_L H / \nu = 1000$, are shown in Fig. 8.8; calculations were done on a finer non-uniform grid as indicated.

Now we can look at the estimation of iteration errors. Several methods were presented in Sect. 5.7. First, an accurate solution was obtained by iterating until the residual norm became negligibly small (of the order of the round-off error in double precision). Then the calculation was repeated and the iteration error was computed as the difference between the converged solution obtained earlier and the intermediate solution.

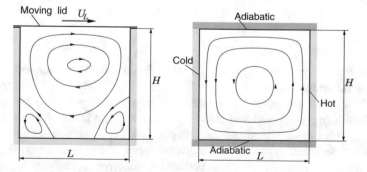

Fig. 8.7 Geometry and boundary conditions for 2D steady flow test cases: lid-driven (left) and buoyancy-driven (right) cavity flows

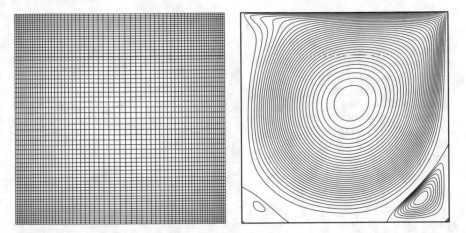

Fig. 8.8 A non-uniform grid with 64 × 64 CV used to solve the cavity flow problems (left) and the streamlines of the lid-driven cavity flow at Re = 1000 (right), calculated on a 256 × 256 CV non-uniform grid (the mass flow between any two adjacent streamlines in one vortex is constant)

Figure 8.9 shows the norm of the iteration error, the estimate obtained from Eqs. (5.89) or (5.96), the difference between two iterates, and the residual. The computation was performed on a 32 × 32 CV grid with under-relaxation factors 0.7 for velocity and 0.3 for pressure (not optimized for efficiency). Because the algorithm needs many iterations to converge, the eigenvalues required by the error estimator were averaged over the latest 50 iterations. The fields were initiated by interpolating the solution from the next coarser grid, which is why the initial error is relatively low.

This figure shows that the error-estimation technique gives good results for the non-linear flow problem. The estimate is not good at the beginning of the solution process, where the error is large. Using the absolute level of either the difference between two iterates or the residuals is not a reliable measure of the iteration error. These quantities do decrease at the same rate as the error, but need to be normalized properly to represent the iteration error quantitatively. Also, they fall very rapidly initially, while the error reduction is much slower. However, after a while all curves become nearly parallel and if one knows roughly the order of the initial error (it is the solution itself if one starts with a zero initial field), then a reliable criterion for stopping the iterations is the reduction of the norm of either the difference between two iterates or the residual by a certain factor, say three or four orders of magnitude. Results similar to those shown in Fig. 8.9 are obtained on other grids and for other flow problems.

We turn next to the estimation of discretization errors. We performed computation on five grids using CDS and UDS discretization; the coarsest had 16 × 16 CVs and the finest had 256 × 256 CVs. Both uniform and non-uniform colocated grids were used, and both SIMPLE and IFSM methods were applied. The strength of the primary vortex, ψ_{min}, which represents the mass flow rate between the vortex center and the

Fig. 8.9 Comparison of the
norms of the exact and
estimated iteration error for
the SIMPLE method, the
difference between two
iterates and the residual for
the solution of lid-driven
cavity flow at $Re = 10^3$ on a
32×32 CV grid, with CDS
discretization for both
convection and diffusion

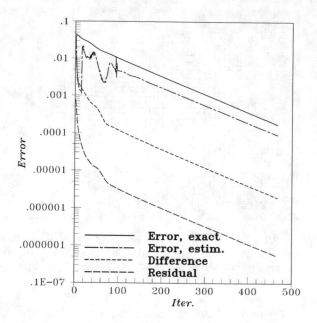

boundary, and the strength of the larger secondary vortex, ψ_{max}, were compared on
all grids. These values were estimated by finding the maximum and minimum value
of the streamfunction in cell corners. The streamfunction values were computed by
summing volume fluxes through cell faces: we set the value at the bottom-left corner
to zero and compute the value at the next vertex by adding the volume flux through
the face connecting the two vertexes. Figure 8.10 shows the variation of computed
vortex strengths as the grid is refined. Results on the four finest grids show monotone
convergence of both quantities towards the grid-independent solution. The results on
the non-uniform grids are obviously substantially more accurate than those from
uniform grids. The expansion ratio for cell growth from the wall towards cavity
center was 1.17166 on the coarsest and 1.01 on the finest grid; the value on the next
finer grid equals the square root of the value from the next coarser grid, in order to
ensure that grid lines from the coarser grid are retained in the finer grid (i.e., each
coarse grid CV contains exactly 4 fine grid CVs).

In order to enable quantitative error estimation, the grid-independent solution was
estimated using the results obtained on the two finest grids and Richardson extrapolation (see Sect. 3.9). These values are: $\psi_{min} = -0.11893$ and $\psi_{max} = 0.00173$.
Note that Richardson extrapolation applied to solutions obtained on either uniform
or non-uniform grids, using either SIMPLE or IFSM, delivered estimates which
were identical to four significant figures; differing by 0.007% for the primary and
0.027% for the secondary vortex (this applies to CDS-discretization). By subtracting
results on a given grid from the reference solution, an error estimate is obtained.
The errors are plotted against average grid size in Fig. 8.11. For both quantities on
both uniform and non-uniform grids, the error reduction expected of a second-order

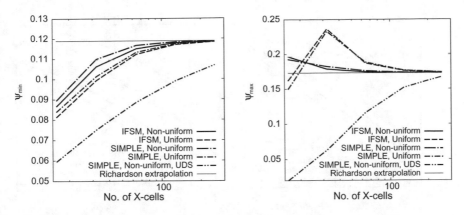

Fig. 8.10 Variation of the strength of primary (ψ_{min}; left) and secondary (ψ_{max}; right) vortexes in a lid-driven cavity flow at Re = 1000, computed using SIMPLE and IFSM solution methods and both uniform and non-uniform grids

method is obtained when CDS-discretization is used; for UDS, the error reduction is only approaching the asymptotic first-order convergence, because the errors are too large. The errors are lower on non-uniform grids, especially for ψ_{max}; because the secondary vortex is confined to a corner, the non-uniform grid, which is much finer there, yields higher accuracy. However, because this quantity is two orders of magnitude smaller than ψ_{min}, error estimates are less reliable (one would have to iterate to a much tighter tolerance to obtain accurate values of ψ_{max}).

Note that errors resulting from using first-order upwind discretization for convection are much higher than errors in solutions obtained using CDS. Even on the grid with 256×256 CVs (which can be considered as very fine) the error for ψ_{min}, using UDS, is around 10%, which is almost two orders of magnitude larger than errors resulting from using CDS; see Fig. 8.11.

The results obtained using SIMPLE and IFSM can only be distinguished from each other on the two coarsest grids; for all finer grids, differences are too small to be seen in graphs. This is expected because in both cases the same discretization is used; the two methods differ only in the pressure-correction equation, which is only expected to affect the convergence rate of the iterative solution process, not the solution itself. We therefore present in Fig. 8.12 only centerline velocity profiles obtained using IFSM. The use of second-order CDS leads to a monotone convergence; the ratio of errors on consecutive grids is a factor of four. The velocity profiles for the two finest grids can hardly be distinguished from each other.

We next investigate the difference between solutions obtained on uniform colocated and staggered grids using CDS discretization. Because the velocity nodes are at different locations on the two grids, staggered values were linearly interpolated to the cell centers (a higher-order interpolation would have been better but linear interpolation is good enough). The average difference for each variable ($\phi = (u_x, u_y, p)$) was determined as:

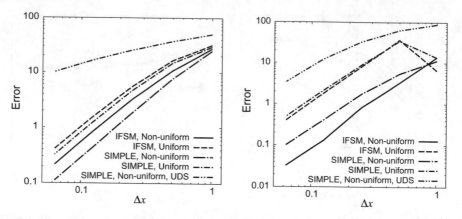

Fig. 8.11 Estimated discretization errors in solutions for the strength of primary ($\psi_{\min}$; left) and secondary ($\psi_{\max}$; right) vortexes in a lid-driven cavity flow at Re = 1000, computed using SIMPLE and IFSM solution methods and both uniform and non-uniform grids

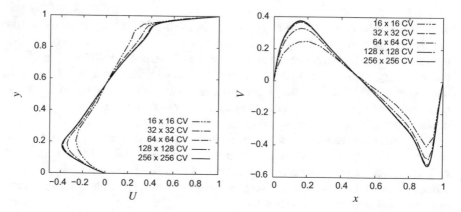

Fig. 8.12 Profiles of u_x velocity along the vertical centerline (left) and of u_y velocity along the horizontal centerline (right) in the lid-driven cavity flow at Re = 1000, calculated on five non-uniform grids using IFSM

$$\epsilon = \frac{\sum_{i=1}^{N} |\phi_i^{\text{stag}} - \phi_i^{\text{col}}|}{N}, \tag{8.86}$$

where N is the number of CVs. For both u_x and u_y, ϵ was on any grid an order of magnitude smaller than discretization errors on the same grid. The differences in pressure were somewhat smaller (no interpolation was necessary).

Convergence properties of the SIMPLE and IFSM methods using CDS and colo-cated non-uniform grids are investigated next. The SIMPLE algorithm has two adjustable parameters: under-relaxation for velocities and under-relaxation for pressure. IFSM does not require under-relaxation for any variable and has only one parameter—the time step (usually expressed in normalized form as a CFL-number,

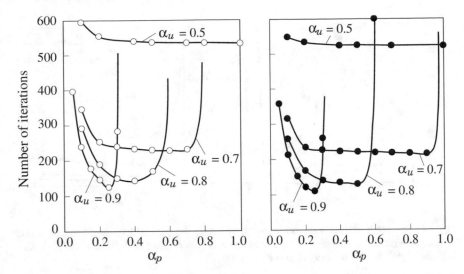

Fig. 8.13 Numbers of outer iterations required to reduce the residual level in all equations three orders of magnitude using SIMPLE with various combinations of under-relaxation parameters and a 32 × 32 CVs uniform grid with staggered (left) and colocated (right) arrangement of variables (lid-driven cavity flow at Re = 1000)

$CFL_x = u_x \Delta t / \Delta x$ or $CFL_y = u_y \Delta t / \Delta y$). We first examine the effect of the under-relaxation parameter for pressure, α_p, see Eq. (7.84), on the convergence of SIMPLE using various under-relaxation parameters for the velocity. Figure 8.13 shows the numbers of outer iterations required to reduce the residual level in all equations three orders of magnitude using various combinations of under-relaxation parameters and a 32 × 32 CVs uniform grid.

This figure shows that the dependence on the under-relaxation parameter α_p is almost the same for the two types of variable arrangements, although the range of good values is somewhat wider for the colocated grid. When the velocity is more strongly under-relaxed, we can use any value of α_p between 0.1 and 1.0, but the method converges slowly. For larger values of α_u, the convergence is faster but the useful range of α_p is restricted. The value of α_p suggested by Eq. 7.89 is nearly optimum; $\alpha_p = 1.1 - \alpha_u$ gives the best results for this flow. Usually, one varies one parameter and determines the other from the above relation.

In Fig. 8.14 we show the effect of the under-relaxation factor for velocity, α_u, on convergence rate of SIMPLE and the effect of CFL-number on the convergence of IFSM for the colocated arrangement of variables and non-uniform grids, in the case of the lid-driven cavity flow at Re = 1000. Similar graphs are obtained for staggered variable arrangement and for uniform grids. For both solution methods, the number of required iterations increases as the grid is refined. In the case of SIMPLE, the optimum value of under-relaxation is close to, but smaller than 1; iterations diverged for $\alpha_u = 0.99$ on all grids, and also for $\alpha_u = 0.98$ on the two finest grids. The finer the grid, the steeper is the increase in the required number of iterations as α_u is reduced

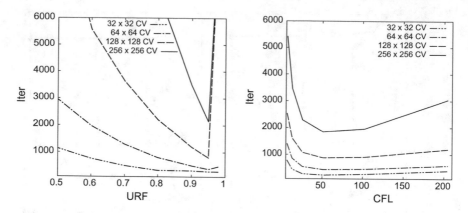

Fig. 8.14 Numbers of outer iterations required to reduce the residual level in all equations four orders of magnitude, as a function of the under-relaxation factor α_u in SIMPLE (left) or CFL-number in IFSM (right); lid-driven cavity flow at Re = 1000

below optimum. IFSM appears to be somewhat less sensitive to the value of CFL-number within a reasonable range: many iterations are required if CFL is below 10, but between 20 and 200, the increase is moderate, especially for coarser grids. We show in Chap. 12 how one can improve the efficiency of computations on fine grids using multigrid methods.

We next investigate the influence of the under-relaxation parameter for velocity in SIMPLE and of CFL-number in IFSM on the solution for the colocated variable arrangement and non-uniform grids. We compare solutions obtained with $\alpha_u = 0.95$ and $\alpha_u = 0.6$ in SIMPLE, and for CFL = 6.4 and CFL = 102.4 in IFSM (the CFL-number was determined using lid velocity and average grid spacing, as would be on a uniform grid; for the non-uniform grids used, the maximum CFL-number found in the whole grid was about 50% higher). The difference in the solutions (after residual levels were reduced over four orders of magnitude, to minimize the effect of iteration errors) was evaluated using the expression (8.86). For SIMPLE, we obtained for velocities ϵ-values around 5×10^{-4} on a grid with 32×32 CV, 8×10^{-5} on a grid with 64×64 CV, and 1.5×10^{-5} on a grid with 128×128 CV. For IFSM, differences in solutions were larger; we obtained for velocities ϵ-values around 2×10^{-2} on a grid with 32×32 CV, 6×10^{-3} on a grid with 64×64 CV, and 1.4×10^{-3} on a grid with 128×128 CV. These differences are much smaller than the discretization errors on corresponding grids (around 10%, 2.5% and 0.6%, respectively) and are reducing with grid refinement at the same rate at which discretization errors are reduced; therefore, they can be neglected.

The dependence of steady-state solution on under-relaxation parameter or time step size stems from the fact that these parameters influence the correction of inter-polated velocities used to compute mass fluxes through cell faces when the colocated arrangement of variables is used. The correction after Rhie and Chow (1983) is more or less the standard approach to avoiding pressure-velocity decoupling and it was

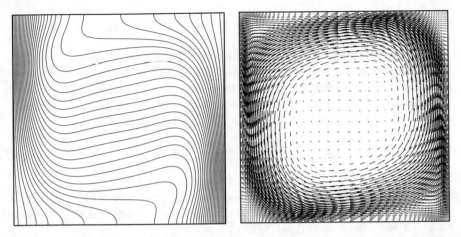

Fig. 8.15 Isotherms (left) and velocity vectors (right) in buoyancy-driven cavity flow at the Rayleigh number Ra = 10^5 and Prandtl number Pr = 0.1 (temperature differences between any two adjacent isotherms are the same)

also used in the computations presented above. However, there are other approaches to avoiding pressure or velocity oscillations on colocated grids (see Armfield and Street 2005). Also, it is possible to fix the interpolation to always correspond to no under-relaxation and be independent of time-step size; see Pascau (2011) for a detailed discussion and one solution approach, and Tukovic et al. (2018) for extension to moving grids. In most applications, such a dependence causes no problems because it is smaller than discretization errors; however, when very small time steps are used while flow is not changing in time, the problems may occur and one of the cures presented in the cited references may become necessary.

We next consider 2D buoyancy-driven flow in a square cavity as shown in Fig. 8.15. We now need to solve the energy equation coupled to Navier–Stokes equations, because the velocity field depends on temperature distribution within the solution domain. The energy equation for an incompressible fluid considered here is represented by the generic scalar transport equation which we considered so far; we only need to set the diffusivity equal to the ratio of viscosity and Prandtl number. The momentum equation for the velocity component in the direction of gravity gains an additional source term; with the Bousinesq-approximation used in this example, this source term is (see an explanation at the end of Sect. 1.4):

$$q_i = \beta \rho_{\text{ref}} g_i (T - T_{\text{ref}}) , \qquad (8.87)$$

where β is the volumetric expansion coefficient, g_i is the gravity component in i-direction and T_{ref} is the reference temperature at which the density ρ_{ref} is computed as a function of temperature. The density is then considered constant, while the above source term in the momentum equations represents the effect of (linearized) density variation around ρ_{ref}. Note that such an approximation is only viable if the density variation is nearly linear; this can be assumed only for a relatively small temperature range when flow in liquids is simulated.

The same five grids used for lid-driven cavity flow computations are re-used in this test case. The cold and hot walls are isothermal. The heated fluid is rising along the hot wall, while cooled fluid is falling along the cold wall. The Prandtl number is 0.1 (meaning that the thermal diffusivity dominates), and the temperature difference and other fluid properties are chosen such that the Rayleigh number is

$$\text{Ra} = \frac{\rho^2 g \beta (T_{\text{hot}} - T_{\text{cold}}) H^3}{\mu^2} \, \text{Pr} = 10^5 \, . \tag{8.88}$$

Here, the following values were used: $T_{\text{hot}} = 10$, $T_{\text{cold}} = 0$, $\rho = 1$, $\beta = 0.01$, $g = 10$, $\mu = 0.001$ and $H = 1$ (SI-units used throughout). For each grid, computations start with initial fields $u_x = 0$, $u_y = 0$ and $T = 6$.

Predicted velocity vectors and isotherms are shown in Fig. 8.15. The flow structure depends strongly on the Prandtl number. A large core of almost stagnant, stably-stratified fluid is formed in the central region of the cavity. Note that isotherms approach adiabatic (top and bottom) walls at a right angle; this must be so at any boundary where heat flux is zero (usually adiabatic walls and symmetry planes). When examining numerical solutions for plausibility, this is one of the features to be checked. We have seen (even in journal publications) results that violate this condition. Where isotherms are more dense, temperature gradient in the direction normal to isotherms is higher and thus the heat flux by conduction is also higher. This is visible in Fig. 8.16, which shows the variation of local heat flux per unit area along the cold wall (heat flux is negative because heat "enters" solution domain): heat transfer is much more intense where hot fluid meets the cold wall near the top (see also isotherms and velocity vectors in Fig. 8.15) than near bottom, where cooled-down fluid leaves the cold wall.

Figure 8.16 shows the dependence of solution on grid fineness. The lines representing solutions on the two finest grids cannot be visually distinguished from each other, indicating a high accuracy. An estimate of discretization errors is presented next.

It is to be expected that non-uniform grids will give more accurate results than uniform grids. This is indeed so. Figure 8.17 shows total heat flux through the isothermal walls and the estimated discretization errors using Richardson extrapolation as a function of grid fineness for both uniform and non-uniform grids, computed using either SIMPLE or IFSM solution methods. Richardson extrapolation yields the same estimate of the grid-independent value to the five significant digits when applied to the results of two finest levels for both grid types and solution methods. This estimate is $Q = 0.39248$, which, when normalized by the heat flux for pure heat conduction, $Q_{\text{cond}} = 0.1$, gives the Nusselt number $\text{Nu} = 3.9248$. By subtracting the solutions on all grids from the estimated grid-independent solution, we obtain an estimate of the discretization error. The errors in predicted total heat flux are plotted in Fig. 8.17. All errors tend asymptotically to the slope expected for second-order schemes (when grid spacing is reduced by one order of magnitude, the error is reduced two orders of

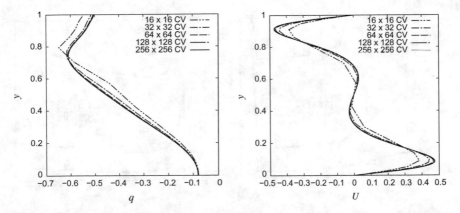

Fig. 8.16 Predicted variation of local heat flux per unit area along cold wall (left) and of u_x velocity along the vertical centerline (right) in the buoyancy-driven cavity flow at Ra $= 10^5$, calculated on five non-uniform grids using IFSM

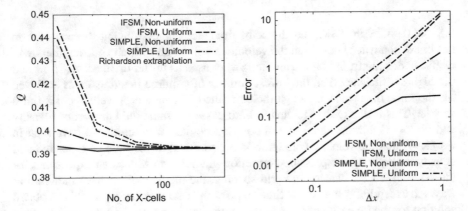

Fig. 8.17 Heat flux through isothermal walls, Q, (left) and the associated discretization error (right) in a buoyancy-driven cavity flow at the Rayleigh number Ra $= 10^5$ and Prandtl number Pr $= 0.1$, as a function of grid fineness (calculation using CDS on both uniform and non-uniform grids, with either SIMPLE or IFSM solution method)

magnitude). The error in the heat flux is substantially smaller on non-uniform than on uniform grids, because these grids resolve the boundary layer better.

Note that, for any given grid, the error in the strength of the primary vortex in lid-driven cavity flow was larger for IFSM than for SIMPLE (see Fig. 8.11), while in the case of buoyancy-driven flow, the error in the total heat flux is smaller for IFSM than for SIMPLE (see Fig. 8.17). The difference between solutions obtained with either method is, however, relatively small in both cases.

We check again the efficiency of the two solution methods and its dependence on under-relaxation factors in SIMPLE and CFL-number in IFSM. As for the lid-driven cavity, the under-relaxation factor for velocity in SIMPLE was varied between

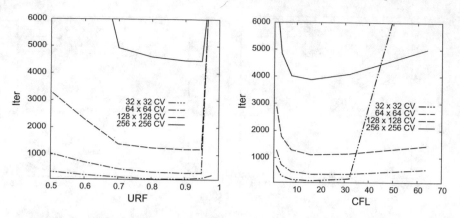

Fig. 8.18 Numbers of outer iterations required to reduce the residual level in all equations four orders of magnitude, as a function of the under-relaxation factor α_u in SIMPLE (left) or CFL-number in IFSM (right); buoyancy-driven cavity flow at Ra $= 10^5$

0.5 and 0.98. With IFSM, the choice of time steps is less obvious because we do not know in advance how high the velocities will be and which CFL-numbers will result. Interestingly, the same time steps were used as for lid-driven cavity flow and the highest efficiency (i.e., the lowest number of required iterations) was achieved for the same time steps, although the computed CFL-numbers were about 3 times smaller. This is due to the fact that in both cases the same fluid properties (density and viscosity) and the same size of solution domain were used. We have seen in Eq. (7.94) that the velocity correction is proportional to the ratio of time step and density multiplying the pressure-correction gradient; thus, one can expect that for the same time step, density and grid spacing, the method will behave in a similar way. An exception is the largest time step for the coarse grids: we could not obtain solution for the time step of 6.4 s in the buoyancy-driven cavity but it was possible for the lid-driven cavity flow.

Figure 8.18 shows the outcome of the efficiency analysis. Because IFSM uses no under-relaxation, the time step (CFL-number) remains the only parameter; on the other hand, in SIMPLE another factor appears—the under-relaxation factor for temperature. If it is set equal to the under-relaxation factor for velocity, convergence becomes very slow. However, we know from experience that the energy equation behaves well in laminar flows and usually requires very little or no under-relaxation. Therefore, here we set the under-relaxation factor for energy equation to 0.99; with this value, the efficiency of SIMPLE is comparable to that of IFSM.

The efficiency of calculating steady incompressible flows using implicit methods like SIMPLE or IFSM can be substantially improved by using the multigrid method for outer iterations, and by starting computations on finer grids with the solution interpolated from a coarser grid, rather than starting with guessed initial fields, as was done here. This will be demonstrated in Chap. 12 for the two test cases studied here.

8.4.2 Unsteady Flow in Square Enclosures

In this section we turn our attention to the prediction of unsteady flows. When marching toward steady-state flow, we could use large time steps and perform only one iteration per time step, because intermediate solutions were of no relevance, as long as the iteration procedure converged. However, when computing unsteady flows we have to ensure that all equations are satisfied to a sufficient degree before proceeding to the next time step. This means that, on one hand, we have to select an appropriate time step which adequately resolves the temporal variation of variables and, on the other hand, we need to perform a sufficient number of iterations per time step in order to ensure that iteration errors are small enough.

For time-accurate flow simulations, the time-integration scheme should be of at least second order. First-order implicit Euler scheme is useful for marching towards steady state (e.g., in IFSM), but in periodic flows it requires too small time steps to obtain satisfactory accuracy. The usual choices are the Crank–Nicolson scheme (which is almost exclusively used in fractional-step methods) and the fully-implicit three-time-level-scheme (which is almost exclusively used in SIMPLE-type methods). Central differences are used for both convection and diffusion terms.

The non-iterative fractional-step method used differs slightly from the method described in Sect. 7.2.1 in that it is based on the fully-implicit three-time-level scheme rather than the usual Crank–Nicolson scheme (the computer code is available in the Internet; see appendix for details):

$$\frac{3(\rho \mathbf{v})^{n+1} - 4(\rho \mathbf{v})^n + (\rho \mathbf{v})^{n-1}}{2\Delta t} + C(\mathbf{v}^{n+1}) = L(\mathbf{v}^{n+1}) - G(p^{n+1}) . \tag{8.89}$$

In order to make the method non-iterative, the non-linear convection fluxes at the time level t_{n+1} are estimated using Adams–Bashforth scheme; the expression is slightly different from the one given in Eq. (7.52):

$$C(\mathbf{v}^{n+1}) \sim 2C(\mathbf{v}^n) - C(\mathbf{v}^{n-1}) + O(\Delta t^2) . \tag{8.90}$$

The rest of the algorithm follows the same route as described in Sect. 7.2.1.

The first flow problem is the lid-driven cavity flow with an oscillating lid: it moves sinusoidally with a velocity

$$u_L = u_{max} \sin(\omega t) , \tag{8.91}$$

where $\omega = 2\pi/P$, with P being the period of oscillation. We set here $u_{max} = 1$ and $P = 10$ s; the cavity height is 1 and viscosity is set to $\mu = 0.001$ (SI units used throughout), leading to the Reynolds number varying between 0 and 1000. The solution is initialized with zero values for both velocities and pressure (fluid at rest). A non-uniform grid with 128×128 CVs was used (the same grid was used to analyze steady-state flows in the previous section).

We first analyze temporal accuracy of implicit Euler and three-time-level schemes by computing the first quarter of lid oscillation (from $t = 0$ to $t = 2.5$ s, corresponding to $\omega t = \pi/2$) with different time step sizes. Fully-implicit methods (SIMPLE and IFSM) allow large time steps to be used and we therefore start with $\Delta t = 0.125$ s (corresponding to 80 time steps per oscillation period of 10 s) and halve it 4 times. The finest time step ($\Delta t = 0.0078125$ s) corresponds then to 1280 time steps per oscillation period. The non-iterative fractional-step method (which we labeled here EFSM, because it is explicit for convection fluxes) cannot work with time steps larger than 0.0125 s, so with this method we used time steps 0.0125 s, 0.00625 s and 0.003125 s.

For assessing discretization errors, it is convenient to use an integral quantity. We chose here to evaluate differences in the strength of the main vortex generated by the moving lid at $t = 2.5$ s. It is computed by integrating mass fluxes to obtain streamfunction values at cell vertexes (starting with value 0 at the bottom-left corner); the minimum value in the solution domain represents the vortex strength for the given direction of lid motion. The grid-independent value was estimated using Richardson extrapolation and the values obtained on the two finest grids; all methods predict this value to be $\psi_{min} = -0.043682$. By subtracting values computed with different methods and time steps from this reference value, discretization errors were estimated.

Figure 8.19 shows the variation of predicted vortex strength and the associated temporal discretization errors as a function of time-step size and the method used. In this example it appears that the results obtained with SIMPLE are slightly more accurate than those obtained with IFSM for each time step; however, this is not always the case—when we repeated this exercise for the 10 times higher viscosity (Reynolds number 10 times smaller), IFSM-results had smaller errors. Both methods show the same behavior, with both time-integration schemes. Obviously, the first-order implicit Euler scheme (here labeled IE) leads to errors which are an order of magnitude larger than when the second-order three-time-level scheme (here labeled TTL) is used. Errors are halved when time step is halved with IE and they are reduced by a factor of four with TTL-scheme.

We forced the codes to reduce residuals four to five orders of magnitude in each time step to ensure that iteration errors are negligible; this is far more than one would normally do, especially when time steps are small, and therefore we have not paid much attention to computational efficiency. However, one can state that IFSM needed slightly less computing time per time step than SIMPLE for reaching the same level of residuals. For the same time step, EFSM needed less computing time than the iterative methods; however, IFSM had the lowest effort required to reach prescribed accuracy. Note that IFSM and SIMPLE already reach a level of discretization error around 0.1% with time steps for which EFSM would not work for stability reasons.

Figure 8.20 shows velocity vectors at times corresponding to $\omega t = \pi/2$ ($u_L = 1$), $\omega t = \pi$ ($u_L = 0$), $\omega t = 3\pi/2$ ($u_L = -1$), and $\omega t = 2\pi$ ($u_L = 0$), created during the fifth oscillation period after the simulation was started. At this time the flow is fully periodic, which can be recognized by the perfect symmetry of solutions which

Fig. 8.19 Strength of the primary vortex at $t = 2.5$ s (upper) and temporal discretization errors (lower) in lid-driven cavity flow with oscillatory lid motion as function of time-step size

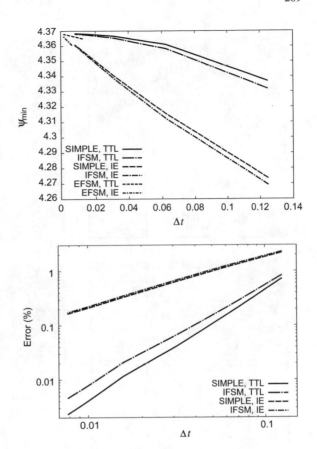

are π-apart in phase. Obviously, a very complicated change in flow pattern takes place during one period and the numerical method needs to capture those changes accurately.

The error analysis presented in Fig. 8.19 suggests that accurate solutions are produced by implicit iterative schemes (SIMPLE and IFSM) with ca. 100 time steps per oscillation period. Indeed, as can be seen in Fig. 8.21, the solution changes slightly only at the sharp peaks as the time step is reduced. The solution obtained using the first-order implicit Euler scheme shows substantial deviation from solutions obtained using the second-order three-time-level scheme, as expected. However, it is remarkable that the second-order scheme reproduces such a complex temporal variation in solution with ca. 60 time steps per oscillation period.

The second unsteady flow which we briefly tackle is buoyancy-driven cavity flow from the previous section, but with a sinusoidally varying hot wall temperature:

$$T_H = T_0 + T_a \sin(\omega t) , \tag{8.92}$$

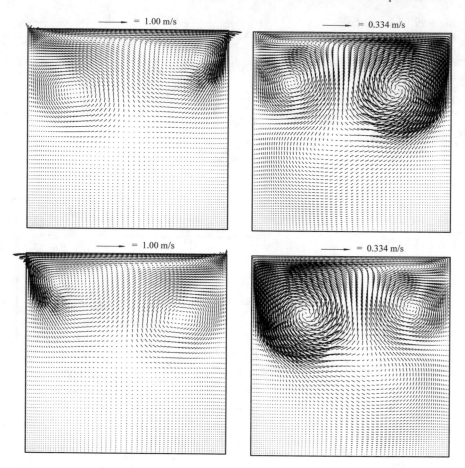

Fig. 8.20 Velocity vectors in the lid-driven cavity flow with an oscillatory lid motion at a quarter, half, three quarters and full period (from left to right, top to bottom, respectively)

where $T_0 = 10$ is the mean hot wall temperature, $T_a = 5$ is the amplitude of temperature fluctuation and $\omega = 2\pi/P$, with $P = 10$ s being the period of oscillation. The fluid is initially at rest and the temperature was set to 5.1, with cold wall temperature being $T_C = 0$ and hot wall temperature being $T_H = 10$. The mean Rayleigh number is the same as in the steady-state flow presented in the preceding section, but due to the periodically varying hot wall temperature, the flow pattern changes significantly over the time. After a few periods, the flow "forgets" the effects of the initialization and becomes fully periodic.

Figure 8.22 shows velocity vectors and Fig. 8.23 isotherms at times corresponding to $\omega t = \pi/2$ ($T_H = 15$), $\omega t = \pi$ ($T_H = 10$), $\omega t = 3\pi/2$ ($T_H = 5$), and $\omega t = 2\pi$ ($T_H = 10$), created when the flow became fully periodic; it is not symmetric (as it was in the case of lid-driven cavity flow), because here only the hot wall temperature

Fig. 8.21 Time history of u_x-velocity at the monitoring location ($x = 0.08672$, $y = 0.90763$) during two oscillation periods: dependence of solution on time-step size (upper; IFSM) and on the scheme used with the largest time step (lower; 62.5 time steps per oscillation period)

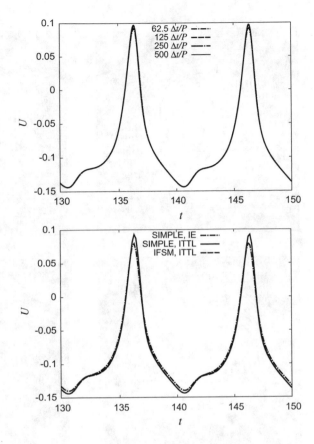

oscillates, but the solutions at times t and $t + P$ are the same. A complicated change in flow pattern takes place during one period, making its prediction a non-trivial task.

Note that isotherms remain always orthogonal to the adiabatic top and bottom boundary, as required for a zero-heat-flux boundary condition. The changing spacing between isotherms shows how the heat flux locally changes: dense isotherms indicate a high temperature gradient in the direction normal to isotherms. In the case of steady-state flow with constant hot and cold wall temperatures, the net heat flux was the same at both isothermal walls. Here the heat flux varies with time and is different at the two walls, because fluid accumulates heat over some part of the period and gives it away during the remaining part. Normally, heat comes in through the hot wall, but due to the oscillatory change of hot wall temperature, there is a short period when the fluid flowing along hot wall is actually hotter than the wall itself, so that the net heat flux changes sign—the cavity looses heat through both walls. This is visible from Fig. 8.24, which shows the variation of total heat flux through the hot wall over two periods: over a short period of time, the heat flux is positive, indicating outgoing flux. Note also that the maximum heat flux through hot wall occurs before the temperature reaches maximum (which happens at the quarter of

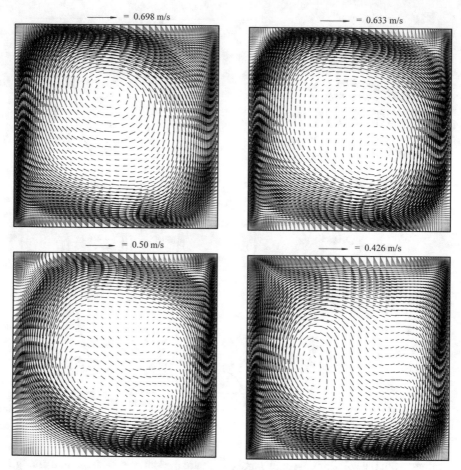

Fig. 8.22 Velocity vectors in the buoyancy-driven cavity flow with an oscillatory hot wall temperature at a quarter, half, three quarters and a full period (from left to right, top to bottom, respectively)

oscillation period, here at $t = 182.5$ and $t = 192.5$ s): there is a phase shift between velocity and temperature field variation.

Figure 8.24 also shows that the second-order time integration scheme already produces a very accurate solution with 62.5 time steps per oscillation period: the three curves corresponding to three different time steps cannot be distinguished in the graph. This is confirmed in Fig. 8.25, which shows u_x-velocity at one monitoring location during two periods, computed using the largest time step with three methods. One can see that there is a significant difference between solutions obtained with first-order implicit Euler scheme and second-order three-time-level scheme: the former smears peaks and does not reach the lowest value due to its first-order error causing numerical diffusion in time. The difference between SIMPLE and IFSM using the second-order scheme is negligible; it becomes even smaller for smaller time steps.

Fig. 8.23 Isotherms in the buoyancy-driven cavity flow with an oscillatory hot wall temperature at a quarter, half, three quarters and a full period (from left to right, top to bottom, respectively)

Fig. 8.24 Heat flux through the hot wall as a function of time during two oscillation periods, computed using IFSM and three-time-level second-order scheme with three different time steps (indicated is the number of time steps per oscillation period)

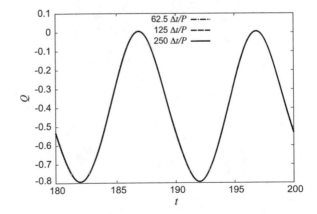

Fig. 8.25 Time history of u_x-velocity at the monitoring location ($x = 0.15479$, $y = 0.83722$) during two oscillation periods: dependence of solution on the time-integration scheme used with the largest time step (62.5 time steps per oscillation period)

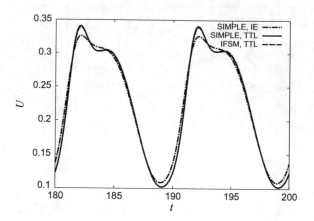

We have dealt so far only with rectangular solution domains and Cartesian grids. Real engineering problems are rarely so simple; in most cases, flow and heat transfer takes place in rather complicated geometries. We explain in the next chapter how the methods described so far can be extended to non-Cartesian structured or arbitrary polyhedral unstructured grids, which can be applied to more complicated solution domain shapes.

Chapter 9
Complex Geometries

Most flows in engineering practice involve complex geometries which are not readily fit with Cartesian grids. Although the principles of discretization and solution methods for algebraic systems described earlier remain valid, there are many different possibilities for their realization. The properties of the solution algorithm depend on the choices of the grid and of the vector and tensor components, and the arrangement of variables on the grid. These issues are discussed in this chapter.

9.1 The Choice of Grid

When the geometry is regular (e.g., rectangular or circular), choosing the grid is simple: the grid lines usually follow the directions defined by boundaries, which coincide with appropriate coordinate directions. In complicated geometries, the choice is not at all trivial. The grid is subject to constraints imposed by the discretization method. If the algorithm is designed for curvilinear orthogonal grids, non-orthogonal grids cannot be used; if the CVs are required to be quadrilaterals or hexahedra, grids consisting of triangles and tetrahedra cannot be used, etc. When the geometry is complex and the constraints cannot be fulfilled, compromises have to be made. However, if the discretization and solution method are both designed for arbitrary polyhedral control volumes, any grid type can be used.

9.1.1 Stepwise Approximation of Curved Boundaries

The simplest computational methods are those that use orthogonal grids (Cartesian or polar-cylindrical). However, in order to apply such a grid to solution domains with inclined or curved boundaries, additional approximations are necessary: either the boundaries have to be approximated by staircase-like steps, or the grid extends

© Springer Nature Switzerland AG 2020
J. H. Ferziger et al., *Computational Methods for Fluid Dynamics*,
https://doi.org/10.1007/978-3-319-99693-6_9

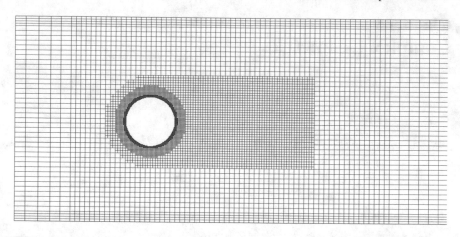

Fig. 9.1 An example of a grid using stepwise approximation of a curved wall boundary, with a local refinement to reduce step height

beyond boundaries and for the cells in their vicinity special approximations have to be used.

The first approach is still sometimes used, but it raises two kinds of problems:

- The number of grid points (or CVs) per grid line is not constant, as it is in a fully regular grid. Thus, although the cells are regular, the major advantage of regular grids is lost as some kind of indirect addressing or special arrays that limit the index range on each grid line have to be used. The computer code may need to be changed for each new problem.
- The approximation of a smooth wall by steps introduces errors into the solution, especially when the grid is coarse. The treatment of the boundary conditions at stepwise walls also requires special attention, because the area influenced by a stepwise approximation is much larger than the immediate area of the step.

An example of such a grid is shown in Fig. 9.1. When the shear forces or heat transfer at walls play an important role (e.g., airfoils, airplane or ship hulls, aerodynamic bodies, turbine blades, etc.), such approach can lead to large errors. It is not recommended, except when the solution algorithm allows local grid refinement near the wall (see Chap. 12 for details of local grid refinement methods) and pressure forces dominate. This is especially true for blunt bodies and when the blockage effects in space occupied by many bodies are more important for the flow analysis than shear stress on individual bodies. This may be the case in civil and environmental engineering (e.g., when studying flow in and around buildings, in rivers and lakes, in the atmosphere, etc.). Theoretically, if the grid spacing at the wall was of the order of natural wall roughness, the stepwise approximation would produce accurate results.

Another possible reason for using this approach is when an existing solution method which contains many physics models (e.g., combustion, multiphase flow, phase change, etc.) cannot be adapted to a grid that fits the boundary better, or when an exact representation of the wall boundary is not essential (e.g., when surfaces are very rough). Indeed, some authors used the locally refined stepwise approximation to model wall roughness effects. Examples are the large-eddy simulation of flow over a wall-mounted hemisphere by Manhart and Wengle (1994) and the simulation of water entry and exit of a circular cylinder by Xing-Kaeding (2006).

As noted above, the use of local refinement removes one of the most attractive features of Cartesian grids, namely the simplicity of neighborhood connections. Either non-refined cells at a refinement interface have to be treated as polyhedra, or one has to use so-called "ghost nodes" (also called "hanging nodes"). These are treated in the same way as at non-conformal or sliding block interfaces, which will be explained in detail in Sect. 9.6.1.

9.1.2 Immersed-Boundary Methods

Immersed-boundary methods use regular (mostly Cartesian) grids and deal with irregular wall boundaries by partially blocking the cells cut by walls. The first publication of such an approach is believed to be the paper by Peskin (1972); in recent years, many variants of immersed-boundary methods have been developed. One of the motivations is studying of flows around moving or deforming bodies, where this approach is believed to be more flexible than moving and deforming body-fitted grids.

There are various flavors of the approach and because the details are only related to the handling of wall boundary conditions while the rest of the method usually corresponds to one of the methods described here, we shall not go into those details. Basically, one first has to identify cells which are cut by wall boundaries (and possibly also their immediate neighbors). A triangulated (stereolithography, STL) description of boundaries is usually used for this purpose rather than their CAD representation. The largest problem is enforcing correct wall boundary conditions at correct locations; this leads to fixing the velocity either in the cut cell, or in the ghost-cell inside of the body, such that the profile fit to a certain number of nodes gives the specified wall velocity where it crosses the wall boundary. When walls are moving, the identification of cut cells and cells outside of the solution domain has to be performed at each time step.

When the grid is fine enough, the methods produce accurate results. However, if high-Reynolds-number flows around aerofoils, turbine blades or other smoothly curved walls are studied, boundary-fitted grids with prism layers along walls are more accurate. Especially when turbulence models with the so-called "low-Reynolds-number" formulation are used, which require very small grid spacing in wall-normal direction, immersed-boundary methods become inefficient because they end up having the grid refined also in wall-tangential direction.

For more details on immersed-boundary methods, see the review article by Peskin (2002) and other publications, e.g., Tseng and Ferziger (2003), Mittal and Iaccarino

(2005), Taira and Colonius (2007), Lundquist et al. (2012), etc. Very detailed descriptions are available in dissertations, e.g., by Peller (2010) and Hylla (2013), among others.

9.1.3   Overlapping Grids

When the geometry is complicated, the problem with structured or block-structured, boundary-fitted grids is that the grid quality usually deteriorates near the wall, where the accuracy should be the highest. Overlapping grids can help to achieve high grid quality near wall by using the best boundary-fitted, prismatic grid near wall and a simple (usually Cartesian) grid away from wall. Methods of this kind are often associated in the literature with names like *Chimera grids* (the Chimera is a mythological creature with lion's head, goat's body, and snake's tail), *overset grids* or *composite grids*. This approach is especially attractive when studying flow around bodies, either in a fixed position or moving. Usually, one creates a *background grid* disregarding solid bodies, which is only fitted to external boundaries (inlet, outlet, symmetry or far-field boundaries etc.) which usually do not move. Grids around bodies are created up to a certain distance from the body and are overlapping the background grid. If bodies are moving, the grids attached to them are moving too (see NASA Chimera Grid Tools User's Manual NASA CGTUM 2010).

Part of the background grid which is covered by the body and the overlapping grid is deactivated, except over a narrow overlap zone where both grids are active. Discretization and solution methods are usually one of those described here; only the coupling of solutions on individual grid is specific to overlapping grids. One possibility is to treat the outer surface of the overlapping grid and the surface of the hole created by deactivating some cells in the background grid as boundaries. On the part of such boundary where the flow enters the grid, inlet (Dirichlet) conditions are prescribed, while on the part where the flow leaves the grid, pressure is specified. Boundary values to be imposed are obtained by interpolation from the other grid. The update of these boundary conditions can be performed after each inner or after each outer iteration; the former is more implicit (as in parallel computing) and leads to better convergence properties.

Another option is to build the coupling of grids into the matrix A of the linear equation system for each equation solved, so that solution is obtained simultaneously on all grids. How this can be done will be explained with reference to Fig. 9.2 and the notation in it. The figure shows active cells on two grids which overlap. Acceptor (ghost, hanging) nodes are defined such that discretization can be performed as if neighbor cells were also active. Variable values at *acceptor* cells are obtained by interpolation from a selected set of *donor* cells from the other grid. This is illustrated in Fig. 9.2: the discretized equation for cell labeled P refers to an acceptor neighbor cell N_a marked by an open circle via fluxes through the face between P and N_a. Wherever in the discretized fluxes the reference is made to the variable value at the center of acceptor cell, it is replaced by an interpolation expression:

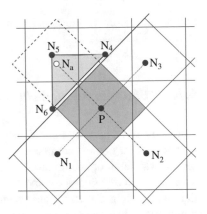

Fig. 9.2 On the definition of acceptor and donor cells on overlapping grids

$$\phi_{N_a} = \alpha_4 \phi_{N_4} + \alpha_5 \phi_{N_5} + \alpha_6 \phi_{N_6} . \tag{9.1}$$

Here α_4, α_5 and α_6 are interpolation factors that add up to unity, obtained by fitting a linear shape function to the three nearest neighbors from the other grid. Thus, in the equations for cell P there will be six off-diagonal coefficients in the linear equation matrix: three neighbors are from the same grid (N_1, N_2 and N_3) and three neighbors come from the overlapping grid (N_4, N_5 and N_6). Depending on the linear equation solver used, some other modifications to the algorithm may be needed. Also, there are many different options for interpolation between donor cells and for the number of donor cells to be used. One possibility is also to use only the variable value and the gradient from the nearest neighbor. A deferred-correction approach is then needed to account for the contribution from the gradient, but that does not represent a big complication.

The main advantages of overlapping grids are:

- One can better optimize the grid quality near a wall than if a single grid is used;
- The grid quality near a wall is preserved if the body moves;
- Parametric studies with bodies at different angles of attack can easily be realized without changing the grid or boundary conditions—one simply rotates the overset grid.
- With overlapping grids, arbitrary body motion can be accounted for, which otherwise may not be possible with other methods.

Note that, with overlapping grids, special interpolation is performed away from the wall, in regions where variation of variables is not too strong, while immersed-boundary methods apply special interpolation directly at the wall, where the highest accuracy is required.

Overlapping grids were used often in the past and recently this feature became available also in commercial codes. There is a special symposium on overset grids that has taken place every two years since 1992; see the official web-site: www.oversetgridsymposium.org for more information. Detailed description of some methods can be found in theses by Hadžić (2005) and Hanaoka (2013); both theses can be downloaded from the Internet. Examples of overset grid application will be presented in Sects. 13.4 and 13.10.1.

9.1.4 Boundary-Fitted Non-orthogonal Grids

Boundary-fitted non-orthogonal grids are most often used to calculate flows in complex geometries (most commercial codes use such grids). They can be structured, block-structured, or unstructured. The advantages of such grids are (1) that they can be adapted to any geometry and (2) that optimum properties are easier to achieve than with orthogonal curvilinear grids. Because CV-faces fall onto solution domain boundaries, the boundary conditions are more easily implemented than with stepwise approximation of curved boundaries. The grid can also be adapted to the flow, i.e., one set of grid lines can be chosen to follow the streamlines (which enhances the accuracy) and the spacing can be made smaller in regions of strong variable variation, especially if block-structured or unstructured grids are used.

Non-orthogonal grids have several disadvantages. If a finite-difference method is used, the transformed equations contain more terms that need to be approximated, thereby increasing both the difficulty of programming and the computing effort per cell for solving the equations. If a finite-volume method is used, one has to create control volumes of acceptable quality, which is a non-trivial task. The grid non-orthogonality may under some conditions cause convergence problems or unphysical solutions. The choice of velocity components and the arrangement of variables on the grid affects the accuracy and efficiency of the algorithm. These issues are discussed further below.

In the remainder of this book we shall assume that the grid is non-orthogonal and unstructured. The principles of discretization and solution methods which we shall present are valid for orthogonal grids as well, because they can be viewed as a special case of a non-orthogonal grid. One section deals with the treatment of block-structured grids, because the same approach is used for non-conformal interfaces in unstructured grids, e.g., at sliding interfaces.

9.2 Grid Generation

The generation of grids for complex geometries is an issue which requires too much space to be dealt with in great detail here. We shall present only some basic ideas and the properties that a grid should have. More details about various methods of grid generation can be found in books and conference proceedings devoted to this topic, e.g., Thompson et al. (1985) and Arcilla et al. (1991).

Even though necessity demands that in complex geometries the grid be non-orthogonal, it is important to make it as nearly orthogonal as possible. In FV-methods orthogonality of grid lines at CV vertexes is unimportant—it is the angle between the cell face surface normal vector and the line connecting the CV-centers on either side of it that matters. Thus, a 2D grid made of equilateral triangles is equivalent to an orthogonal grid, because lines connecting cell centers are orthogonal to cell faces. This will be discussed further in Sect. 9.7.2.

Cell topology is also important. If the midpoint rule integral approximation, linear interpolation, and central differences are used to discretize the equations, then the accuracy will be higher if the CVs are quadrilaterals in 2D and hexahedra in 3D, than if we use triangles and tetrahedra, respectively. The reason is that parts of the errors made at opposite cell faces when discretizing diffusion terms cancel partially (if cell faces are parallel and of equal area, they cancel completely) on quadrilateral and hexahedral CVs. To obtain the same accuracy on triangles and tetrahedra, more sophisticated interpolation and gradient approximations must be used. Especially near solid boundaries it is desirable to have quadrilaterals or hexahedra (prism layer; see Sect. 9.2.3 et seq.), because all quantities vary substantially there and accuracy is especially important in this region.

Accuracy is also improved if one set of grid lines closely follows the streamlines of the flow, especially for the convection terms. This cannot be achieved if triangles or tetrahedra are used, but is possible with quadrilaterals and hexahedra.

Non-uniform grids are the rule rather than exception when complex geometries are treated. However, grid quality becomes an important issue and we have devoted a whole section to it in Chap. 12.

An experienced user may know where strong variation of velocity, pressure, temperature, etc. can be expected; the grid should be fine in these regions because the errors are most likely to be large there. However, even an experienced user will encounter occasional surprises and more sophisticated methods are useful in any event. Errors are convected and diffused across the domain, as discussed in Sect. 3.9, making it essential to achieve as uniform a distribution of truncation error as possible. It is possible, however, to start with a coarse grid and later refine it locally according to an estimate of the discretization error; methods for doing this are called solution-adaptive grid methods and will be described in Chap. 12.

Finally, there is the issue of grid generation. When the geometry is complex, this task usually consumes the largest amount of user time by far; it is not unusual for a designer to spend a week generating a single grid, while the solution takes only a few hours on a parallel computer. Because the accuracy of the solution depends as much (if not more) on the grid quality as on the approximations used for discretization of the equations, grid optimization is a worthwhile investment of time.

Many commercial and public-domain codes for grid generation exist. Automation of the grid generation process, aimed at reducing the user time and speeding up the process is the major goal in this area. It is sometimes easier to generate a set of overlapping grids of good quality than a single block-structured or unstructured grid, but in really complex geometries this approach is difficult due to the existence of too many irregular pieces and the need to use too many grid blocks. However, one can find in publications examples where more than hundred grid blocks are used (e.g., for computing the flow around a military aircraft).

Generation of triangular and tetrahedral meshes is easier to automate, which is one of the reasons for their popularity. However, in most commercial codes, trimmed hexahedral or polyhedral grids are used. When polyhedral grids are generated, the first step is usually to generate a tetrahedral grid and then convert it to a polyhedral one. The grid generation process in industrial applications involves several other steps, which are described in the following sub-sections.

9.2.1 Definition of Flow Domain

For an automatic mesh generation, one needs a closed surface of the solution domain boundary as the starting point. In most cases, the surface is defined in one of the CAD-formats. However, one often obtains CAD-data for solid parts and needs to extract the volume occupied by the fluid inside, outside or between solid bodies. Most commercial software packages for grid generation contain tools that allow Boolean operations with solids and the extraction of the flow volume; sometimes additional steps are needed, such as creation of boundary surfaces for inlets and outlets. An example is shown in Fig. 9.3: the CAD-data contains the ball valve, the handle and the housing. If only the flow around the valve is to be computed, one needs to extract the volume inside valve occupied by fluid and close it by providing inlet and outlet pipe cross-section surfaces; the solid parts can then be discarded. It is also desirable to move outlet boundaries further downstream of valve; if outlet pipes are not included in the CAD-model, one can extrude outlet boundaries in axial direction and thus create adequate pipe sections, as shown in Fig. 9.3.

In the case that heat transfer from hot fluid in the pipe to the environment is of interest, one would keep the solids and also add a larger box around the valve assembly in order to compute the heat transfer from outer wall to the environment. This can be achieved by subtracting the valve assembly from a box. The grid needs then to be generated in both fluid domains and within all solid parts. It is desirable to have conformal grids at solid-fluid interfaces; this will be addressed in more detail in Chap. 12 when grid quality is discussed.

In the case of very complex geometries containing hundreds or even thousands of solid parts (e.g., when studying heat management in a vehicle, which requires that the flow around vehicle, inside the engine compartment, in the passenger cabin etc. as well as heat conduction through solid parts is computed), it may be too complicated to create the flow domain boundary by Boolean operations and imprinting of solid bodies. This is especially true when the assembly of parts contains gaps, intersections or overlapping of surfaces which would require manual repair, or when the imported geometry contains too many details which are not important for the flow and whose inclusion would unnecessarily increase the cell count in the grid. In such situations the so-called "surface-wrapping" tools may be preferred. These tools wrap a surface around all solid parts, like blowing a balloon until it sticks tightly to all walls. The resulting closed surface is provided in discrete form (usually triangulated).

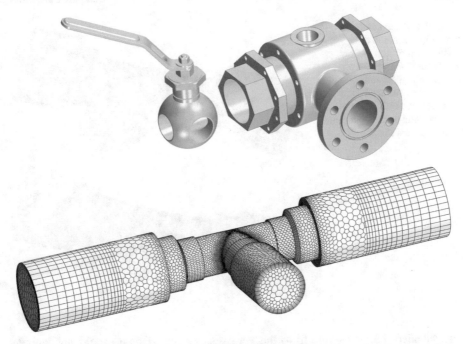

Fig. 9.3 An example of fluid volume extraction and boundary extrusion when simulating flow around a valve: imported CAD-parts (upper) and the extracted flow domain with extruded outlet boundaries (lower)

Surface-wrapping tools usually provide possibilities for the user to control the detail fidelity of the resulting closed surface (e.g., preserving feature lines like sharp edges or material interfaces, resolution of surface curvature, preserving gaps between parts, etc.). It is thus possible to "de-feature" the geometry, i.e., remove the unwanted details (e.g., close holes and gaps, remove small parts, etc.). With tight tolerances, an accurate representation of the geometry without significant loss of detail fidelity can be obtained.

9.2.2 Generation of a Surface Grid

Discrete surface representations obtained by tessellating a CAD representation of geometry are not suitable as the starting point for grid generation: the triangles are representing the geometry accurately, but they usually vary in both shape and size much more than is acceptable for the grid used to simulate fluid flow. For this reason, meshing tools usually first re-create a suitable surface grid which represents the geometry adequately while providing moderate growth rates and being as close as possible to the optimal equilateral triangles (at least the Delaunay-condition should

Fig. 9.4 An example of
initial tessellation of
CAD-description of flow
domain in the valve from
Fig. 9.3 (upper) and
re-meshed surface by
triangulation which serves as
the starting point for volume
mesh generation (lower)

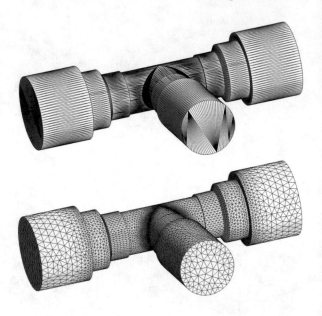

be satisfied, i.e., the circle fit to the vertexes of each triangle should not include
any other vertex—a condition that maximizes the minimum angle of all triangles).
We shall not deal with details of surface grid generation here; interested readers are
encouraged to consult literature on that subject (Frey and George 2008, Chap. 7). We
only want to emphasize the fact that many algorithms for the generation of a volume
grid require as the starting point a closed surface of the solution domain discretized
by an adequate triangular surface grid. Figure 9.4 shows an example of the initial
surface triangulation and the re-meshed surface.

9.2.3 Generation of a Volume Grid

Most tools for the generation of tetrahedral grids start from a triangular surface mesh
on solution domain boundaries. Tetrahedra that have one base on the surface are
generated first and the process is continued inwards as a marching front; the entire
process is something like solving an equation by a marching procedure and, indeed,
some methods are based on the solution of elliptic or hyperbolic partial differential
equations.

Tetrahedral cells are not desirable near walls if the boundary layer needs to be
resolved, because the first grid point must be very close to the wall while relatively
large grid sizes can be used in the directions parallel to the wall. These requirements
lead to long thin tetrahedra, creating problems in the approximation of diffusion
fluxes. For this reason, most grid generation methods first generate prism layers near
solid boundaries, starting with a triangular discretization of the surface and extruding

Fig. 9.5 An example of a
grid made up of prisms near
walls and tetrahedra in the
remaining part of the
solution domain

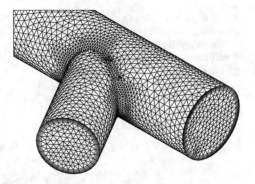

triangles in the direction orthogonal to boundary. Usually, some expansion is applied
to prism layers; factors larger than 1.5 should be avoided, while values up to 1.2
usually provide good results. On top of this layer, a tetrahedral mesh is generated
automatically in the remaining part of the domain. An example of such a grid is shown
in Fig. 9.5. How many prism layers should be created depends on the modeling of
the boundary layer; more details will be given in the next chapter.

This approach enhances grid quality near walls and leads to both more accurate
solutions and better convergence of numerical solution methods; however, it can be
only used if the solution method allows for mixed control volume types. In principle,
any type of method (FD, FV, FE) can be adapted to this kind of grid.

Another approach to automatic grid generation is to cover the solution domain with
a Cartesian grid, and adjust the cells cut by domain boundaries to fit the boundary.
This can be achieved by either projecting the vertexes of cut cells that fall outside
solution domain to the boundary, or by accepting cut cells as they are (polyhedral
CVs). The latter approach requires that the discretization and solution method can
deal with arbitrary polyhedral cells. The problem with this approach is that near walls
irregular cells with poorest quality are generated, where actually the highest grid
quality is required. If this is done on a very coarse level and the grid is then refined
several times, the irregularity is limited to a few locations and may not affect the
accuracy much; however, complex geometries usually impose limits to how coarse
the initial grid can be.

In order to move the irregular trimmed cells further away from walls, one can first
create prism layers along walls; the regular Cartesian grid is then cut by the outer
surface of the near-wall prism layer. An example of such a grid is shown in Fig. 9.6.
This approach allows fast grid generation but requires a solver that can deal with the
polyhedral cells created by cutting regular cells with an arbitrary surface. Also, grid
quality is an issue because cut cells can be both small and of rather irregular shape.
Very small cells between much larger cells are not favorable (too high expansion
ratio). Again, all types of methods can in principle be adapted to this type of grid.

If the discretization and solution method allow the use of an unstructured grid
with cells of arbitrary topology (general polyhedra), the grid generation program is
subject to few constraints. Nowadays, most commercial codes for flow simulation

Fig. 9.6 An example of a trimmed grid created by combining prism layers near a wall and regular Cartesian grid in the bulk of the solution domain, with irregular polyhedral cells along trimming surface

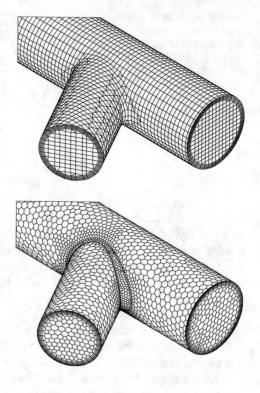

Fig. 9.7 An example of a grid made of polygonal prism layers near wall and arbitrary polyhedra in the bulk of the solution domain

can generate and use arbitrary polyhedral grids. Such grids can be created from tetrahedral grids, as already mentioned above; this is usually the first step in the generation of a polyhedral grid. However, because polyhedral grids do not impose any limits regarding topology (i.e., the number of cell faces and neighbor cells is arbitrary), several optimization steps usually follow. A single cell or a group of polyhedral cells can be modified in many ways to improve the grid quality; this includes moving vertexes, splitting or merging faces or cells, etc. An example of a polyhedral grid is shown in Fig. 9.7.

A polyhedral cell is also created through the cell-wise local grid refinement. A non-refined neighbor cell, although retaining its original shape (e.g., a hexahedron), becomes a logical polyhedron, because one or more faces is replaced by a set of sub-faces.

The solution domain can also be first divided into blocks in which grids with good properties can more easily be created than would be the case for the solution domain as a whole. One has the freedom to choose the best grid topology (structured H-, O-, or C-grid, or unstructured tetrahedral, hexahedral or polyhedral grid) for each block. The grids at block interfaces have then to be imprinted onto each other; the original faces lying in the interface are replaced by several smaller faces, which are uniquely defined as the common surface for two neighbor cells on either side of the interface. The cells on the block interfaces then have faces of irregular shape and must

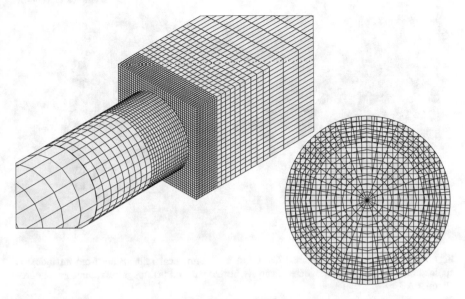

Fig. 9.8 A grid created by combining two blocks with a non-conformal interface and cell-wise local refinement; note the imprinted interface with irregular cell faces

be treated as polyhedra. An example can be seen in Fig. 9.8 which shows a sudden transition from a channel with a rectangular cross-section to a pipe with a circular cross-section. Obviously, a Cartesian grid represents the optimum for the rectangular channel, while a boundary-fitted O-grid is best suited for the circular pipe. Figure 9.8 shows the block interface where the two grids are imprinted onto each other, resulting in irregular cell faces. Alternative strategies for coupling non-conformal grid blocks are also possible; some will be described in Sect. 9.6.1.

The block-wise generation of non-conformal grids in a complex geometry is especially attractive if parts of the geometry are to be varied within a parametric study (e.g., shape optimization). In this case the grid in the part of geometry that is not changing can stay the same and one needs to re-generate the grid only in a small region rather than in the solution domain as a whole. This approach often allows better optimization of grid properties than when a monolithic grid is generated.

If the solution method requires all CVs to be of a particular type, generation of grids for complex geometries may be difficult. It is not unusual for the process of grid generation to take days or even weeks in some industrial applications of CFD. An efficient, automatic grid generator is an essential part of any CFD-software in an industrial environment. The existence of automatic grid generation tools was probably the main reason why CFD has spread so well across many industries, because the tools allow automation of the whole process and enable reduction of the processing time to an affordable scale.

An example of a complex industrial CFD-application is shown in Fig. 9.9, where a section through a trimmed grid applied to the analysis of vehicle thermal man-

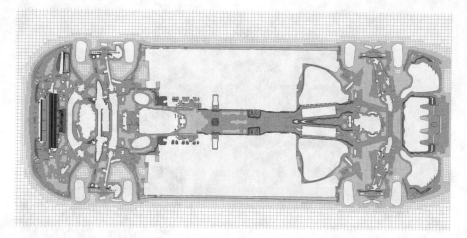

Fig. 9.9 An example of a trimmed Cartesian grid with local refinements from an industrial application—vehicle heat management analysis; horizontal section through the volume grid. Source: Daimler AG

agement is shown. A large portion of computing effort in automotive engineering is spent on ensuring that none of vehicle parts overheats in operation, because that can lead to failure. While car aerodynamics is more influenced by the designer than by engineers, the engine compartment and the under-body are regions where engineers have freedom to optimize shapes and positions within spatial constraints. Figure 9.9, which was created in 2014, indicates the geometrical complexity of solution domains in engineering applications of CFD. With an average cell count around 20–50 million, all relevant details of the geometry can be resolved. Usually, users create templates for grid design (including local grid refinements) for particular applications, thus reducing pre-processing time for subsequent analysis after the first one is carefully prepared.

9.3   The Choice of Velocity Components

In Chap. 1 we discussed issues associated with the choice of components of the momentum. As indicated by Fig. 1.1, only the choice of components in a fixed basis leads to a fully-conservative form of the momentum equations. To ensure momentum conservation, it is desirable to use such a basis and the simplest one is the Cartesian basis. When the flow is three-dimensional, there are no advantages to using any other basis (e.g., grid-oriented, covariant, or contravariant). Only when the choice of another vector basis leads to the simplification of the problem—e.g., by reducing the dimension of the problem—is it worth abandoning the use of Cartesian components. An example of such a case is the flow in a pipe or other axisymmetric geometries. If the flow does not vary in the circumferential direction, the velocity vector has only two non-zero components in the polar-cylindrical basis, but three non-zero Cartesian components. The problem therefore has three dependent variables in terms of the Cartesian components, but only two if the polar-cylindrical components are used, a substantial simplification. If the flow varies in the circumferential direction, the problem is three-dimensional in any case. While there is an apparent advantage in using polar-cylindrical components in axisymmetric geometries because the coordinates are aligned with the flow boundaries, the Cartesian option is superior because the equations are in strong conservative form and we have shown, e.g., in Fig. 9.5, how to handle this type of geometry.

9.3.1 Grid-Oriented Velocity Components

If grid-oriented velocity components are used, non-conservative source terms appear in the momentum equations. These account for the redistribution of the momentum between the components. For example, if polar-cylindrical components are used, the divergence of the convection tensor $\rho \mathbf{vv}$ leads to two such source terms:

- In the momentum equation for the r-component, there is a term $\rho v_\theta^2/r$, which represents the apparent *centrifugal force*. This is not the centrifugal force found in flows analyzed in a rotating coordinate system (e.g., pump or turbine passages)—it is solely due to the transformation from Cartesian to polar-cylindrical coordinates. This term describes the transfer of θ-momentum into r-momentum due to the change of direction of v_θ.
- In the momentum equation for the θ-component, there is a term $-\rho v_r v_\theta/r$, which represents the apparent *Coriolis force*. This term is a source or sink of θ-momentum, depending on the signs of the velocity components.

An example of the action of curvature terms in the momentum equations is visible in Fig. 9.10. It shows how radial and tangential components of velocity vector vary along grid lines even though the velocity field is uniform, i.e., the velocity vector is the same in the whole domain. This change in velocity components happens only because of the change in grid line orientation and has nothing to do with flow physics. While all forms of governing equations are mathematically equivalent, some obviously pose more difficulties when attempting their numerical solution. It is clear that numerical errors would disturb the uniformity of the velocity field in the example shown in Fig. 9.10 if polar-cylindrical or spherical coordinates are used, while the exact solution is trivial to maintain when Cartesian components are used.

In general curvilinear coordinates, there are more such source terms (see books by Sedov 1971; Truesdell 1991, etc). They involve Christoffel symbols (curvature terms, higher-order coordinate derivatives) whose discretization often represents a substantial source of numerical errors. The grid is required to be *smooth*—the change of grid direction from point to point must be small and continuous. Especially on unstructured grids, in which grid lines are not associated with coordinate directions, this basis is difficult to use.

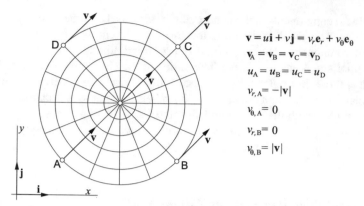

$$v = u\mathbf{i} + v\mathbf{j} = v_r\mathbf{e}_r + v_\theta\mathbf{e}_\theta$$
$$\mathbf{v}_A = \mathbf{v}_B = \mathbf{v}_C = \mathbf{v}_D$$
$$u_A = u_B = u_C = u_D$$
$$v_{r,A} = -|\mathbf{v}|$$
$$v_{\theta,A} = 0$$
$$v_{r,B} = 0$$
$$v_{\theta,B} = |\mathbf{v}|$$

Fig. 9.10 An example of a uniform velocity field and a polar grid, showing variation of radial and circumferential components depending on location in the grid

9.3.2 Cartesian Velocity Components

Recall that we are using Cartesian vector and tensor components exclusively. The discretization and solution techniques remain the same if other components are used but there are more terms to be approximated. The conservation equations in terms of Cartesian components were given in Chap. 1.

If the FD-method is used, one only has to employ the appropriate forms of the divergence and gradient operators for non-orthogonal coordinates (or to transform all derivatives with respect to Cartesian coordinates to the non-orthogonal coordinates). This leads to an increased number of terms, but the conservation properties of the equations remain the same as in Cartesian coordinates, as will be shown below.

In FV-methods there is no need for coordinate transformations. When the gradient normal to the CV surface is approximated, one can use a local coordinate transformation or auxiliary nodes on the line orthogonal to cell face, as will be shown below.

9.4 The Choice of Variable Arrangement

In Chap. 7 we mentioned that apart from colocated variable arrangement, various staggered arrangements are possible. While there were no obvious advantages for one or the other for Cartesian grids, the situation changes substantially when non-orthogonal grids are used.

9.4.1 Staggered Arrangements

The staggered arrangement, presented in Chap. 7 for Cartesian grids, is applicable to non-orthogonal grids only if the grid-oriented velocity components are employed. In

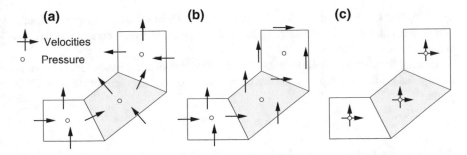

Fig. 9.11 Variable arrangements on a non-orthogonal grid: **a**—staggered arrangement with contravariant velocity components, **b**—staggered arrangement with Cartesian velocity components, **c**—colocated arrangement with Cartesian velocity components

Fig. 9.11 portions of such a grid are shown, in which the grid lines change direction by 90°. In one case, the contravariant, and in the other case, the Cartesian, velocity components are shown at the staggered locations. Recall that the staggered arrangement was introduced in order to achieve strong coupling between the velocities and the pressure gradient. The goal was to have the velocity component normal to cell face lie between the pressure nodes on either side of that face, see Fig. 8.1. For contravariant or covariant grid-oriented components, this goal is also achieved on non-orthogonal grids, see Fig. 9.11a. For Cartesian components, when the grid lines change direction by 90° a situation like the one shown in Fig. 9.11b arises: the velocity component stored at the cell face makes no contribution to the mass flux through that face, as it is parallel to the face. In order to calculate mass fluxes through such CV-faces, one has to use interpolated velocities from surrounding cell faces. This makes the derivation of the pressure-correction equation difficult, and does not ensure the proper coupling of velocities and pressure—oscillations in either may result.

Because, in engineering flows, grid lines often change direction by 180° or more, especially if unstructured grids are used, the staggered arrangement is difficult to use. Some of these problems can be overcome if all Cartesian components are stored at each CV face. However, this becomes complicated in 3D, especially if CVs of arbitrary shape are allowed. To see how this may be done, interested readers may want to look at the paper by Maliska and Raithby (1984).

9.4.2 Colocated Arrangement

It was shown in Chap. 7 that the colocated arrangement is the simplest one, because all variables share the same CV, but it requires more interpolation. It is no more complicated than other arrangements when the grid is non-orthogonal, as can be seen from Fig. 9.11c. The mass flux through any CV-face can be calculated by interpolating the velocities at two nodes on either side of the face; the procedure is the same as on regular Cartesian grids. Most commercial CFD-codes use Cartesian velocity components and the colocated arrangement of variables. We shall concentrate on this arrangement.

In what follows we shall describe the new features of the discretization on non-orthogonal grids, building on what has been done in preceding chapters for Cartesian grids.

9.5 Finite-Difference Methods

Finite-difference methods are not widely used in CFD for engineering applications; however, they find a role in some geophysical applications (see, Sect. 13.7). For those cases, the methods of Chap. 3 are useful. Generally, it is easier to develop methods of higher order (as desired in many geophysical problems) using the finite-difference approach than the finite-volume approach, but the latter is well suited for arbitrary polyhedral control volumes when approximations up to second order are used. For orders higher than second, the complexity increases substantially, as will be shown in a separate section later in this chapter. Accordingly, we give here only a brief introduction to FD-methods for non-orthogonal grids.

9.5.1 Methods Based on Coordinate Transformation

The FD-method is usually used only in conjunction with structured grids, in which case each grid line is a line of constant coordinate ξ_i. The coordinates are defined by the transformation $x_i = x_i(\xi_j)$, $j = 1, 2, 3$, which is characterized by the Jacobian J:

$$J = \det\left(\frac{\partial x_i}{\partial \xi_j}\right) = \begin{vmatrix} \dfrac{\partial x_1}{\partial \xi_1} & \dfrac{\partial x_1}{\partial \xi_2} & \dfrac{\partial x_1}{\partial \xi_3} \\[2mm] \dfrac{\partial x_2}{\partial \xi_1} & \dfrac{\partial x_2}{\partial \xi_2} & \dfrac{\partial x_2}{\partial \xi_3} \\[2mm] \dfrac{\partial x_3}{\partial \xi_1} & \dfrac{\partial x_3}{\partial \xi_2} & \dfrac{\partial x_3}{\partial \xi_3} \end{vmatrix} . \tag{9.2}$$

Because we use the Cartesian vector components, we only need to transform the derivatives with respect to Cartesian coordinates into the generalized coordinates:

$$\frac{\partial \phi}{\partial x_i} = \frac{\partial \phi}{\partial \xi_j} \frac{\partial \xi_j}{\partial x_i} = \frac{\partial \phi}{\partial \xi_j} \frac{\beta^{ij}}{J} , \tag{9.3}$$

where β^{ij} represents the cofactor of $\partial x_i/\partial \xi_j$ in the Jacobian J. In 2D this leads to:

$$\frac{\partial \phi}{\partial x_1} = \frac{1}{J} \left(\frac{\partial \phi}{\partial \xi_1} \frac{\partial x_2}{\partial \xi_2} - \frac{\partial \phi}{\partial \xi_2} \frac{\partial x_2}{\partial \xi_1} \right) . \tag{9.4}$$

The generic conservation equation, which in Cartesian coordinates reads:

$$\frac{\partial (\rho \phi)}{\partial t} + \frac{\partial}{\partial x_j} \left(\rho u_j \phi - \Gamma \frac{\partial \phi}{\partial x_j} \right) = q_\phi , \tag{9.5}$$

transforms to:

$$J \frac{\partial (\rho \phi)}{\partial t} + \frac{\partial}{\partial \xi_j} \left[\rho U_j \phi - \frac{\Gamma}{J} \left(\frac{\partial \phi}{\partial \xi_m} B^{mj} \right) \right] = J q_\phi , \tag{9.6}$$

where

$$U_j = u_k \beta^{kj} = u_1 \beta^{1j} + u_2 \beta^{2j} + u_3 \beta^{3j} \tag{9.7}$$

is proportional to the velocity component normal to the coordinate surface $\xi_j =$ const. The coefficients B^{mj} are defined as:

$$B^{mj} = \beta^{kj} \beta^{km} = \beta^{1j} \beta^{1m} + \beta^{2j} \beta^{2m} + \beta^{3j} \beta^{3m} . \tag{9.8}$$

The transformed momentum equations contain several additional terms that arise because the diffusion terms in the momentum equations contain a derivative not found in the generic conservation equation, see Eqs. (1.16), (1.18) and (1.19). These terms have the same form as the ones shown above and will not be listed here.

Equation (9.6) has the same form as Eq. (9.5), but each term in the latter is replaced by a sum of three terms in the former. As shown above, these terms contain the first derivatives of the coordinates as coefficients. These are not difficult to evaluate numerically (unlike second derivatives). The unusual feature of non-orthogonal grids is that mixed derivatives appear in the diffusion terms. In order to show this clearly, we rewrite the Eq. (9.6) in the expanded form:

$$J \frac{\partial (\rho \phi)}{\partial t} + \frac{\partial}{\partial \xi_1} \left[\rho U_1 \phi - \frac{\Gamma}{J} \left(\frac{\partial \phi}{\partial \xi_1} B^{11} + \frac{\partial \phi}{\partial \xi_2} B^{21} + \frac{\partial \phi}{\partial \xi_3} B^{31} \right) \right] +$$

$$\frac{\partial}{\partial \xi_2} \left[\rho U_2 \phi - \frac{\Gamma}{J} \left(\frac{\partial \phi}{\partial \xi_1} B^{12} + \frac{\partial \phi}{\partial \xi_2} B^{22} + \frac{\partial \phi}{\partial \xi_3} B^{32} \right) \right] + \tag{9.9}$$

$$\frac{\partial}{\partial \xi_3} \left[\rho U_3 \phi - \frac{\Gamma}{J} \left(\frac{\partial \phi}{\partial \xi_1} B^{13} + \frac{\partial \phi}{\partial \xi_2} B^{23} + \frac{\partial \phi}{\partial \xi_3} B^{33} \right) \right] = J q_\phi .$$

All three derivatives of ϕ, which stem from the gradient operator, appear inside each of the outer derivatives, which stem from the divergence operator, see Eq. (1.27). The mixed derivatives of ϕ are multiplied by coefficients B^{mj} with unequal indexes, which become zero when the grid is orthogonal, whether it is rectilinear or curvilinear. If the

Fig. 9.12 On the coordinate
transformation on
non-orthogonal grids

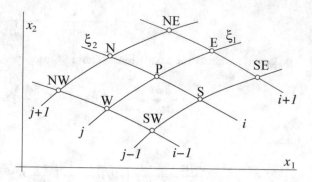

Fig. 9.12 On the coordinate transformation on non-orthogonal grids

grid is non-orthogonal, their magnitudes relative to the diagonal elements B^{ii} depend on the angles between the grid lines and on the grid aspect ratio. When the angle between grid lines is small and the aspect ratio large, the coefficients multiplying mixed derivatives may be larger than the diagonal coefficients, which can lead to numerical problems (poor convergence, oscillations in the solution, etc.). If the non-orthogonality and aspect ratio are moderate, these terms are much smaller than the diagonal ones and cause no problems. The mixed derivative terms are usually treated explicitly, as their inclusion in the implicit computational molecule would make the latter large and solution more expensive. Explicit treatment usually increases the number of outer iterations, but the savings derived from simpler and less expensive inner iterations is far more significant.

The derivatives in Eq. (9.6) can be approximated using one of the FD-approaches described in Chap. 3, see Fig. 9.12. The derivatives along curved coordinates are approximated in the same way as those along straight lines.

Coordinate transformations are often presented as a means of converting a complicated non-orthogonal grid into a simple, uniform Cartesian grid (the spacing in transformed space is arbitrary, but one usually takes $\Delta \xi_i = 1$). Some authors claim that discretization becomes simpler, as the grid in the transformed space appears simpler. This simplification is, however, illusory: the flow does take place in a complex geometry and this fact cannot be hidden by a clever coordinate transformation. Although the transformed grid does look simpler than the non-transformed one, the information about the complexity is contained in the metric coefficients. While the discretization on the uniform transformed mesh is simple and accurate, calculation of the Jacobian and other geometric information is not trivial and introduces additional discretization errors; i.e., it is here that the real difficulty has been hidden.

The mesh spacing $\Delta \xi_i$ need not be specified explicitly. The volume in physical space, ΔV, is defined as:

$$\Delta V = J \Delta \xi_1 \Delta \xi_2 \Delta \xi_3 . \tag{9.10}$$

If we multiply the whole equation by $\Delta \xi_1 \Delta \xi_2 \Delta \xi_3$, and replace $J \Delta \xi_1 \Delta \xi_2 \Delta \xi_3$ everywhere by ΔV, then the mesh spacings $\Delta \xi_i$ disappear in all terms. If central differences are used to approximate the coefficients β^{ij}, e.g., in 2D (see Fig. 9.12):

$$\beta_P^{11} = \left(\frac{\partial x_2}{\partial \xi_2}\right)_P \approx \frac{x_{2,N} - x_{2,S}}{2\,\Delta \xi_2},$$

$$\beta_P^{12} = -\left(\frac{\partial x_2}{\partial \xi_1}\right)_P \approx \frac{x_{2,E} - x_{2,W}}{2\,\Delta \xi_1},$$

$$(9.11)$$

the final discretized terms will involve only the differences in Cartesian coordinates between neighbor nodes and the volumes of imaginary cells around each node. Therefore, all we need is to construct such non-overlapping cells around each grid node and calculate their volume—the coordinates ξ_i need not be assigned any value and the coordinate transformation is hidden.

9.5.2 Methods Based on Shape Functions

Although we are not aware whether anybody has tried it, the FD-method can also be applied to arbitrary unstructured grids. One would have to prescribe a differentiable shape function (probably a polynomial) which describes the variation of the variable ϕ in the vicinity of a particular grid point. The coefficients of the polynomial would be obtained by fitting the shape function to the values of ϕ at a number of surrounding nodes. There would be no need to transform any term in the equation, so the simplest Cartesian form can be used. The shape function can be differentiated analytically to provide expressions for the first and second derivatives with respect to Cartesian coordinates at the grid point in terms of the variable values at surrounding nodes and geometrical parameters. The resulting coefficient matrix would be sparse but it would not have a diagonal structure unless the grid is structured.

One can also allow different shape functions to be used, depending on the local grid topology. This would lead to a different number of neighbors in computational molecules, but a solver that can deal with this complexity can easily be devised (e.g., algebraic multigrid or conjugate-gradient type solvers).

One can also devise a finite-difference method that does not need a grid at all; a set of discrete points adequately distributed over the solution domain is all that is needed. One would then locate a certain number of near neighbors of each point to which one could fit a suitable shape function; the shape function could then be differentiated to obtain approximations of the derivatives at that point. It would be most suitable to use the non-conservative form of differential equation (see Eq. 1.20), because it is simpler to discretize. The method cannot be fully conservative in any event, but this is not a problem if the points are sufficiently densely spaced where required.

It appears easier to distribute points in space than to create suitable control volumes or elements of good quality. One of the greatest problems for FV-methods is the creation of a closed surface of the solution domain, as explained above. The FD-method is more forgiving; any gap between two grid points is simply invisible. In addition, cells should fulfill certain minimum quality requirements, because otherwise either convergence problems, or non-physical solutions, or both can arise. Optimization of CV-shape is complicated and many operations (like imprinting of surfaces with non-conformal grid interfaces) are sensitive to tolerances. On the other hand, it is easier to shift single points in space or introduce new ones as only distances between them play a role—tolerances are not an issue.

The first step would be to place points on the surface, then add points a short distance away in the direction normal to the surface, as when creating prism layers in FV-grids. A second set of points could be regularly distributed in the solution domain (e.g., a Cartesian grid), with higher density near boundaries (like in cell-wise local refinement). Then the two sets of points can be checked for overlap, and, where points are too close to each other, they can be shifted, deleted or merged. Likewise, if gaps are present, new points can easily be inserted. Local refinement is very easy, one needs merely insert more points between the existing ones. If boundaries of solution domain are moving, one would move grid points within certain distance from boundary along with it, while the remaining points could stay where they are.

The only tricky thing would be the derivation of a suitable pressure or pressure-correction equation; however, this could be achieved following the methods presented in the following sections. We hope to see methods of this kind in future editions of this work.

The principles described above apply to all equations. The special features of deriving the pressure or pressure-correction equation or implementing the boundary conditions in FD-methods on non-orthogonal grids will not be dealt with here in detail, as the extension of techniques given so far is straightforward.

9.6 Finite-Volume Methods

The FV method starts from the conservation equation in integral form, e.g., the generic conservation equation:

$$\frac{\partial}{\partial t} \int_V \rho\phi \, dV + \int_S \rho\phi\mathbf{v} \cdot \mathbf{n} \, dS = \int_S \Gamma \mathrm{grad}\phi \cdot \mathbf{n} \, dS + \int_V q_\phi \, dV \ . \qquad (9.12)$$

The principles of FV-methods have been presented in Chap. 4, where rectangular control volumes (CVs) were used to illustrate the frequently used approaches to various approximations. Here we refer to specialties of applying those approximations to CVs of an arbitrary polyhedral shape.

When using Cartesian base vectors, governing equations contain no curvature terms and therefore CVs can be defined by vertexes connected by straight lines. The actual shape of the cell face is not important; because it is bounded by straight line segments, its projections onto Cartesian coordinate surfaces (which represent surface vector components) are the same, whatever the shape.

We shall consider block-structured and unstructured grids and first describe how the necessary grid data can be organized.

9.6.1 Block-Structured Grids

Structured grids are difficult, sometimes impossible, to construct for complex geometries. For example, to compute the flow around a circular cylinder in a free stream, one can easily generate a structured O-type grid around it, but if the cylinder is located in a narrow duct, this is no longer possible. In such a case, block-structured grids provide a useful compromise between (i) the simplicity and wide variety of solvers available for structured grids and (ii) the ability to handle complex geometries that unstructured grids allow.

The idea is to use a regular data structure (lexicographic ordering) within each block while constructing the blocks so as to fill the irregular domain by a very coarse unstructured grid (each block representing one cell).

Many approaches are possible. Some use overlapping blocks (e.g., Hinatsu and Ferziger 1991; Perng and Street 1991; Zang and Street 1995; Hubbard and Chen 1994; 1995). Others rely on non-overlapping blocks (e.g., Coelho et al. 1991; Lilek et al. 1997b). We shall describe one approach that used non-overlapping blocks. It is also well suited for use on parallel computers (see Chap. 12); normally, the computing for each block is assigned to a separate processor.

The solution domain is first subdivided into several sub-domains in such a way that each sub-domain can be fitted with a structured grid with good properties (not too non-orthogonal, individual CV aspect ratios not too large). An example is shown in Fig. 2.2. Within each block the indexes i and j are used to identify the CVs, but we also need a block-identifier. The data is stored in a one-dimensional array. The index of the node (i, j) in block 3 within that one-dimensional array is (see Table 3.2):

$$l = O_3 + (i - 1)N_j^3 + j ,$$

where O_3 is the offset for block 3 (the number of nodes in all preceding blocks, i.e., $N_i^1 N_j^1 + N_i^2 N_j^2$) and N_i^m and N_j^m are the numbers of nodes in the i and j directions in block m.

The grids in two neighboring blocks need not match at the interface; an example is shown in Fig. 2.3. Details of a non-conformal interface are shown in Figs. 9.13 and 9.14. This situation may result for two reasons: (i) the grid in one block is finer than in the other due to accuracy requirements; (ii) the grid in one block moves relative to the other (so-called sliding interface, found typically in simulation of flows around rotating parts).

One possibility of handling such a situation is to use auxiliary nodes on the other side of the interface as if the grid retained the same structure across interface (also called "ghost" or "hanging nodes"), see Fig. 9.13. The variable values at these nodes are then computed using interpolation from surrounding nodes of the neighbor grid. For example, the variable value at the ghost node E for the cell from block A in Fig. 9.13 can be obtained by using the variable value and the gradient from the nearest neighbor node in block B:

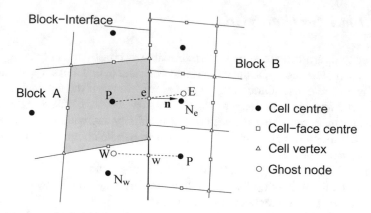

Fig. 9.13 Interface between two blocks with non-conformal grids: non-conservative treatment with ghost (hanging) nodes

$$\phi_E = \phi_{N_e} + (\nabla\phi)_{N_e} \cdot (\mathbf{r}_E - \mathbf{r}_{N_e}) . \tag{9.13}$$

The same treatment can be applied at the ghost node W for the cell from block B. These values are updated either after each inner or after each outer iteration; the treatment is akin to handling of subdomain interfaces in parallel computing using domain decomposition; see Sect. 12.6.2 in Chap. 12 for more details. Another option is to update the nodal value on the right-hand side of Eq. (9.13) after each inner iteration and the remaining part after each outer iteration. There is additional bookkeeping that needs to be implemented, but we shall not deal with it here.

The only problem with this approach is that it is not fully conservative: the sum of fluxes through the faces in block A on one side of the interface may not be equal to the sum of fluxes through the faces in block B on the other side. However, one can devise a correction to enforce conservation (e.g., Zang and Street 1995).

Another approach, which is fully conservative, is to allow CVs along interfaces to have more than four (in 2D) or more than six (in 3D) faces, i.e., to treat them as arbitrary polygonal or polyhedral cells. To this end one has to identify pieces of the interface surface which are common to two cells on either side of the interface. The original cell faces which lie in the interface will not be used for flux approximations—they are only used to perform mapping required to create interface faces. For example, the east face of the shaded cell in block A shown in Fig. 9.14 is not included while working in block A. The coefficient matrix and the source term for this CV will thus be incomplete, because the contribution from its east side is missing; in particular, the coefficient A_E will be zero.

Because the shaded CV in block A of Fig. 9.14 has three neighbors on its east face (which is split into three interface faces), we cannot use the usual notation for structured grids here. In order to treat the irregular cell faces found at block interfaces, we have to use another kind of data structure—one similar to that used when the whole grid is unstructured. Each piece of the interface common to two CVs

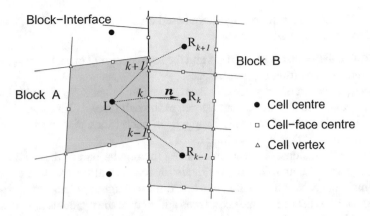

Fig. 9.14 Interface between two blocks with non-conformal grids: conservative treatment with a new data structure

must be identified (by a pre-processing tool) and placed on an interface list together with all of the information needed to approximate the surface integrals:

- the indexes of the left (L) and right (R) neighbor cells,
- the surface vector (pointing from L to R), and
- the coordinates of cell-face center.

With this information, one can use the method used in the interior of each block to approximate the fluxes through these faces. The same approach can be used at the "cuts" that occur in O- and C-type grids; in this case, we are dealing with an interface between two sides of the same block (i.e., A and B are the same block, but the grids may not be conformal at the interface).

Each interface cell face contributes to the source terms for the neighboring CVs (explicit contributions to the convection and diffusion fluxes treated by deferred correction), to the main diagonal coefficient (A_P) of these CVs, and to two off-diagonal coefficients: A_L for node R and A_R for node L. The problem of irregularity of data structure due to having three east neighbors is thus overcome by regarding the contributions to the global coefficient matrix as belonging to the interface cell faces (which always have two neighbor cells) rather than to the CVs. It is then irrelevant how the blocks are ordered relative to each other (the east side of one block can be connected to any side of the other block): one has only to provide the indexes of the neighbor CVs to the interface cell faces.

The contributions from interface cell faces, namely A_L and A_R, make the global coefficient matrix A irregular: neither the number of elements per row nor the bandwidth is constant. However, this is easily dealt with. All we need to do is to modify the iteration matrix M (see Chap. 5) so that it does not contain the elements due to the faces on the block interfaces. This is the same approach used for subdomain interfaces in parallel computing—the contribution from interface faces lags one iteration.

We shall describe a solution algorithm based on an ILU-type of solver; it is easily adapted to other linear equation solvers.

1. Assemble the elements of matrix A and the source term Q in each block, ignoring the contributions from block interfaces.
2. Loop over the list of interface cell faces, updating A_P and Q_P at nodes L and R, and calculate the matrix elements stored at the cell face, A_L and A_R.
3. Calculate elements of matrices L and U in each block disregarding neighbor blocks, i.e., as if they were on their own.
4. Calculate the residuals in each block using the regular part of the matrix A (A_E, A_W, A_N, A_S, A_P, and Q_P); the residuals for the CVs along block interfaces are incomplete, because the coefficients that refer to neighbor blocks are zero.
5. Loop over the list of interface cell faces and update the residuals at nodes L and R by adding the products $A_R \phi_R$ and $A_L \phi_L$, respectively; once all faces have been visited, all the residuals are complete.
6. Compute the variable update at each node in each block and return to step 1.
7. Repeat until the convergence criterion is met.

Because the matrix elements referring to nodes in neighbor blocks do not contribute to the iteration matrix M, one expects that the number of iterations required to converge will be larger than it is in the single block case. This effect can be studied by artificially splitting a structured grid into several subdomains and treating each piece as a block. As already mentioned, this is what is done when implicit methods are parallelized by using domain decomposition in space (see Chap. 12); the degradation of convergence rate of the linear equation solver is then measured by *numerical efficiency*.

Schreck and Perić (1993) and Seidl et al. (1996), among others, have performed numerous tests and found that the performance—especially when conjugate gradient and multigrid solvers are used—remains very good even for a relatively large number of subdomains. When a good structured grid can be constructed, it should be used. Block-structuring does increase the computing effort, but it allows solution of more complex problems and it certainly requires a more complicated algorithm.

An example of the application of this approach at a conformal interface in an O-grid is presented in Sect. 9.12. An implementation of the algorithm for O- and C-type grids is found in the code `caffa.f` in directory `2dgl`; see Appendix A.1. Further details on the implementation for block-structured non-conformal grids are available in Lilek et al. (1997b).

When the grid in one block moves (i.e., we are dealing with a sliding interface), the fact that one block rotates for $\Delta\theta$ within one time step while another block is fixed leads to changing neighborhood connections, irrespective of whether the grid in each block is structured or unstructured. If the above-mentioned conservative approach is used, the mapping of faces at the interface has to be performed at each time step; the interface list from previous time step is cleared and the new one is created. If the hanging-node approach is used, the donor cells and the interpolation stencils need also to be re-defined at each time step.

9.6.2 Unstructured Grids

Unstructured grids allow great flexibility in adapting the grid to domain boundaries. In general, control volumes of arbitrary shape, i.e., with any number of cell faces, can be used. Early versions of unstructured grids used in CFD consisted of cells that had up to six faces, i.e., tetrahedra, prisms, pyramids and hexahedra. They all may be considered special cases of hexahedra so nominally hexahedral grids may include CVs with less than six faces. Each CV was defined by eight vertexes, so the list of CVs also contained a list of associated vertexes. The order of the vertexes in the list represented the relative positions of the cell faces; e.g., the first four vertexes define the bottom face and the last four the top face, see Fig. 9.15. The positions of the six neighbor CVs is also implicitly defined; e.g., the bottom face defined by vertexes 1, 2, 3 and 4 is common to neighbor CV number 1, etc. This was adopted in order to reduce the number of arrays necessary for the definition of connectivity between CVs.

In late 1990s, polyhedral grids were introduced to CFD; this required a change in the data structure, because neither the number of faces per CV nor the number of corners on a face are limited. Modern CFD-codes usually use a data structure similar to the one described below for all cells in the grid, even if they are Cartesian. All data is organized in vertex, face and volume lists.

First, all vertexes are listed with their index and three Cartesian coordinates. Then, a face list is created, in which each face is defined by its index and the indexes of vertexes which define a closed polygon (vertexes are connected by straight line segments in the order they are listed, with the last vertex being connected with the first one to close the polygon). Finally, a list of cells is created, with cell index and the list of faces which enclose it.

Information that is stored in the face list also includes:

- Surface vector components (face projections onto Cartesian coordinate surfaces);
- Coordinates of the face centroid;
- Indexes of the cells on either side of the face (the convention is usually that the surface vector points from the first to the second listed cell);
- Coefficients of the matrix A for cells on each side which multiply neighbor cell variable value.

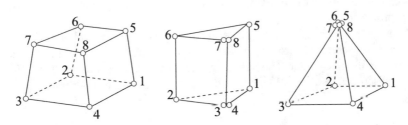

Fig. 9.15 Definition of nominally hexahedral CVs by a list of eight vertexes

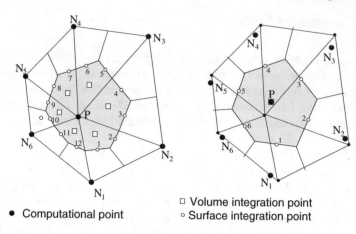

• Computational point

□ Volume integration point
○ Surface integration point

Fig. 9.16 On the construction of polyhedral control volumes from tetrahedral grids (2D illustration examples)

Information that is stored in cell list includes, among others:

- Cell volume;
- Coordinates of the cell centroid;
- Variable values and fluid properties;
- Coefficient A_P for the matrix A.

In recent years, polyhedral grids have become popular. Polyhedra are usually obtained by creating control volumes around vertexes of a tetrahedral grid, that is created first (but there are other approaches as well). This is demonstrated in 2D on the right-hand side in Fig. 9.16. Tetrahedra are split into four hexahedra by connecting midpoints on edges with centroids of faces and volume. The control volume around one tetrahedron vertex is defined by simply joining all hexahedral parts created by splitting tetrahedra which contain that vertex; in this process, faces shared by two hexahedra disappear inside polyhedron. All sub-faces shared by the same two control volumes are merged together, creating one polyhedron face. The computational point is placed in the CV-centroid; the tetrahedral grid is then discarded. This minimizes the number of faces of polyhedral CVs. Usually, some optimization steps follow: vertexes of polyhedral CVs can be moved, faces or even cells can be split or merged to create cells with better properties.

Another approach to converting a tetrahedral grid to a polyhedral one is by cutting the edges of each tetrahedron by a plane orthogonal to it; the cutting point may or may not be the line center. The advantages and disadvantages of various CV-types will be discussed in Chap. 12 where grid quality issues will be addressed.

9.6.3 Grids for Control-Volume-Based Finite-Element Methods

We give here only a short description of the hybrid FE/FV method using triangular elements and linear shape functions. For details on proper finite element methods and their application to Navier–Stokes equations, see the books by Oden (2006), Zienkiewicz et al. (2005) or Fletcher (1991).

For this method, unstructured grids made of triangles (2D) or tetrahedra (3D) are used, but the same principles apply to grids made of quadrilaterals and hexahedra as well (including prisms and pyramids as special cases). The grid represents *elements* which are used to describe the variation of the variables, i.e., to define the shape functions. The computational nodes are located at the element vertexes. Usually, variable ϕ is assumed to vary linearly within the element, i.e., its shape function is (in 2D):

$$\phi = a_0 + a_1 x + a_2 y . \tag{9.14}$$

The coefficients a_0, a_1 and a_2 are determined by fitting the function to the nodal values at the vertexes. They are thus functions of coordinates and variable values at the nodes.

In 2D, the control volumes are formed around each element node by joining the centroids of the elements and midpoints on element edges, thus creating quadrilateral sub-elements around each computational node, as shown on the left-hand side in Fig. 9.16. In 3D, hexahedral sub-elements are created by splitting tetrahedra using midpoints of volume, face and edge, as described above for converting a tetrahedral grid into a polyhedral one. The major difference is that the computational point remains the vertex of the starting grid, which is not the centroid of the control volume as is the case in the method presented in the previous section, see right-hand side of Fig. 9.16. Also, each face resulting from splitting tetrahedra that remained on the surface of the control volume is retained, with an integration point for surface integrals located at its centroid.

The conservation equations in integral form are applied to these CVs of irregular, polyhedral shape with many faces (around 10 in 2D and around 50 in 3D). The surface and volume integrals are calculated *element-wise*, using shape functions defined for each element: for the 2D CV shown in Fig. 9.16, the CV surface consists of 12 sub-faces, and its volume consists of six sub-volumes (from six elements which share the same vertex and thus contribute to the CV). Because the variation of variables over an element is prescribed in form of an analytical function, the integrals can easily be calculated. Usually, the shape function is simply used to compute the variable value at the sub-face or sub-volume centroid and the midpoint-rule approximation is used to evaluate surface and volume integrals, as will be described below.

The algebraic equation for a CV involves the node P and its immediate neighbours (N_1 to N_6 in Fig. 9.16). Even though the 2D grid shown in this figure consists of triangles only, the number of neighbors varies in general from one CV to another, depending on how many triangles share one vertex; this leads to irregular matrix structure. This restricts the range of solvers which can be used; conjugate gradient and Gauss-Seidel solvers, alone or within algebraic multigrid method, are usually employed.

This approach thus uses polyhedral control volumes, although they are never shown; results are presented on the elements defined by the initial grid. Major differences to classical FV-methods applied to polyhedral control volumes—which we shall deal with in the remainder of this chapter—are:

- Even if midpoint-rule approximation for surface and volume integrals is used in both cases, the number of integration points is different: the classical method has a single face between two neighbor cells and a single volume integration points in CV-centroid, while the present method generates multiple faces shared by the same two neighbor cells and evaluates volume integrals for each sub-volume. Thus, computing effort per iteration and storage requirement is lower for the classical FV-method.
- One might expect that the accuracy of the above method could be slightly higher because the integration is performed using a larger number of smaller areas and volumes, but this is questionable because data is interpolated using the same number of computational points and approximations of the same kind.
- In the present method, control volumes are also created around vertexes from the original grid that lie on solution domain boundaries. These CVs require special treatment because there is no equation to be solved there when boundary values are specified.

This approach was followed—although only in 2D and using second-order approximations—by Baliga and Patankar (1983), Schneider and Raw (1987), Masson et al. (1994), Baliga (1997), and others. A 3D version was presented by Raw (1985).

9.6.4 Computation of Grid Parameters

In 3D, the cell faces are not necessarily planar. To calculate cell volumes and cell-face surface vectors, suitable approximations are necessary. A simple method is to represent the cell face by a set of plane triangles. For the hexahedra used in structured grids, Kordula and Vinokur (1983) suggested decomposing each CV into eight tetrahedra (each CV face being subdivided into two triangles) so that no overlapping occurs.

Another way to calculate cell volumes for arbitrary CVs is based on Gauss theorem. By using the identity $1 = \mathrm{div}(x\mathbf{i})$, one can calculate the volume as:

$$\Delta V = \int_V \mathrm{d}V = \int_V \mathrm{div}(x\mathbf{i}) \, \mathrm{d}V = \int_S x\mathbf{i} \cdot \mathbf{n} \, \mathrm{d}S \approx \sum_c x_k \, S_k^x \, , \qquad (9.15)$$

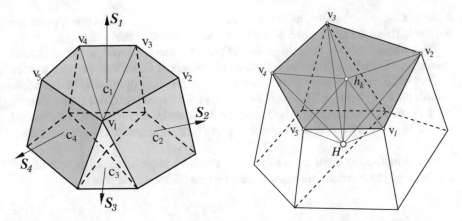

Fig. 9.17 On the calculation of cell volume and surface vectors for arbitrary control volumes: two alternative approaches

where k denotes cell faces and S_k^x is the x-component of the cell face surface vector (see Fig. 9.17):

$$\mathbf{S}_k = S_k \mathbf{n} = S_k^x \mathbf{i} + S_k^y \mathbf{j} + S_k^z \mathbf{k} \,. \tag{9.16}$$

Instead of $x\mathbf{i}$, one can also use $y\mathbf{j}$ or $z\mathbf{k}$, in which case one has to sum the products of $y_k S_k^y$ or $z_k S_k^z$. If each cell face is defined in the same way for both CVs to which it is common, the procedure ensures that no overlapping occurs and that the sum of all CV volumes equals the volume of the solution domain.

An important issue is the definition of the surface vectors at the cell faces. The simplest approach is to decompose them into triangles with one common vertex; see left-hand part of Fig. 9.17. The areas and surface normal vectors of triangles are easily computed. The surface normal vector for the whole cell face is then the sum of surface vectors of all the triangles (see face k in Fig. 9.17):

$$\mathbf{S}_k = \frac{1}{2} \sum_{i=3}^{N_k^v} \left[(\mathbf{r}_{i-1} - \mathbf{r}_1) \times (\mathbf{r}_i - \mathbf{r}_1) \right] \,, \tag{9.17}$$

where N_k^v is the number of vertexes in the cell face and $\mathbf{r}_i$ is the position vector of the vertex i. Note that there are $N_k^v - 2$ triangles. The above expression is correct even if the cell face is twisted or convex. The choice of the common vertex is not important.

The cell face center can be found by averaging the coordinates of the center of each triangle (which is itself the average of its vertex coordinates) weighted by its area. The area of the cell face is equal to the magnitude of its surface vector, e.g.:

$$S_k = |\mathbf{S}_k| = \sqrt{(S_k^x)^2 + (S_k^y)^2 + (S_k^z)^2} \,. \tag{9.18}$$

Note that only projections of the cell face onto Cartesian coordinate planes are required. These are exact when the CV edges are straight, as they are assumed to be.

If, in addition to fluid flow, the motion of particles needs to be computed, one must ensure that the triangulation of cell faces is performed in such a way that surface vectors of all triangles point outwards (i.e., scalar products of all individual surface vectors must be positive). This is not ensured by the method presented above. A more robust method of computing cell-face data is presented with reference to right-hand part of Fig. 9.17, but it creates more triangles and thus leads to a larger computing effort.

In this method an auxiliary hub-point h is defined at each face k; its coordinates can be, e.g., the average coordinates of all vertexes defining the face:

$$\mathbf{r}_{h,k} = \frac{1}{N_k^v} \sum_{i=1}^{N_k^v} \mathbf{r}_{v_i} , \qquad (9.19)$$

where $\mathbf{r}_{v_i}$ represents the ith vertex on face k in the list of vertexes defining the face.

The cell face is then subdivided into N_k^v triangles by connecting each vertex with the hub-point h. The surface vector for mth triangle can be expressed as follows:

$$\mathbf{S}_{k,m} = \frac{1}{2}(\mathbf{r}_{v_{m-1}} - \mathbf{r}_{h,k}) \times (\mathbf{r}_{v_m} - \mathbf{r}_{h,k}) . \qquad (9.20)$$

The surface vector for the whole face equals the sum of surface vectors of all triangles:

$$\mathbf{S}_k = \sum_{m=1}^{N_k^v} \mathbf{S}_{k,m} . \qquad (9.21)$$

The coordinates of the centroid of each triangle m, $\mathbf{r}_{k,m}$, are defined as follows:

$$\mathbf{r}_{k,m} = \frac{1}{3}(\mathbf{r}_{h,k} + \mathbf{r}_{v_m} + \mathbf{r}_{v_{m-1}}) . \qquad (9.22)$$

The coordinates of the face centroid, which are stored for use in the discretization process, are:

$$\mathbf{r}_k = \frac{\sum_{m=1}^{N_k^v} |\mathbf{S}_{k,m}| \mathbf{r}_{k,m}}{\sum_{m=1}^{N_k^v} |\mathbf{S}_{k,m}|} . \qquad (9.23)$$

In order to compute cell volume it is useful to define another auxiliary hub-point H; the coordinates of this point can be defined as the average of coordinates of all cell vertexes. One can now connect vertexes of each triangle in each face with the hub H, thus creating a tetrahedron whose volume can be easily computed:

$$V_{k,m} = \frac{1}{3}\mathbf{S}_{k,m} \cdot (\mathbf{r}_{h,k} - \mathbf{r}_H) . \qquad (9.24)$$

The cell volume can now be computed by summing volumes of all above-defined tetrahedra attached to triangles in all cell faces:

$$V_P = \sum_{k=1}^{N_P^f} \sum_{m=1}^{N_k^v} V_{k,m} \, , \tag{9.25}$$

where N_P^f represents the number of faces enclosing the cell around node P. The coordinates of cell centroid P can be computed by weighting the coordinates of each tetrahedron with its volume:

$$\mathbf{r}_P = \frac{\sum_{k=1}^{N_P^f} \sum_{m=1}^{N_k^v} (\mathbf{r}_{C,m})_k V_{k,m}}{V_P} \, , \tag{9.26}$$

where the coordinates of tetrahedron centroids are defined as the average coordinates of its vertexes:

$$(\mathbf{r}_{C,m})_k = \frac{1}{4}(\mathbf{r}_{h,k} + \mathbf{r}_{v_m} + \mathbf{r}_{v_{m-1}} + \mathbf{r}_H) \, . \tag{9.27}$$

9.7 Approximation of Fluxes and Source Terms

9.7.1 Approximation of Convection Fluxes

We shall use the midpoint rule approximation of the surface and volume integrals exclusively; it is the only second-order approximation that is applicable to integration domains of an arbitrary shape. One only needs to know the coordinates of the face centroid; how these can be computed for arbitrary polygons was described in the preceding section. The necessary steps to develop methods of higher order will be described in Sect. 9.10.

We look first at the calculation of mass fluxes. Only one face, denoted by index k in CVs shown in Fig. 9.18, will be considered; the same approach applies to other faces—only the indexes need be substituted. The CV may have any number of faces.

The midpoint rule approximation of the mass flux through face k leads to:

$$\dot{m}_k = \int_{S_k} \rho \, \mathbf{v} \cdot \mathbf{n} \, \mathrm{d}S \approx (\rho \, \mathbf{v} \cdot \mathbf{n} \, S)_k \, . \tag{9.28}$$

The unit normal vector at the face k is defined by the surface vector $\mathbf{S}_k$ and face area S_k whose projections onto Cartesian coordinate planes are S_k^i:

$$\mathbf{S}_k = \mathbf{n}_k S_k = S_k^i \, \mathbf{i}_i \, , \tag{9.29}$$

and the surface area, S_k, is:

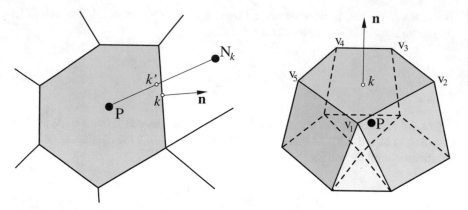

Fig. 9.18 General 2D and 3D control volume and the notation used

$$S_k = \sqrt{\sum_i (S_k^i)^2} \; . \tag{9.30}$$

See Sect. 9.6.4 for details on how surface vector components are computed for arbitrary polyhedral CVs.

With these definitions the expression for the mass flux becomes:

$$\dot{m}_k = \rho_k \sum_i S_k^i u_k^i = \rho_k S_k v_k^n \; , \tag{9.31}$$

where u_k^i are the Cartesian velocity components interpolated to the face centroid k and v_k^n is the velocity component orthogonal to the face k.

The difference between a Cartesian and an arbitrary polyhedral grid is that, in the latter case, the surface vector has components in more than one Cartesian direction and all Cartesian velocity components contribute to the mass flux. Each Cartesian velocity component is multiplied by the corresponding surface vector component (projection of the cell face onto a Cartesian coordinate plane), see Eq. (9.31).

The convection flux of any transported quantity is usually calculated by assuming that the mass flux is known which, with the midpoint rule approximation, leads to:

$$F_k^c = \int_{S_k} \rho \phi \, \mathbf{v} \cdot \mathbf{n} \, \mathrm{d}S \approx \dot{m}_k \phi_k \; , \tag{9.32}$$

where ϕ_k is the value of ϕ at the face centroid k. This represents linearization based on Picard-iteration, as described in Chap. 5.

The variable value at face centroid has to be expressed through nodal values by a suitable interpolation. Several possibilities were presented in Chap. 4, but not all extend easily to arbitrary polyhedral grids, because computational points (CV-centroids) are not located on lines along which interpolation can easily be conducted. For higher-order interpolation, multi-dimensional shape functions are needed.

However, a simple second-order method for arbitrary polyhedral CVs can easily be constructed by using simple geometric data and vector operations. Consider two arbitrary control volumes (CVs) in Fig. 9.18. The following vector is defined based on coordinates of cell and face centroids:

$$\mathbf{d}_k = \mathbf{r}_{N_k} - \mathbf{r}_P \ . \tag{9.33}$$

We also introduce linear interpolation factor along the line connecting cell center P and the center of the neighbor cell N_k as follows:

$$\xi_k = \frac{(\mathbf{r}_k - \mathbf{r}_P) \cdot \mathbf{d}}{\mathbf{d} \cdot \mathbf{d}} \ . \tag{9.34}$$

By projecting the vector $(\mathbf{r}_k - \mathbf{r}_P)$ onto the line connecting cell centers, the auxiliary point k' on this line is defined as:

$$\mathbf{r}_{k'} = \mathbf{r}_{N_k} \xi_k + \mathbf{r}_P (1 - \xi_k) \ . \tag{9.35}$$

Linear interpolation between cell centers on two sides of the face is the simplest second-order approximation, but it leads to the variable value at the location k', which is not the face centroid:

$$\phi_{k'} = \phi_{N_k} \xi_k + \phi_P (1 - \xi_k) \ . \tag{9.36}$$

If the value obtained by the above interpolation is taken to represent the value at the face centroid, the result will not be second-order accurate unless points k' and k almost coincide.

A second-order accurate interpolation for the face centroid location k can be achieved by using the variable value and its gradient from location k' as follows:

$$\phi_k = \phi_{k'} + (\nabla \phi)_{k'} \cdot (\mathbf{r}_k - \mathbf{r}_{k'}) \ . \tag{9.37}$$

The first term on the right-hand side, when expressed via Eq. (9.36), delivers contributions to the coefficient matrix A; the second term is treated explicitly as a deferred correction. The gradient at k' can be computed using interpolation from cell-center values according to expression (9.36). If vector $\mathbf{d}$ passes through face center k, the deferred correction involving gradients will be zero because k' and k then coincide; the face value is then solely computed from cell center values.

Another option is to perform extrapolation from the cell center to the face center from both sides, and then use weighting between the two, depending on distance (equivalent to central-differencing approximation) or flow direction (second-order upwind), or some other criteria:

$$\phi_{k1} = \phi_P + (\nabla \phi)_P \cdot (\mathbf{r}_k - \mathbf{r}_P) \ , \quad \phi_{k2} = \phi_{N_k} + (\nabla \phi)_{N_k} \cdot (\mathbf{r}_k - \mathbf{r}_{N_k}) \ . \tag{9.38}$$

In case of distance weighting, we have:

$$\phi_k = \phi_{k2}\xi_k + \phi_{k1}(1 - \xi_k) \ . \tag{9.39}$$

The individual values ϕ_{k1} and ϕ_{k1} from Eq. (9.38) represent themselves second-order approximations. If one adopts the value obtained by extrapolation from the upstream side, the linear upwind scheme is obtained; it is available in most commercial CFD-codes.

We can also define two additional auxiliary points on the line normal to cell face and passing through its centroid. First we determine the shorter of projections of vectors connecting cell centers with the centroid of the common face onto face normal:

$$a = \min((\mathbf{r}_k - \mathbf{r}_P) \cdot \mathbf{n}, (\mathbf{r}_{N_k} - \mathbf{r}_k) \cdot \mathbf{n}) \ . \tag{9.40}$$

The auxiliary points P' and N'_k are now defined to lie at distance a from face centroid k (one is at the projection of cell center location onto face normal, and the other is brought closer to face so that both are same distance away from face center, see Fig. 9.19):

$$\mathbf{r}_{P'} = \mathbf{r}_k - a\mathbf{n} \ , \quad \mathbf{r}_{N'_k} = \mathbf{r}_k + a\mathbf{n} \ . \tag{9.41}$$

The variable value at the face centroid can now be computed as the average of values at auxiliary nodes P' and N'_k (because these are at equal distance from face center); these can be computed as follows:

$$\phi_{P'} = \phi_P + (\nabla \phi)_P \cdot (\mathbf{r}_{P'} - \mathbf{r}_P) \ , \quad \phi_{N'_k} = \phi_{N_k} + (\nabla \phi)_{N_k} \cdot (\mathbf{r}_{N'_k} - \mathbf{r}_{N_k}) \ . \tag{9.42}$$

$$\phi_k = \frac{1}{2}(\phi_{P'} + \phi_{N'_k}) \ . \tag{9.43}$$

All three options provide a second-order approximation which contains a part referring to variable values at cell centers on either side of the face (which can contribute to the matrix coefficients) and a part that depends on gradients at cell centers (which would normally be treated using deferred-correction approach). The first option reduces to the standard central-differencing approximation (linear interpolation) on Cartesian grids. The second and the third option would reduce to the same approximation if gradients at both cell centers were the same; otherwise, a term proportional to the difference in gradients will remain even if the grid is Cartesian.

Quadratic (3rd-order) and cubic (4th-order) interpolation can easily be constructed by using variable values and gradients at nodes P and N_k, as described in Chap. 4, but they only lead to more accurate approximations at location k'. It would still lead to more accurate solutions when grids are sufficiently fine, but due to the fact that the correction in expression (9.37) and the integral approximation are both of second order, the overall order of convection flux approximation cannot be improved beyond second order; see Sect. 4.7.1 for more details on this issue.

The first-order upwind scheme is straightforward to implement on any grid, see Sect. 4.4.1; however, due to its extreme numerical diffusion, it is usually only used when second-order approximations fail due to extremely bad grid properties. Another area where first-order upwind scheme can be used is to stabilize higher-order schemes in situations where they produce oscillatory solutions. This is achieved by blending two schemes as described in Sect. 5.6; the blending coefficient can either be prescribed by the user or evaluated by the code based on certain criteria.

9.7.2 Approximation of Diffusion Fluxes

The midpoint rule applied to the integrated diffusion flux gives:

$$F_k^d = \int_{S_k} \Gamma \, \nabla \phi \cdot \mathbf{n} \, dS \approx (\Gamma \, \nabla \phi \cdot \mathbf{n})_k S_k = \left(\Gamma \frac{\partial \phi}{\partial n} \right)_k S_k \, . \tag{9.44}$$

The gradient of ϕ at the cell-face center can be expressed either in terms of the derivatives with respect to global Cartesian coordinates x_i or local orthogonal coordinates (n, t, s):

$$\nabla \phi = \frac{\partial \phi}{\partial x_i} \mathbf{i}_i = \frac{\partial \phi}{\partial n} \mathbf{n} + \frac{\partial \phi}{\partial t} \mathbf{t} + \frac{\partial \phi}{\partial s} \mathbf{s} \, , \tag{9.45}$$

where n, t and s represent coordinate directions normal and tangential to the surface, respectively; $\mathbf{n}$, $\mathbf{t}$ and $\mathbf{s}$ are the corresponding unit vectors for each coordinate direction. Note that the coordinate system (n, t, s) is also a Cartesian system, but it is only rotated so that its one coordinate is normal and the other two are tangential to the cell face.

There are many ways to approximate the derivative normal to the cell face or the gradient vector at the cell center; we shall describe only few of them. If the variation of ϕ in the vicinity of the cell face is described by a shape function, it is then possible to differentiate this function at the location k to find the derivatives with respect to the Cartesian coordinates. The diffusion flux is then:

$$F_k^d = \Gamma_k \sum_i \left(\frac{\partial \phi}{\partial x_i} \right)_k S_k^i \, . \tag{9.46}$$

This is easy to implement *explicitly;* an implicit version may be complicated, depending on the order of the shape function and the number of nodes involved.

A simple way to derive a second-order approximation of diffusion flux is obtained when using variable values from auxiliary nodes P' and N_k' introduced in previous section; see Fig. 9.19. The derivative with respect to n approximated as a central difference is simply:

$$\left(\frac{\partial \phi}{\partial n} \right)_k \approx \frac{\phi_{N_k'} - \phi_{P'}}{|\mathbf{r}_{N_k'} - \mathbf{r}_{P'}|} \, . \tag{9.47}$$

Fig. 9.19 On the approximation of diffusion fluxes for arbitrary polyhedral CVs

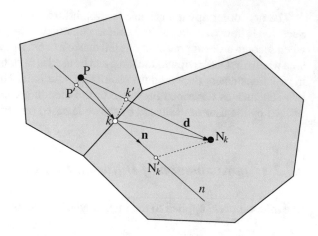

By referring to expressions (9.42), one can re-write the above equation as follows:

$$\left(\frac{\partial \phi}{\partial n}\right)_k \approx \frac{\phi_{N_k} - \phi_P}{|\mathbf{r}_{N_k'} - \mathbf{r}_{P'}|} + \frac{(\nabla \phi)_{N_k} \cdot (\mathbf{r}_{N_k'} - \mathbf{r}_{N_k}) - (\nabla \phi)_P \cdot (\mathbf{r}_{P'} - \mathbf{r}_P)}{|\mathbf{r}_{N_k'} - \mathbf{r}_{P'}|} . \tag{9.48}$$

The first term on the right-hand side represents the implicit part (that contributes to the coefficient matrix) while the second term represents the explicit part, computed using values from previous iteration (deferred correction).

If the face normal passes through cell centers on either side and they are equally away from face center, the deferred correction involving gradients will vanish; the normal derivative will then be solely computed using cell-center values, as is the case on a uniform Cartesian grid. However, if the grid is non-uniform, the above approximation (9.48) would contain a correction term making it second-order accurate at face centroid k.

Another way to calculate derivatives at the cell face is to obtain them first at CV centers, and then interpolate them to the cell faces in the way ϕ_k was determined. However, this can lead to oscillatory solutions and a correction similar to that used to avoid oscillatory pressures on colocated grids needs to be applied. Let us first see how gradients at cell centers can be computed.

A simple way of doing this is provided by the Gauss' theorem; we approximate the derivative at the CV center by the average value over the cell:

$$\left(\frac{\partial \phi}{\partial x_i}\right)_P \approx \frac{\int_V \frac{\partial \phi}{\partial x_i}\, dV}{\Delta V} . \tag{9.49}$$

Then we can consider the derivative $\partial \phi / \partial x_i$ as the divergence of the vector $\phi\, \mathbf{i}_i$ and transform the volume integral in the above equation using Gauss' theorem into a surface integral:

$$\int_V \frac{\partial \phi}{\partial x_i} \, dV = \int_S \phi \, \mathbf{i}_i \cdot \mathbf{n} \, dS \approx \sum_k \phi_k S_k^i \, . \tag{9.50}$$

This shows that one can calculate the derivative of ϕ with respect to x at the CV center by summing the products of ϕ with the x-components of the surface vectors at all faces of the CV and dividing the sum by the CV volume:

$$\left(\frac{\partial \phi}{\partial x_i} \right)_P \approx \frac{\sum_k \phi_k S_k^i}{\Delta V} \, . \tag{9.51}$$

For ϕ_k we can use the values used to calculate the convection fluxes, although one need not necessarily use the same approximation for both terms. For Cartesian grids and linear interpolation, the standard central difference approximation is obtained (only east and west face contribute to the derivative in x-direction because x-components of surface vectors at other faces are zero; the cell volume can be expressed as $\Delta V = S_e \Delta x$):

$$\left(\frac{\partial \phi}{\partial x_i} \right)_P \approx \frac{\phi_E - \phi_W}{2 \, \Delta x} \, . \tag{9.52}$$

Cell-center gradients can also be approximated within second order by using linear shape functions; if we assume linear variation of ϕ between two neighbor cell centers, e.g., P and N_k, we may write:

$$\phi_{N_k} - \phi_P = (\nabla \phi)_P \cdot (\mathbf{r}_{N_k} - \mathbf{r}_P) \, . \tag{9.53}$$

We can write as many such equations as there are neighbors for the cell around node P; however, we need to compute only three derivatives $\partial \phi / \partial x_i$. With the help of least-squares methods, the derivatives can be explicitly computed from such an overde-termined system for arbitrary CV shapes; see Demirdžić and Muzaferija (1995) for more details.

The derivatives calculated in this way can be interpolated to the cell face and the diffusion flux can be calculated from Eq. (9.46). The problem with this approach is that an oscillatory solution may be generated in the course of the iteration procedure and the oscillations will not be detected. How this can be avoided is described below.

For explicit methods, this approach is very simple to implement. It is, however, not suitable for implementation in an implicit method because it produces large computational molecules. The *deferred correction* approach described in Sect. 5.6 offers a way around this problem, but it does not necessarily help to eliminate the oscillations.

A large number of methods to deal with this issue have been developed over the past 20 years; Demirdžić (2015) gives an overview and points out the best one, which is similar to what is obtained by discretizing the diffusion term in non-orthogonal coordinates:

$$F_k^{\mathrm{d}} \approx \Gamma_{\phi,k} S_k \frac{\phi_{N_k} - \phi_{\mathrm{P}}}{(\mathbf{r}_{N_k} - \mathbf{r}_{\mathrm{P}}) \cdot \mathbf{n}_k} + \Gamma_{\phi,k} S_k \underline{\left[(\nabla\phi)_k \cdot \mathbf{n}_k - \overline{(\nabla\phi)}_k \cdot \frac{\mathbf{r}_{N_k} - \mathbf{r}_{\mathrm{P}}}{(\mathbf{r}_{N_k} - \mathbf{r}_{\mathrm{P}}) \cdot \mathbf{n}_k} \right]}.$$

$$(9.54)$$

The first term on the right-hand side is treated implicitly, i.e., it contributes to the coefficient matrix A; the underlined term is calculated using prevailing values of the variables and treated as another deferred correction. In this term, $(\nabla\phi)_k$ is obtained by interpolating gradients from the two CV-centers according to their distance from the face, while $\overline{(\nabla\phi)}_k$ represents the average of gradients at P and N_k. The reason for this distinction is that the first term on the right-hand side, being the central-difference approximation, is second-order accurate at the midpoint between nodes P and N_k, and the last term on the right-hand side should cancel the first term out when the variation of ϕ is smooth, so gradients are interpolated to the midpoint rather than cell-face center.

Note that all terms on the right-hand side of the above equation actually provide approximations at location k' rather than k. Thus, if distance from k' to k is significant, the flux approximation will not be second-order accurate. Note also that approximation of the normal derivative from Eq. (9.54) is similar to the one obtained using auxiliary nodes P$'$ and N$'_k$, Eq. (9.48), but it has been derived using a different approach. The approximation from Eq. (9.48) is second-order accurate at face centroid k even if the line connecting cell centers P and N_k does not pass through face centroid.

In the momentum equations, the diffusion flux contains a few more terms than does the corresponding term in the generic conservation equation, e.g., for u_i:

$$F_k^{\mathrm{d}} = \int_{S_k} \mu \nabla u_i \cdot \mathbf{n} \, \mathrm{d}S + \underline{\int_{S_k} \mu \frac{\partial u_j}{\partial x_i} \mathbf{i}_j \cdot \mathbf{n} \, \mathrm{d}S} . \qquad (9.55)$$

The underlined term is absent in the generic conservation equation. If ρ and μ are constant, the sum of underlined terms over all CV faces is zero by virtue of the continuity equation, see Sect. 7.1. If ρ and μ are not constant, they—except near shocks—vary smoothly and the integral of underlined terms over the whole CV surface is smaller than the integral of the principal term. For this reason, the underlined term is usually treated explicitly. As shown above, the derivatives are easily calculated at the cell face using the gradient vectors from CV-centers. This term causes no oscillations and can be computed using interpolated gradients.

9.7.3 Approximation of Source Terms

The midpoint rule approximates a volume integral by the product of the CV-center value of the integrand and the CV-volume:

$$Q_{\mathrm{P}}^{\phi} = \int_V q_{\phi} \, \mathrm{d}V \approx q_{\phi,\mathrm{P}} \, \Delta V . \qquad (9.56)$$

This approximation is independent of the CV shape and is of second-order accuracy.

Let us look now at the pressure terms in the momentum equations. They can be treated either as conservative forces on the CV-surface, or as non-conservative body forces. In the first case we have (in the equation for u_i):

$$Q_P^p = - \int_S p\, \mathbf{i}_i \cdot \mathbf{n}\, dS \approx \sum_k p_k S_k^i . \tag{9.57}$$

In the second case we get:

$$Q_P^p = - \int_V \frac{\partial p}{\partial x_i}\, dV \approx - \left(\frac{\partial p}{\partial x_i}\right)_P \Delta V . \tag{9.58}$$

The first approach is fully conservative. The second is conservative (and equivalent to the first one) if the derivative $\partial p/\partial x_i$ is calculated using Gauss' theorem. If pressure derivatives at CV-center are computed by differentiating a shape function, this approach is, in general, not conservative.

Other volumetric source terms are approximated in the same way: one first computes the source term at the cell centroid and then simply multiplies it by cell volume. Non-linear source terms need to be linearized; this is done using approaches discussed in Sect. 5.5.

9.8 Pressure-Correction Equation

The SIMPLE algorithm (see Sect. 8.2.1) needs to be modified when the grid is non-orthogonal and/or unstructured. The approach is described in this section. The extension of the implicit fractional-step method presented in the previous Chapter follows the same lines and will not be presented here, because the difference between the two methods is small and was described in detail in Sect. 8.2.3.

For any grid type, the discretized momentum equations have the following form:

$$A_P^{u_i} u_{i,P} + \sum_k A_k^{u_i} u_{i,k} = Q_{i,P} . \tag{9.59}$$

Note that subscript k here denotes cell centers N_k and not face centroids.

The source term $Q_{i,P}$ contains the discretized pressure gradient term. Irrespective of how this term is approximated, one can write:

$$Q_{i,P} = Q_{i,P}^* + Q_{i,P}^p = Q_{i,P}^* - \left(\frac{\delta p}{\delta x_i}\right)_P \Delta V , \tag{9.60}$$

where $\delta p/\delta x_i$ represents the discretized pressure derivative with respect to Cartesian coordinate x_i. If the pressure term is approximated in a conservative way (as a sum of surface forces), the mean pressure gradient over the CV can be expressed as:

$$Q_{i,P}^p = -\int_S p\, \mathbf{i}_i \cdot \mathbf{n}\, dS = -\int_V \frac{\partial p}{\partial x_i}\, dV \quad \Rightarrow \quad \left(\frac{\delta p}{\delta x_i}\right)_P = -\frac{Q_{i,P}^p}{\Delta V}. \qquad (9.61)$$

As always, the correction takes the form of a pressure gradient and the pressure is derived from a Poisson-like equation obtained by imposing the continuity constraint. The objective is to satisfy continuity, i.e., the net mass flux into every CV must be zero. In order to calculate the mass flux, we need velocities at the cell-face centers. In a staggered arrangement these are available. On colocated grids, they are obtained by interpolation.

It was shown in Chap. 7 that, when interpolated velocities at cell faces are used to derive the pressure-correction equation, a large computational molecule results as can oscillations in the pressure and/or velocities. We described a way to modify the interpolated velocity that yields a compact pressure-correction equation and avoids oscillatory solutions. We shall describe briefly an extension of the approach presented in Sect. 8.2.1 to non-orthogonal grids. The method described below is valid for both conservative and non-conservative treatment of the pressure terms in the momentum equations, and with a little modification can be applied to FD schemes on non-orthogonal grids. It is also valid for arbitrarily shaped CVs; we shall consider in the following the face k, see Fig. 9.19.

Following the approach described in Sect. 8.2.1, the interpolated cell-face velocity is corrected by subtracting the difference between the pressure gradient computed at cell face and the interpolated gradient:

$$u_{i,k}^* = \overline{(u_i^*)}_k - \Delta V_k \overline{\left(\frac{1}{A_P^{u_i}}\right)}_k \left[\left(\frac{\delta p}{\delta x_i}\right)_k - \overline{\left(\frac{\delta p}{\delta x_i}\right)}_k\right]^{m-1}, \qquad (9.62)$$

where $*$ denotes velocities at the outer iteration m predicted by solving the momentum equations using the pressure from the previous outer iteration. It was shown in Sect. 8.2.1 that, for a 2D uniform grid, the correction applied to interpolated velocity corresponds to a central-difference approximation to the third derivative of pressure multiplied by $(\Delta x)^2$; it detects oscillations and smooths them out. The correction term may be small and not fulfilling its role if A_P is too large. This can happen when unsteady problems are solved using very small time steps, because A_P contains $\Delta V/\Delta t$, but this problem rarely occurs. The correction term may be multiplied by a constant without affecting the consistency of the approximation. This approach to pressure-velocity coupling on colocated grids was developed in early 1980s and is usually attributed to Rhie and Chow (1983). It is widely used and is employed in most commercial CFD codes.

Only the normal velocity component contributes to the mass flux through a cell face. It depends on the pressure gradient in the normal direction. This allows us to write the following expression for the normal velocity component $v_n = \mathbf{v} \cdot \mathbf{n}$ at a cell face (although we do not solve an equation for this component):

$$v^*_{n,k} = \overline{(v^*_n)}_k - \Delta V_k \overline{\left(\frac{1}{A_P^{v_n}}\right)}_k \left[\left(\frac{\delta p}{\delta n}\right)_k - \left(\frac{\delta p}{\delta n}\right)_k\right]^{m-1} . \tag{9.63}$$

Because $A_P^{u_i}$ is the same for all velocity components in a given CV (except near some boundaries), one can replace $A_P^{u_i}$ by $A_P^{v_n}$.

One can calculate the derivative of pressure in the direction normal to face k at the neighboring CV-centers and interpolate it to the cell face center. Calculation of the normal derivative at the cell face directly would require a coordinate transformation, which is the usual procedure on structured grids. When using CVs of arbitrary shape, we would like to avoid use of coordinate transformations. Using shape functions is a possibility, but it results in a complex pressure-correction equation. The deferred-correction approach could be used to reduce the complexity.

Another approach can be constructed, using auxiliary nodes on face normal shown in Fig. 9.19, as described in Sect. 9.7.2 for diffusion fluxes. The pressure derivative with respect to n can be approximated by a central difference as follows:

$$\left(\frac{\delta p}{\delta n}\right)_k \approx \frac{p_{N'_k} - p_{P'}}{|\mathbf{r}_{N'_k} - \mathbf{r}_{P'}|} . \tag{9.64}$$

The values of pressure at the two auxiliary nodes can be calculated using cell-center values and gradients:

$$\begin{aligned} p_{P'} &\approx p_P + (\nabla p)_P \cdot (\mathbf{r}_{P'} - \mathbf{r}_P) , \\ p_{N'_k} &\approx p_{N_k} + (\nabla p)_{N_k} \cdot (\mathbf{r}_{N'_k} - \mathbf{r}_{N_k}) . \end{aligned} \tag{9.65}$$

With these expressions, Eq. (9.64) becomes:

$$\left(\frac{\delta p}{\delta n}\right)_k \approx \frac{p_{N_k} - p_P}{|(\mathbf{r}_{N'_k} - \mathbf{r}_{P'})|} + \frac{(\nabla p)_{N_k} \cdot (\mathbf{r}_{N'_k} - \mathbf{r}_{N_k}) - (\nabla p)_P \cdot (\mathbf{r}_{P'} - \mathbf{r}_P)}{|(\mathbf{r}_{N'_k} - \mathbf{r}_{P'})|} . \tag{9.66}$$

The second term on the right-hand side disappears when the line connecting nodes P and N_k is orthogonal to the cell face and passes through its center, i.e., when P and P' and N_k and N'_k coincide. If the objective is to just prevent pressure oscillations on colocated grids, it is sufficient to use just the first term on the right-hand side of Eq. (9.66), i.e., one can approximate Eq. (9.63) as:

$$v^*_{n,k} = \overline{(v^*_n)}_k - \frac{\Delta V_k}{|(\mathbf{r}_{N'_k} - \mathbf{r}_{P'})|} \overline{\left(\frac{1}{A_P^{v_n}}\right)}_k \left[(p_{N_k} - p_P) - \overline{(\nabla p)}_k \cdot (\mathbf{r}_{N_k} - \mathbf{r}_P)\right] . \tag{9.67}$$

The correction term in square brackets thus represents the difference between the pressure difference $p_{N_k} - p_P$ and the approximation to it calculated using interpolated pressure gradient, $\overline{(\nabla p)}_k \cdot (\mathbf{r}_{N_k} - \mathbf{r}_P)$. For a smooth pressure distribution, this correction term is small and it tends to zero as the grid is refined. The pressure gradient at the CV centers is available as it was calculated for use in the momentum equations.

Note that interpolated pressure gradient at the face should be computed from values at the two cell centers weighted by 1/2 rather than interpolated according to distances to cell face. The reason is that the gradient computed at the face is second-order accurate not at the face but midway between cell centers; in order to ensure that the correction terms cancels out if pressure variation is smooth, gradients from cell centers should be interpolated to the same location, i.e., simply averaged.

The mass fluxes calculated using the interpolated velocity,

$$\dot{m}_k^* = (\rho v_n^* S)_k \, , \tag{9.68}$$

do not satisfy the continuity requirement, so their sum over all faces of the CV results in a mass source:

$$\sum_k \dot{m}_k^* = \Delta\dot{m} \, , \tag{9.69}$$

which must be reduced to zero. The velocities have to be corrected so that mass conservation is satisfied in each CV. In an implicit method it is not necessary to satisfy mass conservation exactly at the end of each outer iteration. Following the method described earlier, we correct the mass fluxes by expressing the velocity correction through the gradient of the pressure correction, thus:

$$\dot{m}_k' = (\rho v_n' S)_k \approx - (\rho \, \Delta V \, S)_k \overline{\left(\frac{1}{A_P^{v_n}}\right)}_k \left(\frac{\delta p'}{\delta n}\right)_k \approx$$
$$- (\rho \, \Delta V \, S)_k \overline{\left(\frac{1}{A_P^{v_n}}\right)}_k \left[\frac{p'_{N_k'} - p'_P}{|(\mathbf{r}_{N_k'} - \mathbf{r}_{P'})|} - \right. \tag{9.70}$$
$$\left. \frac{(\boldsymbol{\nabla} p')_{N_k} \cdot (\mathbf{r}_{N_k'} - \mathbf{r}_{N_k}) - (\boldsymbol{\nabla} p')_P \cdot (\mathbf{r}_{P'} - \mathbf{r}_P)}{|(\mathbf{r}_{N_k'} - \mathbf{r}_{P'})|}\right] .$$

If the same approximation is applied at the other CV faces, and it is required that the corrected mass fluxes satisfy the continuity equation:

$$\sum_k \dot{m}_k' + \Delta\dot{m} = 0 \, , \tag{9.71}$$

we obtain the pressure-correction equation.

The last term on the right-hand side of Eq. (9.70) leads to an extended computational molecule in the pressure-correction equation. Because this term is small when the non-orthogonality is not severe, it is common practice to neglect it. When the solution converges, the pressure correction becomes zero so the omission of this term does not affect the solution; however, it does affect the convergence rate. For substantially non-orthogonal grids, one has to use a smaller under-relaxation parameter α_p, see Eq. (7.84).

When the above approximation is used, the pressure-correction equation has the usual form; moreover, its coefficient matrix is symmetric so special solvers for symmetric matrices can be used (e.g., the ICCG solver from the conjugate gradients family; see Chap. 5 and the directory `solvers` in the codes repository; see Appendix A.1).

Upon solving the pressure-correction equation, the mass fluxes through cell faces are corrected using Eq. (9.70), leading to the final value at outer iteration m:

$$\dot{m}_k^m = \dot{m}_k^* + \dot{m}_k' \ . \tag{9.72}$$

Cell-center velocities and pressure are corrected by:

$$u_{i,P}^m = u_{i,P}^* - \frac{\Delta V}{A_P^{u_i}} \left(\frac{\delta p'}{\delta x_i}\right)_P \quad \text{and} \quad p_P^m = p_P^{m-1} + \alpha_p p_P' \ . \tag{9.73}$$

The grid non-orthogonality can be taken into account in the pressure-correction equation iteratively, i.e., by using the predictor-corrector approach. One solves first the equation for p' in which the non-orthogonality terms in Eq. (9.70) are neglected. In the second step one corrects the error made in the first step by adding another correction:

$$\dot{m}_k' + \dot{m}_k'' = -(\rho \, \Delta V \, S)_k \overline{\left(\frac{1}{A_P^{v_n}}\right)_k} \left(\frac{\delta p'}{\delta n} + \frac{\delta p''}{\delta n}\right)_k \ , \tag{9.74}$$

which—by neglecting the non-orthogonality terms in the second correction p'' but taking them into account for the first correction p', because p' is now available—leads to the following expression for the second mass-flux correction:

$$\dot{m}_k'' = -(\rho \, \Delta V \, S)_k \overline{\left(\frac{1}{A_P^{v_n}}\right)_k} \left[\frac{p_{N_k}'' - p_P''}{|(\mathbf{r}_{N_k'} - \mathbf{r}_{P'})|} - \frac{(\nabla p')_{N_k} \cdot (\mathbf{r}_{N_k'} - \mathbf{r}_{N_k}) - (\nabla p')_P \cdot (\mathbf{r}_{P'} - \mathbf{r}_P)}{|(\mathbf{r}_{N_k'} - \mathbf{r}_{P'})|} \right] \ . \tag{9.75}$$

The second term on the right-hand side can now be explicitly calculated, because p' is available.

Because the corrected fluxes $\dot{m}^* + \dot{m}'$ were already forced to satisfy the continuity equation, it follows that $\sum_c \dot{m}_c'' = 0$ should now be enforced. This leads to an equation for the second pressure correction p'', which has the same matrix A as the equation for p', but a different right-hand side. This can be exploited in some solvers. The source term of the second pressure correction contains the divergence of the explicit parts of $\dot{m}''$.

The correction procedure can be continued, by introducing third, fourth, etc. corrections. The additional corrections tend to zero; it is rarely necessary to go beyond the two already described as the pressure-correction equation in the SIMPLE

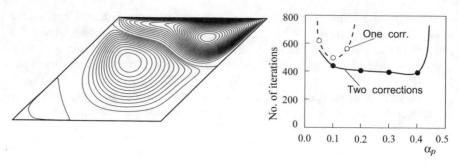

Fig. 9.20 Geometry and predicted streamlines in a lid-driven cavity with side walls inclined at 45°, at Re = 1000 (left), and the numbers of iterations using $\alpha_u = 0.8$ and one or two pressure-correction steps, as a function of α_p (right)

algorithm includes more severe approximations than the non-exact treatment of the effects of grid non-orthogonality.

The inclusion of the second pressure correction has a minor effect on the performance of the algorithm if the grid is nearly orthogonal. However, if the angle between **n** and **d** (see Fig. 9.19) is greater than 45° in much of the domain, convergence may be slow with only one correction. Strong under-relaxation (adding only 5–10% of p' to p^{m-1}) and reduction of the under-relaxation factors for velocity may help but at a cost in efficiency. With two pressure-correction steps, the performance found on orthogonal grids is obtained for non-orthogonal grids as well.

An example of performance degradation on non-orthogonal grids without the second correction is shown in Fig. 9.20. Flow in a lid-driven cavity with side walls inclined at 45° was calculated at Re = 1000; Fig. 9.20 also shows the geometry and computed streamlines. The grid lines are parallel to the walls. With the second pressure correction, the numbers of iterations required for convergence and their dependence on the under- relaxation factor for pressure, α_p, are similar to what is found for orthogonal grids; see Fig. 8.13. If the second correction is not included, the range of usable parameter α_p is very narrow and more iterations are required. Similar results are obtained for other values of the under-relaxation factor for velocity α_u, the differences being greater for larger values of α_u. The range of α_p for which convergence is obtained becomes narrower when the angle between grid lines is reduced and only one pressure correction is calculated.

The method described here is implemented in the code found in the directory 2dgl (see the appendix).

On structured grids, one can transform the normal pressure derivative at cell faces into a combination of derivatives along grid line directions and obtain a pressure-correction equation which involves mixed derivatives; see Sect. 9.5. If the cross-derivatives are treated implicitly, the computational molecule of the pressure-correction equation contains at least nine nodes in 2D and nineteen nodes in 3D. The above two-step procedure results in similar convergence properties as the use of implicitly discretized cross-derivatives (see Perić 1990), but is computationally more efficient, especially in 3D.

9.9 Axisymmetric Problems

Axisymmetric flows are three-dimensional with respect to Cartesian coordinates, i.e., the velocity components are functions of all three coordinates; however, they are only two-dimensional in a cylindrical coordinate system (all derivatives with respect to the circumferential direction are zero, and all three velocity components are functions of only the axial and radial coordinates, z and r). In cases without swirl, the circumferential velocity component is zero everywhere. As it is much easier to work with two independent variables than three, for axisymmetric flows, it makes sense to work in a cylindrical coordinate system rather than a Cartesian one.

In differential form, the 2D conservation equations for mass and momentum, written in a cylindrical coordinate system, read (see, e.g., Bird et al. 2006):

$$\frac{\partial \rho}{\partial t} + \frac{\partial (\rho v_z)}{\partial z} + \frac{1}{r}\frac{\partial (\rho r v_r)}{\partial r} = 0 , \tag{9.76}$$

$$\frac{\partial (\rho v_z)}{\partial t} + \frac{\partial (\rho v_z v_z)}{\partial z} + \frac{1}{r}\frac{\partial (\rho r v_r v_z)}{\partial r} = -\frac{\partial p}{\partial z} + \frac{\partial \tau_{zz}}{\partial z} + \frac{1}{r}\frac{\partial (r\tau_{zr})}{\partial r} + \rho b_z , \tag{9.77}$$

$$\frac{\partial (\rho v_r)}{\partial t} + \frac{\partial (\rho v_z v_r)}{\partial z} + \frac{1}{r}\frac{\partial (\rho r v_r v_r)}{\partial r} = -\frac{\partial p}{\partial r} + \frac{\partial \tau_{rz}}{\partial z} + \frac{1}{r}\frac{\partial (r\tau_{rr})}{\partial r} +$$
$$\frac{\tau_{\theta\theta}}{r} + \frac{\rho v_\theta^2}{r} + \rho b_r , \tag{9.78}$$

$$\frac{\partial (\rho v_\theta)}{\partial t} + \frac{\partial (\rho v_z v_\theta)}{\partial z} + \frac{1}{r}\frac{\partial (\rho r v_r v_\theta)}{\partial r} = -\frac{\rho v_r v_\theta}{r} + \frac{\partial \tau_{\theta z}}{\partial z} +$$
$$\frac{1}{r^2}\frac{\partial (r^2 \tau_{r\theta})}{\partial r} + \rho b_\theta , \tag{9.79}$$

where the non-zero stress tensor components are:

$$\tau_{zz} = 2\mu \frac{\partial v_z}{\partial z} - \frac{2}{3}\mu \nabla \cdot \mathbf{v} , \quad \tau_{rr} = 2\mu \frac{\partial v_r}{\partial r} - \frac{2}{3}\mu \nabla \cdot \mathbf{v} ,$$

$$\tau_{\theta\theta} = -2\mu \frac{v_r}{r} - \frac{2}{3}\mu \nabla \cdot \mathbf{v} , \quad \tau_{rz} = \tau_{zr} = \mu \left(\frac{\partial v_z}{\partial r} + \frac{\partial v_r}{\partial z} \right) , \tag{9.80}$$

$$\tau_{\theta r} = \tau_{r\theta} = \mu r \frac{\partial}{\partial r}\left(\frac{v_\theta}{r} \right) , \quad \tau_{\theta z} = \tau_{z\theta} = \mu \frac{\partial v_\theta}{\partial z} .$$

As discussed in Sect. 9.3.1, the above equations contain two terms which have no analog in Cartesian coordinates: the apparent *centrifugal force* $\rho v_\theta^2 / r$ in the equation for v_r, and the apparent *Coriolis force* $\rho v_r v_\theta / r$ in the equation for v_θ. These terms arise from the coordinate transformation and should not be confused with the centrifugal and Coriolis forces that appear in a rotating coordinate frame. If the swirl velocity v_θ is zero, the apparent forces are zero and the third equation becomes redundant.

When a FD-method is used, the derivatives with respect to both axial and radial coordinates are approximated in the same way as in Cartesian coordinates; any method described in Chap. 3 can be used.

Finite volume methods require some care. The conservation equations in integral form given earlier (e.g., (8.1) and (8.2)) remain the same, with the addition of apparent forces as source terms. These are integrated over the volume as described in Sect. 4.3. The CV size in the θ-direction is unity, i.e., one radian. Care is needed with pressure terms. If these are treated as body forces and the pressure derivatives in z and r directions are integrated over the volume as shown in Eq. (9.58), no additional steps are necessary. However, if the pressure is integrated over the CV surface as in Eq. (9.57), it is not sufficient to integrate only over north, south, west and east cell face, as was the case in plane 2D problems—one has to consider the radial component of pressure forces onto the front and back surfaces.

Thus, we have to add these terms, which have no counterparts in planar 2D problems, to the momentum equation for v_r:

$$Q^r = -\frac{2\mu\,\Delta V}{r_P^2}v_{r,P} + p_P\Delta S + \left(\frac{\rho v_\theta^2}{r}\right)_P \Delta V \,, \tag{9.81}$$

where ΔS is the area of the front face.

In the equation for v_θ, if it needs to be solved, one has to include the source term (the apparent Coriolis force):

$$Q^\theta = -\left(\frac{\rho v_r v_\theta}{r}\right)_P \Delta V \,. \tag{9.82}$$

The only other difference compared to planar 2D problems is the calculation of cell face areas and volumes. The areas of cell faces 'n', 'e', 'w' and 's' are calculated as in plane geometry, see Eq. (9.29), with the inclusion of a factor of r_k (where k denotes the cell face center). The areas of the front and back faces are calculated in the same way as the volume in plane geometry (where the third dimension is unity). The volume of axisymmetric CVs with any number of faces is:

$$\Delta V = \frac{1}{6}\sum_{i=1}^{N_v}(z_{i-1} - z_i)(r_{i-1}^2 + r_i^2 + r_i\,r_{i-1}) \,, \tag{9.83}$$

where N_v denotes the number of vertexes, counted counter-clockwise, with $i = 0$ corresponding to $i = N_v$.

An important issue in axisymmetric swirling flows is the coupling of radial and circumferential velocity components. The equation for v_r contains v_θ^2, and the equation for v_θ contains the product of v_r and v_θ as source terms; see above. The combination of the sequential (decoupled) solution procedure and Picard linearization may prove inefficient. The coupling can be improved by using the multigrid method for the outer iterations, see Chap. 12, by a coupled solution method, or by using more implicit linearization schemes, see Sect. 5.5.

If the coordinates z and r of the cylindrical coordinate system are replaced by x and y, the analogy with the equations in Cartesian coordinates becomes obvious. Indeed, if r is set to unity and v_θ and $\tau_{\theta\theta}$ are set to zero, these equations become identical to those in Cartesian coordinates, with $v_z = u_x$ and $v_r = u_y$. Thus the same computer code can be used for both plane and axisymmetric 2D flows; for axisymmetric problems, one sets $r = y$ and includes $\tau_{\theta\theta}$ and, if the swirl component is non-zero, the v_θ equation.

9.10 Higher-Order Finite-Volume Methods

It is worth noting that the derivation of high order FV-methods is more difficult than construction of FD methods of high order. In FD-methods, we only have to approximate the first and the second derivatives at a grid point with higher-order approximations, which is relatively easy to do on structured grids (see Chap. 3). In FV-methods, there are three kinds of approximation:

- Approximation of surface and volume integrals;
- Interpolation of variable values to locations other than CV center;
- Approximation of first derivatives at cell center and all cell faces.

The second-order accuracy of the midpoint rule is the highest accuracy achievable with single-point approximations. Any higher-order FV-scheme requires interpolation of higher order to more than one cell-face location as well as more sophisticated multi-point integral approximations to compute convection fluxes. For diffusion fluxes, one needs, in addition, to approximate the derivatives at multiple locations within cell faces with higher order. This is manageable on structured grids, but rather difficult on unstructured grids, especially those made of arbitrary polyhedral CVs. For the sake of simplicity of implementation, extension, debugging and maintenance, second-order accuracy appears to be the best compromise between accuracy and efficiency.

Only when very high accuracy is required (discretization errors below 1%) do higher-order methods become cost-effective. One also has to bear in mind that higher-order methods produce more accurate results than a second-order method only if the grid is *sufficiently fine*. If the grid is not fine enough, higher-order methods may produce oscillatory solutions, and the average error may be higher than for a second-order scheme. Higher-order schemes also require more memory and computing time per grid point than second-order schemes. For industrial applications, for which errors of the order of 1% are acceptable, a second-order scheme coupled with local grid refinement offers the best combination of accuracy, simplicity of programming and code maintenance, robustness, and efficiency.

As already mentioned, higher-order methods can be easier realized using FD- or FE-approach than using FV-method. Finite-element methods of high order are quite standard in structural mechanics, especially for linear problems. The continuity equation for incompressible flows causes trouble, which is why FE-methods for CFD usually use unequal order of approximations for momentum and continuity equations. FD-methods for unstructured grids are also not common nowadays, but we expect to see higher-order methods of this kind in the near future.

9.11 Implementation of Boundary Conditions

The implementation of boundary conditions on non-orthogonal grids requires special attention because the boundaries are usually not aligned with the Cartesian velocity components. The FV-method requires that the boundary fluxes either be known or expressed in terms of known quantities and interior nodal values. Of course, the number of CVs must match the number of unknowns.

We shall often refer to a local coordinate system (n, t, s), which is a rotated Cartesian frame with n being the outward normal to the boundary and t and s tangential to the boundary.

9.11.1 Inlet

Usually, at an inlet boundary, all quantities have to be prescribed. If the conditions at the inlet are not well known and approximations to variable profiles need to be made, it is useful to move the boundary upstream as far from the region of interest as possible. Because the velocity and other variables are given, all convection fluxes can be directly calculated. The diffusion fluxes are usually not known, but they can be approximated using known boundary values of the variables and one-sided finite-difference approximations for the gradients.

If velocity is specified at a boundary, then it does not need to be corrected during iterations. If the velocity correction at a boundary is equal to zero, that translates to the zero-gradient condition for pressure correction in the SIMPLE algorithm; see Eq. (9.70). Thus, the pressure-correction equation has Neumann boundary conditions at all boundaries where velocity is specified.

9.11.2 Outlet

At the outlet we usually know little about the flow. For this reason, these boundaries should be as far downstream of the region of interest as possible. Otherwise, errors may propagate upstream. The flow should be directed out of the domain over the entire outlet cross-section, and if possible, be parallel and orthogonal to the outlet boundary. In high Reynolds-number flows, upstream propagation of errors—at least in steady flows—is weak so it is easy to find suitable approximations for boundary conditions. Usually one extrapolates along grid lines from the interior to the boundary (or, better, along streamlines). The simplest approximation is that of zero gradient along grid lines. For the convection flux this means that a first-order upwind approximation is used. The condition of zero gradient along a grid line can be easily implemented implicitly. For example, at the east face of a structured 2D grid, the first-order backward approximation gives $\phi_E = \phi_P$. When we insert this expression into the discretized equation for the CV next to boundary, we have:

$$(A_P + A_E)\phi_P + A_W\phi_W + A_N\phi_N + A_S\phi_S = Q_P , \qquad (9.84)$$

so the boundary value ϕ_E does not appear in the equation. This does not mean that the diffusion flux is zero at the outlet boundary, except when the grid is orthogonal to the boundary.

If higher accuracy is required, one has to use higher-order, one-sided finite-difference approximations of the derivatives at the outlet boundary. Both convection and diffusion fluxes have to be expressed in terms of the variable values at inner nodes.

If velocities are extrapolated to outlet boundary, one usually corrects them—if the flow is assumed to be incompressible—so that the outflow mass flux matches the inflow mass flux. The continuity equation is then globally satisfied, i.e., when mass conservation equations for all CVs are summed, the mass fluxes over all inner cell faces cancel out and when boundary fluxes match, the net mass flux is zero. The consequence of correcting boundary velocities in this way is that they can be treated as fixed for the current outer iteration, thus leading to the zero-gradient condition for the pressure-correction equation. If Neumann conditions apply at all boundaries, one has to ensure that the algebraic sum of source terms in the pressure-correction equation is equal to zero—otherwise the formulation is not well posed. The correction of outlet velocities described above guarantees that this condition is satisfied. However, when Neumann conditions apply to all boundaries, the solution of the pressure-correction equation is not unique—one can add a constant to all values and the equation is still satisfied. For this reason one usually keeps pressure fixed at one reference location and corrects pressure by adding the difference between the pressure correction computed at given grid point and the pressure correction computed at the reference location.

When the flow is unsteady, especially when turbulence is directly simulated, care is needed to avoid reflection of errors at the outlet boundary. These issues are discussed in Sects. 10.2 and 13.6.

9.11.3 Impermeable Walls

At an impermeable wall, the following condition applies:

$$u_i = u_{i,\text{wall}} . \tag{9.85}$$

This condition follows from the fact that viscous fluids stick to solid boundaries (no-slip condition).

Because there is no flow through the wall, convection fluxes of all quantities are zero. Diffusion fluxes require some attention. For scalar quantities, such as thermal energy, they may be zero (adiabatic walls), they may be specified (prescribed heat flux), or the value of the scalar may be prescribed (isothermal walls). If the flux is known, it can be inserted into the conservation equation for the near-wall CVs, e.g., for the boundary face labeled 's':

$$F_s^d = \int_{S_s} \Gamma \, \nabla \phi \cdot \mathbf{n} \, dS = \int_{S_s} \Gamma \left(\frac{\partial \phi}{\partial n} \right) dS = \int_{S_s} f \, dS \approx f_s S_s , \tag{9.86}$$

where f is the prescribed flux per unit area. If the value of ϕ is specified at the wall, we need to approximate the normal gradient of ϕ using one-sided differences. From such an approximation we can also calculate the value of ϕ at the wall when the flux is prescribed. Many possibilities exist; one is to calculate the value of ϕ at an auxiliary point P′ located on the normal n, see Fig. 9.21, and use the approximation:

$$\left(\frac{\partial \phi}{\partial n} \right)_s \approx \frac{\phi_{P'} - \phi_S}{\delta n} , \tag{9.87}$$

where $\delta n = (\mathbf{r}_S - \mathbf{r}_{P'}) \cdot \mathbf{n}$ is the distance between points P′ and S. If the non-orthogonality is not severe, one can use ϕ_P instead of $\phi_{P'}$. Shape functions or extrapolated gradients from cell centers can also be used. The flux can then be approximated using the midpoint rule as:

$$F_s^d \approx \Gamma_s \left(\frac{\partial \phi}{\partial n} \right)_s S_s \approx \Gamma_s \frac{\phi_{P'} - \phi_S}{\delta n} S_s . \tag{9.88}$$

Diffusion fluxes in the momentum equations require special attention. If we were solving for the velocity components v_n, v_t and v_s, we could use the approach described in Sect. 7.1.6. The viscous stresses at a wall are:

$$\tau_{nn} = 2\mu \left(\frac{\partial v_n}{\partial n} \right)_{\text{wall}} = 0 , \quad \tau_{nt} = \mu \left(\frac{\partial v_t}{\partial n} \right)_{\text{wall}} . \tag{9.89}$$

Here we assume that the coordinate t is in the direction of the shear force at the wall, so $\tau_{ns} = 0$. This force is parallel to the projection of the velocity vector onto the wall (s is orthogonal to it). This is equivalent to the assumption that the velocity vector

Fig. 9.21 On the implementation of boundary conditions at a wall

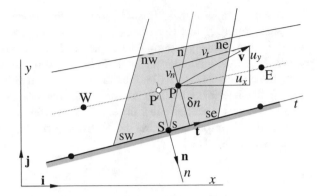

does not change its direction between the first grid point and the wall, which is not quite true but is a reasonable approximation which becomes more accurate as the distance of the first grid point to wall reduces.

Both v_t and v_n can easily be calculated at node P. In 2D, the unit vector $\mathbf{t}$ is easily obtained from coordinates of the corners 'se' and 'sw', see Fig. 9.21. In 3D, we have to determine the direction of the vector $\mathbf{t}$. From the velocity parallel to the wall we can define unit vector $\mathbf{t}$ as follows:

$$\mathbf{v}_t = \mathbf{v} - (\mathbf{v} \cdot \mathbf{n})\mathbf{n} \quad \Rightarrow \quad \mathbf{t} = \frac{\mathbf{v}_t}{|\mathbf{v}_t|} . \tag{9.90}$$

The velocity components needed to approximate the stresses are then:

$$v_n = \mathbf{v} \cdot \mathbf{n} = un_x + vn_y + wn_z , \quad v_t = \mathbf{v} \cdot \mathbf{t} = ut_x + vt_y + wt_z . \tag{9.91}$$

The derivatives can be calculated as in Eq. (9.87).

One could transform the stress τ_{nt} to obtain τ_{xx}, τ_{xy} etc., but this is not necessary. The surface integral of τ_{nt} results in a force:

$$\mathbf{f}_{\text{wall}} = \int_{S_s} \mathbf{t}\tau_{nt} \, dS \approx (\mathbf{t}\tau_{nt} S)_s , \tag{9.92}$$

whose x, y and z components correspond to the integrals needed in the discretized momentum equations; e.g., in the equation for u_x:

$$f_x = \int_{S_s} (\tau_{xx}\mathbf{i} + \tau_{yx}\mathbf{j} + \tau_{zx}\mathbf{k}) \cdot \mathbf{n} \, dS = \mathbf{i} \cdot \mathbf{f}_{\text{wall}} \approx (t_x \tau_{nt} S)_s . \tag{9.93}$$

Alternatively we can use the velocity gradients at cell centers (calculated, e.g., using the Gauss theorem, see Eq. (9.49)), extrapolate them to the center of wall cell face, calculate the shear stresses τ_{xx}, τ_{xy} etc., and calculate the shear force components from the above expression.

We, thus, replace the diffusion fluxes in the momentum equations at walls by the shear force. If this force is calculated explicitly using values from the previous iteration, convergence may be impaired. If the force is written as a function of Cartesian velocity components at node P, part of it can be treated implicitly. In this case the coefficients A_P will not be the same for all velocity components (as is the case for the interior cells). This is undesirable, because the coefficients A_P are needed in the pressure-correction equation, and if they differ, we would have to store all three values. It is therefore best to use the deferred-correction approach, as in the interior; we approximate

$$f_i^i = \mu S \frac{\delta u_i}{\delta n} , \tag{9.94}$$

implicitly and add the difference between the implicit approximation and the force calculated using one of the above-mentioned approaches to the right-hand side of the equation. Here δn is the distance of node P from wall. The coefficient A_P is then the same for all velocity components, and the explicit terms partially cancel out. The rate of convergence is almost unaffected.

Because velocities are specified at walls, the pressure-correction equation has Neumann conditions at wall boundaries.

9.11.4 Symmetry Planes

In many flows there are one or more symmetry planes. When the flow is steady, there is a solution which is symmetric with respect to this plane (in many cases, e.g., diffusers or channels with sudden expansions, there exist also asymmetric steady solutions, which are usually more stable than the symmetric solution). The symmetric solution can be obtained by solving the problem in part of the solution domain only, using symmetry conditions.

At a symmetry plane the convection fluxes of all quantities are zero. Also, the normal gradients of the velocity components parallel to symmetry plane and of all scalar quantities are zero there. Thus, diffusion fluxes of all scalar quantities are zero at symmetry planes. The normal velocity component is zero, but its derivative in the normal direction is not; thus, the normal stress τ_{nn} is non-zero. The surface integral of τ_{nn} results in a force:

$$\mathbf{f}_{\text{sym}} = \int_{S_s} \mathbf{n} \tau_{nn} \, \mathrm{d}S \approx (\mathbf{n} \tau_{nn} S)_s . \tag{9.95}$$

When the symmetry boundary does not coincide with a Cartesian coordinate plane, the diffusion fluxes of all three Cartesian velocity components will be non-zero. These fluxes can be calculated by obtaining first the resultant normal force from (9.95) and an approximation of the normal derivative as described in the preceding section, and splitting this force into its Cartesian components. Alternatively, one can extrapolate the velocity gradients from interior to the boundary and use an expression similar to (9.93), e.g., for the u_x component at the face 's' (see Fig. 9.21):

$$f_x = \int_{S_s} (\tau_{xx}\mathbf{i} + \tau_{yx}\mathbf{j} + \tau_{zx}\mathbf{k}) \cdot \mathbf{n}\, dS = \mathbf{i} \cdot \mathbf{f}_{\text{sym}} \approx (n_x \tau_{nn} S)_s \,. \tag{9.96}$$

As in the case of wall boundaries, one can split the diffusion fluxes at a symmetry boundary into an implicit part, involving the velocity components at the CV-center (which contributes to the coefficient A_P) or use the deferred-correction approach to keep A_P same for all velocity components.

Again, the pressure-correction equation has Neumann boundary conditions at symmetry boundaries due to the fact that the normal velocity component is prescribed.

9.11.5 Specified Pressure

In incompressible flows one usually specifies the mass flow rate at inlet and uses extrapolation at the outlet. However, there are situations in which the mass flow rate is not known, but the pressure drop between inlet and outlet is prescribed. Also, pressure is sometimes specified at a far-field boundary.

When the pressure is specified at a boundary, velocity cannot be prescribed—it has to be extrapolated from the interior using the same approach as for cell faces between two CVs, see Eq. (9.63); the only difference is that now the locations of the cell face and one neighbor node coincide. The pressure gradient at the boundary is approximated using one-sided differences; for example, at the 'e' face, one can use the following expression, which is a first-order backward difference:

$$\left(\frac{\partial p}{\partial n}\right)_e \approx \frac{p_E - p_P}{(\mathbf{r}_E - \mathbf{r}_P) \cdot \mathbf{n}} \,. \tag{9.97}$$

The boundary velocities determined in this way need to be corrected to satisfy the mass conservation; the mass flux corrections $\dot{m}'$ are not zero at boundaries where the pressure is specified. However, boundary pressure is not corrected, i.e., $p' = 0$ at the boundary. This is used as a Dirichlet boundary condition in the pressure-correction equation. More details about the implementation of boundary conditions when static pressure is specified at a boundary can be found in Chap. 11.

If the Reynolds number is high, the solution process will converge slowly if the above approach is applied when the inlet and outlet pressures are specified. Another possibility is to guess first the mass flow rate at the inlet and treat it as prescribed for one outer iteration, and consider the pressure to be specified only at the outlet. The inlet velocities should then be corrected by trying to match the extrapolated pressure at the inlet boundary with the specified pressure. An iterative correction procedure is used to drive the difference between the two pressures to zero.

9.12 Examples

In this section we present examples of computing laminar flows in geometries which require body-fitted grids. Two examples deal with steady-state flows and one with unsteady flow. In one case a structured grid and a code which can be downloaded from the Internet are used; for the other two examples, a commercial CFD software is used. The aim of these examples is to demonstrate how such flow problems can be solved, how to analyze the accuracy of solutions and how different grid types affect the computing effort and quality of results.

9.12.1 Flow Around Circular Cylinder at Re = 20

As an example we consider first laminar 2D flow around a circular cylinder in an infinite environment exposed to a uniform cross-flow at Reynolds number Re = 20. The Reynolds number is based on the uniform stream velocity U_∞, fluid viscosity μ and the cylinder diameter D. The solution domain is finite and extends $16D$ upstream and downstream of the cylinder, as well above and below it, see Fig. 9.22 which shows the whole solution domain and the boundary-fitted, structured O-type grid used for these computations. The two computer codes that were used—one based on the SIMPLE algorithm and one on the implicit fractional-step method (labeled IFSM)—are available on the Internet; see Appendix A.1 for details.

Five systematically refined grids (i.e., each coarser grid CV is split into four finer-grid CVs) were used in order to be able to estimate discretization errors; the coarsest grid had 24×16 CVs and the finest had 384×256 CVs (the number of cells around

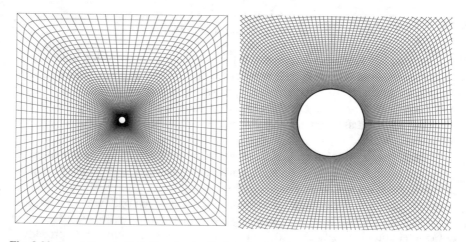

Fig. 9.22 Level-three grid (left) and a detail of the level-four grid around a cylinder (right), used to compute steady and unsteady 2D flow around a circular cylinder

Fig. 9.23 Predicted
streamlines in cylinder
vicinity at Re = 20

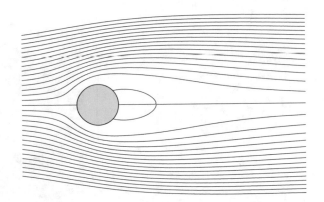

cylinder × the number of cells in radial direction). Cells are uniformly distributed around the cylinder, while they expand in the radial direction; the expansion factor is 1.25 on the coarsest grid and for every finer grid, the expansion factor equals the square root of the expansion factor on the preceding grid. Thus, on the finest grid, the expansion factor is only 1.014044, due to the fact that now 256 cells are fit within the same distance from the cylinder to the outside boundary. Because grids are of O-type, they have a seam where west and east boundary meet each other: it is the horizontal centerline behind the cylinder, as indicated by the thicker line in Fig. 9.22 for the level-four grid. At the left boundary, velocity $U_\infty = 1$ m/s was specified (cylinder diameter is $D = 1$ m); at the downstream side velocities are extrapolated with zero-gradient condition; at the top and bottom boundary, symmetry conditions are specified. Note that, in the O-grid used, all these segments are portions of the south boundary, while the cylinder surface represents the north boundary. A second-order CDS was used for spatial discretization.

The flow at this Reynolds number is steady; it separates from cylinder surface and forms two weak recirculating vortexes behind cylinder, as can be seen in the plot of streamlines in Fig. 9.23 and velocity vectors in Fig. 9.24. The uniform incoming flow is deflected by the cylinder in a large zone around it; the distance of $16D$ from cylinder to the outer boundary of solution domain is probably not sufficient to represent the truly undisturbed far-field condition of uniform flow, but it is expected that the effect of the outer boundary on the flow around cylinder is not too strong. In any case, when we determine the discretization errors below, they will be valid only for the flow subject to the specified boundary conditions.

Figure 9.24 shows also pressure contours; they clearly indicate that the presence of cylinder is felt even at a relatively large distance, because pressure gradients are non-zero. The highest pressure is found at the front stagnation point, as expected. Dense isobars indicate a rapid pressure decrease over both upper and lower part of cylinder surface, until the minimum is reached shortly after the equator. From there onwards, pressure starts increasing again and this adverse pressure gradient leads to flow separation and formation of recirculation zone behind the cylinder.

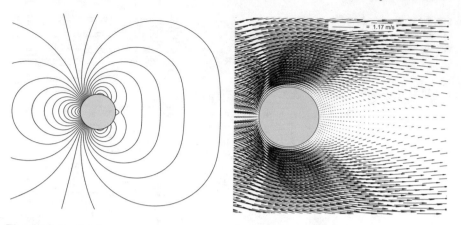

Fig. 9.24 Predicted pressure contours (left) and velocity vectors (right) in cylinder vicinity at Re = 20; only every fourth vector is plotted. The maximum velocity vector is shown on the inset for scale

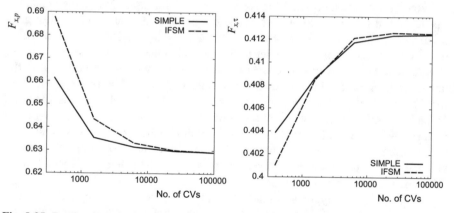

Fig. 9.25 Predicted pressure (left) and shear (right) force on cylinder at Re = 20

In an inviscid flow, pressure contours would be fully symmetric around the vertical centerline, with pressure at the rear stagnation point being the same as at the front, thus resulting in zero drag force and no recirculation. Due to fluid viscosity, both pressure (normal to wall) and shear (tangential to wall) forces, when integrated over cylinder surface, lead to a non-zero net force component in the flow direction. Because the steady flow is symmetric around the horizontal centerline, the lift force is equal to zero. The convergence of pressure and shear drag component towards a grid-independent solution is shown in Fig. 9.25.

On each grid, outer iterations were performed until residual norms (the sum of absolute values of residuals at all CVs) were reduced seven orders of magnitude; this is more than necessary, but we wanted to be sure that iteration errors were negligible when determining discretization errors. Results from both the SIMPLE and IFSM

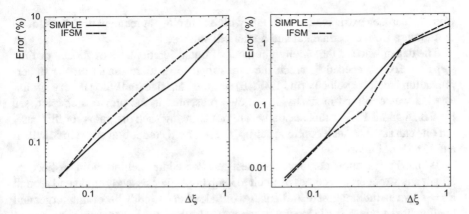

Fig. 9.26 Discretization errors in pressure force (left) and in shear force (right) on cylinder at $Re = 20$, estimated using the Richardson extrapolation technique

methods converge towards the same grid-independent solution, but the errors on coarse grids are smaller when the SIMPLE algorithm is used. It is interesting to note in Fig. 9.25 that pressure force is overestimated on coarse grids, while the shear force is under-estimated. Therefore, the relative error in the total force is lower than errors in component forces, because those are of the opposite sign and thus partially cancel out. It is not unusual to find out that errors from different sources cancel out, but they may augment as well. It is, therefore, always important to check the grid dependence of the solution.

Figure 9.26 shows estimated discretization errors using Richardson extrapolation (see Sect. 3.9 for details). Second-order convergence towards a grid-independent solution is obtained, as expected. On the coarsest grid, both pressure and shear force are substantially in error (around 10% and 3%, respectively); on the finest grid, the errors are quite low (0.03% and 0.007%, respectively). The level-four grid delivers solutions with errors of the order of 0.1% which would be sufficiently small for most applications.

For flows around bodies, the drag and lift coefficients are defined as:

$$C_D = \frac{F_x}{\frac{1}{2}\rho U_\infty^2 S} , \quad C_L = \frac{F_y}{\frac{1}{2}\rho U_\infty^2 S} , \tag{9.98}$$

where F_x and F_y are the x and y component of the force exerted by the fluid on the body and S is the cross-sectional body area perpendicular to flow direction. In a 2D flow simulation, the dimension in z-direction is assumed to be unity, and because in our simulation $D = 1$ was used, the area is also equal to $1\,\text{m}^2$. The velocity of undisturbed flow was also taken to be $U_\infty = 1\,\text{m/s}$. Therefore, if the fluid density $\rho = 1\,\text{kg/m}^3$, the drag and lift coefficients are obtained by simply multiplying the computed forces by 2. The drag coefficient thus obtained from the total force resulting from Richardson extrapolation is 2.083, which is in close agreement with data found in literature from experimental and numerical studies.

The dependence of the required number of outer iterations (in SIMPLE) or time steps (in IFSM) needed to reach the prescribed level of residuals on the under-relaxation factor for velocity (in SIMPLE) or time-step size (in IFSM) is very similar to what was observed in steady-state flows presented in the previous chapter, see Figs. 8.14 and 8.18. For this reason we are not showing such diagrams for this case, but one can say that, under optimal set-up, the required computing effort in SIMPLE and IFSM is similar.

When the Reynolds number increases, the two eddies behind cylinder become longer and stronger; it becomes difficult to keep both eddies exactly equal and around Re = 45, even the smallest disturbances would lead to one eddy becoming larger and the flow loses symmetry. Once the symmetry is broken, the flow becomes unsteady; eddies start outgrowing each other and detaching alternately from each side of the cylinder, leading to the well-known von Karman vortex street. Both experiments and simulations show that around Re = 200 the flow becomes three-dimensional and no longer perfectly periodic; finally, at still larger Reynolds numbers, the wake becomes turbulent. This, of course, is only seen in 3D simulations. In the next section, we take a closer look at the unsteady flow around cylinder at Re = 200.

9.12.2   Flow Around Circular Cylinder at Re = 200

We present here some results from simulations performed in 2D for Re = 200, using the commercial flow solver STAR-CCM+ and three systematically refined grids (in each refinement step, the grid spacing was halved in both directions). Around the cylinder 10 prism layers are created; the outer surface of the prism layer cuts the Cartesian grid, which fills the remaining space. The domain size is the same as in the previous example: it is rectangular and extends 16 D in both positive and negative x- and y-directions away from cylinder center. The Cartesian grid was locally refined in 4 steps so that the cell size is 6.25% of the 'base size' in a rectangular zone around cylinder and in its wake; Fig. 9.27 shows the part of the coarsest grid which includes the finest zone. The coarsest grid had 6,196 CV; the medium grid had 21,744 CV; the finest grid had 109,320 CV. The thickness of the prism layer was proportional to the base size and was thus halved each time the grid was refined. On the finest grid, there were around 220 cells along cylinder perimeter. The simulation file and a more detailed report about the simulation are available in the Internet; see the Appendix for details.

Only the second-order time discretization (three-time-level scheme; see Sect. 6.3.2.4) was used; it was demonstrated in Sect. 6.4 that first-order time advancing schemes are not suitable for simulation of unsteady problems when a time-accurate solution is required. Four time steps were used to test time-step dependence of solution: 0.04 s, 0.02 s, 0.01 s and 0.005 s, corresponding to ca. 63, 126, 253 and 506 time steps per period of drag oscillation, respectively.

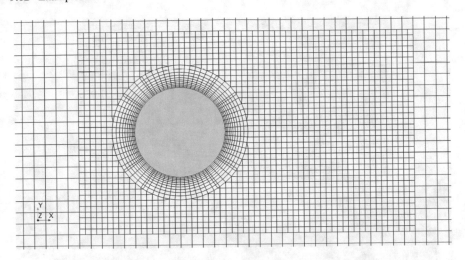

Fig. 9.27 Detail of the coarsest grid around cylinder used to simulate the flow at Re = 200

The flow around cylinder at Re = 200 is highly unsteady, with strong vortexes shedding off of the cylinder at a regular frequency. On the finest grid, this frequency was determined to be $f = 0.1977$, which corresponds to the dimensionless Strouhal-number (because in our case both D and U_∞ are equal to 1):

$$\text{St} = \frac{fD}{U_\infty} = \frac{D}{U_\infty P} ,\tag{9.99}$$

where P is the period of oscillation of the lift force, found to be equal to 5.059 s. This value corresponds well to data found in the literature. The drag force oscillates with a double frequency, because it has one maximum and one minimum per each vortex shedding, while the maximum lift force occurs when a vortex sheds on one side and minimum when the next vortex sheds on the other side.

Figure 9.28 shows instantaneous velocity vectors and pressure contours computed on the finest grid using the smallest time step. One can see one big vortex just shed off the lower side; another vortex is starting to form at the upper side of cylinder, which will grow while the other vortex is moving away.

The distribution of isobars on the front side of cylinder is similar to that at Re = 20, cf. Fig. 9.24, except that the locations of minimum pressure have now moved upstream and are found before the equator and that the distribution on the upper and lower side of cylinder surface is not symmetric. Also, there are more contours between the front stagnation point (where maximum pressure is found) and the minimum pressure location, suggesting a higher pressure gradient. Indeed, as can be seen by comparing the plots of velocity vectors from Figs. 9.24 and 9.28, the fluid acceleration is much stronger at Re = 200 than at Re = 20. While at Re = 20 the maximum velocity in the field was 1.17 m/s (17% higher than the free stream

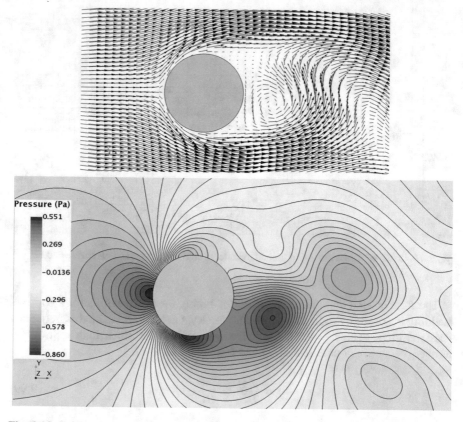

Fig. 9.28 Predicted instantaneous velocity vectors (upper) and isobars (lower) in the unsteady flow around circular cylinder at Re = 200; interpolated velocity vectors are shown on a uniform presentation grid

velocity), at Re = 200 the maximum velocity is 1.47 m/s (47% higher than the free stream velocity).

The SIMPLE algorithm was used to solve the Navier–Stokes equations. As noted earlier, SIMPLE requires two under-relaxation factors to be chosen. For steady-state flows, one typically chooses 0.8 for velocities and 0.2 for pressure, but when the flow is unsteady and small time steps are used, both under-relaxation factors can be increased. To be on the safe side, we used here 0.8 for velocities and 0.5 for pressure on all grids and for all time steps and forced the code to perform 10 iterations per time step. For the two smallest time steps one could increase the under-relaxation factor for velocity to 0.9 and instead of a fixed number of outer iterations per time step, one could use a suitable criterion to stop outer iterations when the criterion is satisfied (e.g., by specifying a level of residuals which should be reached). Figure 9.29 shows how the residual norms in the continuity and momentum equations varied with outer iterations within 4 time steps on the finest grid with the smallest time step. Note

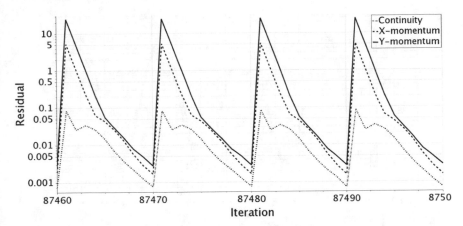

Fig. 9.29 Reduction of residuals during outer iterations in several time steps

that residuals in momentum equations drop more than 3 orders of magnitude in 10 iterations, while the residual norm in the continuity equation reduces about 2 orders of magnitude. The reason is that momentum equations are non-linear and a larger disturbance to the balance is invoked when advancing to the next time level than in the linear mass conservation equation.

The analysis of discretization errors when computing unsteady flows is more difficult than in the case of steady-state flows. When the unsteadiness is imposed by boundary conditions, as in the cases studied in the previous chapter (see Sect. 8.4.2), the situation is simpler because the period of oscillation is prescribed. In the present case the boundary conditions are steady and the flow unsteadiness results solely from the inherent instability; also, the cylinder surface is smooth and the flow separation point is not fixed, as would be the case for a cylinder with a rectangular cross-section. Thus, variation of both time step and grid size causes changes in all flow features.

Figure 9.30 shows the dependence of predicted drag and lift force on the grid fineness. Results for all three grids are presented; the time step was held constant at 0.01 s (ca. 253 time steps per period of drag oscillation). From this figure it is obvious that the difference between solutions from the coarse and medium grid is much larger than the difference between solutions on medium and fine grid. One expects, with a second-order discretization, that the difference between solutions on consecutive grids reduces by a factor 4 if the grid spacing is halved, which is the case here.

Figure 9.31 shows the dependence of solution on the time-step size for the fixed (fine) grid. The simulation was run with the largest time step (0.04 s) until a periodic state was reached; it was then saved and used as the starting point for subsequent simulations over 10 periods of lift force oscillation with all four time steps. Over a few periods, the solution adjusted to the new time step size and a periodic state was again reached. The figure shows clearly that a second-order convergence towards a time-step independent solution is obtained: the two curves corresponding to the two smallest time steps cannot be distinguished from each other, with the difference

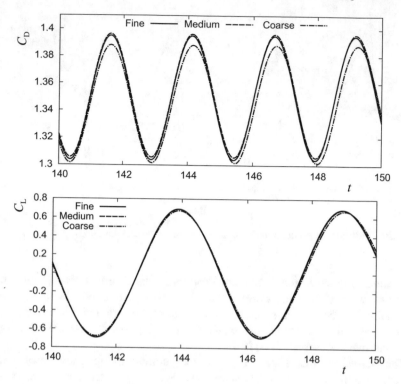

Fig. 9.30 Grid-dependence analysis: variation of drag (upper) and lift (lower) over two periods of lift force oscillation, computed on the three grids with the same time step, $\Delta t = 0.01\,\text{s}$

to the next coarser grid increasing by a factor of 4, as expected of a second-order time-discretization scheme.

The analysis of discretization errors in unsteady flow simulations requires choosing an adequate initial grid and time step size (based on varying one at a time), and then simultaneous refinement of both the grid and time step size.

9.12.3   Flow Around Circular Cylinder in a Channel at Re = 200

We also performed calculations of 3D laminar flow around a circular cylinder mounted between two walls in a duct with a square cross-section. Although it would be still possible to generate a decent block-structured grid for this geometry, we use this example to demonstrate the application of unstructured grids and commercial software for both grid generation and flow computation. Three grid types are created and used for flow analysis: trimmed Cartesian (as in the previous example), tetrahe-

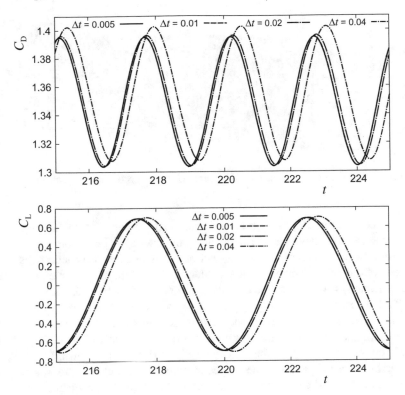

Fig. 9.31 Time-step-dependence analysis: variation of drag (upper) and lift (lower) over two periods of lift force oscillation, computed on the finest grid with four different time steps

dral and polyhedral. Prism layers along cylinder and channel walls were created in an analogous way in all three cases. The duct axis points in the x-direction and the cylinder axis points in the y direction (horizontal).

Figure 9.32 shows the geometry of the solution domain and the coarse polyhedral grid on the boundaries; Fig. 9.33 shows a longitudinal section at $y = 0$ through all three kinds of grid. The duct extensions are (in meters): $-1.75 \leq x \leq 3.25$, $-0.5 \leq y \leq 0.5$ and $-0.5 \leq z \leq 0.5$; the cylinder is positioned at the coordinate system origin and has a diameter of 0.4 m. The cylinder thus blocks 40% of the duct cross-section. The hypothetic fluid has a density of 1 kg/m^3 and a viscosity of 0.005 Pa·s. At the inlet ($x = -1.75$ m), a uniform velocity field with $u_x = 1$ m/s was specified, while at the outlet ($x = 3.25$ m), a constant pressure was prescribed. The Reynolds number based on duct height is Re $= 200$. The velocity field was initialized with a constant velocity in the x-direction equal to the inlet velocity. As shown in the previous example, the flow around a cylinder in an infinite environment would be unsteady under the same conditions, but due to the confinement in the duct, the laminar flow is still steady in this case.

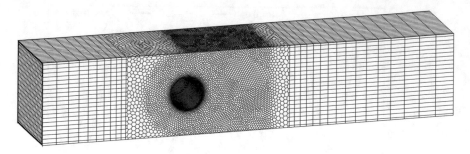

Fig. 9.32 Cylinder in a duct: geometry and a coarse polyhedral grid on solution domain boundaries

We generated three types of grid in a shorter duct segment around the cylinder and performed the so-called *grid extrusion* along the duct axis, both from the upstream and from the downstream duct cross-section. This leads to prismatic cells in the portions of the duct downstream of the inlet and upstream of the outlet, in which a gradual expansion or contraction of cells in x-direction is achieved, as can be seen in Fig. 9.32, where prisms with a polygonal base are created. In the case of tetrahedral grid, the extruded sections are made of prisms with triangular base, while in the case of trimmed hexahedral grid, the extruded regions contain elongated hexahedra. Local grid refinement around the cylinder, and to some extent downstream of it, was invoked by specifying two volume shapes for which a smaller cell size was required: a larger cylindrical shape around cylinder and a block on the downstream side. The grid generation software generates a regular-shape polyhedra (dodecahedra) and tetrahedra within regular refinement zones, as can be seen in Fig. 9.33.

Computations were performed using STAR-CCM+ software from Siemens; it is based on the FV-method and midpoint-rule approximations of all integrals. A second-order upwind scheme is used for convection (linear extrapolation to the cell-face center using the variable value and the gradient at the upstream cell center), while linear shape functions are used for approximation of gradients (leading to central differences on Cartesian grids). The SIMPLE algorithm is used with under-relaxation factors of 0.8 for velocities and 0.2 for pressure. Figure 9.34 shows the variation of residuals with outer iterations on a medium-size polyhedral grid (ca. 1.9 million cells). The residuals fall one order of magnitude very quickly, but iteration errors usually do not follow residuals from the very beginning (see Fig. 8.9 in the previous chapter). A safe way to estimate iteration errors from a residual plot is to extrapolate backward from the final state. For example, we see in Fig. 9.34 that the residual norm for the u_x-velocity is around 1e-6 at iteration 800. Projecting backward along the mean slope we reach iteration 0 at the level of around 0.01, thus indicating that the true reduction of iteration errors is around 4 orders of magnitude. Indeed, a closer look at monitoring values of velocities and pressure shows no variation in the first 4 significant digits, which is another useful check to make sure that the solution has really converged.

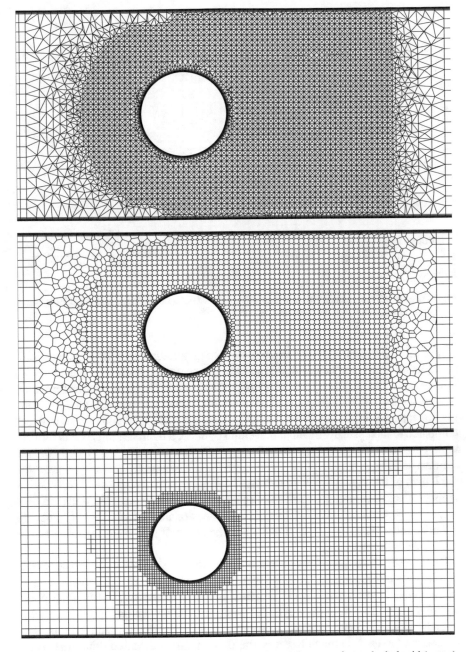

Fig. 9.33 Computational grid in the longitudinal symmetry plane $y = 0$: tetrahedral grid (upper), polyhedral grid (middle) and trimmed hexahedral grid (lower)

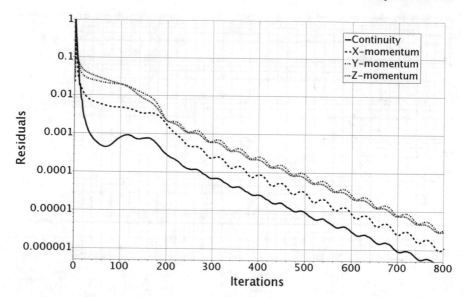

Fig. 9.34 Variation of residuals with increasing number of outer iterations in the SIMPLE algorithm for the 3D laminar flow around a circular cylinder in a duct

Figure 9.35 shows velocity vectors in the two longitudinal symmetry planes, computed using a trimmed hexahedral grid. One can see in the horizontal plane ($z = 0$) how velocity vectors turn backward at both cylinder ends; that indicates the formation of a horse-shoe-vortex. The vertical cross-section shows a strong acceleration of the flow as it passes around cylinder; velocity vectors double in length around cylinder compared to their size upstream of it. Behind cylinder a recirculation zone is formed, which is almost two diameters long. Because the flow is steady, the velocity field is symmetric in both section planes, due to geometric symmetry.[1] However, note that the length of the recirculation zone behind cylinder varies in the lateral direction: it is longest at the center, becomes smaller as one moves towards side walls but then increases again close to walls. Thus, three-dimensionality effects are not limited to cylinder ends where they meet side walls—the effect of side walls is visible across the whole cylinder span.

Figure 9.36 shows pressure distribution in the vertical symmetry plane. The isobars around cylinder are similar to those seen in 2D-computations for a cylinder in an infinite environment: the highest pressure is found at the front stagnation point, where the flow impinges onto cylinder wall, and the lowest pressure is located at cylinder sides. Pressure recovers to some extent on the downstream side, but due to viscous losses, it is much lower at the downstream stagnation point than at the upstream one. One closed pressure contour downstream of cylinder indicates the end

---

[1]The steady-state flow does not have to be symmetric when the geometry is symmetric; in some symmetric geometries—such as diffusers and sudden channel expansions—an asymmetric steady-state flow may be obtained, both in experiments and in simulations.

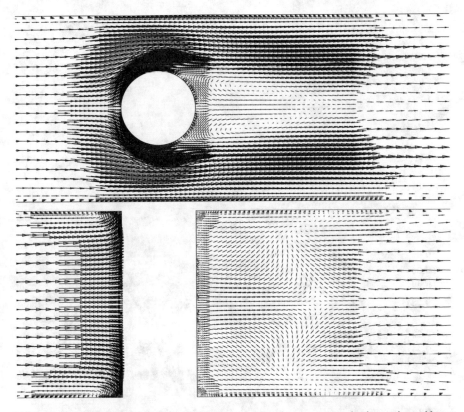

Fig. 9.35 Velocity vectors in two longitudinal symmetry planes, computed using a trimmed hexahedral grid: $y = 0$ (upper) and $z = 0$ (lower)

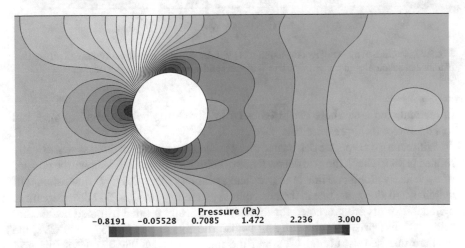

Pressure (Pa)

−0.8191   −0.05528   0.7085   1.472   2.236   3.000

Fig. 9.36 Pressure distribution in the longitudinal symmetry plane $y = 0$, computed using a trimmed hexahedral grid (flow from left to right)

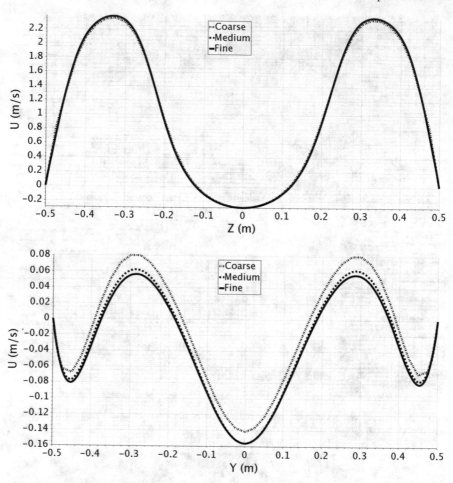

Fig. 9.37 Profiles of u_x-velocity downstream of cylinder, computed on three systematically refined trimmed hexahedral grids: $x = 0.35$, $y = 0$ (upper) and $x = 0.75$, $z = 0$ (lower)

of recirculation zone: here two streams (from above and below cylinder) meet and this leads to a local pressure rise.

With each grid type, we performed computations on a series of three systematically refined grids to check the grid dependence of solutions. Figure 9.37 shows the profiles of u_x velocity along one line in the vertical symmetry plane 0.875 D downstream of cylinder, and along one line in the horizontal symmetry plane 1.875 D downstream of cylinder. Results are presented for three systematically refined trimmed hexahedral grids with five prism layers along all walls. The coarse grid had 377,006 CVs, the medium one had 1,611,904 CVs and the fine grid had 8,952,321 CVs (the grid spacing was halved with each refinement). For the profile along the vertical line, the differences between solutions obtained on all three grids are very small; the peak

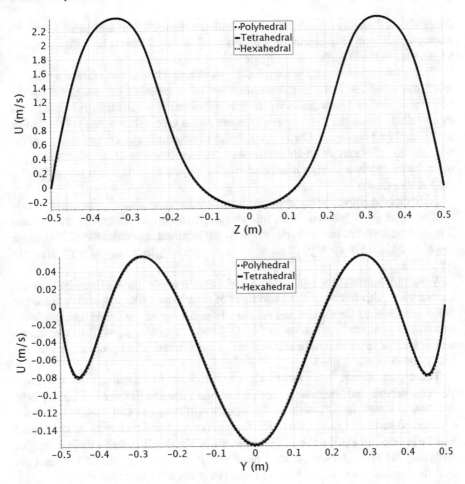

Fig. 9.38 Profiles of u_x-velocity downstream of cylinder, computed on three different grid types: $x = 0.35$, $y = 0$ (upper) and $x = 0.75$, $z = 0$ (lower)

values are under-predicted by the coarse grid by ca. 2%, while the profiles from medium and fine grid can hardly be distinguished in the graph. The differences are better visible in the horizontal section further downstream from cylinder: the peak values are here one order of magnitude smaller than the mean velocity in the duct and thus even smaller differences are clearly distinguishable. The difference between the coarse and medium grid is about 4 times larger than the difference between the medium and fine grid, as expected of a second-order method. One can thus estimate that the average discretization error on the finest grid is of the order of 0.1% of the mean duct velocity.

Figure 9.38 shows the comparison of the same velocity profiles computed on fine grids of different type. The differences are smaller than the difference between medium and fine grids of the same type, see Fig. 9.37. This confirms that, when the grid is sufficiently fine, the same grid-independent solution will be obtained, no matter which type of computational grid is used. What differs is the effort required to generate the grid and solve the Navier–Stokes equations.

When using commercial software such as that used here, the effort required to reach the same level of discretization errors is, in general, the lowest when the trimmed hexahedral grids are used. The least efficient are tetrahedral grids: one needs more cells to reach the same level of discretization errors than when polyhedral or hexahedral grids are used. On a grid with the same number of control volumes, the convergence of iterations under the same conditions (number of inner iterations per outer iteration when solving linearized equation systems, under-relaxation factors, etc.) is also slower.

Note that the above statements are valid when the same kind of approximations is applied on all grids (here: midpoint rule for integral approximations, linear interpolation or extrapolation, linear shape functions for gradient approximation). The ratios might be different if one applied discretizations specially tuned to one particular grid type.

We do not present here a quantitative comparison of computing times because we did not vary grids to ensure the same level of discretization errors, and because the ratios are problem-dependent; the statements given above are based on experience from many industrial applications of CFD. The comparison also differs depending on the kind of finite approximations used in the discretization process, and on the linear equation solver used.

Three-dimensional flows are much more difficult to visualize than 2D flows. Velocity vectors and streamlines, which are often used in 2D, are difficult to both draw and interpret in 3D problems. Presentation of contours and vector projections on selected surfaces (planes, iso-surfaces of some quantity, boundary surfaces, etc.) and the possibility to view them from different directions is perhaps the best way of analyzing 3D flows. Unsteady flows require animation of the results. We shall not deal further with these issues here, but want to stress their importance.

Chapter 10
Turbulent Flows

10.1 Introduction

Most flows encountered in engineering practice are turbulent (Pope 2000, and Jovanović 2004) and therefore require different treatment compared to laminar flows studied so far. Turbulent flows are characterized by the following properties:

- Turbulent flows are highly unsteady. A plot of the velocity as a function of time at most points in the flow would appear random to an observer unfamiliar with these flows. The word 'chaotic' could be used but it has been given another definition in recent years.
- They are three-dimensional. The time-averaged velocity may be a function of only two coordinates, but the instantaneous field fluctuates rapidly in all three spatial dimensions.
- They contain a great deal of vorticity. Indeed, vortex stretching is one of the principal mechanisms by which the intensity of turbulence is increased.
- Turbulence increases the rate at which conserved quantities are stirred. Stirring is a process in which parcels of fluid with differing concentrations of at least one of the conserved properties are brought into contact. The actual *mixing* is accomplished by diffusion. Nonetheless, this process is often called *turbulent diffusion*.
- By means of the processes just mentioned, turbulence brings fluids of differing momentum content into contact. The reduction of the velocity gradients due to the action of viscosity reduces the kinetic energy of the flow; in other words, mixing is a *dissipative* process. The lost kinetic energy is irreversibly converted into internal energy of the fluid.
- It has been shown in recent years that turbulent flows contain *coherent structures*— repeatable and essentially deterministic events that are responsible for a large part of the mixing. However, the random component of turbulent flows causes these events to differ from each other in size, strength, and time interval between occurrences, making study of them very difficult.
- Turbulent flows fluctuate on a broad range of length and time scales. This property makes direct numerical simulation of turbulent flows very difficult. (See below.)

© Springer Nature Switzerland AG 2020
J. H. Ferziger et al., *Computational Methods for Fluid Dynamics*,
https://doi.org/10.1007/978-3-319-99693-6_10

All of these properties are important. The effects produced by turbulence may or may not be desirable, depending on the application. Intense mixing is useful when chemical mixing or heat transfer are needed; both of these may be increased by orders of magnitude by turbulence. On the other hand, increased mixing of momentum results in increased frictional forces, thus increasing the power required to pump a fluid or to propel a vehicle; again, an increase by an order of magnitude is not unusual. Engineers need to be able to understand and predict these effects in order to achieve good designs. In some cases, it is possible to control the turbulence, at least in part.

In the past, the primary approach to studying turbulent flows was experimental. Overall parameters such as the time-averaged drag or heat transfer are relatively easy to measure but as the sophistication of engineering devices increases, the levels of detail and accuracy required also increase, as does cost and the expense and difficulty of making measurements. To optimize a design, it is usually necessary to understand the source of the undesired effects; this requires detailed measurements that are costly and time-consuming. Some types of measurements, for example, the fluctuating pressure within a flow, are almost impossible to make at the present time. Others cannot be made with the required precision. As a result, numerical methods have an important role to play, and that role is perhaps nowhere more advanced than in the design and optimization of commercial and military aircraft and ships. There CFD-analysis for components (e.g., airfoils, propellers, turbines, etc.) as well as whole body configurations is routinely undertaken. However, the numerical method requirements vary depending on what one wants to analyze in the flow and we shall return to this principle in Sect. 12.1.1. As an example, in time-averaged flow, the determination of forces, heat fluxes, etc. pose less stringent requirements than if one wants to look at Reynolds stresses or if even the more problematic triple correlations are of interest (both in terms of accuracy and the duration of simulation).

Before proceeding to the discussion of numerical methods for these flows, it is helpful to summarize the approaches to predicting turbulent flows. Bardina et al. (1980) had a list of six categories on which we base our summary as follows:

- The first approach involves the use of *correlations* such as ones that give the friction factor as a function of the Reynolds number or the Nusselt number of heat transfer as a function of the Reynolds and Prandtl numbers. This method, which is usually taught in introductory courses, is very useful but is limited to simple types of flows, ones that can be characterized by just a few parameters. As its use does not require CFD, we shall say no more about it here.
- The second approach uses *integral equations* which can be derived from the equations of motion by integrating over one or more coordinates. Usually this reduces the problem to one or more ordinary differential equations which are easily solved. The methods applied to these equations are those for ordinary differential equations which are discussed in Chap. 6.
- The third approach is based on equations obtained by decomposing the equations of motion into mean and fluctuating components (Pope 2000). Unfortunately, these decomposed equations do not form closed sets (see Sect. 10.3.5.1) so these meth-

ods require the introduction of approximations (*turbulence models*). Some of the turbulence models in common use today and a discussion of the problems associated with the numerical solution of equations containing turbulence models are presented later in this chapter. There we will focus on so-called *one-point closures*.[1]

The actual approach to handling the turbulence models is dictated by the nature of the process used to obtain the mean and fluctuating equations, leading to subcategories of this third approach as follows:

- We obtain a set of partial differential equations called the *Reynolds-averaged Navier–Stokes* (or RANS) equations if the process to create the mean is averaging the equations of motion over time or over an ensemble of realizations (an imagined set of flows in which all controllable factors are kept fixed). The resulting equations may represent either a time-dependent or steady flow as we discuss below.
- We obtain a set of equations called the *large-eddy simulation (LES)* equations when the mean is achieved by averaging (or filtering) over finite volumes in space.[2] LES solves then for an accurate representation of the largest scale motions of the flow while approximating or modeling the small scale motions. It can be regarded as a kind of compromise between RANS (see above) and direct numerical simulation (see below).

• Finally, the fourth approach is *direct numerical simulation (DNS)* in which the Navier–Stokes equations are solved for all of the motions in a turbulent flow.

As one progresses down this list, more and more of the turbulent motions are computed and fewer are approximated by models. This makes the methods close to the bottom more exact but the computation time is increased considerably.

All of the methods described in this chapter require the solution of some form of the conservation equations for mass, momentum, energy, or chemical species. The major difficulty is that turbulent flows contain variations on a much wider range of length and time scales than laminar flows. So, even though they are similar to the laminar flow equations, the equations describing turbulent flows are usually much more difficult and expensive to solve.

---

[1] There are also *two-point closures* which use equations for the correlation of the velocity components at two spatial points or, more often, the Fourier transform of these equations. These methods are not widely used in practice (Leschziner 2010) and most often are used in pure research, so we shall not consider them further. However, Lesieur (2010, 2011) presents a charming review with good insight to their history and state-of-the-art.

[2] By averaging over 'relatively' large volumes in space, one obtains *very large-eddy simulation (VLES)*. We will discuss later (Sects. 10.3.1; 10.3.7) how to define 'relatively.'

10.2   Direct Numerical Simulation (DNS)

10.2.1   Overview

The most accurate approach to turbulence simulation is to solve the Navier–Stokes equations without averaging or approximation other than numerical discretizations whose errors can be estimated and controlled. It is also the simplest approach from the conceptual point of view. In such simulations, all of the motions contained in the flow are resolved. The computed flow field obtained is equivalent to a single realization of a flow or a short-duration laboratory experiment; as noted above, this approach is called direct numerical simulation (DNS).

   In a direct numerical simulation, in order to assure that all of the significant structures of the turbulence have been captured, the domain on which the computation is performed must be at least as large as the physical domain to be considered or the largest turbulent eddy. A useful measure of the latter scale is the integral scale (L) of the turbulence which is essentially the distance over which the fluctuating component of the velocity remains correlated. Thus, each linear dimension of the domain must be at least a few times the integral scale. A valid simulation must also capture all of the kinetic energy dissipation. This occurs on the smallest scales, the ones on which viscosity is active, so the size of the grid must be on the order of a viscously-determined scale, called the Kolmogoroff scale, η. Usually the resolution requirement is stated as

$$k_{max}\eta = \frac{\pi}{\Delta}\eta \geq 1.5 \qquad (10.1)$$

so the grid size $\Delta \leq 2\eta$. Schumacher et al. (2005) point out that dissipation occurs over a range of scales (the peak of the dissipation spectrum is at $k\eta \sim 0.2$) and locally scales smaller than η appear in the flow; their simulations were run for $k_{max}\eta$ up to 34. In order to capture the essence of their flows, Kaneda and Ishihara (2006) and Bermejo-Moreno et al. (2009) used values of $k_{max}\eta$ of up to 2 and 4, respectively.

   For homogeneous isotropic turbulence, the simplest type of turbulence, there is no reason to use anything other than a uniform grid. In this case, the argument just given shows that the number of grid points in each direction must be on the order of L/η; it can be shown (Tennekes and Lumley 1976) that this ratio is proportional to $\mathrm{Re}_L^{3/4}$.[3] Here Re_L is a Reynolds number based on the magnitude of the velocity fluctuations and the integral scale; this parameter is typically about 0.01 times the macroscopic Reynolds number engineers use to describe a flow. Because this number of points must be employed in each of the three coordinate directions, and the time step is related to the grid size, the cost of a simulation scales as Re_L^3. In terms of the

---

[3] It could be worse! The NSF Report on Simulation-Based Engineering Science (2006) actually notes that the *tyranny of scales* dominates simulation efforts in many fields, including fluid mechanics. So, even for $\mathrm{Re}_L \sim 10^7$, the length scale ratio is on the order of 2×10^5. However, they point out that for protein folding the time scale ratio is $\sim 10^{12}$ while for advanced material design the spatial scale ratio is $\sim 10^{10}$!

Reynolds number that an engineer would use to describe the flow, the scaling of the cost may be somewhat different.

Because the number of grid points that can be used in a computation is limited by the processing speed and memory of the machine on which it is carried out, direct numerical simulation is typically done in geometrically simple domains. On present machines, it is possible to make direct numerical simulations of homogeneous flows at turbulent Reynolds numbers up to the order of 10^5 using as many as 4096^3 grid points (Ishihara et al. 2009). As noted in the preceding paragraph, this corresponds to overall flow Reynolds numbers about two orders of magnitude larger and allows DNS to reach the low end of the range of Reynolds numbers of engineering interest, making it a useful method in some cases. In other cases, it may be possible to extrapolate from the Reynolds number of the simulation to the Reynolds number of actual interest by using some kind of extrapolation. For further details about DNS, see Pope (2000) and Moin and Mahesh (1998) for overviews and the specific papers of Ishihara et al. (2009), who use a Fourier spectral method (Sect. 3.11) to study isotropic flows, and Wu and Moin (2009), who employ a fractional-step (Sect. 7.2.1) finite-difference method to study a boundary layer flow.

10.2.2 Discussion

The results of a DNS contain very detailed information about the flow. This can be very useful but, on the one hand, it is far more information than any engineer needs and, on the other, DNS is expensive and not often used as a design tool. One must then ask what DNS can be used for. With it, we can obtain detailed information about the velocity, pressure, and any other variable of interest at a large number of grid points. These results are the equivalent of experimental data and can be used to produce statistical information or to create a 'numerical flow visualization.' From the latter, one can learn a great deal, for example, about the coherent structures that exist in the flow (Fig. 10.1). This wealth of information can then be used to develop a deeper understanding of the physics of the flow or to construct a quantitative model, perhaps of the RANS or LES type, which will allow other, similar, flows to be computed at a lower cost, making such models useful as engineering design tools.

Some examples of kinds of uses to which DNS has been put are:

- Understanding the process for laminar to turbulent transition, as well as the mechanisms of turbulence production, energy transfer, and dissipation in turbulent flows;
- Simulation of the production of aerodynamic noise;
- Understanding the effects of compressibility on turbulence;
- Understanding the interaction between combustion and turbulence;
- Controlling and reducing drag on a solid surface.

Other applications of DNS have already been made and many others will undoubtedly be carried out in the future.

Fig. 10.1 Hairpin forests in the fully turbulent region of a flat-plate boundary layer. Iso-surfaces of the second invariant of the velocity gradient tensor shown; surfaces colored based on local values of dimensionless distance from the wall. Values greater than 300 are in red (flow from left to right). From: Wu and Moin 2011

The increasing speed of computers has made it possible to carry out DNS of simple flows at low Reynolds numbers on workstations. By simple flows, we mean any homogeneous turbulent flow (there are many), channel flow, free shear flows, and a few others. On large parallel computers, as noted above, DNS is done with 4096^3 ($\sim 6.9 \times 10^{10}$) or more grid points. The computation time depends on the machine and the number of grid points used so no useful estimate can be given. Indeed, one usually chooses the flow to simulate and the number of grid points to fit the available computer resources. A complete state-of-the-art simulation may employ hundreds of thousands of cores on a computer system and consume millions of core-hours. As computers become faster and memories larger, more complex and higher Reynolds number flows will be simulated.

A wide variety of numerical methods can be employed in direct numerical simulation and large-eddy simulation. Almost any method described in this book can be used; however, for truly large-scale computation on parallel systems, essentially explicit codes based on CDS or spectral schemes are most often used; we note below that in some cases, implicit methods are used for certain terms in the equations. Because these methods have been presented in earlier chapters, we shall not give a lot of detail here. However, there are important differences between DNS and LES and simulations of steady flows and it is important that these be discussed.

The most important requirements placed on numerical methods for DNS and LES arise from the need to produce an accurate realization of a flow that contains a wide range of length and time scales. Because an accurate time history is required, techniques designed for steady flows are inefficient and should not be used without considerable modification. The need for accuracy requires the time step to be small and, obviously, the time-advance method must be stable for the time step selected. In most cases, explicit methods that are stable for the time step demanded by the accuracy requirement are available so there is no reason to incur the extra expense associated with implicit methods; most simulations have therefore used explicit time-advance methods. A notable (but not the only) exception occurs near solid surfaces. The important structures in these regions are of very small size and very fine grids must be used, especially in the direction normal to the wall. Numerical instability may arise from the viscous terms involving derivatives normal to the wall so these are often treated implicitly. In complex geometries, it may be necessary to treat still more terms implicitly.

The time advance methods most commonly used in DNS and LES are of second to fourth-order accuracy; Runge–Kutta methods have been used most commonly but others, such as the Adams–Bashforth and leapfrog methods have also been used. In general, for a given order of accuracy, Runge–Kutta methods require more computation per time step. Despite this, they are preferred because, for a given time step, the errors they produce are much smaller than those of the competing methods. Thus, in practice, they allow a larger time step for the same accuracy and this more than compensates for the increased amount of computation. The Crank–Nicolson method is often applied to the terms that must be treated implicitly, e.g., viscous terms or wall-normal convection terms.

A difficulty with time-advance methods is that ones of accuracy higher than first order require storage of data at more than one time step (including intermediate time steps). Thus, there is an advantage to designing and using methods which demand relatively little storage. Leonard and Wray (1982) presented a third-order Runge–Kutta method which requires less storage than the standard Runge–Kutta method of that accuracy (see an application of this method in Bhaskaran and Lele 2010). On the other hand, for compressible flow and computational acoustics, different characteristics are desired, e.g., low dissipation and low dispersion Runge–Kutta scheme (Hu et al. 1996, as applied in Bhagatwala and Lele 2011).

A further issue of importance in DNS is the need to handle a wide range of length scales; this requires a change in the way one thinks about discretization methods. The most common descriptor of the accuracy of a spatial discretization method is its order, a number that describes the rate at which the discretization error decreases when the grid size is reduced. Returning to the discussion in Chap. 3, it is useful to think in terms of the Fourier decomposition of the velocity field. We showed (Sect. 3.11) that, on a uniform grid, the velocity field can be represented in terms of a Fourier series:

$$u(x) = \sum \tilde{u}(k) \, e^{ikx} \, . \tag{10.2}$$

The highest wavenumber k that can be resolved on a grid of size Δx is $\pi/\Delta x$, so we consider only $0 < k < \pi/\Delta x$. The series (10.2) can be differentiated term by term. The exact derivative of e^{ikx}, ike^{ikx} is replaced by $ik_{\text{eff}}e^{ikx}$ where k_{eff} is the effective wavenumber defined in Sect. 3.11 when a finite difference approximation is used. The plot of k_{eff} given in Fig. 3.12 shows that central differences are accurate only for $k < \pi/2\Delta x$, the first half of the wavenumber range of interest.

The difficulty for turbulent flow simulations that is not encountered in steady flow simulations is that turbulence spectra (the distributions of turbulence energy over wavenumber or inverse length scale) are usually large over a significant part of the wavenumber range $\{0, \pi/\Delta x\}$ so the order of the method and its behavior in the context of Fig. 3.12 may become significant. In that figure, we see that ideally the numerical scheme would have an effective wavenumber approaching the exact and spectral line. A fourth-order CDS is certainly an improvement, but not in any way ideal. Compact schemes (Lele 1992, and Mahesh 1998; see Sect. 3.3.3) can yield higher-order schemes that are spectral-like. In compressible flow simulations and computational aeroacoustics such schemes are popular, e.g., see the fourth-order compact scheme of Kim (2007) or the implementation of a sixth-order scheme by Kawai and Lele (2010). However, because grid resolution is critical in defining the smallest structures resolvable in DNS, impressive results can be obtained with second-order CDS as well if the grid is fine enough (Wu and Moin 2009).

It is also useful to reiterate the importance of using an energy-conservative spatial differencing scheme. Many methods, including all upwind ones, are dissipative; that is, they include as part of the truncation error a diffusion term that dissipates energy in a time-dependent calculation. Their use has been advocated because the dissipation they introduce often stabilizes numerical methods. When these methods are applied to steady problems, the dissipative error may not be too large in the steady-state result (although we showed in earlier chapters that these errors may be quite large). When these methods are used in DNS, the dissipation produced is often much greater than that due to the physical viscosity and the results obtained may have little connection to the physics of the problem. For a discussion of energy conservation, see Sect. 7.1.3. Also, as demonstrated in Chap. 7, energy conservation prevents the velocity from growing without bound and thus maintains stability.

The methods and step sizes in time and space need to be related. The errors made in the spatial and temporal discretizations should be as nearly equal as possible, i.e., they should be balanced. This is not possible point-by-point and for every time step but, if this condition is not satisfied in an average sense, one is using too fine a step in one of the independent variables and the simulation could be made at lower cost with little loss of accuracy.

Accuracy is difficult to measure in DNS and LES. The reason is inherent in the nature of turbulent flows. A small change in the initial state of a turbulent flow is amplified exponentially in time and, after a relatively short time, the perturbed flow hardly resembles the original one. This is a physical phenomenon that has nothing to do with the numerical method. Because any numerical method introduces some error and any change in the method or the parameters will change that error, direct comparison of two solutions with the goal of determining the error is not possible. Instead, one can repeat the simulation with a different grid (which should differ considerably from the original one) and statistical properties of the two solutions can be compared. From the difference, an estimate of the error can be found. Unfortunately, it is difficult to know how the error changes with the grid size, so this type of estimate can only be an approximation. A simpler approach, which has been used by most people who compute simple turbulent flows, is to look at the spectrum of the turbulence. If the energy in the smallest scales is sufficiently smaller than that at the peak in the energy spectrum, it is probably safe to assume that the flow has been well resolved.

The accuracy requirement makes use of spectral methods common in DNS and LES, where the domain configuration and boundary conditions permit. These methods were described briefly earlier, in Sect. 3.11. In essence, they use Fourier series as a means of computing derivatives. The use of Fourier transforms is feasible only because the fast Fourier transform algorithm (Cooley and Tukey 1965; Brigham

1988) reduces the cost of computing a Fourier transform to $n \log_2 n$ operations. Unfortunately, this algorithm is applicable only for equi-spaced grids and a few other special cases. A number of specialized methods of this kind have been developed for solving the Navier–Stokes equations; more details of spectral methods are given in Sect. 3.11 and Canuto et al. (2007).

Rather than directly approximating the Navier–Stokes equations, an intriguing application of the spectral method is to multiply them by a sequence of 'test or basis functions', integrate over the entire domain, and then find a solution that satisfies the resulting equations (see Sect. 3.11.2.1). Functions which satisfy this form of the equations are known as 'weak solutions'. One can represent the solution of the Navier–Stokes equations as a series of vector functions, each of which has zero divergence. This choice removes the pressure from the integral form of the equations, thereby reducing the number of dependent variables that need to be computed and stored. The set of dependent variables can be further reduced by noting that, if a function has zero divergence, its third component can be computed from the other two. The result is that only two sets of dependent variables need to be computed, reducing the memory requirements by half. As these methods are quite specialized and their development requires considerable space, they are not given in detail here; the interested reader is referred to the paper by Moser et al. (1983) or to Sect. 3.4.2 of Canuto et al. (2007).

10.2.3 Initial and Boundary Conditions

Another difficulty in DNS is that of generating initial and boundary conditions. The former must contain all the details of the initial three-dimensional velocity field. Because coherent structures are an important component of the flow, it is difficult to construct such a field. Furthermore, the effects of initial conditions are typically 'remembered' by the flow for a considerable time, usually a few 'eddy-turnover times.' An eddy-turnover time is essentially the integral time scale of the flow or the integral length scale divided by the root-mean-square velocity (q). Thus the initial conditions have a significant effect on the results. Frequently, the first part of a simulation that is started with artificially constructed initial conditions must be discarded because it is not faithful to the physics. The question of how to select initial conditions is as much art as science and no unique prescriptions applicable to all flows can be given but we shall give some examples.

For homogeneous isotropic turbulence, the simplest case, periodic boundary conditions are used and it is easiest to construct the initial conditions in Fourier space, i.e., we need to create $\hat{u}_i(\mathbf{k})$. This is done by giving the spectrum which sets the amplitude of the Fourier mode, i.e., $|\hat{u}_i(\mathbf{k})|$. The requirement of continuity $\mathbf{k} \cdot \hat{u}_i(\mathbf{k})$ places another restriction on that mode. This leaves just one random number to be chosen to completely define $\hat{u}_i(\mathbf{k})$; it is usually a phase angle. The simulation must then be run for about two eddy-turnover times before it can be considered to represent real turbulence.

The best initial conditions for other flows are obtained from the results of previous simulations. For example, for homogeneous turbulence subjected to strain, the best initial conditions are taken from developed isotropic turbulence. For channel flow, the best choice has been found to be a mixture of the mean velocity, instability modes (which have nearly the right structure), and noise. For a curved channel, one can take the results of a fully developed plane channel flow as the initial condition.

Similar considerations apply to the boundary conditions where the flow enters the domain (inflow conditions). The correct conditions must contain the complete velocity field on a plane (or other surface) of a turbulent flow at each time step which is difficult to construct. As an example, one way this can be done for the developing flow in a curved channel is to use results for the flow in a plane channel. A simulation of a plane channel flow is made (either simultaneously or in advance) and the velocity components on one plane normal to the main flow direction provide the inflow condition for the curved channel. Chow and Street (2009) used such a strategy for an inflow condition for flow over a hill in Scotland with a code used to predict atmospheric mesoscale flows.[4]

As already noted, for flows which do not vary (in the statistical sense) in a given direction, one can use periodic boundary conditions in that direction. These are easy to use, fit especially well with spectral methods, and provide conditions at the nominal boundary that are as realistic as possible.

Outflow boundaries are less difficult to handle. One possibility is to use extrapolation conditions which require the derivatives of all quantities in the direction normal to the boundary be zero:

$$\frac{\partial \phi}{\partial n} = 0 \,, \tag{10.3}$$

where ϕ is any of the dependent variables. This condition is often used in steady flows but is not satisfactory in unsteady flows. For the latter, it is better to replace this condition by an unsteady convective condition. A number of such conditions have been tried but one that appears to work well is also one of the simplest:

$$\frac{\partial \phi}{\partial t} + U \frac{\partial \phi}{\partial n} = 0 \,, \tag{10.4}$$

where U is a velocity that is independent of location on the outflow surface and is chosen so that overall conservation is maintained, i.e., it is the velocity required to make the outflow mass flux equal to the incoming mass flux. This condition appears to avoid the problem caused by pressure perturbations being reflected off the outflow boundary back to the interior of the domain. Another option is to use the forcing technique described in Sect. 13.6 to damp velocity fluctuations in directions perpendicular to the mean flow direction over some distance towards outlet boundary. In this way all vortexes disappear before the boundary is reached and reflections

---

[4]"Mesoscale" refers to weather systems with horizontal dimensions generally ranging from around 5 km to several hundred or perhaps 10^3 km.

are avoided. However, there remains the issue of determining the optimum forcing parameter; this issue will be discussed in more detail in Sect. 13.6.

On smooth solid walls, no slip boundary conditions, which have been described in Chaps. 8 and 9, may be used. One must bear in mind that at boundaries of this type the turbulence tends to develop small but very important structures ('streaks') that require very fine grids especially in the direction normal to the wall and, to a lesser extent, in the spanwise direction (the direction normal to both the wall and the principal flow direction). An alternative approach is to employ an immersed boundary method (IBM) for complex shapes (Kang et al. 2009; Fadlun et al. 2000; Kim et al. 2001; Ye et al. 1999) or rough walls (Leonardi et al. 2003; Orlandi and Leonardi 2008). In this method, the governing equations are discretized and solved on a regular grid, but the boundary condition is imposed, e.g., via body forces on the actual boundary which leads to appropriate forcing at relevant points of the regular grid, even though they are near, but not on, boundary. Also, interpolations of velocities from the desired conditions on the actual boundary to the regular grid boundary can be used, and mass conservation must be maintained. Most of these IBM methods employ a fractional-step method (cf. Sect. 7.2.1 and 8.3) and the Kim and Ye works use FV, while the others use FD schemes.

Symmetry boundary conditions, which are often used in RANS computations to reduce the size of the domain are usually not applicable in DNS or LES because, although the mean flow may be symmetric about some particular plane, the instantaneous flow is not and important physical effects may be removed by application of conditions of this type. Symmetry conditions have, however, been used to represent free surfaces.

Despite all attempts to make the initial and boundary conditions as realistic as possible, a simulation must be run for some time before the flow develops all of the correct characteristics of the physical flow. This situation derives from the physics of turbulent flows so there is little one can do to speed up the process; one possibility is mentioned below. As we have noted, the eddy-turnover time scale is the key time scale of the problem. In many flows, it can be related to a time scale characteristic of the flow as a whole, i.e., a mean flow time scale. However, in separated flows, there are regions that communicate with the remainder of the flow on a very long time scale and the development process can be very slow, making very long run times necessary.

The best way to ascertain that flow development is complete is to monitor some quantity, preferably one that is sensitive to the parts of the flow that are slow to develop; the choice depends on the flow being simulated. As an example, one might measure a spatial average of the skin friction in the recirculating region of a separated flow as a function of time. Initially, there is usually a systematic increase or decrease of the monitored quantity; when the flow becomes fully developed, the value will show statistical fluctuations with time. After this point, statistical average results (for example, for the mean velocity or its fluctuations) may be obtained by averaging over time and/or a statistically homogeneous coordinate in the flow. In so doing it is important to remember that, because turbulence is not purely random, the sample size is not the same as the number of points used in the averaging process. A conservative estimate is to assume that each volume of diameter equal to the integral scale (and each time period equal to the integral time scale) represents only a single sample.

The development process can be sped up by using a coarse grid initially. When the flow is developed on that grid, the fine grid can be introduced. If this is done, some waiting is still necessary for the flow to develop on the fine grid but it may be smaller than the time that would have been required had the fine grid been used throughout the simulation.

The paper by Wu and Moin (2009) about zero-pressure-gradient boundary-layer simulations is a *tour de force* on the set up of initial and boundary conditions for a problem that is very sensitive to them and how the physics of the flow can be a good guide on how to proceed.

10.2.4   Examples of DNS Application

10.2.4.1   Spatial Decay of Grid Turbulence

As an illustrative example of what can be accomplished with DNS, we shall take a deceptively simple flow, the flow created by an oscillating grid in a large body of quiescent fluid. The oscillation of the grid creates turbulence which decreases in intensity with distance from the grid. This process of energy transfer away from the oscillating grid is usually called *turbulent diffusion*; energy transfer by turbulence plays an important role in many flows so its prediction is important but it is surprisingly difficult to model. Briggs et al. (1996) made simulations of this flow and obtained good agreement with the experimentally determined rate of decay of the turbulence with distance from the grid. The energy decays approximately as $x^{-\alpha}$ with $2 < \alpha < 3$; determination of the exponent α is difficult both experimentally and computationally because the rapid decay does not provide a large enough region to allow one to compute its value accurately.

Using visualizations based on simulations of this flow, Briggs et al. (1996) showed that the dominant mechanism of turbulent diffusion in this flow is the movement of energetic parcels of fluid through the undisturbed fluid. This may seem a simple and logical explanation but is contrary to earlier proposals. Figure 10.2 shows the

Distance from
the source

↓

Fig. 10.2 Contours of the kinetic energy on a plane in the flow created by an oscillating grid in a quiescent fluid; the grid is located at the top of the figure. Energetic packets of fluid transfer energy away from the grid region. From Briggs et al. (1996)

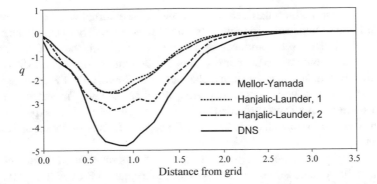

Fig. 10.3 The profile of the flux of turbulent kinetic energy, q, compared with the predictions of some commonly used turbulence models (Mellor and Yamada 1982; Hanjalić and Launder 1976, 1980); from Briggs et al. (1996)

contours of the kinetic energy on one plane in this flow. One sees that the large energetic regions are of approximately the same size throughout the flow but there are fewer of them far from the grid. The reasons are that those parcels that propagate parallel to the grid do not move very far in the direction normal to the grid and that small 'blobs' of energetic fluid are quickly destroyed by the action of viscous diffusion.

The results were used to test turbulence models. A typical example of such a test is shown in Fig. 10.3 in which the profile to the flux of turbulent kinetic energy is given and compared with the predictions of some commonly used turbulence models. It is clear that the models do not work very well even in a flow as simple as this one. The probable reason is that the models were designed to deal with turbulence generated by shear which has a character very different from that created by the oscillating grid.

The simulation used a code that was designed for the simulation of homogeneous turbulence (Rogallo 1981). Periodic boundary conditions are applied in all three directions; this implies that there is actually a periodic array of grids but this causes no problem so long as the distance between neighboring grids is sufficiently greater than the distance required for the turbulence to decay. The code uses the Fourier spectral method and a third-order Runge–Kutta method in time.

These results illustrate some important features of DNS. The method allows one to compute statistical quantities that can be compared with experimental data to validate the results. It also allows computation of quantities that are difficult to measure in the laboratory and that are useful in assessing models. At the same time, the method yields visualizations of the flow that can provide insight into the physics of the turbulence. It is rarely possible to obtain both statistical data and visualizations of the same flow in a laboratory. As the example above shows, the combination can be very valuable.

In direct numerical simulations, one can control the external variables in a manner that is difficult or impossible to implement in the laboratory. There have been several cases in which the results produced by DNS disagreed with those of experiments and the former turned out to be more nearly correct. One example is the distribution of turbulence statistics near a wall in a channel flow; the results of Kim et al. (1987) proved to be more accurate than the experiments when both were repeated with more care. An earlier example was provided by Bardina et al. (1980) which explained some apparently anomalous results in an experiment on the effects of rotation on isotropic turbulence.

DNS makes it possible to investigate certain effects much more accurately than would be otherwise possible. It is also possible to try methods of control that cannot be realized experimentally. The point of doing so is to provide insight into the physics of the flow and thus to indicate possibilities that may be realizable (and to point the direction toward realizable approaches). An example is the study of drag reduction and control on a flat plate conducted by Choi et al. (1994). They showed that, by using controlled blowing and suction through the wall (or a pulsating wall surface), the turbulent drag of a flat plate could be reduced by 30%. Bewley et al. (1994) used optimal control methods to demonstrate the possibility that the flow could be forced to relaminarize at low Reynolds number and that reduction in the skin friction is possible at high Reynolds numbers.

10.2.4.2   Flow Around a Sphere at Re = 5,000

A sphere is a body of a very simple shape, but the fluid flow around it is very complicated. DNS of such a flow is possible at moderate Reynolds numbers; Seidl et al. (1998) presented such a simulation using second-order time and space discretization and unstructured hexahedral grids with local refinement. We shall not go into details of the flow physics at this stage (more is to come after the LES-approach is described) but we want to introduce a method of testing the realizability of the simulation and analyzing the variation of turbulence structure in different zones of the flow.

Figure 10.4 shows flow pattern from simulation and experiment. Although the dye distribution (coming from two holes, one on the front of the sphere ahead of the separation line and one on the rear side, after separation) does not correspond exactly to the azimuthal component of vorticity, it is obvious that both figures agree when it comes to identifying the main features of the flow. The flow separation takes place shortly before the equator; the shear layer becomes unstable and rolls up, creating vortex rings; these eventually break down into isotropic turbulence further downstream. The back flow within the recirculation zone is turbulent, as both experiment and simulation indicate.

Another interesting way of analyzing flow structure and at the same time checking the realizability of the solution is to plot the invariants of the anisotropy tensor computed at various points in the flow field in a map introduced first by Lumley (1979). The components of the anisotropy tensor are defined as follows:

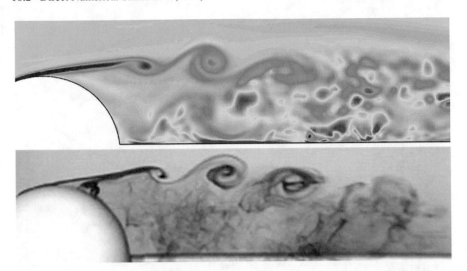

Fig. 10.4 Comparison of flow pattern from DNS and experiment: azimuthal component of computed instantaneous vorticity (upper) and the snapshot of dye distribution in an experiment (from Seidl et al. 1998)

$$b_{ij} = \frac{\overline{u_i u_j}}{q} - \frac{1}{3}\delta_{ij} , \qquad (10.5)$$

where u_i are the fluctuating components of the velocity vector, q stands for the turbulence intensity, $q = \overline{u_i u_i}$ and δ_{ij} is the Kronecker-delta. The two non-zero invariants of this tensor, II and III, are:

$$II = -\frac{1}{2}b_{ij}b_{ji} , \quad \text{and} \quad III = \frac{1}{3}b_{ij}b_{jk}b_{ki} . \qquad (10.6)$$

Lumley (1979) showed that the possible states of turbulence in the parameter space of $-II$ and III are bound by three lines, as shown in Fig. 10.5. The upper-right corner represents one-component turbulence; the bottom corner represents isotropic turbulence, while the corner on the left-hand side represents isotropic two-component turbulence. The two limiting lines emerging from the bottom corner represent an axisymmetric turbulence state, and the third line represents the planar two-component turbulence. Figure 10.5 shows the values of the two invariants in the map for the set of points selected along a closed streamline of the steady-state mean flow. One can see from the map that the state of turbulence varies significantly along the selected streamline. The starting point (marked by a larger symbol) is relatively close to the top-right corner, indicating that one component of the Reynolds-stress dominates. Further on, along the outer path, we are moving close to the axisymmetric limit towards full isotropy, which is the state found near the reattachment point and within the larger part of the recirculation zone. The last part, corresponding to points closer to the sphere surface, indicates that the state of turbulence is dominated by two components.

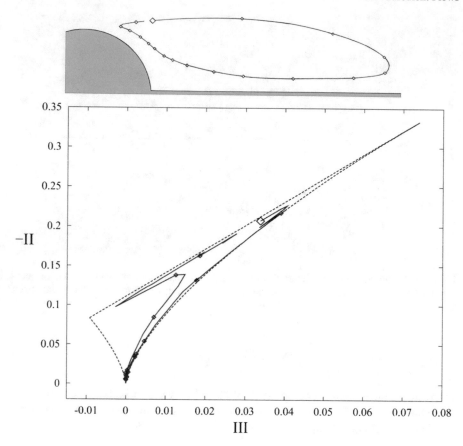

Fig. 10.5 Points on a mean flow streamline at which the invariants of the anisotropy tensor are evaluated (upper) and the locations of those points in the invariant map (lower); from Seidl et al. (1998)

All points from the streamline lie properly inside the triangle bounded by the limiting lines, indicating that the simulation has not violated physics laws—the solution is realizable. Because the flow upstream of the sphere is laminar and the separation zone behind it is not closed (meaning that 'fresh' fluid enters and 'older' fluid leaves this zone continuously), an interesting question arises: how does a fluid element from a non-turbulent upstream zone enter the turbulent state inside the map? From Fig. 10.5 (lower), it appears that the only possibility is at the top corner of the triangle: fluid elements in the vicinity of the separation streamline experience first one-dimensional fluctuations before those grow to multi-dimensional turbulence. Fluid elements leaving the recirculation zone remain in the turbulent wake where the turbulence slowly decays further downstream.

10.2.5 Other Applications of DNS

Flow behind sphere is a challenging research subject because it contains all types of turbulence states: by scanning the turbulent wake one finds points corresponding to all locations within the invariant map. Many other flows that are commonly studied using DNS—such as fully-developed flow in a plane channel or pipe—cover only a small portion of the map. However, because one needs a very fine grid only within the turbulent zone, solution methods which allow local grid refinement are desirable. Such methods are usually based on the finite-volume method and use unstructured grids with second-order discretization, thus requiring finer grids to obtain satisfactory accuracy than higher-order methods which are used for simpler geometries. With structured grids, like those used by Pal et al. (2017) in their study of stratified flow past a sphere at a subcritical[5] Reynolds number of 3700 and moderate Froude number, one wastes a lot of grid points in zones where a fine grid is not needed; this becomes especially critical at higher Reynolds numbers.

The advantage of specialized solution methods for simple geometries is that they allow for flows at higher Reynolds numbers to be adequately resolved. Lee and Moser (2015) in their DNS of plane channel flow reached $Re = 2.5 \times 10^5$, based on mean velocity and channel width. The studied flow exhibits characteristics of high Reynolds number wall-bounded turbulent flows and allows a more detailed analysis of the boundary layer, thus delivering invaluable information for developing new or tuning existing turbulence models. See Sect. 10.3.5.5 for a discussion of wall functions relying on the so-called log-law, where the DNS suggests that some of the "law"—parameters are not as universal as was so far believed.

With advancing computing technology, DNS will continue to be applied to more complex geometries and higher Reynolds numbers, thus providing data which either cannot be measured at all or not with the desired accuracy, but which will be extremely useful in both science and engineering.

10.3 Simulation of Turbulence with Models

10.3.1 Model Categories

As noted above, when the Navier–Stokes equations are averaged in some way, the result is that the equations are not closed because of the nonlinear convection terms. This means that the mean equations contain nonlinear correlation terms involving the unknown fluctuating variables. One can derive exact equations for these correlations, but they contain yet more complex unknown correlation terms which are not determinable from mean variables and the terms for which equations are written.

---

[5]Subcritical flow past a sphere is a low Reynolds number regime in which the drag coefficient is not a function of Reynolds number and there is laminar boundary layer separation.

Thus, one needs to construct *models*, i.e., approximate expressions or equations, for these unknown correlations. Leschziner (2010) presents a useful list of constraints and desirable traits for models; we paraphrase as follows: Models should be:

- based on rational principles and concepts of physics, rather than intuition;
- constructed from appropriate mathematical principles, such as dimensional homogeneity, consistency, and frame invariance;
- constrained to yield physically realizable behavior;
- widely applicable;
- mathematically simple;
- built from variables with accessible boundary conditions;
- computationally stable.

This and the following sections are devoted to methods for generating models. We can categorize the methods as follows:

1. RANS, unsteady RANS (URANS), and transient RANS (TRANS): When Osborne Reynolds (1895) presented the later-called Reynolds-averaged Navier–Stokes equations, they were averaged over a spatial volume, not time. However, the traditional averaging in the literature is over time or ensembles.

   a. Averaging over all time: If the average is over all time, then the resulting equations are the traditional *steady* RANS equations, i.e., the mean flow defined by the average is steady.

   b. Averaging over a time which is *long compared to the time scale of, say, the integral scale of the turbulence* (see Chen and Jaw 1998): If the average is over a time that is long compared to the time scales of turbulence, then the resulting equations will be unsteady, being a temporal low-pass filtered version of the original equations. This approach is rarely, if ever, actually used, in part, because then the model for the high frequency turbulence would have to depend on the filter scale as discussed below.

   c. Averaging over ensembles: If the equations are averaged over ensembles (a set of statistically identical realizations), then the averaging removes all turbulent (random) eddies because they have random phases between different members of the ensemble. If the ensembles are statistically stationary, then the mean flow will be steady. However, the resulting mean flow equations may be unsteady if coherent, deterministic elements are present in the motions under study,[6] i.e., the ensembles are not stationary. An example is flows driven by a periodic motion of boundaries, i.e., flow in a cylinder of an internal combustion engine, driven by the motion of a piston and valves with a repeatable pattern. Then, the equations are often called *unsteady* RANS or URANS (Durbin 2002; Iaccarino et al. 2003; Wegner et al. 2004). Hanjalić (2002) presents a construct similar to these which he calls transient RANS (TRANS) (see Sect. 10.3.7). In either case, the averaged equations are still unclosed so models are needed for the turbulence effects. As a consequence

---

[6]Chen and Jaw (1998) illustrate the creation of an ensemble average in their Fig. 1.8.

of the averaging process, Wyngaard (2010) reminds us "that the ensemble-averaged field is unlikely to exist in any realization of a turbulent flow, even for an instant."

The literature is often not precise on the averaging or filtering process. In reality, for all of the methods listed here the averaging or filtering is rarely explicitly done; neither the averaging-time nor number of required ensembles is defined. Thus, the averaged equations look essentially the same in all cases, and, typically, the same turbulence models are applied. This approach is likely correct for RANS and for the equations from ensemble averaging (URANS and TRANS) because in those cases, it is assumed that all of the turbulence (the random motions) has been averaged out. Then, the turbulence model is expected to represent the effect of the missing parts on the remaining mean flow, even if it is unsteady and supports, for example, deterministic motions. On the other hand, the temporally-filtered RANS equations solve for the low frequency motions of the flow while the high-frequency or small-time-scale motions are modeled. Accordingly, in such a simulation, the turbulence model must depend on the time scale of the highest resolved frequency in a pattern consistent with the LES approach (below); few, if any, studies actually take this approach, so that virtually all simulations are either RANS [long-time averaged and steady] or URANS/TRANS [ensemble-averaged and perhaps unsteady] simulations, but no turbulence is resolved.

2. LES, VLES: For large-eddy simulations (LES) and very-large-eddy simulations (VLES), the Navier–Stokes and scalar equations are filtered (averaged) over space. The resulting equations are essentially identical to the URANS equations except that the models for the unclosed terms have a different meaning and (perhaps) form; it is important to note that *while most of the turbulent energy is in the subfilter scale for URANS, LES resolves most of the turbulence energy*. Thus, LES and VLES resolve all unsteady features that are larger than the filter size, while the small-scale (subfilter-scale) features are modeled. There are various thoughts about the scales involved. Pope (2000) offers some guidelines, viz., the simulation is LES if "the filter and grid are sufficiently fine to resolve 80% of the energy everywhere." The definition has caveats near walls. The simulation is VLES if "the filter and grid are too coarse to resolve 80% of the energy." As imagined, this puts more weight on having a subfilter model containing as much physics as possible. Both Bryan et al. (2003) and Wyngaard (2004) pose that the grid scale of an LES should be much less than the length scale of the peak of the energy spectrum. Matheou and Chung (2014) used the Kolmogorov energy spectrum to quantify these criteria for flows in the atmospheric boundary layer. They postulate that to resolve 80% of the turbulent kinetic energy, the grid scale should be less than 1/12th of the peak spectrum scale, while to resolve 90% of the TKE, the grid scale should be less than 1/32nd of the peak spectrum scale. Their experience was that the 90% level was needed for reasonable convergence.

3. ILES: This method actually generates the models implicitly, and so is termed, implicit large-eddy simulation. We shall begin with it in the next section.

10.3.2   Implicit Large-Eddy Simulation (ILES)

Implicit large-eddy simulation (ILES) operates on the concept that one can apply a numerical method directly to the Navier–Stokes equations and then tailor the numerical scheme so that the resulting method is both eddy-resolving and properly dissipative, i.e., it is a large-eddy simulation, but with no explicit model for the subgrid-scale fluctuations; such fluctuations cannot be represented by the solution because the grid is a spatial filter (the highest wavenumber that can be resolved on a grid is $\pi/\Delta x$).

One easy way to understand the concept of ILES is to use the modified equation approach (see Sect. 6.3.2.2), wherein we seek to examine the truncation error terms in the numerical method. Rider (2007) does this analysis for compressible turbulence, and one can see from the analysis what the truncation terms are and how they constitute an effective subgrid-scale model. The monograph edited by Grinstein et al. (2007) provides a broad overview as well as a number of specific examples on ILES implementation. In that book, Smolarkiewicz and Margolin (2007) provide convincing evidence that their MPDATA is an ILES code that can accurately simulate geophysical flows, particularly, atmospheric circulation and boundary-layer motions. The heart of MPDATA is an iterative finite-difference approximation for the advective terms; it is second-order accurate and conservative. The iteration uses an upstream differencing term first, followed by a second pass to increase the accuracy. The authors state that MPDATA is in the class of non-oscillatory Lax-Wendroff schemes.

Aspden et al. (2008) give a different and insightful view of ILES based on scaling analysis. For computation, they employ the Lawrence Berkeley NL's CCSE IAMR code which is an incompressible, variable-density FV fractional-step method, second-order accurate in space and time, using an unsplit Godunov method (Colella 1990) and a monotonicity-limited fourth-order CD slope approximation (Colella 1985).

Any method of this type should satisfy the following requirement: if we keep refining the grid, more turbulence should be resolved and less modeled, until we reach the grid fineness at which all of turbulence would be resolved (i.e., we have a DNS) and the contribution of the subgrid-scale model becomes negligible. ILES obviously satisfies this requirement.

10.3.3   Large-Eddy Simulation (LES)

10.3.3.1   Large-Eddy Simulation (LES) Equations

Turbulent flows contain a wide range of length and time scales; the range of eddy sizes that might be found in a flow is shown schematically on the left-hand side of Fig. 10.6. The right-hand side of this figure shows the time history of a typical velocity component at a point in the flow; the range of scales on which fluctuations occur is obvious.

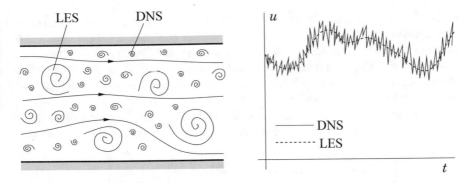

Fig. 10.6 Schematic representation of turbulent motion (left) and the time-dependence of a velocity component at a point (right)

The large scale motions are generally much more energetic than the small scale ones; their size and strength make them by far the most effective transporters of the conserved properties. The small scales are usually much weaker and provide little transport of these properties. A simulation which treats the large eddies more exactly than the small ones may make sense; large-eddy simulation is just such an approach. *Large-eddy simulations are three-dimensional and time-dependent.* Although expensive, they are much less costly than DNS of the same flow. In general, because it is more accurate, DNS is the preferred method whenever it is feasible. LES is the preferred method for flows in which the Reynolds number is too high or the geometry is too complex to allow application of DNS (see, e.g., Rodi et al. 2013 or Sagaut 2006). For example, LES has become a dominant tool in the atmospheric sciences, touching on clouds, precipitation, transport of pollution, wind flow in valleys (e.g., Chen et al. 2004b; Chow et al. 2006; Shi et al. 2018b,a); in the planetary boundary layer community physical processes dealt with in LES include buoyancy, rotation, entrainment, condensation, and interaction with the rough ground and ocean surfaces (Moeng and Sullivan 2015). LES has even found its way into the design of high-pressure gas turbines (Bhaskaran and Lele 2010) and into a dominant role in aeroacoustics simulations (e.g., Bodony and Lele 2008; Brès et al. 2017).

It is essential to define the quantities to be computed precisely. We need a velocity field that contains only the large scale components of the total field. This is best produced by filtering the velocity field (**?**); in this approach, the large or resolved scale field, the one to be simulated, is essentially a local average of the complete field. We shall use one-dimensional notation; the generalization to three dimensions is straightforward. The filtered velocity is defined by:

$$\overline{u}_i(x) = \int G(x, x')\, u_i(x')\, \mathrm{d}x' , \tag{10.7}$$

where $G(x, x')$, the filter kernel, is a localized function.[7] Filter kernels which have been applied in LES include a Gaussian, a box filter (a simple local average) and a cutoff (a filter which eliminates all Fourier coefficients belonging to wavenumbers above a cutoff). Every filter has a length scale associated with it, Δ. Roughly, eddies of size larger than Δ are large eddies while those smaller than Δ are small eddies, the ones that need to be modeled.

When the Navier–Stokes equations with constant density (incompressible flow) are filtered,[8] one obtains a set of equations very similar to the URANS equations:

$$\frac{\partial(\rho\bar{u}_i)}{\partial t} + \frac{\partial(\rho\overline{u_i u_j})}{\partial x_j} = -\frac{\partial\bar{p}}{\partial x_i} + \frac{\partial}{\partial x_j}\left[\mu\left(\frac{\partial\bar{u}_i}{\partial x_j} + \frac{\partial\bar{u}_j}{\partial x_i}\right)\right]. \tag{10.8}$$

Because the continuity equation is linear, filtering does not change it:

$$\frac{\partial(\rho\bar{u}_i)}{\partial x_i} = 0. \tag{10.9}$$

It is important to note that, because

$$\overline{u_i u_j} \neq \bar{u}_i \bar{u}_j \tag{10.10}$$

and the quantity on the left side of this inequality is not easily computed, a modeling approximation for the difference between the two sides of this inequality,

$$\tau_{ij}^s = -\rho(\overline{u_i u_j} - \bar{u}_i \bar{u}_j) \tag{10.11}$$

must be introduced. Thus, the momentum equations solved in LES are:

$$\frac{\partial(\rho\bar{u}_i)}{\partial t} + \frac{\partial(\rho\bar{u}_i\bar{u}_j)}{\partial x_j} = \frac{\partial\tau_{ij}^s}{\partial x_j} - \frac{\partial\bar{p}}{\partial x_i} + \frac{\partial}{\partial x_j}\left[\mu\left(\frac{\partial\bar{u}_i}{\partial x_j} + \frac{\partial\bar{u}_j}{\partial x_i}\right)\right]. \tag{10.12}$$

In the context of LES, τ_{ij}^s is called the *subgrid-scale Reynolds stress*. The name 'stress' stems from the way in which it is treated rather than its physical nature. It is in fact the large scale momentum flux caused by the action of the small or unresolved scales. The name 'subgrid scale' is also somewhat of a misnomer. The width of the filter, Δ, need not have anything to do with the grid size, h, other than the obvious condition that $\Delta \geq h$ as noted before. Traditionally, authors made such a connection

[7] The reader may note that there is no mention of the grid here. The filter size is at least as large as the grid size and often significantly larger. In traditional LES, the filter width is assumed to be the grid size and for now we will follow that practice. In Sects. 10.3.3.4 and 10.3.3.7, we will explore the relationship between grid and filter widths and how that affects the modeling.

[8] As is the case for the time and ensemble averaged equations above, the filter operation is defined, but the equations are, in general, not explicitly filtered. The equations are essentially the result of an implicit and unknown filter in traditional LES. Bose et al. (2010) have shown the value of explicit filtering.

and their nomenclature has stuck. Now, the models used to approximate the subgrid-scale Reynolds stress (10.11) are called *subgrid-scale (SGS)* or *subfilter-scale (SFS) models* depending on the model used as discussed below.

The subgrid-scale Reynolds stress contains local averages of the small scale field so models for it should be based on the local velocity field or, perhaps, on the past history of the local fluid motions. The latter can be accomplished by using a model that solves partial differential equations to obtain the parameters needed to determine the SGS Reynolds stress.

10.3.3.2 Smagorinsky and Related Models

The earliest and most commonly used subgrid-scale model is one proposed by Smagorinsky (1963). It is an eddy-viscosity model. All such models are based on the notion that the principal effects of the SGS Reynolds stress are increased transport and dissipation. As these phenomena are due to the viscosity in laminar flows, it seems reasonable to assume that a feasible model might be:

$$\tau_{ij}^{s} - \frac{1}{3}\tau_{kk}^{s}\delta_{ij} = \mu_{t}\left(\frac{\partial \overline{u}_i}{\partial x_j} + \frac{\partial \overline{u}_j}{\partial x_i}\right) = 2\mu_{t}\overline{S}_{ij} , \qquad (10.13)$$

where μ_t is the eddy viscosity and $\overline{S}_{ij}$ is the strain rate of the large scale or resolved field. Wyngaard (2010) shows how Lilly, in 1967, formally derived this model using the evolution equation for τ_{ij}^{s}. Similar models are also often used in connection with the RANS equations; see below.

The form of the subgrid-scale eddy viscosity can be derived by dimensional arguments and is:

$$\mu_{t} = C_{S}^{2}\rho\Delta^{2}|\overline{S}| , \qquad (10.14)$$

where C_S is a model parameter to be determined, Δ is the filter length scale, and $|\overline{S}| = (\overline{S}_{ij}\overline{S}_{ij})^{1/2}$. This form for the eddy viscosity can be derived in a number of ways. Theories provide estimates of the parameter. Most of these methods apply only to isotropic turbulence for which they all agree that $C_S \approx 0.2$. Unfortunately, C_S is not constant; it may be a function of Reynolds number and/or other non-dimensional parameters and may take different values in different flows.

The Smagorinsky model, although relatively successful, is not without problems, and its use has declined in favor of the more sophisticated models described below. It is not recommended, but its use persists. To simulate channel flow with it, several modifications are required. The value of the parameter C_S in the bulk of the flow has to be reduced from 0.2 to approximately 0.065, which reduces the eddy viscosity by almost an order of magnitude. Changes of this magnitude are required in all shear flows. In regions close to the surfaces of the channel, the value has to be reduced even further. One successful recipe is to borrow the van Driest damping that has long been used to reduce the near-wall eddy viscosity in RANS models:

$$C_S = C_{S0} \left(1 - e^{-n^+/A^+} \right)^2 , \tag{10.15}$$

where n^+ is the distance from the wall in viscous wall units ($n^+ = nu_\tau/\nu$, where u_τ is the shear velocity, $u_\tau = \sqrt{\tau_{\text{wall}}/\rho}$, and τ_{wall} is the shear stress at the wall) and A^+ is a constant usually taken to be approximately 25. Although this modification produces the desired results, it is difficult to justify in the context of LES.

A further problem is that, near a wall, the flow structure is very anisotropic. Regions of low and high speed fluid (streaks) are created; they are approximately 1000 viscous units long and 30–50 viscous units wide in both the spanwise and normal directions. Resolving the streaks requires a highly anisotropic grid and the choice of length scale, Δ, to use in the SGS model is not obvious. The usual choice is $(\Delta_1 \Delta_2 \Delta_3)^{1/3}$ but $(\Delta_1^2 + \Delta_2^2 + \Delta_3^2)^{1/2}$ is possible and others are easily constructed; here Δ_i is the width associated with the filter in the ith coordinate direction.

In a stably-stratified fluid, it is necessary to reduce the Smagorinsky parameter. Stratification is common in geophysical flows; the usual practice is to make the parameter a function of a Richardson or Froude number. These are related non-dimensional parameters that represent the relative importance of stratification and shear. Similar effects occur in flows in which rotation and/or curvature play significant roles. Typically, the Richardson number is based on the properties of the mean flow field. In some codes, the transport of SGS turbulent kinetic energy $k = \frac{1}{2}\overline{u_i' u_i'}$ (where $u_i' = u_i - \overline{u}_i$) is computed (Pope 2000) and used in defining the coefficient as

$$\mu_{t,j} \sim k^{\frac{1}{2}} l_j ,$$

where the vertical component of the turbulence length scale l_3 is adjusted for buoyancy; $l_1 = l_2 = \Delta$ or h (i.e., nominally equal to the grid or filter scale); see, e.g., Xue (2000).

Thus there are many difficulties with the Smagorinsky model. If we wish to simulate more complex and/or higher Reynolds number flows, it may be important to have a more accurate model (see, e.g., Shi et al. 2018b,a). Indeed, detailed tests based on results derived from DNS data, show that the Smagorinsky model is quite poor in representing the details of the subgrid-scale stresses. In particular, the eddy-viscosity model forces the stress τ_{ij}^s to be aligned with the strain rate $\overline{S}_{ij}$, which it is not in reality!

10.3.3.3   Dynamic Models

The smallest scales that are resolved in a simulation are similar in many ways to the still smaller scales that are treated via the model. This idea leads to an alternative subgrid-scale model, the *scale-similarity model* (Bardina et al. 1980). The principal argument is that the important interactions between the resolved and unresolved scales involve the smallest eddies of the former and the largest eddies of the latter, i.e., eddies that are a little larger or a little smaller than the length scale, Δ, associated with the filter. Arguments based on this concept lead to the following model:

$$\tau_{ij}^{s} = -\rho(\overline{\overline{u}_i \overline{u}_j} - \overline{\overline{u}}_i \overline{\overline{u}}_j) \,, \tag{10.16}$$

where the double overline indicates a quantity that has been filtered twice. This model correlates very well with the actual SGS Reynolds stress, but dissipates hardly any energy and cannot serve as a 'stand alone' SGS model. It transfers energy more or less equally to the smallest scales from the larger scales (*forward scatter*) and from the smallest resolved scales to larger scales (*back scatter*), which is useful. To correct for the lack of dissipation, one may combine the Smagorinsky and scale similarity models to produce a 'mixed' model. This model improves the quality of simulations. For further details, see Sagaut (2006).

The concept underlying the scale similarity model, namely, that the smallest resolved scale motions can provide information that can be used to model the largest subgrid-scale motions, can be taken a step further, leading to the dynamic model or procedure (Germano et al. 1991). This procedure is based on the assumption that one of the models described above is an acceptable representation of the small scales.

The essence of the seminal Germano procedure lies in the idea of scale invariance (Meneveau and Katz 2000): "Scale invariance means that certain features of the flow remain the same in different scales of the motion." Here, it is assumed that the coefficient in and the form of an SGS model remain the same at the grid scale and at scales which are some multiple of the grid scale. Accordingly, while the original filtering above at a scale Δ yielded

$$\tau_{ij}^{s} = -\rho(\overline{u_i u_j} - \overline{u}_i \overline{u}_j) \,,$$

and, using the Smagorinsky model,

$$\tau_{ij}^{s} - \frac{1}{3}\tau_{kk}^{s}\delta_{ij} = 2C_S^2 \rho \Delta^2 |\overline{S}|\overline{S}_{ij} \,,$$

filtering the filtered equations (10.8) again at a "test filter" scale $\hat{\Delta}$ produces

$$T_{ij}^{s} = -\rho(\widehat{\overline{u_i u_j}} - \hat{\overline{u}}_i \hat{\overline{u}}_j) \,,$$

and

$$T_{ij}^{s} - \frac{1}{3}T_{kk}^{s}\delta_{ij} = 2C_S^2 \rho \hat{\Delta}|\hat{\overline{S}}|\hat{\overline{S}}_{ij} \,,$$

where now the ratio $\hat{\Delta}/\Delta$ becomes an adjustable constant of the method; various choices have been made, but typically a ratio of ~ 2 is used. Germano created what is now known as the Germano Identity, $\mathcal{L}_{ij} = T_{ij}^{s} - \hat{\tau}_{ij}^{s}$, from which it is straightforward to show

$$\mathcal{L}_{ij} = \rho\left(\widehat{\overline{u_i}\,\overline{u_j}} - \hat{\overline{u}}_i \hat{\overline{u}}_j\right) = 2C_S^2 \rho\left(\hat{\Delta}|\hat{\overline{S}}|\hat{\overline{S}}_{ij} - \Delta^2\widehat{|\overline{S}|\overline{S}_{ij}}\right) + \frac{1}{3}\delta_{ij}\mathcal{L}_{kk} \,, \tag{10.17}$$

where $\mathcal{L}_{kk}$ contains the isotropic terms.[9] In this Eq. (10.17) the coefficient C_S^2 is the only unknown and can be determined in several ways. The isotropic terms cancel out, e.g., see Sagaut (2006), Lilly (1992) or Germano et al. (1991), because $S_{ii} = 0$ in incompressible flow and the isotropic terms appear in the actual solution method in the form of $S_{ij}\delta_{ij}\mathcal{L}_{kk}$.

The essential ingredients of the dynamic model are (1) that it uses information from the smallest eddies in the resolved field to calculate the coefficient and (2) that the same model with the same value of the coefficient is assumed applicable to both the actual LES and the coarser-scale filtered equations. The dynamic process gives the model coefficient as the ratio of two quantities, and the coefficient is computed, at every spatial grid point and every time step, directly from the current resolved variables. We shall not present the actual process here; however, note that Eq. (10.17) produces six equations for one unknown and so is overdetermined. The dynamic process then sets up schemes to minimize the error

$$\epsilon_{ij} = \mathcal{L}_{ij} - 2C_S^2\rho \left(\hat{\Delta}|\hat{\bar{S}}|\hat{\bar{S}}_{ij} - \Delta^2\widehat{|\bar{S}|\bar{S}_{ij}} \right) - \frac{1}{3}\delta_{ij}\mathcal{L}_{kk} . \tag{10.18}$$

Germano et al. (1991) contracted the error with the strain rate, i.e, $\epsilon_{ij}\bar{S}_{ij}$ generates a single equation for C_S^2. Two significant improvements to the original model proposed by Germano et al. (1991) were (1) the least-squares procedure for evaluating the coefficient suggested by Lilly (1992) and (2) the introduction by Wong and Lilly (1994) of a new base model to replace the Smagorinsky model; their model is based on Kolomogorov scaling and does not require calculation of the strain rate during dynamic procedures, making wall boundary conditions less critical. Their eddy viscosity is

$$\mu_t = C^{2/3}\rho\Delta^{4/3}\epsilon^{1/3} = C_\epsilon\rho\Delta^{4/3} . \tag{10.19}$$

C_ϵ is the coefficient determined in the dynamic process, and there is no requirement that dissipation rate be equal to SGS energy production rate.

The dynamic procedure with the Smagorinsky or Wong–Lilly model as its basis removes many of the difficulties described earlier:

- In shear flows, the Smagorinsky model parameter needs to be much smaller than in isotropic turbulence. The dynamic model produces this change automatically.
- The model parameter has to be reduced even further near walls. The dynamic model automatically decreases the parameter in the correct manner near the wall.
- The definition of the length scale for anisotropic grids or filters is unclear. This issue becomes moot with the dynamic model because the model compensates for any error in the length scale by changing the value of the parameter.

---

[9]The discerning reader will notice that we quietly extracted the coefficient C_S^2, which is a function of space and time, from under the test-filter average in Eq. (10.17); this is equivalent to assuming that it is constant over the volume of the test filter. This is a convenient choice, but not the only possible one.

Although it is a considerable improvement on the Smagorinsky model, there are problems with the dynamic procedure. The model parameter it produces is a rapidly varying function of the spatial coordinates and time so the eddy viscosity takes large values of both signs. Although a negative eddy viscosity has been suggested as a way of representing energy transfer from the small scales to the large ones (this process is called *back scatter*), if the eddy viscosity is negative over too large a spatial region or for too long a time, numerical instability can and does occur. One cure is to reset any eddy viscosity $\mu_t < -\mu$ equal to $-\mu$, i.e., equal to the negative of the molecular viscosity; this is called *clipping*. Another useful alternative is to employ averaging in space or time. For details, the reader is referred to the papers cited above. These techniques produce further improvements but are still not completely satisfactory; finding more robust models for the subgrid scale is the subject of current research as is shown below.

The arguments on which the dynamic model is based are not restricted to the Smagorinsky model. One could, instead, use the mixed Smagorinsky–scale-similarity model. The mixed model has been used by Zang et al. (1993) and Shah and Ferziger (1995) with considerable success.

Finally, we note that other versions of the dynamic procedure have been devised to overcome the difficulties with the simplest form of the model. One of the better of these is the Lagrangian dynamic model of Meneveau et al. (1996). In this model, the terms in the numerator and denominator of the expression for the model parameter of the dynamic procedure are averaged along flow trajectories. This is done by solving partial differential 'relaxation-transport' equations for integrals of the terms over the Lagrangian tracks. An application of this model is given in Sect. 10.3.4.3.

10.3.3.4 Models Using Reconstruction of Subfilter Scales

Carati et al. (2001) derived governing equations for LES which lead naturally to both SFS and SGS mixed models and distinguish between explicit filtering of the Navier–Stokes equations and discretization for numerical computation. While the explicit filter for LES is given by Eq. (10.7), the effect of the discretization is also an (unfortunately) unknown filter (cf. the ILES method of Sect. 10.3.2). The governing equations on a discrete grid can be obtained by applying an explicit filter (overbar) and a discretization operator (tilde) to the incompressible and constant-density Navier–Stokes and continuity equations to produce:

$$\frac{\partial(\widetilde{\bar{u}_i})}{\partial t} + \frac{\partial(\widetilde{\overline{\bar{u}_i \bar{u}_j}})}{\partial x_j} = -\frac{1}{\rho}\frac{\partial \widetilde{\bar{p}}}{\partial x_i} + \frac{\partial}{\partial x_j}\left[\nu\left(\frac{\partial \widetilde{\bar{u}_i}}{\partial x_j} + \frac{\partial \widetilde{\bar{u}_j}}{\partial x_i}\right)\right] - \frac{\partial}{\partial x_j}\left[\frac{\widetilde{\tau}_{ij}}{\rho}\right] \quad (10.20)$$

and

$$\frac{\partial(\overline{\bar{u}_i})}{\partial x_i} = 0 . \quad (10.21)$$

The total SFS stress[10] is defined as

$$\tau_{ij} = \rho(\overline{u_i u_j} - \overline{u}_i \overline{u}_j) \,, \tag{10.22}$$

which leads directly to a very insightful decomposition, namely,

$$\tau_{SFS} = \tau_{ij} = \rho\left(\overline{u_i u_j} - \overline{\overline{u}_i \overline{u}_j}\right) + \rho\left(\overline{\overline{u}_i \overline{u}_j} - \overline{u}_i \overline{u}_j\right) = \tau_{SGS} + \tau_{RSFS} \,. \tag{10.23}$$

We observe now that the flow field is effectively divided into three domains: the resolved field $\overline{u}_i$, the resolved subfilter-scale (RSFS) field $(\tilde{u}_i - \overline{\tilde{u}}_i)$, and the subgrid-scale (SGS) field $(u_i - \tilde{u}_i)$. The resolved field is that computed by the code; the RSFS field exists on the grid and so can actually be reconstructed to some level of approximation from the resolved data as was done by Stolz et al. (2001) and Chow et al. (2005); and the SGS field effects must be modeled. We can see how this plays out in the SFS stress terms which, once they are defined, become the "model" that is used in the LES.

The Resolved Subfilter-Scale (RSFS) Stress

The RSFS stress can be actually calculated without modeling by use of the deconvolution method of Stolz et al. (2001) which approximately reconstructs the original velocity field from the resolved field (see also Carati et al. 2001; Chow et al. 2005). Reconstruction uses the van Cittert iterative series expansion method and provides an estimate of the unfiltered velocity $(\tilde{u}_i)$ in terms of the filtered velocity $(\overline{\tilde{u}}_i)$ as

$$\tilde{u}_i^{\star} \sim \overline{\tilde{u}}_i + (I - G) * \overline{\tilde{u}}_i + (I - G) * [(I - G) * \overline{\tilde{u}}_i] + \cdots \,, \tag{10.24}$$

where I is the identity operator, G is the filter in Eq. (10.7), and $*$ signifies a convolution. Then, we observe that the approximate velocity $\tilde{u}_i^{\star}$ can be inserted into the first term of the RSFS stress, thereby making it possible to calculate the stress without a model, viz.,

$$\tau_{RSFS} = \rho\left(\overline{\tilde{u}_i \tilde{u}_j} - \overline{\tilde{u}}_i \overline{\tilde{u}}_j\right) \approx \rho\left(\overline{\tilde{u}_i^{\star} \tilde{u}_j^{\star}} - \overline{\tilde{u}}_i \overline{\tilde{u}}_j\right) \,. \tag{10.25}$$

The expansion velocity can also be substituted into the other RSFS term, but it is not essential to do so; note that the RSFS term $\rho \overline{\tilde{u}_i^{\star} \tilde{u}_j^{\star}}$ must actually be explicitly filtered in the calculation (Chow et al. 2005; Gullbrand and Chow 2003). The user can choose the level of reconstruction. Level 0 uses one term and yields an approximation essentially equivalent to the scale-similarity model of Bardina et al. (1980); level 1 uses two terms, has a significant impact on the results, and is probably a useful compromise

---

[10]Please note that Carati et al. (2001) and Chow et al. (2005), as well as many others, use a stress definition as the velocity product without including the density.

between cost and physics in mixed models.[11] Stolz et al. (2001) used this reconstruction as a full model and, in effect, neglected the SGS portion of the stress; however, the reconstruction term needed to be supplemented by an energy drain term added to the right side of the filtered Navier–Stokes equations. The reconstruction terms allow back scatter of energy from the subfilter to the resolved scales (see Chow et al. 2005).

The Subgrid-Scale (SGS) Stress

The SGS stress

$$\tau_{SGS} = \rho \left(\overline{u_i u_j} - \overline{\tilde{u}_i \tilde{u}_j} \right) \tag{10.26}$$

contains a correlation of the exact (non-filtered) velocities and so must be modeled. Any SGS model can be used, including those mentioned above and the algebraic stress models described below.

The Carati et al. (2001) approach leads, then, to a mixed expression for the total SFS stress. Chow et al. (2005) created a dynamic-reconstruction model (DRM) by using reconstruction plus a dynamic SGS model (the Wong and Lilly 1994, dynamic model), along with a near-wall model (which has a role similar to that of the van Driest model described in Sect. 10.3.3.2); Ludwig et al. (2009) compared the performance of that model to others and found that the DRM was superior to Smagorinsky and TKE models. Zhou and Chow (2011) applied the DRM strategy to the stable atmospheric boundary layer.

On the other hand, one can neglect either the RSFS or SGS terms and create a functional model, e.g., the Stolz et al. (2001) approximate deconvolution using only RSFS terms and an energy sink, a dynamic Smagorinsky alone, or the algebraic stress model of Enriquez et al. (2010) described next. A rationale for the stand-alone SGS models is to assume that the filter width Δ and the grid size h are equal so the RSFS zone vanishes. This is not strictly correct because the explicit and (implicit) grid filters are different, but it suffices. In general, for the mixed models, following the advice of Chow and Moin (2003) regarding error and the size of the SGS force, one sets $2 \leq \Delta/h \leq 4$.

10.3.3.5 Explicit Algebraic Reynolds Stress Models (EARSM)

One of the problems with the above SGS models is that they do not allow for the normal-stress anisotropy that is observed near the ground/walls in experiments and field studies (see Sullivan et al. 2003). Alternatives include:

---

[11]Accordingly, level n has $n + 1$ terms from the expansion (10.24); see the Appendix of Shi et al. (2018b) for a discussion and an application.

- Nonlinear SGS models such as that of Kosović (1997), which allows back scatter of energy and normal SGS stress anisotropy (cf. reconstruction models which also allow back scatter and are anisotropic at the RSFS).
- Deriving the transport equations for the SGS stresses and then retaining additional terms beyond those which yield the Smagorinsky model. Wyngaard (2004) and Hatlee and Wyngaard (2007) show those equations and suggest new anisotropic SGS models with improved performance. Extending this approach, Ramachandran and Wyngaard (2010) solve a truncated version of the partial differential transport equations for the SGS stresses. They did this for the moderately convective atmospheric boundary layer which matches the HATS experiments reported in Sullivan et al. (2003).
- Generating algebraic stress models from the SGS stress transport equations to capture new physics at moderate cost. We comment briefly on two algebraic stress models.

Generally, explicit algebraic stress approaches aim to produce a model not requiring solution of differential equations and dispensing with the eddy viscosity. Rodi (1976) created such a model for URANS and Wallin and Johansson (2000) have extended that (see Sect. 10.3.5.4). Marstorp et al. (2009) have built then an EARSM for LES of rotating channel flows. Working along the same lines as Wallin and Johansson, they either truncate or model the terms in the transport equations for the SGS stress anisotropy tensor; they obtain an explicit algebraic model for the SGS stress with two model parameters, i.e., the SGS kinetic energy and the SGS time scale. They offer a dynamic and a non-dynamic procedure to evaluate them. Their simulation results agree well with DNS data and both the resolved Reynolds stress anisotropy and the SGS stress anisotropy are improved as a consequence of avoiding the use of an eddy viscosity.

Most recently, Rasam et al. (2013) combined the explicit algebraic scalar flux model of Wallin and Johansson (2000) and the EARSM for LES of Marstorp et al. (2009), yielding a model capable of producing SGS scalar flux anisotropy and predicting scalar profiles reasonably well, as compared with filtered DNS data.

Findikakis and Street (1979) described an EARSM for the SGS terms in a LES based on ideas from the RANS turbulence models of Launder et al. (1975); see Sect. 10.3.5.4 below. Working in the context of the Carati et al. (2001) decomposition (described above), Enriquez et al. (2010) built on the Findikakis and Street work and that of Chow et al. (2005) to create a linear algebraic subgrid-scale stress model combined with reconstruction of the RSFS stresses for application to the neutral atmospheric boundary layer. The transport equations for the SGS stresses have been simplified to represent only the production, dissipation, and pressure redistribution terms in those equations, yielding a set of six linear algebraic equations, which are inverted at each grid point in each time step. There are no parameters to be determined beyond the constants implicit in the Launder et al. (1975) models which have been adopted. This model was embedded in the ARPS mesoscale, LES code (Xue et al. 2000; Chow et al. 2005). Convection, diffusion and viscosity are neglected in the SGS Reynolds-stress equations, but dissipation is modeled in the normal stress equations and the turbulent kinetic energy is obtained from solution of a partial differential transport equation. Results reported show that, when used without reconstruction, this EARSM produces results that are superior to Smagorinsky simulations and equivalent to dynamic Wong–Lilly results, and it reproduces the observed normal stress anisotropy. Applications to neutral, stable and convective atmospheric boundary layers are contained in Enriquez (2013).

10.3.3.6 Boundary Conditions for LES

The boundary conditions and numerical methods used for LES are very similar to those used in DNS. The most important difference is that, when LES is applied to flows in complex geometries, some numerical methods (for example, spectral methods) become difficult to apply. In these cases, one is forced to use finite difference, finite volume, or finite element methods. In principle, any method described earlier in this book could be used, but it is important to bear in mind that structures that challenge the resolution of the grid may exist almost anywhere in the flow. For this reason, it is important to employ methods of the highest accuracy possible.

As we noted in Sects. 10.2, and 10.3.5.5, behavior of flow near walls may require special treatment of boundary conditions or use of wall functions in the near wall region. Sagaut (2006) presents a discussion of alternatives, as does Rodi et al. (2013), which also includes alternatives for rough walls. Stoll and Porté-Agel (2006) describe a number of rough-wall models, but here we describe one of the most simple alternatives, as it is adequate in almost all cases. This model arises from the atmospheric sciences and oceanography, where the solid boundaries are often rough and the no-slip boundary condition is not appropriate, because either the roughness cannot be accurately represented and/or the roughness elements penetrate the mean viscous sublayer and the forcing includes both pressure and viscous effects. Then, a free-slip condition is applied where the velocity is not constrained and a relationship between the velocity and the momentum flux (wall stress) is established at the boundary. This mirrors what is often done for RANS flows which have wall models (cf. Sect. 10.3.5.5), i.e., a logarithmic velocity profile is assumed to exist near the boundary (or if the flow is not neutrally stable, a Monin–Obukhov similarity profile may be used (Porté-Agel et al. 2000; Zhou and Chow 2011)). So, instead of resolving the flow which requires a very fine grid normal to the wall,[12] a boundary condition forcing this logarithmic behavior is used (Rodi et al. 2013; Porté-Agel et al. 2000; Chow et al. 2005) and the wall stress is given by

$$(\tau_x)_{\text{wall}} = \rho C_D u_1 \sqrt{u_1^2 + v_1^2} \quad , \qquad (\tau_y)_{\text{wall}} = \rho C_D u_2 \sqrt{u_1^2 + v_1^2} \quad , \qquad (10.27)$$

where C_D is a drag coefficient and u_1 and v_1 are the streamwise and transverse velocity components at the first grid point off the wall from the profile of the velocity

---

[12]N.B.: A fine grid is needed also along the wall for DNS, but LES often has a rather high (horizontal to normal) grid aspect ratio near walls or the ground.

in the wall-normal direction.[13] Typically, when $(\tau_x)_{\text{wall}} = \rho u_\tau^2$ and the log-law for a rough wall is

$$\frac{u_1}{u_\tau} = \frac{1}{\kappa} \ln \frac{z_1 + z_0}{z_0} , \qquad (10.28)$$

it follows that

$$C_D = \left[\frac{1}{\kappa} \ln \left(\frac{z_1 + z_0}{z_0} \right) \right]^{-2} , \qquad (10.29)$$

where z_0 is the roughness length and z_1 is the distance to the first point off the wall; cf. Eq. (10.63).

In LES, it is possible to use wall functions of the kind used in RANS modeling (see Sect. 10.3.5) for smooth walls in order to avoid the fine grids otherwise required. Piomelli and Balaras (2002) and then Piomelli (2008) present reviews of a range of wall-layer models, the latter concluding that "... there is no single method that is clearly superior to the others."

10.3.3.7 Truncation Error and Computational Mixing

The accuracy of numerical large-eddy simulations can be affected by, among other things, the size of the numerical truncation error compared to the subgrid- or subfilter-scale forcing, the form of the SFS/SGS models, the numerical algorithm, and computational mixing. The last is a typical feature of geophysical flow simulations where high-order (e.g., 5th and 6th) advection schemes are used; in that context, Xue (2000) states that "most numerical models employ numerical diffusion or computational mixing to control small-scale (near two grid intervals in wavelength) noise that can arise from numerical dispersion, nonlinear instability, discontinuous physical processes, and external forcing."

The following sections suggest some guidelines regarding error, mixing effects and simulation quality.

Error and accuracy: Chow and Moin (2003) conducted a study of the balance between numerical errors and the SFS forcing. Using a direct numerical simulation (DNS) dataset of stably stratified shear flow to perform a priori tests, Chow and Moin (2003) compare the numerical error from several finite difference schemes to the magnitude of the SFS force. To ensure that the SFS terms are larger than the numerical errors contained in the numerical scheme for the nonlinear convection terms, they found:

1. for a second-order finite difference scheme, the filter size should be at least four times the grid spacing;
2. for a sixth-order Padé scheme, the filter size needs to be only at least twice as large as the grid spacing.

---

[13]For Coriolis-influenced flows, the velocity rotates with distance from the boundary and both components may be non-zero even if the flow is uni-directional far from the boundary; see Sect. 10.3.4.3.

However, Celik et al. (2009) argue that "This is nearly impossible to achieve in engineering LES applications." The consequence of not satisfying the above criteria is that LES then contains, in addition to the modeled SGS stresses, an unknown—but possibly significant—contribution from numerical errors.

Mixing Effects: The computational mixing terms typically are 4th or higher-order hyperviscosity terms which are applied to momentum and scalar equations. For commercial or professionally developed codes guidelines suggest coefficient values so as to give the minimum amount of mixing required for stability. However, this may still have a significant impact on results. The Bryan et al. (2003) test case (see their Appendix) of buoyant convection, with sixth-order convection and a sixth-order explicit filter, showed that for energy spectra only information at wavelengths greater than six times the grid spacing represents a physical solution and that it was possible to create an arbitrary spectral slope for scales less than six times the grid size by changing the computational mixing coefficient. Using a mesoscale code, Michioka and Chow (2008) conducted a high-resolution LES of scalar transport over complex terrain and demonstrated that:

1. Large computational mixing coefficients may not dramatically change the simulated mesoscale flow fields (because the computational mixing damps only near grid scale noise).
2. Computational mixing can noticeably impact turbulent fluctuations and near-ground velocity profiles in high-resolution simulations, because small-scale motions are part of the resolved turbulent flow field.
3. Using the smallest feasible computational mixing coefficient was essential to achieving a good match between simulation and measured data for the maximum ground concentration of the scalar.

Simulation Quality: The verification and/or validation of LES is difficult because as, for example, the grid size is reduced, both the numerical truncation error and the SGS or SFS model errors decrease and the fraction of flow resolved changes. Thus, in some sense the LES converges towards DNS and as Celik et al. (2009) state "there is no such thing as grid-independent LES." It is, of course, reasonable to use a sequence of smaller grid sizes to test for convergence (by which we mean that the represented flow changes from one grid to another are small enough for the desired purpose). Sullivan and Patton (2011) present a detailed evaluation of convergence with grid size reduction, noting that the convergence of some measures of performance is slower than others and that as grid size is reduced, it is possible that the representation of the physics at interfaces and other zones of sharp gradients may change.

Meyers et al. (2007) do an error assessment on a homogeneous isotropic flow to demonstrate a method in which they systematically change the Smagorinsky coefficient and use different discretizations, thus attempting to isolate the numerical and model error impacts. Following a series of antecedent papers, Celik et al. (2009) offer strategies for assessment of engineering LES applications. They too suggest systematic grid and model variation, which requires several code runs, and make application to a number of cases, presenting graphic evidence. However, Sullivan

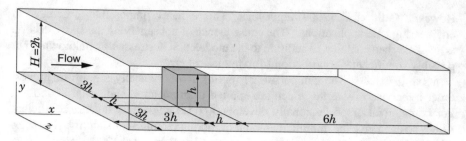

Fig. 10.7 The solution domain for the flow over a cube mounted on a channel wall; from Shah and Ferziger (1997)

and Patton (2011) caution that, while Celik et al. (2009) often rely on DNS as a testing base, it is not available for the high-Reynolds number planetary boundary layer.

10.3.4 Examples of LES Application

10.3.4.1 Flow Over a Wall-Mounted Cube

As an example of the method, we shall use the flow over a cube mounted on one wall of a channel. The geometry is shown in Fig. 10.7. For the simulation shown, which was made by Shah and Ferziger (1997), the Reynolds number based on the maximum velocity at the inflow and the cube height is 3200. The inflow is fully developed channel flow and was taken from a separate simulation of that flow, the outlet condition was the convective condition given by Eq. 10.4. Periodic boundary conditions were used in the spanwise direction and no-slip conditions at all wall surfaces.

The LES used a grid of $240 \times 128 \times 128$ control volumes with second-order accuracy. The time advancement method was of the fractional-step type. The convection terms were treated explicitly by a third-order Runge–Kutta method in time while the viscous terms were treated implicitly. In particular, the method used for the latter was an approximate factorization of the Crank–Nicolson method. The pressure was obtained by solving a Poisson equation with the multigrid method.

Figure 10.8 gives the streamlines of the time-averaged flow in the region close to the wall; a great deal of information about the flow can be discerned from this plot. The incoming flow does not separate in the traditional sense but reaches a stagnation or saddle point (marked by A on the figure) and goes around the body. Some of the flow further above the lower wall hits the front face of the cube; about half of it flows downwards and creates the region of reversed flow in front of the body. As the flow down the front face of the cube nears the lower wall, there is a secondary separation and a reattachment line (marked by B in the figure) just ahead of the cube. On each

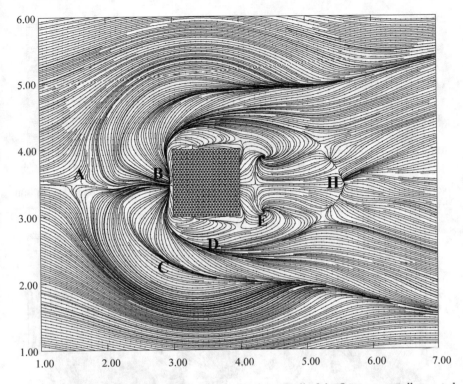

Fig. 10.8 The streamlines in the region close to the lower wall of the flow over a wall-mounted cube; from Shah and Ferziger (1997)

side of the cube, one finds a region of converging streamlines (marked as C) and another of diverging streamlines (marked D); these are the traces of the horseshoe vortex (about which more is said below). Behind the body one finds two areas of swirling flow (marked E) which are the footprints of an arch vortex. Finally, there is a reattachment line (marked H) further downstream of the body.

Figure 10.9 shows the streamlines of the time-averaged flow in the center-plane of the flow. Many of the features described above are clearly seen including the separation zone in the upstream corner (F), which is also the head of the horseshoe vortex, the head of the arch vortex (G), the reattachment line (H), and the recirculation zone (I) above the body which does not reattach on the upper surface.

Finally, Fig. 10.10 gives a projection of the streamlines of the time-averaged flow on a plane parallel to the back face of the cube just downstream of the body. The horseshoe vortex (J) is clearly seen as are smaller corner vortexes.

It is important to note that the instantaneous flow looks very different than the time-averaged flow. For example, the arch vortex does not exist in an instantaneous sense; there are vortexes in the flow but they are almost always asymmetric on the two sides of the cube. Indeed, the near-symmetry of Fig. 10.8 is an indication that the averaging time is (almost) long enough.

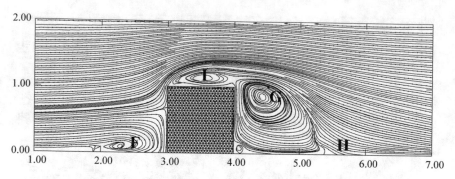

Fig. 10.9 The streamlines in the vertical center-plane of the flow over a wall-mounted cube; from Shah and Ferziger (1997)

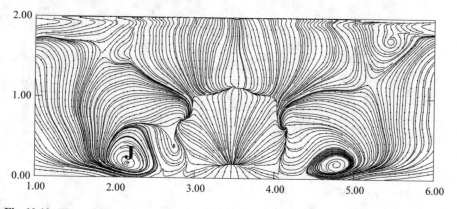

Fig. 10.10 The projection of streamlines of the flow over a wall-mounted cube onto a plane parallel to the back face, 0.1 step hight behind the cube; from Shah and Ferziger (1997)

It is clear from these results that an LES (or DNS for simpler flows) provides a great deal of information about a flow. Performance of such a simulation has more in common with doing an experiment than it does to the types of simulations described in Sect. 10.3.5 and the qualitative information gained from it can be extremely valuable.

10.3.4.2   Flow Around a Sphere at Re = 50,000

In Sect. 10.2.4.2 we briefly discussed a DNS of flow around smooth sphere at Re = 5,000; for the flow at Re = 50,000 one would need an extremely fine grid to be able to perform a proper DNS. The logical step is to switch to LES, which resolves the large scales and models the subgrid-scale turbulence. Even for LES we have had to make a finer grid than the one used for DNS at the lower Reynolds number: around 40 million cells were used in an unstructured grid made of Cartesian cells with local refinement in the wake and trimming by a prism layer along walls. Figure 10.11

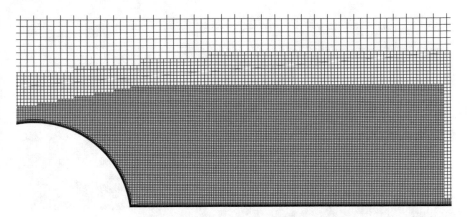

Fig. 10.11 Longitudinal section through the solution domain at $y = 0$ for the smooth sphere, showing the grid structure around sphere (in the finest grid the spacing is twice smaller in all directions)

shows the structure of the grid in the sphere wake; so the grid structure can be seen, this grid is twice coarser than the grid used to produce the results presented below.

The flow around sphere at Re = 50,000 has been studied experimentally by Bakić (2002); he used a billiard ball with a diameter $D = 61.4$ mm, held by a stick with a diameter $d = 8$ mm in a small wind tunnel with a rectangular cross-section of 300 × 300 mm. He also performed measurements for a sphere with a trip-wire attached circumferentially at a location 75° from the front stagnation point; the diameter of the trip-wire was 0.5 mm.

The solution domain in the simulation matches the experimental geometry, with the inlet boundary being 300 mm upstream and the outlet boundary being 300 mm downstream of the sphere center. A uniform velocity of 12.43 m/s was specified at the inlet, the fluid being air with density $\rho = 1.204$ kg/m^3 and viscosity $\mu = 1.837 \times 10^{-5}$ Pa·s. Simulations were performed for both a smooth sphere and a sphere with a trip-wire. The grid spacing in the wake of the smooth sphere (see Fig. 10.11) was 0.265 mm in all three directions, which is approximately $D/232$; this spacing was kept over a distance of approximately $1.7D$ downstream of the rear stagnation point and the diameter of the finest grid zone was almost $1.5D$. The grid was coarsened in several steps away from the finest zone shown in Fig. 10.11; no attempt was made to resolve the boundary layers growing along the wind tunnel walls, but prism layers along the sphere and the holding stick surface started with the thickness of the next-to-wall cell being 0.03 mm ($D/2407$).

For the case with trip-wire, the grid design had to be changed. In order to capture laminar boundary layer separation at the trip-wire surface, the cell size in the zone extending ca. half a trip-wire diameter upstream and above and 4 diameters downstream of trip-wire was 0.041667 mm ($D/1474$). The zone with twice that spacing extended along the sphere surface all the way until boundary-layer separation on the downstream side of the sphere; see Fig. 10.12, which shows the detail of the

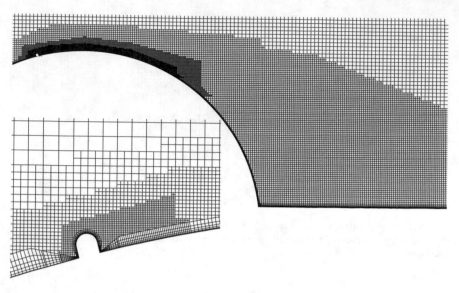

Fig. 10.12 Longitudinal section through the solution domain at $y = 0$ for the sphere with a trip wire, showing the grid structure around the sphere and trip wire

actual grid around trip wire and in immediate sphere wake. In the largest part of sphere wake, the grid size was 0.3333 mm ($D/184$). The thickness of the first prism layer next to wall was 0.02 mm ($D/3070$). The wake is much narrower in the case with trip-wire, so the fine-grid zone behind the sphere is accordingly made smaller compared to the case of a smooth sphere.

The time step was 10 μs in both cases, leading to an average Courant-number of 0.5 based on mean velocity and grid spacing in the finest zone for the smooth sphere. The Courant-numbers were larger in the finest grid around the trip wire. Second-order schemes were used in both space and time (central-differencing for convection and diffusion and quadratic backward interpolation in time). The commercial code STAR-CCM+ was the solver (fully-implicit in time, i.e., convection fluxes, diffusion fluxes and source terms are computed at the new time level; the SIMPLE algorithm is used for pressure-velocity coupling; the flow is assumed to be incompressible). The under-relaxation parameters were 0.95 for velocity and 0.75 for pressure, with 5 outer iterations performed per time step to update the non-linear terms. Simulations were performed also on one coarser grid and with three different subgrid-scale models; this is part of an ongoing study and we shall not go into all details here. The one important message is that, by using a locally refined unstructured grid (which could be also polyhedral), we can resolve the turbulent wake while not wasting cells in zones where the flow is not turbulent and the variables do not vary so rapidly in space. A visual inspection of vorticity structures in Fig. 10.13 and 10.14 suggests that the grid used might be of an adequate resolution for LES, because the structures are substantially larger than the fine grid spacing.

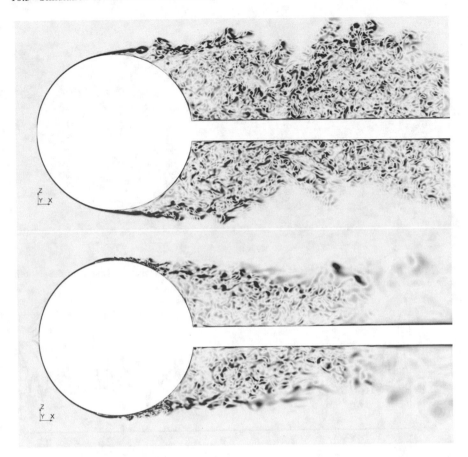

Fig. 10.13 Instantaneous contours of vorticity component ω_y in the plane $y = 0$ for the smooth sphere (upper) and for the sphere with a trip-wire (lower)

This subcritical flow cannot be predicted well using RANS models; we shall not show any results of such computations but only state that all models tested significantly under-predict the drag and over-predict the length of the recirculation zone behind sphere (significantly meaning 25% or more). However, LES using a dynamic Smagorinsky subgrid-scale model predicts the mean drag coefficient for a smooth sphere to be around 0.48, which is close to experimental data found in the literature. The trip-wire was expected to delay the separation and reduce the size of the recirculation zone behind sphere, leading to a significantly lower drag; that was indeed the case. As shown in Fig. 10.15, the predicted drag fluctuates around a value close to 0.175, which is almost 3 times lower than for a smooth sphere. This demonstrates the power of CFD: when using an adequate grid and turbulence-modeling approach, one can predict the effects of small geometry changes on the flow.

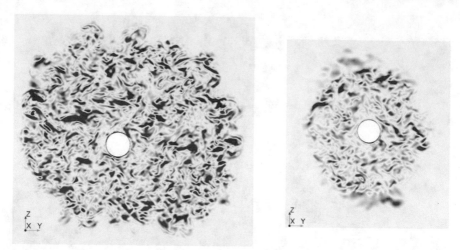

Fig. 10.14 Instantaneous contours of vorticity component ω_x in the plane $x/D = 1$ for the smooth sphere (left) and for the sphere with a trip-wire (right)

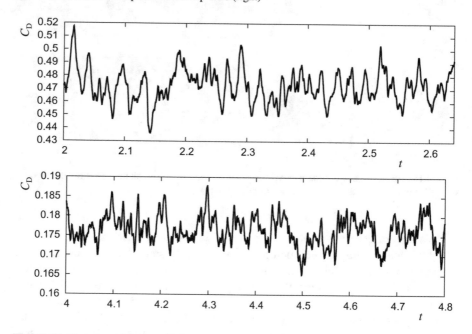

Fig. 10.15 Variation of drag coefficient for the smooth sphere (upper) and for the sphere with a trip-wire (lower) during the last 65,000 time steps

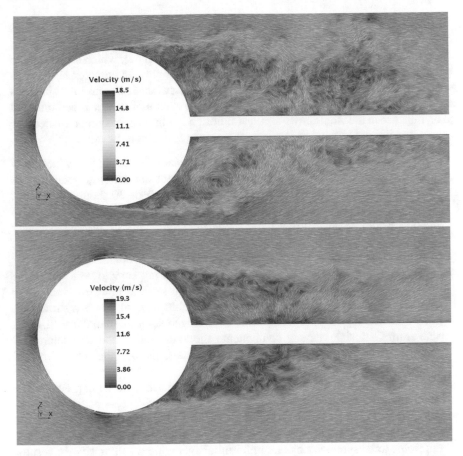

Fig. 10.16 Instantaneous flow patterns for the smooth sphere (upper) and for the sphere with a trip-wire (lower)

LES provides insight into flow behavior, which is important for engineers who may need to improve the design. Animation of flow can easily be performed using pictures from LES created at a prescribed time increment; we cannot reproduce an animation in this book, but we show in Figs. 10.13, 10.14, and 10.16 the instantaneous flow patterns for the sphere with both smooth surface and surface-mounted trip-wire, revealing details of both flow separation and wake behavior. These figures clearly show the effect of trip wire on the flow: the boundary layer becomes turbulent following reattachment after separation from the trip wire, leading to a delayed main separation and a much narrower wake. Every engineer would immediately know that the consequence of such a change in flow behavior will lead to a significant reduction of drag.

Engineers are often interested in mean quantities and rms-values (root-mean-square) of their fluctuation. This information can easily be obtained for integral quantities like drag of a body, e.g., by processing signals like those shown in Fig.

10.15. Obtaining the mean velocity and pressure distribution is less trivial. In LES of channel or pipe flows, averaging is usually performed in both time and one spatial direction (spanwise for channel and circumferential for pipe), which allows one to obtained steady values with a smaller number of samples. In complex geometries spatial averaging is usually not possible; in the present case, because the sphere is mounted in a wind tunnel with a rectangular cross-section, the flow is certainly not axisymmetric in the whole cross-section (although it might be axisymmetric near the sphere and the rod holding it). Three problems are encountered here:

- Without spatial averaging, the required number of samples for obtaining the steady time-averaged flow becomes very large. In the present simulations, averaging was performed over the last 65 thousand (smooth sphere) or 70 thousand samples (sphere with trip-wire), but the time-averaged flow is not axisymmetric; many more samples would be needed to achieve the perfect symmetry of averaged velocity profiles. In experiments, one often assumes symmetry and measures the profiles along one line from the symmetry axis toward the outer radius.
- When using unstructured grids, spatial averaging is not easy to perform. In the present case, cells are Cartesian with variation in size; in order to average the solution in the circumferential direction, we would have to create a structured, polar-cylindrical grid, interpolate the time-averaged solution onto this grid, compute from Cartesian velocity components the axial, radial and circumferential components on the structured grid and then average these in the circumferential direction.
- The assumption that the mean flow is axisymmetric may be wrong, even if the geometry were fully axisymmetric. Indeed, there is evidence—both from simulations (e.g., by Constantinescu and Squires 2004) and experiments (e.g., Taneda 1978)—that the wake of a sphere tends to tilt with respect to the flow direction. Enforcing axial symmetry by circumferential averaging would falsify the results.

We show the time-averaged flow patterns for both spheres in Fig. 10.17. While the mean flow field for the smooth sphere appears nearly symmetric, for the sphere with a trip-wire the flow pattern is closely symmetrical only in the plane $z = 0$; in the plane $y = 0$, the flow is highly asymmetric. As we will show in Sect. 10.3.6, some RANS models also predict asymmetric steady-state flow around smooth sphere, so this appears to be the feature of the flow rather than a deficiency of the simulation. The length of the recirculation zone agrees well in both cases with the experiments of Bakić, who found that the recirculation zone ends at $x/D = 1.43$ for the smooth sphere and at $x/D = 1$ for the sphere with the trip wire.

We note again that the time-averaged flow is only a construct obtained by averaging of flow realizations at different times; it never exists in reality, not even for an instant of time. While it can be obtained by averaging of measured data as well as simulation data, it cannot be observed in nature. A long-time exposure in an experiment would produce a different picture because the spreading around the mean does not cancel out positive and negative values as happens in a mathematical averaging. Interestingly, the instantaneous plot of vorticity component ω_x in the plane at $x/D = 1$ showed in Fig. 10.14 appears almost symmetric for the sphere with a trip-wire, while the

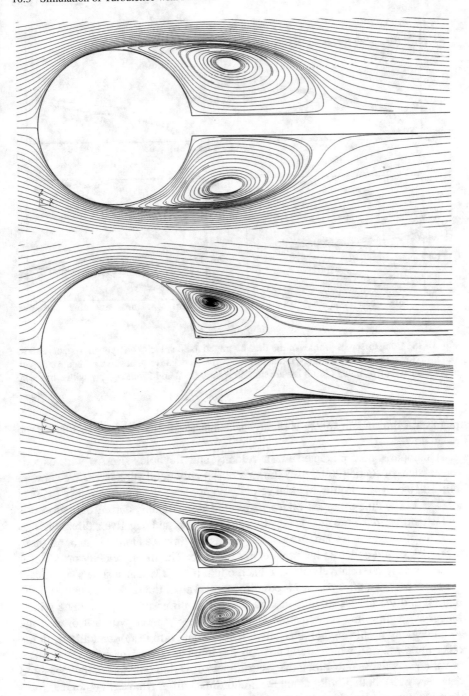

Fig. 10.17   Time-averaged flow pattern for the smooth sphere in the plane $y = 0$ (upper) and for the sphere with a trip-wire in the plane $y = 0$ (middle) and $z = 0$ (lower)

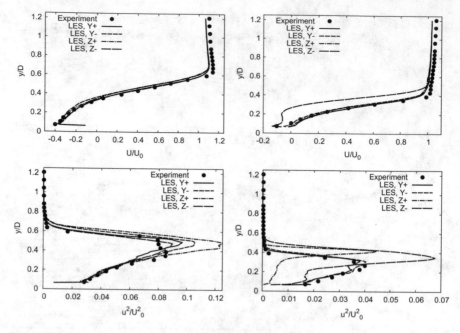

Fig. 10.18 Comparison of profiles of the time-averaged axial velocity component (upper) and its variance (lower) at $x/D = 1$ along 4 radial lines (positive y, negative y, positive z, negative z) with experimental data of Bakić (2002) for the smooth sphere (left) and for the sphere with a trip-wire (right)

picture for the smooth sphere is highly asymmetric—contrary to the results for the mean flow.

A comparison of predicted mean velocity and Reynolds-stresses with experimental data of Bakić (2002) also shows a reasonably good agreement. We show in Fig. 10.18 only the profiles of axial velocity component U_x and its variance $(u'_x)^2$ at $x/D = 1$. The time-averaged values are plotted along 4 lines, starting at the rod axis and extending in the positive y, negative y, positive z, and negative z directions.

If the flow was axisymmetric and the averaging time was long enough, all these 4 profiles would collapse; here this is not the case. The mean velocity profiles for smooth sphere are relatively close to each other, but the variance shows significant variation between all 4 profiles around the peak value; the difference between the highest and the lowest value is around 30%. This difference might reduce with a longer averaging time (more samples). In the case of sphere with a trip wire, the mean velocity profile along the negative z-coordinate differs substantially from the other three profiles, which are almost on top of each other. However, the profiles of variance collapse only along the y-coordinate direction (and match the measured data very well); both profiles along the z-coordinate differ from each other and from the other two profiles. Here the flow appears to have a tilted wake, as discussed above.

More details from these simulations (including grid- and SGS-model-dependence as well as experimental data) will be presented in future publications.

10.3.4.3 Renewable Energy Simulation: Large Wind-Turbine Farm

Jacobson and Delucchi (2009) outlined "a path to sustainable energy by 2030." Their plan called for solar, water, geothermal, and wind-based sources, among which would be 3.8 million large wind turbines. Indeed, wind energy is a fast growing electric power source already (e.g., in 2016, 12.3% of all consumed electric energy in Germany was produced by wind turbines, with the tendency increasing). Sta. Maria and Jacobson (2009) give an overview and elementary analysis of the effect on energy in the atmosphere of large groups of wind turbines, i.e., *wind farms*. Increased near-surface turbulence due to a wind farm (from turbulence downstream of the rotors) may affect heat and vapor fluxes there, and some studies have shown increased mixing in the boundary layer. In addition, in wind farms, the turbines interact among themselves.

Current and immediate future technology points to wind turbines with rotor diameters of more than 150 m, hub heights (the axis level of the rotor at the top of the support tower) around 200 m, and power outputs around 10 MW. Lu and Porté-Agel (2011) carried out a three-dimensional LES of a very large wind farm in a stable atmospheric boundary layer. Their rotor diameter is 112 m with a hub height of 119 m. The study is summarized below.

Flow domain: The domain and physical set-up were taken from a well-known stable boundary layer (SBL) case with a boundary layer height of approximately 175 m, i.e., the intercomparison study of Beare et al. (2006) based on the Global Energy and Water Cycle Experiment Atmospheric Boundary Layer Study (GABLS). Thus, the flow domain and simulation had been completely tested and the numerical code verified for the flow without the wind turbines, making assessment of the turbine impacts easily observed and quantified.

Figure 10.19 shows the domain on the left. The basic idea is that a single wind turbine is placed in a domain with periodic lateral boundary conditions, thereby creating an infinite wind farm with specified turbine placements. The dimensions of the domain are vertical height $L_z = 400$ m and lateral width $L_y = 5D = 560$ m, where $D = 112$ m is the rotor diameter mentioned above. There are two simulation configurations, one with $L_x = 5D$ (the 5D case) and one with $L_x = 8D$ (the 8D case). With this configuration, a given turbine is affected by its neighbors and conversely, as in a real wind farm.

Wind turbine: Figure 10.19 shows the three-blade turbine on the right. It is located $x_c = 80$ m from the left edge of the domain at a height $z_c = 119$ m and along the centerline of the domain (at $y_c = 260$ m). The turbines are spaced, then, 5 or 8 diameters apart in the streamwise direction and 5 diameters apart laterally. These are considered typical wind farm spacings. The turbine rotates at 8 rpm, consistent with the chosen turbine, its power generation, blades, etc.

In the calculation, the turbine is parameterized with the actuator-line method (ALM) in which the actual motion of the blades is accounted for (Ivanell et al. 2009). The forces on the blades are represented on lines along the blade axis. Body forces (caused by the lift and drag at each point along the line, on each blade, and at each time step) are calculated using the local angle of attack of a blade element

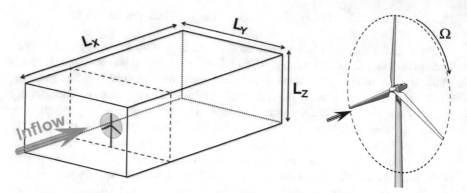

Fig. 10.19 Domain layout with turbine sketched (left) and three-blade wind turbine, generator, and tower (right) in Lu and Porté-Agel (2011). (Reproduced with the permission of AIP Publishing)

which depends on the blade form, the instantaneous resolved flow velocity, and actual measured airfoil performance data for that blade. This produces unsteady, spatially-varying forces on the plane of the turbine that are transferred into the flow momentum equations via a Gaussian smoothing to connect to the nearby grid points. The flow results feed back into the ALM and affect the computed lift and drag along the blades. The details are found in the cited ALM paper and the Lu and Porté-Agel paper and its references.

The LES code and model: The code used is a modified version of the LES code by Porté-Agel et al. (2000), which is pseudo-spectral in the horizontal and uses a 2nd-order CDS on a staggered grid in the vertical direction. The continuity equation, momentum conservation equations with a wind-induced forcing term from the ALM, and a transport equation for potential temperature[14] are solved for incompressible flow in the Boussinesq approximation (Sect. 1.7.5) and include Coriolis forcing. At the top of the domain a stress-free/zero-gradient boundary condition is used. At the ground, the instantaneous wall stress is related to the horizontal velocities at the first vertical node through the application of the Monin–Obukhov similarity theory for rough walls which is applicable because the flow is stratified (see Eqs. (10.27) and (10.29) above). A similar boundary condition is used for heat flux.

The grid has uniformly-spaced points with a grid spacing of $\sim$3.3 m, with 270 $\times$ 168 $\times$ 120 points for the 8D case, and 168 $\times$ 168 $\times$ 120 points for the 5D case. Several tests were run to validate these choices. A Courant limit of $C = 0.06$ was used with a 2nd-order accurate Adams–Bashforth time-stepping scheme. The spectral results were de-aliased according to the standard 3/2 rule. Before data is harvested, the code is run until quasi-statistically-steady conditions are obtained; recall that this

---

[14]The potential temperature is used extensively in meteorology because it has properties that suit it to the study of stratified flow in a compressible medium (air); it is the temperature that a parcel of air at a height z would have after it was moved to the ground without exchanging heat with its surroundings (adiabatically); so $\Theta(z) = T(z)[p_{\text{ground}}/p(z)]^{0.286}$.

flow is three-dimensional and unsteady with both resolved turbulent motions and coherent motions induced by the turbines.

The SGS model for fluxes of momentum and heat is a variant of the dynamic model of Sect. 10.3.3.3 above and is described in Stoll and Porté-Agel (2008); Porté-Agel et al. (2000) and subsequent papers. Two important features are added, and they improve the behavior of the dynamic method significantly: (1) to minimize the error in the dynamic process the total error to be minimized is generated by accumulating the local error along Lagrangian tracks (pathlines of fluid particles in the flow) (Meneveau et al. 1996) and (2) the scale similarity is relaxed so the model form and coefficient are scale-dependent. The base filter/grid width is computed as $(\Delta x \Delta y \Delta z)^{1/3}$. The grid filter and the two test filters are two-dimensional low-pass filters in the horizontal directions; there is no filtering in the vertical FD direction. The variables are filtered in Fourier space with a sharp cutoff filter, which removes all wavenumbers larger than the filter scale. The structure of the dynamic equations is essentially the same as above; however, because scale invariance is not forced, a form for the change of the coefficient with grid spacing is assumed and two test filters are used, one at twice the grid size and one at four times the grid size. Then, using the same minimizing strategy as in the traditional dynamic model, the unknowns can be determined. Of course, here there is also a dynamic procedure for the SGS potential temperature. Note that, because the dynamic process uses the smallest resolved scales, it is not necessary to include a correction for stability in the SGS model because that effect is already in the small resolved scales (cf. Sect. 10.3.3.2).

Simulation results: At the scale of the wind farm and because of the stable (thermally-stratified) conditions, vertical shear due to the velocity profile changes with height, as does the horizontal shear due the Coriolis force effect (Fig. 10.20). These cause significant asymmetric loadings on the turbines. Here we can see the power of the LES as it actually resolves the tip vortices from the moving turbine blades. Also the Coriolis force not only causes additional lateral shearing loads on wind turbines but also drives part of the turbulence energy away from the center of the wind-turbine wakes. The wind-turbine motions enhance the vertical mixing of heat, resulting in increased air temperatures in the wind-turbine wakes and lowered surface heat flux, thus affecting the thermal energy budget.

We show three figures from the study. Figure 10.21 shows clearly that the mean vertical profiles (averaged over space and time) of the horizontal velocity components and of the potential temperature are heavily impacted by the presence of the turbine. The jet at the top of the boundary layer is removed by the energy extraction of the turbine, and the Coriolis effect is altered when the turbine is present. The difference between the two $S_x = L_x/D$ cases is not large. The increase in the depth of the mixed layer is evident in the potential temperature.

Figure 10.22 shows the evolution of the mean streamwise velocity profile at various S_x distances downwind. Note the turbine is at $S_{\text{turbine}} \sim 0.7$. More energy is extracted from the flow when the turbines are closer together.

Finally, we see the impact of the turbine on the energy spectra of the flow in Fig. 10.23. **N** is the Brunt–Väisälä or *stratification frequency* of natural oscillations in stable stratification. Other than the obvious increase in turbulent energy downwind

Fig. 10.20 Visualization of tip vortices at $t = 150$ s caused by moving turbine blades, using iso-surfaces of vorticity $\omega (\sim 0.3|\omega|)$. Reproduced from Lu and Porté-Agel (2011) with the permission of AIP Publishing

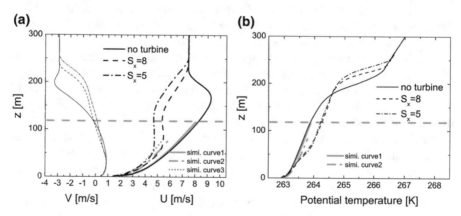

Fig. 10.21 Vertical mean profiles of velocity and potential temperature. Dashed horizontal line is turbine hub height; light lines in background on curves are M-O similarity curves. Reproduced from Lu and Porté-Agel (2011) with the permission of AIP Publishing

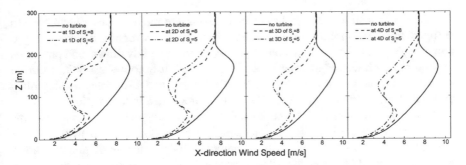

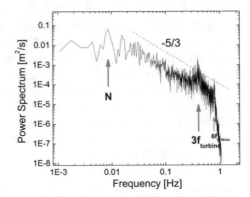

Fig. 10.22 Downwind variation of streamwise (axial) profile on the centerline of domain. Reproduced from Lu and Porté-Agel (2011) with the permission of AIP Publishing

Fig. 10.23 Wind turbine effect on the turbulence energy spectra at $x = 3D$ downstream of turbine plane, y_c, and $z = z_c + D/2$ (the height of the top tip of the blade). Reproduced from Lu and Porté-Agel (2011) with the permission of AIP Publishing

of the turbine shown here, the paper gives an in-depth discussion of turbulent fluxes under the influence of the turbine and interprets them for meteorological situations.

Lu and Porté-Agel state that their results show "that large-eddy simulation can provide valuable three-dimensional high-resolution velocity and temperature fields needed for the quantitative description of wind-turbine wakes and their effects on the turbulent fluxes of heat and momentum inside and above wind farms." Here, the surface momentum flux was reduced by more than 30%, and the surface buoyancy flux was reduced by more than 15%. The wind farm had a strong effect on vertical turbulent fluxes of momentum and heat, which could impact local meteorology.

10.3.5 Reynolds-Averaged Navier–Stokes (RANS) Simulations

Traditionally, engineers were normally interested in knowing just a few quantitative properties of a turbulent flow, such as the average forces on a body (and, perhaps, their distribution), the degree of mixing between two incoming streams of fluid, or the amount of a substance that has reacted. Using the methods described above to compute these quantities is, to say the least, overkill. However, things have changed and today's problems are more complex, designs more tightly drawn, and processes depend more on details of what is occurring. Then, it becomes essential to make judicious use of the methods contained in the DNS and LES schemes described above and intelligent use of the RANS methods described in this section.

As noted above, due to Osborne Reynolds over a century ago, the methods of this section are called the *Reynolds-averaged* methods. Leschziner (2010) wrote "..., at the time of writing, the large majority of computational predictions of industrial flows are based on the RANS equations." In Reynolds-averaged approaches to turbulence, the governing equations are averaged in some way as described in Sect. 10.3. As we look at the various segments below we need to keep alert to the implications of these averaging strategies. Recall that the practical strategies were:

1. **Steady flow**: All of the unsteadiness is averaged out, i.e., all unsteadiness is regarded as part of the turbulence. The result is that the mean flow equations are steady. These are then the *Reynolds-averaged Navier–Stokes* (RANS) equations.
2. **Unsteady flow**: The equations are averaged over a complete set of statistically identical realizations of the flow (ensembles).[15] As a consequence, all of the random fluctuations are averaged out and so are implicitly part of the 'turbulence'. However, if there are deterministic and coherent structures in the flow, they should survive the averaging. These ensemble-averaged equations may have an unsteady mean. Such flows were defined above as unsteady RANS or URANS (Durbin 2002) or transient RANS or TRANS (Hanjalić 2002).

Again, we recall that, on averaging, the non-linearity of the Navier–Stokes equations gives rise to terms that must be modeled, just as they did earlier. The complexity of turbulence, which was discussed briefly above, makes it unlikely that any single Reynolds-averaged model will be able to represent well all turbulent flows, so turbulence models should be regarded as engineering approximations rather than scientific laws. Hanjalić (2004) presents a comprehensive review of RANS and its turbulence models.

10.3.5.1   Reynolds-Averaged Navier–Stokes (RANS) Equations

In a statistically steady flow, every variable can be written as the sum of a time-averaged value $\overline{\phi}$ and a fluctuation about that value ϕ':

$$\phi(x_i, t) = \overline{\phi}(x_i) + \phi'(x_i, t) , \tag{10.30}$$

where

$$\overline{\phi}(x_i) = \lim_{T \to \infty} \frac{1}{T} \int_0^T \phi(x_i, t) \, dt . \tag{10.31}$$

---

[15]Weather forecasters, in particular, use a finite set of simulations (which may not be statistically identical) and ensemble-average them to achieve improved predictions.

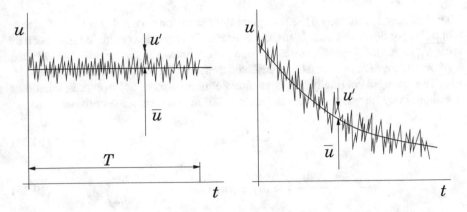

Fig. 10.24 Time-averaging for a statistically steady flow (left) and ensemble-averaging for an unsteady flow (right)

Here t is the time and T is the averaging interval. This interval must be large compared to the typical time scale of the fluctuations; thus, we are interested in the limit of $T \to \infty$, see Fig. 10.24. If T is large enough, $\overline{\phi}$ does not depend on the time at which the averaging is started.

If the flow is unsteady, time-averaging cannot be used unless the accompanying turbulence model's time scale is adjusted for the averaging time period $\Delta T < \infty$. In most cases, unsteady flow is dealt with by ensemble-averaging. This concept was discussed earlier and is illustrated in Fig. 10.24[16]:

$$\overline{\phi}(x_i, t) = \lim_{N \to \infty} \frac{1}{N} \sum_{n=1}^{N} \phi(x_i, t) , \qquad (10.32)$$

where N is the number of members of the ensemble and must be large enough to eliminate the effects of the turbulent (random) fluctuations. This type of averaging can be applied to any flow. We use the term *Reynolds-averaging* to refer to any of these averaging processes; applying it to the Navier–Stokes equations yields the Reynolds-averaged Navier–Stokes (RANS) equations for the steady case and URANS or TRANS for unsteady cases.

From Eq. (10.31), it follows that $\overline{\phi'} = 0$. Thus, averaging any linear term in the conservation equations simply gives the identical term for the averaged quantity. From a quadratic nonlinear term we get two terms, the product of the average and a covariance:

$$\overline{u_i \phi} = \overline{(\overline{u}_i + u_i')(\overline{\phi} + \phi')} = \overline{u}_i \overline{\phi} + \overline{u_i' \phi'} . \qquad (10.33)$$

---

[16]For the unsteady case with persistent structures, the flow may not change monotonically and the result might look much like Fig. 10.6 if the LES line is imagined to represent the coherent structures in the ensemble-averaged flow. See Fig. 1.8 in Chen and Jaw (1998).

The last term is zero only if the two quantities are uncorrelated; this is rarely the case in turbulent flows and, as a result, the conservation equations contain terms such as $\rho \overline{u'_i u'_j}$, called the *Reynolds stresses*,[17] and $\rho \overline{u'_i \phi'}$, known as the *turbulent scalar flux*, among others. These cannot be represented uniquely in terms of the mean quantities.

The averaged continuity and momentum equations can, for incompressible flows without body forces, be written in tensor notation and Cartesian coordinates as:

$$\frac{\partial(\rho \overline{u}_i)}{\partial x_i} = 0 \,, \tag{10.34}$$

$$\frac{\partial(\rho \overline{u}_i)}{\partial t} + \frac{\partial}{\partial x_j}\left(\rho \overline{u}_i \overline{u}_j + \rho \overline{u'_i u'_j}\right) = -\frac{\partial \overline{p}}{\partial x_i} + \frac{\partial \overline{\tau}_{ij}}{\partial x_j} \,, \tag{10.35}$$

where the $\overline{\tau}_{ij}$ are the mean viscous stress tensor components:

$$\overline{\tau}_{ij} = \mu \left(\frac{\partial \overline{u}_i}{\partial x_j} + \frac{\partial \overline{u}_j}{\partial x_i}\right) \,. \tag{10.36}$$

Finally the equation for the mean of a scalar quantity can be written:

$$\frac{\partial(\rho \overline{\phi})}{\partial t} + \frac{\partial}{\partial x_j}\left(\rho \overline{u}_j \overline{\phi} + \rho \overline{u'_j \phi'}\right) = \frac{\partial}{\partial x_j}\left(\Gamma \frac{\partial \overline{\phi}}{\partial x_j}\right) \,. \tag{10.37}$$

The presence of the Reynolds stresses and turbulent scalar flux in the conservation equations means that these equations are not closed, that is to say, they contain more variables than there are equations. Closure requires use of some approximations, which usually take the form of prescribing the Reynolds stress tensor and turbulent scalar fluxes in terms of the mean quantities.

It is possible to derive equations for the higher-order correlations, e.g., for the Reynolds stress tensor, but these contain still more (and higher-order) unknown correlations that require modeling approximations. These equations will be introduced later but the important point is that it is impossible to derive a closed set of exact equations. The approximate *turbulence models* that we use in engineering are often called *parameterizations* in the geosciences.

10.3.5.2   Simple Turbulence Models and Their Application

To close the equations we must introduce a turbulence model. To see what a reasonable model might be, we note, as we did in the preceding section, that in laminar flows, energy dissipation and transport of mass, momentum, and energy normal to the streamlines are mediated by the viscosity, so it is natural to assume that the effect of turbulence can be represented as an increased viscosity. This leads to the

---

[17]Note the similarity to the subgrid-scale Reynolds stresses, Eq. 10.11.

eddy-viscosity model for the Reynolds stress:

$$- \rho \overline{u_i' u_j'} = \mu_t \left(\frac{\partial \overline{u}_i}{\partial x_j} + \frac{\partial \overline{u}_j}{\partial x_i} \right) - \frac{2}{3} \rho \delta_{ij} k \ , \tag{10.38}$$

and the eddy-diffusion model for a scalar:

$$- \rho \overline{u_j' \phi'} = \Gamma_t \frac{\partial \overline{\phi}}{\partial x_j} \ . \tag{10.39}$$

In Eq. (10.38), k is the turbulent kinetic energy:

$$k = \frac{1}{2} \overline{u_i' u_i'} = \frac{1}{2} \left(\overline{u_x' u_x'} + \overline{u_y' u_y'} + \overline{u_z' u_z'} \right) \ , \tag{10.40}$$

where μ_t is the turbulent (or eddy) viscosity and Γ_t is the turbulent diffusivity. The equations are still not closed, but the number of additional unknowns is reduced from 9 (6 components of the Reynolds stress tensor and 3 components of the turbulent flux vector) to two (μ_t and Γ_t).

The last term in Eq. (10.38) is required to guarantee that, when both sides of the equation are contracted (the two indices are set equal and summed over), the equation remains correct. Although the eddy-viscosity hypothesis is not correct in detail, it is easy to implement and, with careful application, can provide reasonably good results for many flows.

In the simplest description, turbulence can be characterized by two parameters: its kinetic energy, k, or a velocity, $q = \sqrt{2k}$, and a length scale, L. Dimensional analysis shows that:

$$\mu_t = C_\mu \rho q L \ , \tag{10.41}$$

where C_μ is a dimensionless constant whose value will be given later.

In the simplest practical models, mixing-length models, k is determined from the mean velocity field using the approximation $q = L \, \partial u / \partial y$ and L is a prescribed function of the coordinates. Accurate prescription of L is possible for simple flows but not for separated or highly three-dimensional flows. Mixing-length models can therefore be applied only to relatively simple flows; they are also known as zero-equation models.

The difficulty in prescribing the turbulence quantities suggests that one might use partial differential equations to compute them. Because a minimum description of turbulence requires at least a velocity scale and a length scale, a model which derives the needed quantities from two such equations is a logical choice. In almost all such models, an equation for the turbulent kinetic energy, k, determines the velocity scale. The exact equation for this quantity is not difficult to derive:

$$\frac{\partial(\rho k)}{\partial t} + \frac{\partial(\rho \overline{u}_j k)}{\partial x_j} = \frac{\partial}{\partial x_j}\left(\mu \frac{\partial k}{\partial x_j}\right) - \frac{\partial}{\partial x_j}\left(\frac{\rho}{2}\overline{u'_j u'_i u'_i} + \overline{p' u'_j}\right) -$$

$$\rho \overline{u'_i u'_j}\frac{\partial \overline{u}_i}{\partial x_j} - \mu \overline{\frac{\partial u'_i}{\partial x_k}\frac{\partial u'_i}{\partial x_k}} . \tag{10.42}$$

For details of the derivation of this equation, see Pope (2000), Chen and Jaw (1998) or Wilcox (2006). The terms on the left-hand side of this equation and the first term on the right-hand side need no modeling. The last term represents the product of the density ρ and the dissipation, ε, the rate at which turbulence energy is irreversibly converted into internal energy. We shall give an equation for the dissipation below.

The second term on the right-hand side represents *turbulent diffusion* of kinetic energy (which is actually transport of velocity fluctuations by the fluctuations themselves); it is almost always modeled by use of a gradient-diffusion assumption:

$$-\left(\frac{\rho}{2}\overline{u'_j u'_i u'_i} + \overline{p' u'_j}\right) \approx \frac{\mu_t}{\sigma_k}\frac{\partial k}{\partial x_j}, \tag{10.43}$$

where μ_t is the eddy viscosity defined above and σ_k is a *turbulent Prandtl number* whose value is approximately unity. One of the great weaknesses of the eddy viscosity is that it is a scalar which greatly limits its ability to represent general turbulent processes. In more complex models, that will not be described here, the eddy viscosity can be made into a tensor, or better yet, the model can be created without the presence of the eddy viscosity (see EARSMs in Sects. 10.3.5.4 and 10.3.3.5).

The third term of the right-hand side of Eq. (10.42) represents the *rate of production* of turbulent kinetic energy by the mean flow, a transfer of kinetic energy from the mean flow to the turbulence. If we use the eddy-viscosity hypothesis (10.38) to estimate the Reynolds stress, it can be written:

$$P_k = -\rho\overline{u'_i u'_j}\frac{\partial \overline{u}_i}{\partial x_j} \approx \mu_t\left(\frac{\partial \overline{u}_i}{\partial x_j} + \frac{\partial \overline{u}_j}{\partial x_i}\right)\frac{\partial \overline{u}_i}{\partial x_j} \tag{10.44}$$

and, as the right-hand side of this equation can be calculated from quantities that will be computed, the development of the turbulent kinetic energy equation is complete.

As mentioned above, another equation is required to determine the length scale of the turbulence. The choice is not obvious and a number of equations have been used for this purpose. The most popular one is based on the observations that the dissipation is needed in the energy equation and, in so-called equilibrium turbulent flows, i.e., ones in which the rates of production and destruction of turbulence are in near-balance, the dissipation, ε, and k and L are related by[18]:

---

[18]This relationship plays a large role in LES using a TKE-based SGS model; in that case, $L = \Delta$ and a constant of proportionality of $O(1)$ is used.

$$\varepsilon \approx \frac{k^{3/2}}{L} .$$

(10.45)

This idea is based on the fact, that at high Reynolds numbers, there is a cascade of energy from the largest scales to the smallest ones and that the energy transferred to the small scales is dissipated. Equation (10.45) is based on an estimate of the inertial energy transfer.

Equation (10.45) allows one to use an equation for the dissipation as a means of obtaining both ε and L. No constant is used in Eq. (10.45) because the constant can be combined with others in the complete model.

Although an exact equation for the dissipation can be derived from the Navier–Stokes equations, the modeling applied to it is so severe that it is best to regard the entire equation as a model. We shall therefore make no attempt to derive it. In its most commonly used form, this equation is:

$$\frac{\partial(\rho\varepsilon)}{\partial t} + \frac{\partial(\rho u_j \varepsilon)}{\partial x_j} = C_{\varepsilon 1} P_k \frac{\varepsilon}{k} - \rho C_{\varepsilon 2} \frac{\varepsilon^2}{k} + \frac{\partial}{\partial x_j}\left(\frac{\mu_t}{\sigma_\varepsilon}\frac{\partial\varepsilon}{\partial x_j}\right) .$$

(10.46)

In this model, the eddy viscosity is expressed as:

$$\mu_t = \rho C_\mu \sqrt{k} L = \rho C_\mu \frac{k^2}{\varepsilon} ,$$

(10.47)

where Eq. (10.45) was used to determine L.

The model based on Eqs. (10.42) and (10.46) is called the k–ε model and has been widely used. This model contains five parameters; the most commonly used values for them are:

$$C_\mu = 0.09; \quad C_{\varepsilon 1} = 1.44; \quad C_{\varepsilon 2} = 1.92; \quad \sigma_k = 1.0; \quad \sigma_\varepsilon = 1.3 .$$

(10.48)

The implementation of this model in a computer code is relatively simple to carry out. The RANS equations have the same form as the laminar equations provided the molecular viscosity, μ, is replaced by the effective viscosity $\mu_{eff} = \mu + \mu_t$. The most important difference is that two new partial differential equations need to be solved and that μ_t usually varies by several orders of magnitude within the flow domain. Because the time scales associated with the turbulence are much shorter than those connected with the mean flow, the equations with the k–ε model (or any other turbulence model) are much stiffer than the laminar equations. Thus, there is little difficulty in the discretization of these equations other than one to be discussed below but the solution method has to take the increased stiffness into account.

For this reason, in the numerical solution procedure, one first performs an outer iteration of the momentum and pressure-correction equations in which the value of the eddy viscosity is based on the values of k and ε at the end of the preceding iteration. After this has been completed, an outer iteration of the turbulent kinetic energy and dissipation equations is made. Because these equations are highly nonlinear, they have to be linearized prior to iteration. After completing an iteration of the turbulence model equations, we are ready to recalculate the eddy viscosity and start a new outer iteration.

The stiffness of equations that use eddy-viscosity models requires either time-marching or under-relaxation to achieve converged steady-state solutions. Too large a time step (or under-relaxation factors in an iterative method) can lead to negative values of either k of ε (especially near walls) and thus to numerical instability. Even when using time-marching approach, under-relaxation may still be necessary to enhance numerical stability. Typical values of the under-relaxation parameters are similar to the ones used in the momentum equations (0.6–0.9, depending on time-step size, grid quality and flow problem; higher values are applicable to high-quality grids and small time steps).

A number of other *two-equation* models have been proposed; we shall describe just one of them. An obvious idea is to write a differential equation for the length scale itself; this has been tried but has not met with much success. The second most commonly used model is the k–ω model, originally introduced by Saffman but popularized by Wilcox. In this model, use is made of an equation for an inverse time scale ω; this quantity can be given various interpretations but they are not very enlightening so they are omitted here. The k–ω model uses the turbulent kinetic energy equation (10.42) but it has to be modified a bit:

$$\frac{\partial(\rho k)}{\partial t} + \frac{\partial(\rho \overline{u}_j k)}{\partial x_j} = P_k - \rho \beta^* k\omega + \frac{\partial}{\partial x_j}\left[\left(\mu + \frac{\mu_t}{\sigma_k^*}\right)\frac{\partial \omega}{\partial x_j}\right]. \tag{10.49}$$

Nearly everything that was said about it above applies here. The ω equation as given by Wilcox (2006) is:

$$\frac{\partial(\rho \omega)}{\partial t} + \frac{\partial(\rho \overline{u}_j \omega)}{\partial x_j} = \alpha \frac{\omega}{k} P_k - \rho \beta \omega^2 + \frac{\partial}{\partial x_j}\left[\left(\mu + \frac{\mu_t}{\sigma_\omega^*}\right)\frac{\partial \omega}{\partial x_j}\right]. \tag{10.50}$$

In this model, the eddy viscosity is expressed as:

$$\mu_t = \rho \frac{k}{\omega} . \tag{10.51}$$

The coefficients that go into this model are a bit more complicated than those in the k–ε model. They are:

$$\alpha = \frac{5}{9} , \quad \beta = 0.075 , \quad \beta^* = 0.09 , \quad \sigma_k^* = \sigma_\omega^* = 2 , \quad \varepsilon = \beta^* \omega k . \tag{10.52}$$

The numerical behavior of this model is similar to that of the k–ε model.

The reader interested in knowing more about these models is referred to the book by Wilcox (2006). A popular variant of this model was introduced in 1993 by Menter (see Menter 1994); his shear-stress transport turbulence model is used (sometimes along with detached eddy simulation; see Sect. 10.3.7) in aerodynamic studies of airplanes, trucks, etc. (Menter et al. 2003).[19]

10.3.5.3   The v2f Model

As should be clear from the above, a major problem with turbulence models is that the proper conditions to be applied near walls are not known. The difficulty comes from the fact that we simply do not know how some of these quantities behave near a wall. Also, the variation of the turbulent kinetic energy and, even more so, the dissipation are very rapid near a wall. This suggests that it is not a good idea to try to prescribe conditions on these quantities in that region. Another major issue is that, despite years of effort devoted to it, the development of 'low Reynolds number' models designed to treat the near-wall region, relatively little success has been achieved.

Durbin (1991) suggested that the problem is not that the turbulent Reynolds number is low near a wall (although viscous effects are certainly important). The impermeability condition (zero normal velocity at wall) is far more important. This suggests that instead of trying to find low Reynolds number models, one should work with a quantity that becomes very small near a wall due to the impermeability condition. Such a quantity is the normal velocity (usually called v by engineers) and its fluctuations (v'^2) and so Durbin introduced an equation for this quantity. It was found that the model also required a damping function f, hence the name v^2-f (or $v2f$) model. It appears to give improved results at essentially the same cost as the $k-\varepsilon$ model. Iaccarino et al. (2003) used the $v2f$ model in a very successful unsteady RANS (URANS) simulation of unsteady separated flow.

To remedy the problems encountered near walls, Durbin suggested the use of elliptic relaxation. The idea is the following. Suppose that ϕ_{ij} is some quantity that is modeled. Let the value predicted by a model be ϕ_{ij}^{m}. Instead of accepting this value as the one to be used in the model, we solve the equation:

$$\nabla^2 \phi_{ij} - \frac{1}{L^2} \phi_{ij} = \phi_{ij}^{m} , \qquad (10.53)$$

where L is the length scale of the turbulence, usually taken to be $L \approx k^{3/2}/\varepsilon$. The introduction of this procedure appears to relieve a great deal of the difficulty. More details on these and other similar models can be found in a recent book by Durbin (2009); see also Durbin and Pettersson Reif (2011).

---

[19]The NASA Turbulence Modeling Resource (NASA TMR 2019) provides documentation for RANS turbulence models, including the latest (often corrected) versions of Spalart-Allmaras, Menter, Wilcox and other models and verification and validation test cases, grids, and databases.

10.3.5.4   Reynolds-Stress and Algebraic Reynolds-Stress Models

Eddy-viscosity models have significant deficiencies; some are consequences of the eddy-viscosity assumption, Eq. (10.38), not being valid. In two dimensions, there is always a choice of the eddy viscosity that allows this equation to give the correct profile of the shear stress (the 1–2 component of τ_{ij}). In three-dimensional flows, the Reynolds stress and the strain rate may not be related in such a simple way. This means that the eddy viscosity may no longer be a scalar; indeed, both measurements and simulations show that it becomes a tensor quantity.

Anisotropic (tensor) models based on using the k and ε equations have been proposed. Abe et al. (2003) describe a non-linear eddy-viscosity model that is particularly successful at capturing the highly anisotropic turbulence near walls. Leschziner (2010) describes a wide range of models, including linear eddy-viscosity, non-linear eddy-viscosity, and Reynolds-stress models.

The most complex models in common use today are Reynolds-stress models which are based on dynamic equations for the Reynolds-stress tensor $\tau_{ij} = \rho\overline{u_i'u_j'}$ itself. These equations can be derived from the Navier–Stokes equations and are:

$$\frac{\partial \tau_{ij}}{\partial t} + \frac{\partial(\overline{u}_k \tau_{ij})}{\partial x_k} = -\left(\tau_{ik}\frac{\partial \overline{u}_j}{\partial x_k} + \tau_{jk}\frac{\partial \overline{u}_i}{\partial x_k}\right) + \rho\varepsilon_{ij} - \prod_{ij} +$$

$$\frac{\partial}{\partial x_k}\left(\nu\frac{\partial \tau_{ij}}{\partial x_k} + C_{ijk}\right) . \tag{10.54}$$

Because the tensor is symmetric, only six equations need to be solved. The first two terms of the right-hand side are the production terms and require no approximation or modeling.

The other terms are:

$$\prod_{ij} = \overline{p'\left(\frac{\partial u_i'}{\partial x_j} + \frac{\partial u_j'}{\partial x_i}\right)} , \tag{10.55}$$

which is often called the *pressure-strain* term. It redistributes turbulent kinetic energy among the components of the Reynolds-stress tensor but does not change the total kinetic energy. The next term is:

$$\rho\varepsilon_{ij} = 2\mu\overline{\frac{\partial u_i'}{\partial x_k}\frac{\partial u_j'}{\partial x_k}} , \tag{10.56}$$

which is the dissipation tensor. The last term is:

$$C_{ijk} = \rho\overline{u_i'u_j'u_k'} + \overline{p'u_i'}\delta_{jk} + \overline{p'u_j'}\delta_{ik} \tag{10.57}$$

and is often called the *turbulent diffusion*.

The dissipation, pressure-strain, and turbulent diffusion terms cannot be computed exactly in terms of the other terms in the equations and therefore must be modeled. The simplest and most common model for the dissipation term treats it as isotropic:

$$\varepsilon_{ij} = \frac{2}{3}\varepsilon\delta_{ij} \, . \tag{10.58}$$

This means that an equation for the dissipation must be solved along with the Reynolds-stress equations. Typically, this is taken to be the dissipation equation used in the $k-\varepsilon$ model. More sophisticated (and therefore more complex) models have been suggested.

The simplest model for the pressure-strain term is one that assumes that the function of this term is to attempt to make the turbulence more isotropic. This model has not met with great success. The most successful models are based on decomposing the pressure-strain term into a 'rapid' part, which involves interactions between the turbulence and the mean-flow gradients, and a 'slow' part involving only interactions among turbulent quantities (this part is typically modeled with a return-to-isotropy term). See Launder et al. (1975) or Pope (2000) for more details.

The turbulent diffusion terms are usually modeled using a gradient-diffusion type of approximation. In the simplest case, the diffusivity is assumed to be isotropic and is simply a multiple of the eddy viscosity used in the models discussed earlier. In recent years, anisotropic and nonlinear models have been suggested. Again, no attempt is made to discuss them in detail here.

In three dimensions, Reynolds-stress models require the solution of seven partial differential equations in addition to the equations for the mean flow. Still more equations are needed when scalar quantities need to be predicted. These equations are solved in a manner similar to that for the $k-\varepsilon$ equations. The only additional issue is that, when the Reynolds-averaged Navier–Stokes equations are solved together with a Reynolds-stress model, they are even stiffer than those obtained with the $k-\varepsilon$ equations and even more care is required in their solution and the calculations usually converge more slowly. The usual approach in application is to compute the flow using the $k-\varepsilon$ turbulence model first, estimate the initial values of the Reynolds-stress components from the eddy-viscosity hypothesis and then continue the computation using the Reynolds-stress model. This usually helps because it provides more reasonable starting fields of all variables than would be the case if starting the computation with the Reynolds-stress model and a simple initialization of variables. At the same time, in this way one obtains solutions with two turbulence models and the comparison of the two solutions also represents a useful information.

The Reynolds-stress models, by eliminating the eddy viscosity, are able to resolve anisotropy, yet they require solution of additional differential equations which raises costs. But, there is no doubt that Reynolds-stress models can represent turbulent flow phenomena more correctly than the two-equation models (see Hadžić 1999, for some illustrative examples). Particularly good results have been obtained for some flows in which $k-\varepsilon$ models perform badly (e.g., swirling flows, flows with stagnation points or lines, flows with strong curvature and with separation from curved surfaces, etc.).

Which model is best for which kind of flow (none is expected to be good for all flows) is not yet clear; however, Leschziner (2010) gives a good account of what is available, while Hanjalić (2004) covers many of the details. Because one cannot always be sure about the choice and/or performance of a turbulence model, it is important to ensure that any differences between solutions are due to model differences and not to numerical errors. This is one reason why numerical accuracy is emphasized in this book; its importance cannot be overemphasized and constant attention to it is required.

As noted above in Sect. 10.3.3.5 for LES, explicit algebraic Reynolds-stress models (EARSMs) are an attractive alternative in which the differential equation load of the Reynolds-stress models is reduced, but the essential ability to resolve anisotropy and perhaps other important physics is retained. The EARSM by Wallin and Johansson (2000), which is aimed at both incompressible and compressible, rotational flows, describes anisotropy via the mean strain rate and the mean rotation rate tensors. It uses the Rodi (1976) postulate that convection and diffusion of the Reynolds stress can be represented in terms of the Reynolds stress, the turbulent energy production, and the turbulent energy dissipation. This leads them to a nonlinear implicit model, which they simplify to achieve an EARSM. The method requires transport equations for the turbulent kinetic energy and dissipation rate and solution of a linear equation set and one nonlinear algebraic equation for the ratio of production to dissipation. This EARSM correctly accounts for rotation and has been shown to be superior to classical eddy-viscosity-based models.

There are many versions of all models described above. The modifications are aimed to correct various deficiencies of the basic models, like not accounting for the anisotropy of turbulence, inadequately modeling the effects of stagnation or separation, adverse or favorable pressure gradient, streamline curvature, damping of turbulence near solid walls, or transition from laminar to turbulent flow. It is beyond the scope of this book to go into all these details here; interested readers will find sufficient information in references cited above and in particular in Patel et al. (1985) and Wilcox (2006). Commercial CFD-codes usually have more than 20 turbulence model versions to choose from. Almost all models come in two variants, depending on how the wall boundary conditions are treated: "high-Re" version, which assumes that all computational points are within turbulent part of the boundary layer, and "low-Re" version, which assumes that the near-wall grid resolves the viscous sublayer and provides damping of turbulence in this region. Wall treatment is described in more detail in the next section.

10.3.5.5   Boundary Conditions for RANS-Computations

The application of boundary conditions at inlet, outlet and symmetry planes remains for RANS-computations the same as in the case of laminar flows, so we shall not repeat any details here; see Sect. 9.11 for further information. It is, however, worth noting that at an inflow boundary, k and ε are often not known; if they are available, the known values should, of course, be used in the same way as described in previous chapters for the generic scalar variable. If k is not known, it is usually estimated from an assumed turbulence intensity, $I_t = \sqrt{\overline{u'^2}}/\overline{u}$. For example, by specifying $I_t = 0.01$ (low turbulence intensity of 1%) and assuming that $\overline{u_x'^2} = \overline{u_y'^2} = \overline{u_z'^2} = I_t^2 \overline{u}^2$, one obtains $k = \frac{3}{2} I_t^2 \overline{u}^2 = 1.5 \times 10^{-4} \overline{u}^2$. The value of ε should be selected so that the length scale derived from Eq. (10.45) is approximately one-tenth of the width of a shear layer or the domain size. If the Reynolds stresses and mean velocities are measured at inlet, ε can be estimated using the assumption of local equilibrium; this leads to (in a cross-section $x = $ const.):

$$\varepsilon \approx -\overline{uv}\frac{\partial \overline{u}}{\partial y} . \tag{10.59}$$

The velocity field itself is often not precisely known at the inlet (especially in the case of internal flows). The flow rate is usually known, and when the inlet cross-section is known, one can compute the mean velocity. The simplest approximation is then to prescribe the constant velocity at inlet. If it is possible to make a reasonable approximation of velocity variation across the inlet boundary, then this should be done. For example, if the inlet represents a section across a duct, pipe or annulus, one can prescribe velocity profiles for a fully-developed flow in such a geometry. The fully-developed flow can easily be computed by using a single layer of cells, applying periodic boundary conditions with prescribed flow rate at inlet and outlet; the result of such a calculation can then be used to prescribe the variable values at inlet of a more complicated solution domain.

If the distribution of variable values needs to be approximated at the inlet, one should try to move the inlet boundary as far upstream of the region of interest as possible. When the geometry of the upstream flow path is not available, one can, as an approximation, extrude the inlet cross-section in the upstream direction to allow the approximations made at inlet boundary to develop into reasonable profiles by the time the flow reaches the zone of interest. Most commercial grid generation tools allow for such extrusions at both inlet and outlet. It is also advisable to move the outlet cross-section as far downstream from the region of interest as possible, in order to minimize the effect of approximations made there on the flow in the important part of the solution domain. In addition, one usually uses coarser grids or gradually reduces the order of approximation for convection terms towards the outlet, in order to increase the numerical diffusion and avoid reflection of disturbances at the outlet boundary.

Before describing wall boundary conditions for RANS-computations of turbulent flows we first recognize that, immediately next to a wall, the effects of turbulence are relatively small and the flow is basically laminar. This part of the wall boundary layer is called *viscous sublayer* and both experiments and DNS indicate that the velocity component parallel to wall varies there linearly with wall distance (which we denote here with the local coordinate in wall-normal direction, n; see Fig. 9.2.1). If this sublayer is resolved by the numerical grid, then the boundary conditions for momentum equations are the same as in laminar flows; see Sect. 9.11 for details. We recall that the normal viscous stress τ_{nn} is equal to zero because the derivative of the wall-normal velocity component v_n with respect to n must be zero at the wall, and that the shear stress is equal to the product of molecular viscosity and the derivative of wall-parallel velocity component v_t with respect to n:

$$\tau_{nn} = 2\mu \left(\frac{\partial v_n}{\partial n} \right)_{\text{wall}} = 0 \, , \quad \tau_{nt} = \mu \left(\frac{\partial v_t}{\partial n} \right)_{\text{wall}} . \tag{10.60}$$

Because the profile of v_t is linear near wall, even the simplest one-sided forward or backward difference across half cell is accurate.

The problem is that, if one wants to resolve the viscous sublayer in the case of high Reynolds number 3D flows, the cells near wall must be extremely thin. On the one hand, this requires many prismatic cell layers next to a wall, and on the other hand, this makes the aspect ratio of cells very high. Because the coefficients resulting from discretized Laplace-operator (diffusion terms in transport equations, the corresponding terms in the pressure or pressure-correction equation) are proportional to the aspect ratio squared, this means that the coefficients in the wall-normal direction are several orders of magnitude larger than in other directions. This makes the equations stiff, requires higher arithmetic precision and makes the solution of discretized equations more difficult. In addition, if the wall is curved, thin cells next to the wall may become too warped unless the grid is also substantially refined in the tangential direction. This is why the so called low-Re turbulence models with grids that resolve the viscous sublayer are only used for flows at moderate Reynolds numbers (e.g., for comparisons with experiments at model scale). For flows at very high Reynolds numbers, e.g., around a ship, an airplane or other large-scale objects, we need an alternative, cheaper approach.

Engineers always try to find generalized scaling laws and it was recognized some 100 years ago that the velocity profile across a boundary layer at different Reynolds numbers can be made to collapse if the so-called *shear velocity* u_τ (also called *friction velocity*) is used for scaling as follows:

$$u_\tau = \sqrt{\frac{\tau_{\text{wall}}}{\rho}} \, , \tag{10.61}$$

where τ_{wall} is the magnitude of wall shear stress (if the local wall-tangential coordinate t is aligned with the direction of shear stress vector, then $\tau_{\text{wall}} = |\tau_{nt}|$). The velocity is scaled with u_τ to obtain u^+ and plotted against dimensionless distance from wall, n^+, as follows:

$$u^+ = \frac{\overline{v}_t}{u_\tau} \, , \quad n^+ = \frac{\rho u_\tau n}{\mu} . \tag{10.62}$$

Traditionally, y^+ is used in literature to denote the dimensionless distance from the wall, because early computations were two-dimensional and y was the coordinate normal to the wall. However, this is meaningless in complex geometries, and we therefore prefer to denote the wall distance by n, which is a local coordinate normal to the wall surface.

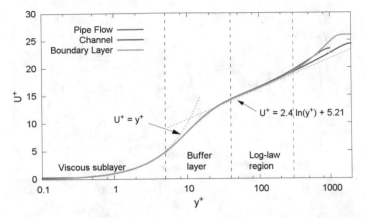

Fig. 10.25 Normalized velocity profiles across a boundary layer for turbulent flows in a plane channel, a pipe with a circular cross-section, and the boundary layer on a flat plate, obtained from DNS data of Lee and Moser (2015), El Khoury et al. (2013) and Schlatter and Örlü (2010), respectively

Figure 10.25 shows normalized velocity profiles for three different turbulent flows at different Reynolds numbers (these are 2D-flows in the x-direction, hence $n^+ = y^+$); they indeed fall on top of each other, except further away from the wall. Three distinct regions can be identified: in addition to the already mentioned viscous sublayer immediately next to wall, there is an important part with a logarithmic variation, and a buffer layer between linear and logarithmic zones. When the Reynolds number is increased, the logarithmic range extends to higher values of n^+; see e.g., Wosnik et al. (2000) or Lee and Moser (2015). The so-called *logarithmic law of the wall* has been confirmed both experimentally and by DNS data; however, latest research work suggests that the law is not as universal as once was thought. There are apparently important variations, depending on Reynolds number and the geometry of flow domain, but we shall not go into these details; interested readers may consult recent literature on the subject, e.g., Smits et al. (2011) and Smits and Marusic (2013). Because the purpose of this section is to demonstrate how to implement turbulence models into a CFD-code, we shall stick to the classical method describing the velocity profile in the logarithmic range as:

$$u^+ = \frac{1}{\kappa} \ln n^+ + B \, , \tag{10.63}$$

where κ is called the von Karman constant and B is an empirical constant. It is usually taken that $\kappa = 0.41$ and $B \approx 5.2$ but these constants are also not strictly universal. Using their DNS of plane channel flow at Re $= 2.5 \times 10^5$ (based on mean velocity and channel width), Lee and Moser (2015) derived lower values, namely $\kappa = 0.384$ and $B = 4.27$; for rough walls, smaller values for B are obtained, which means that the profile as shown in Fig. 10.25 is shifted downward.

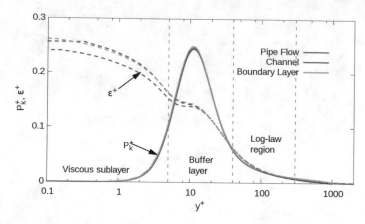

Fig. 10.26 The variation of normalized production and dissipation of turbulent kinetic energy across boundary layer in a plane channel, a pipe and a flat plate boundary layer flow (see Fig. 10.25 for data sources)

The computational effort can be substantially reduced if the first computational point can be placed in the logarithmic range rather than in the viscous sublayer. We need the velocity gradient at the wall to compute the wall shear stress, but one cannot obtain it with sufficient accuracy from higher-order polynomials, as in the case of a laminar flow. However, from the log-law and a few other assumptions, it is possible to derive a relationship between wall shear stress and the velocity at a point in the logarithmic part of the profile. This is what the so-called *wall functions* do.

Launder and Spalding (1974) proposed what became known as "high-Re wall functions".[20] Two assumptions in addition to that of a logarithmic velocity profile are made:

- The flow is assumed to be in local equilibrium, meaning the production and dissipation of turbulence are nearly equal.
- The total shear stress (i.e., the sum of viscous and turbulent contribution) is constant between the wall and the first computational point and equal to wall shear stress τ_{wall}.

Figure 10.26 shows the variation of normalized production and dissipation of turbulent kinetic energy across the boundary layer in a plane channel, a pipe and a flat plate boundary layer flow, obtained from DNS data. This data, as well as numerous measurements, do indicate that, within the logarithmic region, the production and dissipation are indeed nearly balanced, at least in these relatively simple flows. The second assumption is also supported by DNS and measurement data: near wall, where the velocity varies linearly, the viscous stress is obviously constant and across the

---

[20]Note that the statements "high-Re" and "low-Re" have nothing to do with the actual Reynolds number for the particular flow problem—they are related to how close to wall the computational points reach.

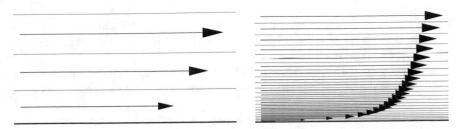

Fig. 10.27 Velocity vectors near a wall in a channel flow at Reynolds number 5×10^5 (based on the mean velocity and the distance between channel walls), computed using wall functions and a coarser grid near the wall (left) and a fully-resolved boundary layer (right); thin lines between vectors represent CV-boundaries (grid lines)

buffer layer the viscous part reduces while the turbulent part increases, keeping the sum nearly constant.

Under these assumptions, one can show that the following relation holds:

$$u_\tau = C_\mu^{1/4}\sqrt{k} \,. \tag{10.64}$$

From this equation and Eq. (10.63) we can derive an expression connecting the velocity at the first computational point above the wall and the wall shear stress[21]:

$$\tau_w = \rho u_\tau^2 = \rho C_\mu^{1/4}\kappa\sqrt{k}\frac{\bar{v}_t}{\ln(n^+ E)} \,, \tag{10.65}$$

where $E = e^{\kappa B}$. This allows the computation of the wall shear stress from the values of the wall-parallel velocity component and turbulent kinetic energy at the center of the near-wall CV. The product of wall shear stress and face area delivers a force which, when projected onto the Cartesian coordinate directions provides the necessary contributions to the discretized momentum equations for the near-wall CV, in the same way as was described for laminar flows in Sect. 9.11.

Figure 10.27 shows velocity vectors near the bottom wall of a plane channel flow at the Reynolds number 5×10^5 when computed using wall functions on a coarser grid and using a fine grid which resolves the viscous sublayer. The channel is 100 mm wide, and in the case of wall functions the cell next to the wall was 0.315 mm thick ($n^+ = 31$ at cell center). There were 15 prism layers next to the wall up to a distance of 10 mm, expanding with a factor of 1.1; for the "low-Re" approach, there were 40 prism layers over the same distance, the expansion factor was also the same, and the first cell next to the wall was 0.013 mm thick (23.4 times thinner than with wall functions; $n^+ = 1.32$ at the cell center). Obviously, there are no problems with very thin cells in the case of a plane channel wall, but if the wall was curved, one would also need refinement in tangential direction to avoid excessive cell warping

---

[21]If the wall is rough, we may use the condition derived in Sect. 10.3.3.3 and given in Eqs. (10.27) and (10.29).

(see Chap. 12 for a discussion of grid quality issues). This figure emphasizes the substantial variation of velocity between the wall and the first computational point when wall functions are used.

In many practical applications of CFD, it is difficult to keep the computational points next to the wall within the viscous sublayer everywhere and at all times (for the "low-Re wall treatment") or within the logarithmic range (for "high-Re wall functions"). The presence of flow separation, stagnation and reattachment zones leads inevitably to a large variation of wall shear stress and places where it becomes practically zero. The assumptions on which wall functions are based may not be valid, or computational points may locally fall outside of the viscous sublayer. Many researchers therefore tried to develop more general versions of wall boundary conditions for turbulent flows. Jakirlić and Jovanović (2010) proposed an approach which allows the first computational point to be close to the edge of the viscous sublayer, thus relaxing the requirements for near-wall grid fineness of the standard "low-Re wall treatment". Other researchers used modifications of Reichardt's law (Reichardt 1951) to provide the so-called "all-y^+ wall treatment": if the grid is fine enough, the linear velocity profile is assumed, and if the near-wall computational point is within the logarithmic layer, standard wall functions are applied. When the computational point is in between, the two approaches are blended. We will not go into further details here; in Chap. 13 this issue will be addressed again from the practical application point of view.

The profiles of the turbulent kinetic energy and its dissipation rate are typically much more peaked near the wall than the mean velocity profile. These peaks are difficult to capture; one should probably use a finer grid for the turbulence quantities than for the mean flow but this is rarely done. If the same grid is used for all quantities, the resolution may be insufficient for the turbulence quantities and there is a chance that the solution will contain wiggles if higher-order schemes are used, which can lead to negative values of these quantities in this region. This possibility can be avoided by locally blending the central difference scheme with a low-order upwind discretization for the convection terms in the k and ε equations. This, of course, decreases the accuracy to which these quantities are calculated but is necessary if the grid is not fine enough.

Wall boundary conditions for the model equations also require special attention. One possibility is to solve the equations accurately right up to the wall, when the grid resolves the viscous sublayer. In the k–ε model, it is appropriate to set $k = 0$ at the wall but the dissipation is not zero there; instead one can use the condition:

$$\varepsilon = v \left(\frac{\partial^2 k}{\partial n^2} \right)_{\text{wall}} \quad \text{or} \quad \varepsilon = 2v \left(\frac{\partial k^{1/2}}{\partial n} \right)^2_{\text{wall}} . \tag{10.66}$$

As noted above, it is necessary to modify the model itself near the wall, because its presence damps turbulent fluctuations and results in the existence of the practically laminar viscous sublayer.

When *wall function* type boundary conditions are used, the diffusive flux of k through the wall is usually taken to be zero, yielding the boundary condition that the normal derivative of k is zero. The dissipation boundary condition is derived by assuming equilibrium, i.e., balance of production and dissipation in the near wall region. The production in wall region is computed from:

$$P_k \approx \tau_w \frac{\partial \bar{v}_t}{\partial n},$$ (10.67)

which is an approximation to the dominant term of Eq. (10.44); it is valid near the wall because the shear stress is nearly constant in this region and velocity derivatives in directions parallel to the wall are much smaller than the derivative in the direction normal to the wall. The velocity derivative at cell center required in the above equation can be derived from the logarithmic velocity profile (10.63):

$$\left(\frac{\partial \bar{v}_t}{\partial n} \right)_P = \frac{u_\tau}{\kappa n_P} = \frac{C_\mu^{1/4} \sqrt{k_P}}{\kappa n_P}.$$ (10.68)

When the above approximations are used, the discretized equation for ε is not applied in the control volume next to the wall; instead, ε is at the CV center set equal to:

$$\varepsilon_P = \frac{C_\mu^{3/4} k_P^{3/2}}{\kappa n_P}.$$ (10.69)

This expression is derived from Eq. (10.45) using the approximation for the length scale

$$L = \frac{\kappa}{C_\mu^{3/4}} n \approx 2.5 \, n,$$ (10.70)

which is valid near the wall under the conditions used to derive the wall functions.

It should be noted that the above boundary conditions are valid when the first grid point is within the logarithmic region, i.e., when $n_P^+ > 30$. Problems arise in separated flows; within the recirculation region and, especially, in the separation and reattachment regions, the above conditions are not satisfied. Usually the possibility that wall functions may not be valid in these regions is ignored and they are applied everywhere. However, if the above conditions are violated over a large portion of the solid boundaries, serious modeling errors may result. Alternative wall functions and so-called "all-y^+" models have been proposed and are available in most commercial codes; they cannot eliminate modeling errors completely but can help to minimized them.

Three papers that may be of use in constructing useful wall treatments are as follows. First, Durbin (2009) provides an insightful review for applied turbulence modeling at walls, including some constraints, wall functions and an elliptic relaxation model. Second, Popovac and Hanjalić (2007) showcase wall boundary conditions for turbulent flows and heat transfer. Third, Billard et al. (2015) introduce a robust formulation of adaptive wall functions for use in heat transfer calculations in the context of the near-wall elliptic-blending eddy-viscosity model.

At computational boundaries far from walls (far-field or free-stream boundaries), the following boundary conditions can be used:

- If the surrounding flow is turbulent:

$$\bar{u}\frac{\partial k}{\partial x} = -\varepsilon \; ; \quad \bar{u}\frac{\partial \varepsilon}{\partial x} = -C_{\varepsilon 2}\frac{\varepsilon^2}{k} \; . \tag{10.71}$$

- In a free stream:

$$k \approx 0 \; ; \quad \varepsilon \approx 0 \; ; \quad \mu_t = C_\mu \rho \frac{k^2}{\varepsilon} \approx 0 \; . \tag{10.72}$$

Boundary conditions for Reynolds stresses are even more complicated. We shall not go into details here but note that, in general, the model must provide an approximation of the variation of each variable solved for in the vicinity of each boundary type. In most cases, the conditions reduce to the specified boundary value (Dirichlet-condition) or the specified gradient in boundary-normal direction (Neumann-condition). These conditions can be implemented into the discretized equations for cells next to boundary using methods described for the generic scalar variable for laminar flows; see Sect. 9.11 for further information.

10.3.6   Example of RANS Application: Flow Around a Sphere at Re = 500,000

We noted earlier that RANS models do not predict well the flow around a sphere at the subcritical Reynolds numbers, where laminar separation is followed by a turbulent wake; LES is ideally suited for that class of flows. At supercritical Reynolds numbers, LES would require a very fine grid and small time steps, making the simulation cost very high. For this class of flows, a RANS or URANS approach is usually used. We consider here the flow around a sphere at a Reynolds number Re = 500,000. Because the Reynolds number is not too high (as would be the case in a flow around a vehicle, ship or airplane), we can afford to resolve the boundary layer and thus created a grid with 15 prism layers, with the first layer next to wall being 0.01 mm thick; outside prism layers the grid size in the wake is 0.4375 mm ($D/140$). The same kind of grid (trimmed Cartesian with local refinement) as in the LES-case at 10 times lower Reynolds number and the same commercial flow solver (STAR-CCM+) was used. The grid around the sphere and in its wake is shown in Fig. 10.28. The solution domain geometry is also the same—a smooth sphere with a diameter $D = 61.4$ mm held by a rod with a diameter $d = 8$ mm in a wind tunnel with a cross-section of 300×300 mm. In the simulation the same velocity and viscosity were used as in the

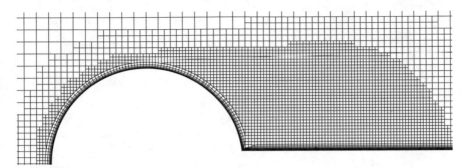

Fig. 10.28 The coarse grid around a sphere and in its wake, used for RANS computation of flow at Re = 500,000 (the grid spacing was reduced by a factor 1.5 each time the grid was refined)

previous case, only the density was increased by a factor of 10 (from 1.204 to 12.04 kg/m^3) to obtain the 10 times higher Reynolds number.

The flow around the sphere has several features with which RANS models usually do not deal well; the most important one is the separation from a smooth, curved surface. We tried here four RANS models, out of many versions available in the commercial flow solver: (i) the standard low-Re version of the k-ε model; (ii) the SST-version of k-ω model; (iii) the lag-EB k-ε model; (iv) the Reynolds-stress model. Only the lag-EB k-ε model produced results which are close to experimental data from the literature; all other models led to much too high drag and a too large recirculation zone.

The lag-EB k-ε model is a relatively new addition to the class of k-ε models in the code; details can be found in Lardeau (2018). The "lag" in the name means that the model accounts for the fact that the mean stress and strain are not always aligned (one lags behind the other); 'EB' stands for elliptic blending. In addition to k and ε, transport equations are solved for two additional variables. Computations with this model were performed on three grids; the first prism layer next to wall was the same for all three grids, as was the total thickness of all prism layers; however, in the coarse grid there were 10 prism layers, in the medium grid 12, and in the finest grid 15. Outside of the prism layer, the grid spacing was increased by a factor of 1.5 from fine to medium and from medium to coarse grid; the fine-grid values were given above. The coarse grid had 728,923 control volumes, the medium one had 2,131,351 and the fine one 6,591,260. The computed drag values were: 0.0679 on the coarse, 0.0638 on the medium and 0.0619 on the fine grid. Richardson extrapolation leads to an estimate of the grid-independent value of 0.0604, which is close to data from the literature. Other turbulence models were only used with the finest grid.

The low-Re k-ε model does not converge to a steady-state solution; when the computation is continued in transient mode, the drag coefficient oscillates between 0.108 and 0.146, which is substantially higher than values from experiments reported in literature (e.g., Achenbach 1972). The SST k-ω and the lag-EB k-ε model converge to a nearly steady-state solution; Fig. 10.29 shows residuals from the computation

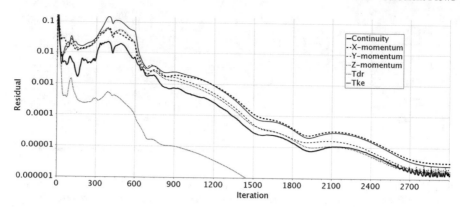

Fig. 10.29 Residual history for the computation of flow around sphere at Re = 500,000 on the fine grid using the lag-EB k-ε model

with lag-EB k-ε model. Residuals drop almost five orders of magnitude and then remain at that level, with small oscillations; the drag value does not change any more on the five most significant digits. Continuing computation in a transient mode does not lead to any significant change in the flow; for practical purposes it can be considered converged to a steady state. The drag coefficient obtained in the computation using the SST k-ω model is about twice as high as in experiments (0.146). Computations with the Reynolds-stress model were started from the solution obtained with the lag-EB k-ε model; the convergence is oscillatory and very slow, but the drag coefficient does not vary much and tends to a mean value of ca. 0.108, which is also substantially higher than in experiments.

Leder (1992) gives the length of the recirculation zone behind sphere for Reynolds numbers between 150,000 and 300,000 as constant at about $0.2D$. All computations predict longer recirculation zones; the smallest value results from computations with the lag-EB k-ε model (somewhat longer than $0.4D$). Figure 10.30 shows flow patterns in section planes at $y = 0$ and $z = 0$ for the SST k-ω and lag-EB k-ε model. One can see that, although the solution is practically steady, the flow is not axisymmetric; especially, the solution obtained using the SST k-ω model is highly asymmetric. As noted earlier, asymmetric wakes were observed in other numerical and experimental studies (Constantinescu and Squires 2004; Taneda 1978).

Although the geometry of the flow domain is quite simple, flow around sphere is very difficult to predict with RANS models. As will be shown in the following chapters, a much better agreement between experiments and RANS-computations is obtained in most flows of industrial relevance than in the present test case (e.g., ship resistance is usually predicted to within 2% of the experimental value; in many cases the agreement is even better, and only rarely is the discrepancy larger than 2%). The reason is that, in the case of sphere, flow features at which RANS models are not so good dominate; in a more complex geometry, other features come into play which RANS models can predict much better.

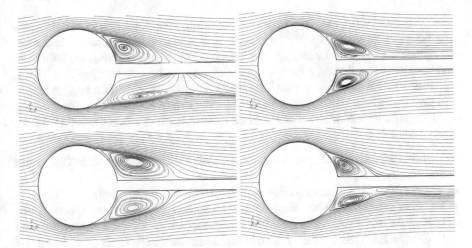

Fig. 10.30 The flow pattern in the section at $y = 0$ (upper) and $z = 0$ (lower), computed using the SST k-ω model (left) and the lag-EB k-ε model (right)

10.3.7 Very-Large-Eddy Simulation/TRANS/DES

The objective of flow simulation is usually to obtain information about particular properties of the flow at minimum cost. It is wise to use the simplest tool that will provide the desired results, but it is not easy to know in advance how well each method will work. We are dealing with a hierarchy, from simple to complex, namely, steady RANS, unsteady RANS, VLES, LES, and DNS. We add to that set an interesting blend of LES and URANS called *Detached Eddy Simulation* (DES), available in several variants under acronyms DDES or IDDES.

Clearly, if RANS methods are successful, there is no reason to use LES, etc. On the other hand, when a RANS variant does not work, it may be a good idea to try a URANS or LES-type of simulation. However, recall that URANS is an ensemble construct so all of the turbulence energy is likely in the modeled scales. LES on the other hand resolves a large fraction of the turbulence energy.

One way to use simulation methods is to perform unsteady RANS, LES and/or DNS of 'building block' flows, ones that are structurally similar to those of actual interest. From the results, RANS models that can be applied to more complex flows can be validated and improved. RANS computations can then be the everyday tool. LES may need be performed only when there are significant changes in the design. There are areas of simulation where the use of LES has become dominant, however, e.g., mesoscale and atmospheric boundary layer simulations for research and for weather prediction, or aero-acoustics simulations in engineering.

It appears that we have to either use RANS, which is affordable, or LES, which is more accurate but rather expensive. It is natural to ask whether there are methods that provide the advantages of both RANS and LES while avoiding the disadvantages?

A straightforward way to proceed is to examine the definition of the domains of LES given in Sect. 10.3. By increasing grid size in a given domain and flow, we reduce the amount of energy in the resolved flow below the nominal 80% level and have what is called *very-large-eddy simulation* (VLES). Accordingly, in VLES, one uses an LES code to compute the unsteady flow, but employs a large grid size and perhaps more sophisticated turbulence model to properly represent the modeled motions, e.g., Reynolds stress or algebraic Reynolds stress models (Sect. 10.3.5.4). Examples include:

- VLES of the unsteady flow in draft tubes of hydroelectric power plants (Gyllenram and Nilsson 2006).
- VLES of flow in a direct injection combustor (Shih and Liu 2009);
- VLES of gravity-wave-breaking induced turbulence in a deep atmosphere. Smolarkiewicz and Prusa 2002 employ a non-oscillatory forward-in-time method (NFT) (an ILES) as a VLES and so compute explicitly the large coherent eddies that are resolvable on the grid.

On the other hand, flows over bluff bodies usually produce strong vortexes in their wakes. These vortexes produce fluctuating forces on the body in both the streamwise and spanwise directions whose prediction is very important. These include flows over buildings (wind engineering), ocean platforms, and vehicles, among others. If the vortexes are sufficiently larger than the bulk of the motions that constitute the 'turbulence', the turbulence model may remove only the smaller-scale motions, and one may convert an aperiodic flow into a periodic one, which may have significant consequences. Durbin (2002), Iaccarino et al. (2003), and Wegner et al. (2004) demonstrate that the standard ensemble-averaged URANS approach will accomplish this goal if the remaining structures are periodic; if they are not, a long-term simulation will lead to a steady flow. Iaccarino et al. (2003) show that URANS can provide both quantitative and qualitative agreement with experimental data for flow which is not statistically stationary.

However, in the presence of dominant forcing from buoyancy, for example, a construct called transient RANS (TRANS) (Hanjalić and Kenjereš 2001; Kenjereš and Hanjalić 1999) has been successful, also. In their TRANS simulation of the coherent eddy structure in flows driven by thermal buoyancy and the Lorentz force, they used an explicit triple decomposition of the original motions into (1) a steady long-time average, (2) a quasi-periodic (coherent structure) component, and (3) random (stochastic) fluctuations. The large coherent-eddy-structures are fully resolved by time integration of the three-dimensional ensemble-averaged Navier–Stokes equations. The incoherent part is modeled by a RANS-type closure. In this particular case, there was strong buoyant forcing that produced well-defined structures which were only pseudo-periodic; the authors appealed then to the concept of a spectral gap to allow decomposition of the flow. Hanjalić (2002) shows a convincing result for Rayleigh-Bénard convection, and Kenjereš and Hanjalic (2002) use their method to show the effects of terrain and thermal stratification in mesoscale modeling of diurnal cycles in the atmospheric boundary layer. Strong destabilizing forces or periodic forcing appear to distinguish the TRANS simulations from the URANS ones described above (see Durbin 2002).

Created in 1997, Detached-Eddy Simulation (DES) was aimed at predicting massively separated flows. Recently, Spalart et al. (2006) presented a new version called Delayed DES (DDES), which corrects some faults of the original. In its main domain of separated flow, the accuracy of DES predictions is typically superior to that of steady or unsteady RANS methods. In DES, the 'mean' flow equations are the same throughout the flow domain; however, in the near wall region, the turbulence model reduces to a RANS formulation, while away from the wall an LES subgrid model is applied. The transition distance is computed by the algorithm under specific rules. Most formulations use, or are derived from, the Spalart and Allmaras (1994) RANS model. A number of interesting applications have been made, especially in engineering design. As examples:

1. Viswanathan et al. (2008) used DES to study the massively-separated flow around an aerodynamic body at an angle of attack using a finite-volume, unstructured grid code with second-order accuracy in time and space. Motions of the body are permitted.
2. Konan et al. (2011) applied DES and Lagrangian particle tracking in rough-wall turbulent channel flow in order to investigate the effect of the wall roughness on the dispersed phase in the flow.

Nowadays, most commercial CFD-codes offer all flavors of approaches to compute turbulent flows, from RANS, over URANS and DES to LES and even DNS. Although the discretization methods in commercial codes are usually limited to second order in space and time (with possibly higher-order interpolation for convection terms), reasonably good results have been reported for a range of practical applications.

Chapter 11
Compressible Flow

11.1 Introduction

Compressible flows are important in aerodynamics and turbomachinery among other applications. In high speed flows around aircraft, the Reynolds numbers are extremely high and turbulence effects are confined to thin boundary layers. The drag consists of two components, frictional drag due to the boundary layer and pressure or form drag which is essentially inviscid in nature; there may also be wave drag due to shocks which may be computed from the inviscid equations provided that care is taken to assure that the second law of thermodynamics is obeyed. If frictional drag is ignored, these flows may be computed using the inviscid momentum *Euler* equations.

Due to the importance of compressible flow in civilian and military applications, many methods of solving the equations of compressible flow have been developed. Among these are special methods for the Euler equations such as the method of characteristics and numerous methods that may be capable of extension to viscous flows. Most of these methods are specifically designed for compressible flows and become very inefficient when applied to incompressible flows. A number of variations on the reason for this can be given. One is that, in compressible flows, the continuity equation contains a time derivative which drops out in the incompressible limit. As a result, the equations become extremely stiff in the limit of weak compressibility, necessitating the use of very small time steps or implicit methods. Another version of the argument is that the compressible equations support sound waves which have a definite speed associated with them. As some information propagates at the flow velocity, the larger of the two velocities determines the allowable time step in an explicit method. In the low speed limit, one is forced to take a time step inversely proportional to the sound speed for any fluid velocity; this step size may be much smaller than the one a method designed for incompressible flows might allow.

Discretization and solution of the compressible flow equations can be carried out with methods already described. For example, to solve the time-dependent equations, one can use any of the time-advance methods discussed in Chap. 6. As the effect of diffusion is usually small in compressible flows because the Reynolds numbers

© Springer Nature Switzerland AG 2020
J. H. Ferziger et al., *Computational Methods for Fluid Dynamics*,
https://doi.org/10.1007/978-3-319-99693-6_11

are high, there may be discontinuities, e.g., shocks, in the flow. Special methods for producing smooth solutions near shocks have been constructed. These include simple upwind methods, flux blending methods, essentially non-oscillatory (ENO) methods, and total variation diminishing (TVD) methods. These are described in Sects. 4.4.6 and 11.3 and may be found in a number of other books, e.g., Tannehill et al. (1977) and Hirsch (2007).

We first describe in the next section how methods originally designed for incompressible flows and presented in Chaps. 7–9 can be extended to handle also compressible flows. In Sect. 11.3 we then discuss some aspects of methods designed specifically for computing compressible flows, including extension in the other direction—from compressible to incompressible flows.

11.2   Pressure-Correction Methods for Arbitrary Mach Number

To compute compressible flows, it is necessary to solve not only the continuity and momentum equations but also a conservation equation for the thermal energy (or one for the total energy) and an equation of state. The latter is a thermodynamic relation connecting the density, temperature, and pressure. The energy equation was given in Chap. 1; for incompressible flows it reduces to a scalar transport equation for the temperature and only the convection and heat conduction are important. In compressible flows, viscous dissipation may be a significant heat source and conversion of internal energy to kinetic energy (and vice versa) by means of flow dilatation is also important. All terms in the equations must then be retained. In integral form the energy equation is:

$$\frac{\partial}{\partial t} \int_V \rho h \, dV + \int_S \rho h \mathbf{v} \cdot \mathbf{n} \, dS = \int_S k \nabla T \cdot \mathbf{n} \, dS +$$
$$\int_V \left[\mathbf{v} \cdot \nabla p + \mathsf{S} : \nabla \mathbf{v} \right] dV + \frac{\partial}{\partial t} \int_V p \, dV \; . \tag{11.1}$$

Here h is the enthalpy per unit mass, T is the absolute temperature (K), k is the thermal conductivity and S is the viscous part of the stress tensor, $\mathsf{S} = \mathsf{T} + p\mathsf{I}$. For a perfect gas with constant specific heats, c_p and c_v, the enthalpy becomes $h = c_p T$, allowing the energy equation to be written in terms of the temperature. Furthermore, under these assumptions, the equation of state is:

$$p = \rho R T \; , \tag{11.2}$$

where R is the gas constant. The set of equations is completed by adding the continuity equation:

$$\frac{\partial}{\partial t} \int_V \rho \, dV + \int_S \rho \mathbf{v} \cdot \mathbf{n} \, dS = 0 \tag{11.3}$$

and the momentum equation:

$$\frac{\partial}{\partial t} \int_V \rho \mathbf{v} \, dV + \int_S \rho \mathbf{v} \mathbf{v} \cdot \mathbf{n} \, dS = \int_S \mathsf{T} \cdot \mathbf{n} \, dS + \int_V \rho \mathbf{b} \, dV \;, \qquad (11.4)$$

where T is the stress tensor (including pressure terms) and $\mathbf{b}$ represents body forces per unit mass; see Chap. 1 for a discussion of various forms of these equations.

It is natural to use the continuity equation to compute the density and to derive the temperature from the energy equation. This leaves role of determining the pressure to the equation of state. We thus see that the roles of the various equations are quite different from the ones they play in incompressible flows. Also note that the nature of the pressure is completely different. In incompressible flows there is only the dynamic pressure whose absolute value is of no consequence; for compressible flows, it is the thermodynamic pressure whose absolute value is of critical importance.

The discretization of the equations can be carried out using the methods described in Chaps. 3 and 4. The only changes required involve the boundary conditions (which need to be different because the compressible equations are hyperbolic in character), the nature and treatment of the coupling between the density and the pressure, and the fact that shock waves, which are very thin regions of extremely large change in many of the variables, may exist in compressible flows. Below we shall extend the pressure-correction methods to flows at arbitrary Mach number, following the approach of Demirdžić et al. (1993). Similar methods have been published by Issa and Lockwood (1977), Karki and Patankar (1989) and Van Doormal et al. (1987).

As mentioned above, the discretization of the compressible momentum equations is essentially identical to that employed for the incompressible equations, see Chaps. 7 and 8, so we shall not repeat it here. We shall limit the discussion to the implicit pressure-correction methods described in Chap. 7, but the ideas can be applied to other schemes as well.

To obtain the solution at the new time level using an implicit method, several outer iterations are performed; see Sects. 7.2.1 and 7.2.2 for a detailed description of the fractional-step and SIMPLE scheme for incompressible flows. If time step is small, only a few outer iterations per time step are necessary. For steady problems, the time step may be infinite and the under-relaxation parameter acts like a pseudo-time step. We consider only the segregated solution method, in which the linearized (around values from the previous outer iteration) equations for velocity components, pressure correction, temperature and other scalar variables are solved in turn. While solving for one variable, other variables are treated as known. The steps necessary to extend the methods described in Sects. 7.2.1 and 7.2.2 to be able to compute flows at any Mach number are described in the following two sections.

11.2.1   Implicit Fractional-Step Method for All Flow Speeds

The treatment of momentum equations is basically the same as for incompressible flows, cf. Eqs. (7.62)–(7.64); the only thing to note is that the density in terms computed at outer iteration m of the new time level is taken from the previous iteration, $m - 1$. The continuity equation represented only a constraint on velocity field (it had to be divergence-free at all times) in incompressible flows; now it contains a time derivative and must be treated consistently with other transport equations. For the sake of completeness, we give here all the steps, including those that do not require any change. The algorithm, applicable to both compressible and incompressible flows, is as follows:

1. At the mth iteration within the new time step, solve the momentum equation of the following form for the estimate of the new time level solution, using the 2nd-order, fully-implicit, three-time-level scheme for time integration (see Sect. 6.3.2.4):

$$\frac{3(\rho^{m-1}\mathbf{v}^*) - 4(\rho\mathbf{v})^n + (\rho\mathbf{v})^{n-1}}{2\Delta t} + C(\rho^{m-1}\mathbf{v}^*) = L(\mathbf{v}^*) - G(p^{m-1}) \,, \quad (11.5)$$

where $\mathbf{v}^*$ is the predictor value for $\mathbf{v}^m$; it needs to be corrected to enforce continuity. Because particular spatial discretization schemes are not important here, we use symbolic notation for convection fluxes (C), diffusion fluxes (L) and gradient operator (G). Note that density from previous outer iteration is used in convection terms; also, if viscosity depends on temperature or other variables, we use in viscous terms the values from the previous iteration.

2. Require that the corrected velocity and pressure satisfy this form of the momentum equation:

$$\frac{3(\rho^{m-1}\mathbf{v}^m) - 4(\rho\mathbf{v})^n + (\rho\mathbf{v}^{n-1}}{2\Delta t} + C(\rho^{m-1}\mathbf{v}^*) = L(\mathbf{v}^*) - G(p^m) \,. \quad (11.6)$$

By subtracting Eq. (11.5) from Eq. (11.6) we obtain the following relation between velocity and pressure correction:

$$\frac{3}{2\Delta t}\left[(\rho^{m-1}\mathbf{v}^m) - (\rho^{m-1}\mathbf{v}^*)\right] = -G(p') \quad \Rightarrow \quad \rho^{m-1}\mathbf{v}' = -\frac{2\Delta t}{3}G(p') \,. \quad (11.7)$$

Here $\mathbf{v}' = \mathbf{v}^m - \mathbf{v}^*$ and $p' = p^m - p^{m-1}$.

3. The discretized continuity equation will not be satisfied with ρ^{m-1} and $\mathbf{v}^*$—a mass-imbalance results:

$$\frac{3\rho^{m-1} - 4\rho^n + \rho^{n-1}}{2\Delta t} + D(\rho^{m-1}\mathbf{v}^*) = \Delta\dot{m} \,. \quad (11.8)$$

4. Require that the continuity equation is satisfied by the corrected density ρ^* and velocity $\mathbf{v}^m$ fields:

$$\frac{3\rho^* - 4\rho^n + \rho^{n-1}}{2\Delta t} + D(\rho^*\mathbf{v}^m) = 0 \,. \quad (11.9)$$

Here ρ^* is an estimate of the density at iteration m; the final value will be computed at the end from the equation of state after T^m is computed. We introduce the density correction $\rho' = \rho^* - \rho^{m-1}$ and expand the product of corrected density and velocity as follows:

$$\rho^* \mathbf{v}^m = (\rho^{m-1} + \rho')(\mathbf{v}^* + \mathbf{v}') = \rho^{m-1}\mathbf{v}^* + \rho^{m-1}\mathbf{v}' + \rho'\mathbf{v}^* + \underline{\rho'\mathbf{v}'} . \quad (11.10)$$

The underlined term, being the product of two corrections, tends to zero faster than other terms and is neglected from here on. Using Eqs. (11.8) and (11.10), we can re-write Eq. (11.9) as:

$$\frac{3\rho'}{2\Delta t} + \Delta \dot{m} + D(\rho^{m-1}\mathbf{v}') + D(\rho'\mathbf{v}^*) = 0 . \quad (11.11)$$

5. In order to obtain a pressure-correction equation from Eq. (11.11), we need to express velocity and density corrections through the pressure correction. For velocity correction, this is already done in Eq. (11.7): it is proportional to the gradient of pressure correction, as is the case in incompressible flows. For the link between density correction and pressure correction, we need to refer to the equation of state:

$$\rho = f(p, T) \quad \Rightarrow \quad \rho' = \frac{\partial \rho}{\partial p}p' = \frac{\partial f(p, T)}{\partial p}p' = C_\rho p' . \quad (11.12)$$

With these expressions, one can re-write Eq. (11.11) as the pressure-correction equation for all flow speeds:

$$\frac{3C_\rho p'}{2\Delta t} + D(C_\rho \mathbf{v}^* p') = \frac{2\Delta t}{3}D(G(p')) - \Delta \dot{m} . \quad (11.13)$$

6. Upon solving the above pressure-correction equation, velocity, pressure and density are corrected to obtain $\mathbf{v}^m$, p^m and ρ^*; these values are used in the next step to solve the energy equation, from which the updated temperature T^m is obtained. Finally, the new density is computed from the equation of state, $\rho^m = f(p^m, T^m)$. After a sufficient number of iterations is performed, all corrections become negligible and we can set $\mathbf{v}^{n+1} = \mathbf{v}^m$, $p^{n+1} = p^m$, $T^{n+1} = T^m$ and $\rho^{n+1} = \rho^m$, and proceed to the next time step.

In the case of incompressible flow, the left-hand side of the pressure-correction equation (11.13) becomes zero and we recover the Poisson-equation that we had earlier, when the implicit fractional-step method was introduced in Sect. 7.2.1. In the case of compressible flow, the pressure-correction equation (11.13) looks like any other transport equation: it has the rate-of-change and convection term on the left-hand side and the diffusion and source term on the right-hand side. The properties of this equation will be discussed further below, after the same kind of modification is introduced into the SIMPLE algorithm.

11.2.2 SIMPLE Method for All Flow Speeds

As shown in Sect. 7.2.2, the only important difference between the implicit fractional-step method and SIMPLE is that the former corrects the velocity in the transient term, while the latter applies correction to all terms contributing to the main diagonal of the coefficient matrix (transient term and parts of convection and diffusion term). We therefore only briefly summarize the steps in the SIMPLE-Method extended to all flow speeds.

1. In the first step of the new outer iteration m, the momentum equation is solved for $\mathbf{v}^*$, whereby the density, pressure and all fluid properties are taken from the previous iteration $m - 1$ (the superscript indicating this is omitted for clarity, except where necessary). The coefficient matrix is split into the main diagonal A_D and off-diagonal part, A_OD:

$$(A_\mathrm{D} + A_\mathrm{OD})u_i^* = Q - G_i(p^{m-1}) , \qquad (11.14)$$

where G denotes the discretized gradient operator; the particular discretization scheme is not important here, which is why we use symbolic notation. The velocity field obtained by solving this equation, $\mathbf{v}^{m*}$, will in general not satisfy the continuity equation; for this density and pressure also need to be updated. However, we first introduce velocity correction due to pressure correction, while keeping the density at the previous iteration level:

$$p^* = p^{m-1} + p' \quad , \quad u_i^{**} = u_i^* + u_i' . \qquad (11.15)$$

2. The relation between velocity and pressure correction is obtained by requiring that the corrected velocity and pressure satisfy the following simplified version of Eq. (11.14):

$$A_\mathrm{D}u_i^{**} + A_\mathrm{OD}u_i^* = Q - G_i(p^*) . \qquad (11.16)$$

Now by subtracting Eq. (11.14) from Eq. (11.16), we obtain the following relation between velocity and pressure corrections:

$$A_\mathrm{D}u_i' = -G_i(p') \quad \Rightarrow \quad u_i' = -(A_\mathrm{D})^{-1}G_i(p') . \qquad (11.17)$$

3. We now turn to the mass conservation equation. The following two steps are identical to steps 3 and 4 in the implicit fractional-step method just described (assuming that the same time-integration scheme is used—here the fully-implicit scheme with three time levels), so we shall not repeat them here; see Eqs. (11.8)–(11.11).

4. The relation between density and pressure correction is the same as before, cf. Eq. (11.12); the relation between velocity and pressure correction is given by Eq. (11.17). By inserting these relations into Eq. (11.11), the following form of the pressure-correction equation is obtained:

$$\frac{3C_\rho p'}{2\Delta t} + D(C_\rho \mathbf{v}^* p') = D[\rho^{m-1}(A_D)^{-1}G(p')] - \Delta\dot{m} . \tag{11.18}$$

For the sake of simplicity, we assumed here that the main diagonal coefficient is the same for all three velocity components; this is usually the case, but if not, the difference is easily taken into account.

By comparing the pressure-correction equations for SIMPLE, (11.18), with the corresponding equation for IFSM, (11.13), we see that only the first term on the right-hand side (the one that resembles the discrete Laplace-operator) is different. This is due to the different expressions for the link between velocity and pressure corrections, cf. Eq. (11.7) for IFSM and Eq. (11.17) for SIMPLE.

11.2.3 Properties of the Pressure-Correction Equation

The coefficient C_ρ in Eq. (11.12) is determined from the equation of state; for a perfect gas:

$$C_\rho = \left(\frac{\partial f(p, T)}{\partial p}\right)_T = \frac{1}{RT} . \tag{11.19}$$

For other gases and when liquids are considered compressible, the derivative may need to be computed numerically. The converged solution is independent of this coefficient because all corrections are then zero; only the intermediate results are affected. It is important that the connection between the density and pressure corrections be qualitatively correct and the coefficient can, of course, influence the convergence rate of the method.

The coefficients in the pressure-correction equation depend on the approximations used for the gradients and cell-face values of the pressure correction. The part which stems from the velocity correction is identical to that for the incompressible case; it requires an approximation of derivative of pressure correction in the direction normal to cell face which should be approximated in the same way as the pressure terms in momentum equations. The part which stems from density correction corresponds to the convection flux in other conservation equations. It requires an approximation of pressure correction at the cell-face center; see Chaps. 4 and 7 for examples of various commonly used approximations.

Despite the similarity in appearance to the pressure-correction equation for incompressible flows, there are important differences. The incompressible equation is a discretized Poisson equation, i.e., the coefficients represent an approximation to the Laplacian operator. In the compressible case, there are contributions that represent the fact that the equation for the pressure in a compressible flow contains convection and unsteady terms, i.e., it is actually a convected wave equation. For an incompressible flow, if the mass flux is prescribed at the boundary, the pressure may be indeterminate to within an additive constant. The presence of convection terms in the compressible pressure-correction equation makes the solution unique.

The relative importance of the terms resulting from velocity and density correction depends on the type of flow. The diffusion term is of order $1/Ma^2$ relative to the convection term so the Mach number is the determining factor. At low Mach numbers, the Laplacian term dominates and we recover the Poisson equation. On the other hand, at high Mach number (highly compressible flow), the convection term dominates, reflecting the hyperbolic nature of the flow. Solving the pressure-correction equation is then equivalent to solving the continuity equation for density. Thus the pressure-correction method automatically adjusts to the local nature of the flow and the same method can be applied to the entire flow region, even if it contains both high and low Mach number zones (e.g., in a flow around bluff body).

For the approximation of the Laplacian, central-difference approximations are always applied. On the other hand, for the approximation of convection terms a variety of approximations may be used, just as is the case for the convection terms in the momentum equations. If higher-order approximations are used, the 'deferred correction' method may be used. On the left-hand side of the equation, the matrix is constructed on the basis of the first-order upwind approximation while the right-hand side contains the difference between the higher-order approximation and the first-order upwind approximation, assuring that the method converges to the solution belonging to the higher-order approximation; see Sect. 5.6 for details. Also, if the grid is severely non-orthogonal, deferred correction can be used to simplify the pressure-correction equation as described in Sect. 9.8.

These differences are reflected in the pressure-correction equation in another way. Because the equation is no longer a pure Poisson equation, the central coefficient A_P is not the negative of the sum of the neighbor coefficients. Only when $\text{div} \mathbf{v} = 0$, is this property obtained. Also, while the pressure-correction equation for incompressible flow has a symmetric coefficient matrix and allows that some special solvers can be used, the equation looses this property in the case of compressible flow due to contribution from convection terms.

11.2.4 Boundary Conditions

For incompressible flows the following boundary conditions are usually applied:

- Prescribed velocity and temperature on inflow boundaries;
- Zero gradient normal to the boundary for all scalar quantities and the velocity component parallel to the surface on a symmetry plane; zero velocity normal to such a surface;
- No-slip (zero relative velocity) conditions, zero normal stress and prescribed temperature or heat flux on a solid surface;
- Prescribed gradient (usually zero) of all quantities on an outflow surface.

These boundary conditions also hold for compressible flow and are treated in the same way as in incompressible flows. However, in compressible flows there are further possible boundary conditions:

- Prescribed total pressure;
- Prescribed total temperature;
- Prescribed static pressure on the outflow boundary[1];
- At a supersonic outflow boundary, zero gradients of all quantities are usually specified.

The implementation of these boundary conditions is described below.

11.2.4.1 Prescribed Total Pressure on the Inflow Boundary

The implementation of these boundary conditions will be described for the west boundary of a two-dimensional domain with the aid of Fig. 11.1.

One possibility is to note that, for isentropic flow of an ideal gas, the total pressure is defined as:

$$p_t = p \left(1 + \frac{\gamma - 1}{2} \frac{u_x^2 + u_y^2}{\gamma RT} \right)^{\frac{\gamma}{\gamma - 1}}, \tag{11.20}$$

where p is the static pressure and $\gamma = c_p/c_v$. The flow direction must be prescribed; it is defined by:

$$\tan \beta = \frac{u_y}{u_x}, \quad \text{i.e.,} \quad u_y = u_x \tan \beta . \tag{11.21}$$

These boundary conditions can be implemented by extrapolating the pressure from the interior of the solution domain to the boundary and then calculating the velocity there with the aid of Eqs. (11.20) and (11.21). These velocities can be treated as known within an outer iteration. The temperature can be prescribed or it can be calculated from the total temperature:

$$T_t = T \left(1 + \frac{\gamma - 1}{2} \frac{u_x^2 + u_y^2}{\gamma RT} \right) . \tag{11.22}$$

This treatment leads to slow convergence of the iterative method as there are many combinations of pressure and velocity that satisfy Eq. (11.20). One must implicitly take into consideration the influence of the pressure on the velocity at the inflow. One way of doing this is described below.

At the beginning of an outer iteration the velocities at the inflow boundary (side 'w' in Fig. 11.1) must be computed from Eqs. (11.20) to (11.21) and the prevailing values of the pressure; they will then be treated as fixed during the outer iteration of the momentum equation. The mass fluxes at the inflow are taken from the preceding outer iteration; they should satisfy the continuity equation. From the solution of the

---

[1]For incompressible flows, the static pressure can also be prescribed on either the inflow or outflow boundary. As the mass flux is a function of the difference in pressure between the inflow and outflow, the velocity at the inflow boundary cannot be prescribed if the pressure is prescribed at both inflow and outflow boundaries.

Fig. 11.1 A control volume
next to an inlet boundary
with a prescribed flow
direction

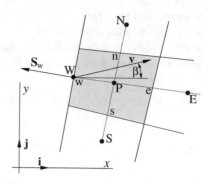

momentum equation, (u_x^*, u_y^*), a new mass flux $\dot{m}^*$ is computed. The 'prescribed' velocities on the inflow boundary are used to compute the mass flux there. In the following correction step, the mass flux (including its value at the inflow boundary) is corrected and mass conservation is enforced. The difference between the mass flux correction on the boundary and that at interior control volume faces is that, at the boundary, only the velocity (and not the density) is corrected. The velocity correction is expressed in terms of the pressure correction (and not its gradient):

$$u'_{x,w} = \left(\frac{\partial u_x}{\partial p}\right)_w p'_w = C_u p'_w \; ; \quad u'_{y,w} = u'_{x,w} \tan \beta \; . \tag{11.23}$$

The coefficient C_u is determined with the aid of Eq. (11.20):

$$C_u = -\frac{\gamma \, R T^{m-1}}{p_t u_x^{m*} \gamma \, (1 + \tan^2 \beta) \left[1 + \dfrac{\gamma - 1}{2} \dfrac{(u_x^*)^2 (1 + \tan^2 \beta)}{\gamma \, R T^{m-1}}\right]^{\frac{1-2\gamma}{\gamma-1}}} \; . \tag{11.24}$$

The correction of the mass flux at the inflow boundary is expressed as:

$$\dot{m}'_w = [\rho^{m-1} u'_x (S^x + S^y \tan \beta)]_w = \left[\rho^{m-1} C_u (S^x + S^y \tan \beta)\right]_w \overline{(p')}_w \; . \tag{11.25}$$

The pressure correction at the boundary, $\overline{(p')}_w$, is expressed by means of extrapolation from the center of the neighboring control volume i.e., as a linear combination of p'_P and p'_E. From the above equation we obtain a contribution to the coefficients A_P and A_E in the pressure-correction equation for the control volume next to the boundary. Because the density is not corrected at the inflow, there is no convection contribution to the pressure-correction equation there so the coefficient A_W is zero.

After solution of the pressure correction equation, the velocity components and the mass fluxes in the entire domain including the inflow boundary are corrected. The corrected mass fluxes satisfy the continuity equation within the convergence tolerance. These are used to compute the coefficients in all of the transport equations for the next outer iteration. The convective velocities at the inflow boundary are computed from Eqs. (11.20) to (11.21). The pressure adjusts itself so that the velocity satisfies the continuity equation and the boundary condition on the total pressure. The temperature at the inflow is calculated from Eq. (11.22), and the density from the equation of state (11.2).

11.2.4.2 Prescribed Static Pressure

In subsonic flows, the static pressure is usually prescribed on the outflow boundary. Then the pressure correction on this boundary is zero (this is used as a boundary condition in the pressure correction equation) but the mass flux correction is non-zero. The velocity components are obtained by extrapolation from the neighboring control volume centers, in a way similar to calculating cell-face velocities on colocated grids, e.g., for the 'e' face and mth outer iteration:

$$v^*_{n,e} = \overline{(v^*_n)}_e - \Delta V_e \overline{\left(\frac{1}{A^u_P}\right)}_e \left[\left(\frac{\delta p^{m-1}}{\delta n}\right)_e - \overline{\left(\frac{\delta p^{m-1}}{\delta n}\right)}_e\right], \qquad (11.26)$$

where v^*_n is the velocity component in the direction normal to outflow boundary, which is easily computed from Cartesian components obtained by solving the momentum equations, u^*_i, and the known components of the unit outward normal vector, $v^*_n = \mathbf{v}^* \cdot \mathbf{n}$. The only difference from the calculation of the velocity at inner cell faces is that here the overbar denotes extrapolation from inner cells, rather than interpolation between cell centers on either side of the face. At high flow speeds, if the outflow boundary is far downstream, one can usually use the simple upwind scheme, i.e., use the cell-center values (node P) in place of values denoted by overbar; linear extrapolation from W and P is also easily implemented on structured grids.

The mass fluxes constructed from these velocities do not, in general, satisfy the continuity equation and must therefore be corrected. Normally, both the velocity and density need to be corrected, as described above. The velocity correction is:

$$v'_{n,e} = -\Delta V_e \overline{\left(\frac{1}{A^u_P}\right)}_e \left(\frac{\delta p'}{\delta n}\right)_e . \qquad (11.27)$$

The convection (density) contribution to the mass flux correction would turn out to be zero (because $\rho'_e = (C_\rho p')_e$, and $p'_e = 0$ because the pressure is prescribed); however, although pressure is prescribed, the temperature is not fixed (it is extrapolated from inside), so the density does need to be corrected. The simplest approximation is the first-order upwind approximation, i.e., taking $\rho'_e = \rho'_P$. The mass flux correction is then:

$$\dot{m}'_e = (\rho^{m-1}v'_n + \rho'v^*_n)_e S_e . \qquad (11.28)$$

Note, however, that the density correction is not used to correct the density at the outflow boundary—it is calculated always from the equation of state once the pressure and temperature are calculated. The mass flux, on the other hand, has to be corrected using the above expression, because only the correction used to derive the pressure-correction equation does ensure mass conservation. Because, at convergence, all corrections go to zero, the above treatment of density correction is consistent with other approximations and does not affect the accuracy of the solution, only the rate of convergence of the iterative scheme. The coefficient for the boundary node in the pressure-correction equation contains no contribution from the convection term (due to upwinding)—its contribution goes to the central coefficient A_P. The pressure derivative in the normal direction is usually approximated as:

$$\left(\frac{\delta p'}{\delta n}\right)_e \approx \frac{p'_E - p'_P}{(\mathbf{r}_E - \mathbf{r}_P) \cdot \mathbf{n}} , \qquad (11.29)$$

following the approach described in Sect. 9.8 and Eq. (9.66).

The coefficient A_P in the pressure-correction equation for the control volume next to the boundary thus changes compared to those at inner CVs. Due to the convection term in the pressure-correction equation and the Dirichlet boundary condition where static pressure is specified, it usually converges faster than for incompressible flow (where Neumann boundary conditions are usually applied at all boundaries and the equation is fully elliptic).

11.2.4.3   Non-reflecting and Free-Stream Boundaries

At some portions of the boundary the exact conditions to be applied may not be known, but pressure waves and/or shocks should be able to pass through the boundary without reflection. Usually, one-dimensional theory is used to compute the velocity at boundary, based on the prescribed free-stream pressure and temperature. If the free stream is supersonic, shocks may cross the boundary and one makes then a distinction between the parallel velocity component, which is simply extrapolated to the boundary, and the normal component, which is computed from theory. The latter condition depends on whether compression or expansion (Prandtl–Meyer) waves hit the boundary. The pressure is usually extrapolated from the interior to the boundary, while the normal velocity component is computed using the extrapolated pressure and the prescribed free-stream Mach number.

There are many schemes designed to produce non-reflecting and free-stream boundaries. Their derivation relies on the outgoing characteristics computed via one-dimensional theory; the implementation depends on the discretization and the solution method. A detailed discussion of these (*numerical*) boundary conditions can be found in Hirsch (2007) and Durran (2010).

If no shocks are crossing the free-stream or outlet boundary, one can prescribe static pressure there as outlined above and apply the solution forcing technique (see Sect. 13.6) to avoid reflection of pressure waves from these boundaries. This is especially practical for weakly compressible flows in which acoustic pressure waves are captured (aero-acoustics or hydro-acoustics analysis); their reflection at boundaries must be avoided. More details about this topic can be found in Perić (2019).

11.2.4.4 Supersonic Outflow

If the flow at the outflow is supersonic, all of the variables at the boundary must be obtained by extrapolation from the interior, i.e., no boundary information needs to be prescribed. The treatment of the pressure-correction equation is similar to that in the case in which the static pressure is prescribed. However, because the pressure at the boundary is not prescribed but is extrapolated, the pressure correction also needs to be extrapolated—it is not zero as in the above case. Because p'_E is expressed as a linear combination of p'_P and p'_W (if the pressure gradient can be neglected, one may also set $p'_E = p'_P$), the node E does not occur in the algebraic equation, so $A_E = 0$. The coefficients of nodes appearing in the approximation of the mass-flux correction through the boundary are different from those in the interior region.

Some examples of application of the pressure-correction scheme to solving compressible flow problems are presented below. More examples can be found in Demirdžić et al. (1993), Lilek (1995) and Riahi et al. (2018).

11.2.5 Examples

We present below the results of the solution of the Euler equations for a flow over a circular arc bump. Figure 11.2 shows the geometry and the predicted isolines of Mach number for the subsonic, transonic and supersonic conditions. The thickness-to-chord ratio of the circular arc is 10% for subsonic and transonic cases and 4% for the supersonic case. Uniform inlet flow at Mach numbers Ma = 0.5 (subsonic), 0.675 (transonic) and 1.65 (supersonic) is specified. Because Euler equations are solved, viscosity is set to zero and slip conditions are prescribed at walls (flow tangency, as for symmetry surfaces). These problems were the test cases in a workshop in 1981 (see Rizzi and Viviand 1981) and are often used to assess the accuracy of numerical schemes.

For subsonic flow, because the geometry is symmetric and the flow is inviscid, the flow is also symmetric. The total pressure should be constant throughout the solution domain, which is useful in assessing numerical error. In the transonic case, one shock is obtained on the lower wall. When the oncoming flow is supersonic, a shock is generated as the flow reaches the bump. This shock is reflected by the upper wall; it crosses another shock, which issues from the end of the bump, where another sudden change in wall slope is encountered.

Figure 11.3 shows distribution of Mach number along lower and upper walls for the three cases, respectively. The solution error is very small on the finest grid and subsonic flow; this can be seen from the effects of grid refinement, as well as from the

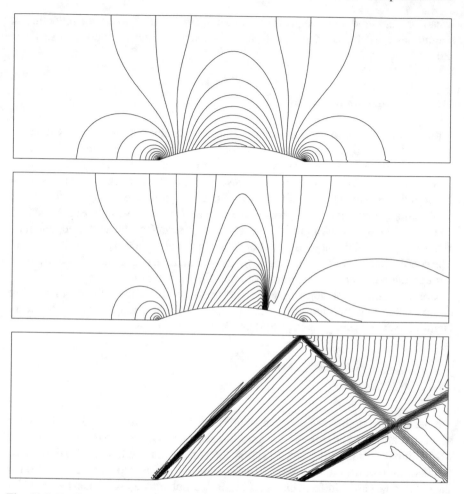

Fig. 11.2 Predicted Mach number contours for inviscid flow through a channel with a circular arc bump in the lower wall: subsonic flow at $Ma_{in} = 0.5$ (above), transonic flow at $Ma_{in} = 0.675$ (middle), and supersonic flow at $Ma_{in} = 1.65$ (bottom); from Lilek (1995)

fact that the Mach numbers at both walls at the outlet are identical and equal to the inlet value. The total pressure error was below 0.25%. In the transonic and supersonic cases, grid refinement affects only the steepness of the shock; it is resolved within three grid points. If central differencing is used for all terms in all equations, strong oscillations at the shocks make solution difficult. In the calculations presented here, 10% of UDS and 90% of CDS were used to reduce the oscillations; they are still present, as can be seen from Fig. 11.3, but they are limited to two grid points near the shock. It is interesting to note that the position of the shocks does not change with grid refinement—only the steepness is improved (this was observed in many applications). The conservation properties of the FV method used and the dominant role of CDS approximations is probably responsible for this feature.

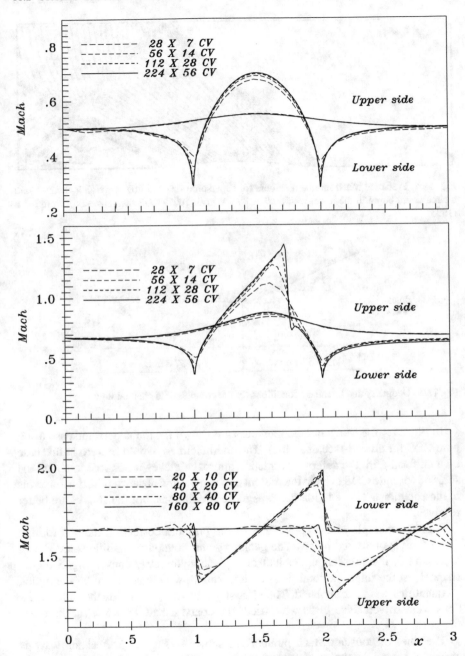

Fig. 11.3 Predicted Mach number profiles along lower and upper wall for inviscid flow through a channel with a circular arc bump in lower wall: subsonic flow at $Ma_{in} = 0.5$ (above; 95% CDS, 5% UDS), transonic flow at $Ma_{in} = 0.675$ (middle; 90% CDS, 10% UDS), and supersonic flow at $Ma_{in} = 1.65$ (bottom; 90% CDS, 10% UDS); from Lilek (1995)

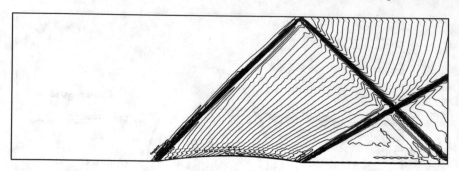

Fig. 11.4 Predicted Mach number contours for supersonic inviscid flow through a channel with a circular arc bump in the lower wall (160 × 80 CV grid, 100% CDS discretization); from Lilek (1995)

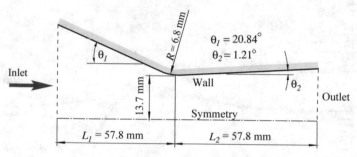

Fig. 11.5 Geometry and boundary conditions for the compressible channel flow

In Fig. 11.4 the Mach number contours are shown for the supersonic case using pure CDS for all cell-face quantities. The coefficient A_P would be zero in this case on a uniform grid; the deferred correction approach makes it possible to obtain the solution for pure CDS even in the presence of shocks and absence of diffusion terms in the equations. The solution contains more oscillations, but the shocks are better resolved.

Another example of the application of the pressure-correction method to high speed flow is presented below. The geometry and boundary conditions are shown in Fig. 11.5. It represents upper half of a plane, symmetric converging/diverging channel. At the inlet, the total pressure and enthalpy were specified; at the outlet, all quantities were extrapolated. The viscosity was set to zero, i.e., the Euler equations were solved. Five grids were used: the coarsest had 42 ×5 CVs, the finest 672 × 80 CVs.

The lines of constant Mach number are shown in Fig. 11.6. A shock wave is produced behind the throat, because the flow cannot accelerate due to the change in geometry. The shock wave is reflected from the wall and the symmetry plane twice before it exits through the outlet cross-section.

In Fig. 11.7 the computed pressure distribution along the channel wall is compared with experimental data of Mason et al. (1980). Results on all grids are shown. On the coarsest grid, the solution oscillates; it is fairly smooth on all other grids. As in

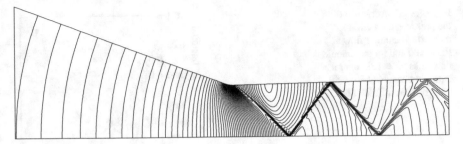

Fig. 11.6 Mach number contours in the compressible channel flow (from minimum Ma $= 0.22$ at inlet to maximum Ma $= 1.46$, step 0.02); from Lilek (1995)

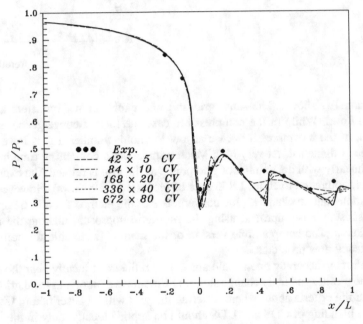

Fig. 11.7 Comparison of predicted (Lilek 1995) and measured (Mason et al. 1980) distribution of pressure along the channel wall

the previous example, the locations of the shocks do not change with grid refinement but the steepness is improved as the grid is refined. The numerical error is low everywhere except near the exit, where the grid is relatively coarse; the results on the two finest grids can hardly be distinguished. Agreement with the experimental data is also quite good.

The solution method presented in this section tends to converge faster as the Mach number is increased (except when the CDS contribution is so large that strong oscillations appear at the shocks; in most applications it was about 90–95%). In Fig. 11.8 the convergence of the method for the solution of laminar incompressible flow at Re $= 100$ and for the supersonic flow at Ma $= 1.65$ over a bump (see Fig. 11.4)

Fig. 11.8 Convergence of
the pressure-correction
method for laminar flow at
Re = 100 and for supersonic
flow at $Ma_{in} = 1.65$ over a
bump in channel (160×80
CV grid); from Lilek (1995)

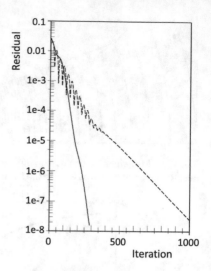

in a channel is shown. The same grid and under-relaxation parameters are used
for both flows. While in the compressible case the rate of convergence is nearly
constant, in the incompressible case at low Reynolds number it gets lower as the
tolerance is tightened. At very high Mach numbers, the computing time increases
almost linearly with the number of grid points as the grid is refined (the exponent is
about 1.1, compared to about 1.8 in case of incompressible flows). However, as we
shall demonstrate in Chap. 12, the convergence of the method for elliptic problems
can be substantially improved using the multigrid approach, making the method
very efficient. The compressible version of the method is suitable for both steady
and unsteady flow problems.

For ultimate accuracy one should apply grid refinement locally near the shocks,
where the profiles suddenly change slope. The methods of applying local grid refinement and the criteria about where to refine the grid will be described in Chap. 12.
Also, the blending of CDS and UDS should be applied locally, only in the vicinity
of shocks, and not globally, as in the above applications. The criteria for decision
where and how much of UDS to blend with CDS can be based on a monotonicity
requirement on the solution, on total variation diminishing (TVD, see next section)
or on other suitable requirements.

11.3 Methods Designed for Compressible Flow

11.3.1 Introduction

The method described above is an adaptation of methods designed for computing
incompressible flows to the treatment of compressible flows. It was mentioned several
times in this book that there are methods specifically designed for the solution of

compressible flows. In particular, these methods can be used in conjunction with the artificial compressibility methods described in Chap. 7. In this section, we briefly describe some of these methods. The purpose is to give enough information about these methods to allow comparison with the methods described above. We shall not present them in sufficient detail to allow the reader to develop codes based on them. The latter task requires a separate volume; readers interested in such a treatment are referred to the texts by Hirsch (2007) and Tannehill et al. (1977).

Historically, the development of methods for the computation of compressible flows proceeded in stages. Initially (until about 1970), only the equations for linearized potential flow were solved. Later, as computer capacity increased, interest moved progressively to the non-linear potential flow equations and, in the 1980s, to the Euler equations. Methods for the viscous flow or Navier–Stokes equations (in most cases the RANS-equations, because the high Reynolds numbers assure that the flows are turbulent) have become the focus of research over the past 20 years.

If there is a major theme running through these methods, it is explicit recognition that the equations are hyperbolic and, thus, have real characteristics along which information about the solution travels at finite speeds. The other essential issue (which arises from the existence of characteristics) is that the compressible flow equations support shock waves and other kinds of discontinuities in the solutions. The discontinuities are sharp in inviscid flows, but have finite width when the viscosity is non-zero. Respecting these properties is important, so it is explicitly taken into account in most methods.

These methods are mainly applied to the aerodynamics of aircraft, rockets, and turbine blades. Because the speeds are high, explicit methods would need to use very small time steps and would be very inefficient. Consequently, implicit methods would be useful and have been developed. However, many of the methods used are explicit.

11.3.2   Discontinuities

The need to treat discontinuities raises another set of issues. We have seen that, in the attempt to capture any kind of rapid change in a solution, discretization methods are likely to produce results that contain oscillations or 'wiggles'. This is especially so when non-dissipative discretizations (which includes essentially all central-difference schemes) are used. A shock (or any other discontinuity) represents the extreme of a rapidly varying solution and therefore presents the ultimate challenge to discretization methods. It can be shown that no discretization method of order higher than first can guarantee a monotonic solution when the solution contains discontinuities (Hirsch 2007, Sect. 8.3). Because accuracy is best obtained through the use of central-difference methods (or their equivalents in finite volume methods), many modern methods for compressible flow use central differences everywhere except near the discontinuities where special upwind methods are applied.

These are the issues that must be faced in the design of numerical methods for compressible flows. We now look, in a very general and superficial way, at some of the methods that have been proposed to deal with them.

The earliest schemes were based on explicit methods and central differencing. One of the most notable of these is the method of MacCormack (2003) (reprint from 1969) which is still used. To avoid the problem of oscillations at shocks in this type of method, it is necessary to introduce artificial dissipation into the equations. The usual second-order dissipation (equivalent to ordinary viscosity) would smooth the solution everywhere so a term more sensitive to the rapid variation at the shock is needed. A fourth-order dissipative term, i.e., an added term that contains fourth derivatives of the velocity, is the most common addition but higher-order terms have also been used.

The first effective implicit methods were developed by Beam and Warming (1978). Their method is based on approximate factorization of the Crank-Nicolson method and can be considered an extension of the ADI method presented in Chaps. 5 and 8 to compressible flow. As with the ADI method, this method has an optimum time step for convergence to a steady solution. The use of central differences again requires addition of an explicit fourth-order dissipation term to the equations.

More recently, there has been an interest in upwind schemes of greater sophistication. The objective is always to produce a well-defined discontinuity without introducing an undue amount of error into the smooth part of the solution. One scheme for accomplishing this is the flux-vector-splitting method of Steger and Warming (1981) to which a number of modifications and extensions have been suggested. The idea is to locally split the flux (because the application is to the Euler equations, this means the convection flux of momentum) into components that flow along the various characteristics of the equations. In general, these fluxes flow in different directions. Each flux is then treated by an upwind method appropriate to the direction in which it flows. The resulting method is fairly complex but the upwinding provides stability and smoothness at discontinuities.

11.3.3 Limiters

Next, we mention a class of schemes that use limiters to provide smooth and accurate solutions. The earliest of these (and one of the easiest to explain) is the flux-corrected transport (FCT) method of Boris and Book (1973); see also Kuzmin et al. (2012). In a one-dimensional version of the method, one might compute the solution using a simple first-order upwind method. The diffusive error in the solution can be estimated (one way is to use a higher-order scheme and take the difference). This estimated error is then subtracted from the solution (a so-called anti-diffusive step) but only to the extent that it does not produce oscillations. In Kuzmin et al. (2012) it is noted that Zalesak (1979) extended FCT to multidimensional form and Parrott and Christie (1986) generalized FCT to finite elements on unstructured grids.

Still more sophisticated methods are based on similar ideas and are generally referred to as *flux limiters*; see Sect. 4.4.6 where we give a number of alternatives. As a reminder, the concept is to limit the flux of the conserved quantity into a control volume to a level that will not produce a local maximum or minimum of the profile of that quantity in that control volume. In total variation diminishing (TVD) schemes, one of the most popular types of these methods, the idea is to reduce the total variation of the quantity q defined by:

$$TV(q^n) = \sum_k |q_k^n - q_{k-1}^n|, \tag{11.30}$$

where k is a grid point index, by limiting the flux of the quantity through the control volume faces.

These methods have been demonstrated to be capable of producing very clean shocks in one-dimensional problems. The obvious way of applying them in multi-dimensional problems is to use the one-dimensional version in each direction. This is not entirely satisfactory, for reasons similar to those that make upwind methods for incompressible flows inaccurate in more than one dimension; this issue was discussed in Sect. 4.7.

TVD schemes reduce the order of approximation in the vicinity of a discontinuity. They become first order at the discontinuity itself because this is the only approximation that is guaranteed to yield a monotonic solution. The first-order nature of the scheme means that a great deal of numerical dissipation is introduced. Another class of schemes, called *essentially non-oscillatory* (ENO) schemes has been developed (see context in Sect. 4.4.6). They do not demand monotonicity and, instead of reducing the order of the approximation, they use different computational molecules or shape functions near a discontinuity; one-sided stencils are used to avoid interpolation across a discontinuity.

In weighted ENO-schemes (WENO), several stencils are defined and checked for the oscillations they produce; depending on the kind of detected oscillations, weight factors are used to define the final shape function (usually called a *reconstruction polynomial*). For computational efficiency, the stencils should be few in number and compact, but to avoid oscillations while keeping a high order of approximation requires that a large number of neighbors be used in the scheme. Studies on atmospheric boundary-layer flows and shocks have shown that, because WENO schemes are dissipative in the presence of large gradients, changing the advection scheme can lead to significant changes in results. Fu et al. (2016, 2017) tackle this issue with methods for so-called "targeted ENO schemes." These produce optimized higher-order (e.g., sixth- or eighth-order) schemes with lower dissipation and better near-shock performance, for example.

Sophisticated non-oscillatory methods for unstructured adaptive grids are described by Abgrall (1994), Liu et al. (1994), Sonar (1997), and Friedrich (1998), among others. These schemes are difficult to implement in implicit methods; in explicit methods they increase the computing time per time step, but the accuracy and lack of oscillations usually compensates for the higher cost.

Finally, we mention that, although they were designed for the solution of elliptic equations, multigrid methods have been applied with great success to compressible flow problems.

11.3.4   Preconditioning

It will also be noted that most of the recent methods just described are explicit. This means that there are limitations on the time steps (or effective equivalent) that can be used with them. As usual, the limitation takes the form of a Courant condition but, due to the presence of sound waves, it has the modified form:

$$\frac{|u \pm c| \Delta t}{\Delta x} < \alpha , \tag{11.31}$$

where c is the sound speed in the gas and, as usual, α is a parameter that depends on the particular time-advancement method used.

For flows that are only slightly compressible i.e., $\mathrm{Ma} = u/c \ll 1$, this condition reduces to:

$$\frac{c \, \Delta t}{\Delta x} < \alpha , \tag{11.32}$$

which is much more restrictive than the Courant condition:

$$\frac{u \Delta t}{\Delta x} < \alpha \tag{11.33}$$

that is usually applicable in incompressible flows. Thus, methods for compressible flows tend to become very inefficient in the limit of slightly compressible flow. The pressure-correction methods presented above seem to be fairly efficient for both incompressible and compressible, steady and unsteady flows. This is why they are mostly used in general-purpose commercial codes, aimed at a wide range of applications from incompressible to highly compressible flow.

There are also methods that were originally developed for compressible flows and then extended to handle flows at all Mach numbers. As noted in Sect. 11.1, the governing equations become numerically very stiff at low Mach numbers. To counter that we can use preconditioning (the concept was discussed in Sect. 5.3). We present briefly the main ideas from one such method, which is implemented in two commercial codes; more details can be found in Weiss and Smith (1995) and Weiss et al. (1999). The method uses preconditioning of the time derivative, in order to be applicable to incompressible flows; this approach was first suggested by Turkel (1987). The coupled system of conservation equations is considered, with pressure p, the Cartesian velocity components u_i and temperature T as the solution variables. The preconditioning matrix K, which multiplies the terms with a time derivative in the coupled system of equations, is defined as:

$$K = \begin{bmatrix} \theta & 0 & 0 & 0 & \dfrac{\partial \rho}{\partial T} \\[2mm] \theta u_x & \rho & 0 & 0 & u_x \dfrac{\partial \rho}{\partial T} \\[2mm] \theta u_y & 0 & \rho & 0 & u_y \dfrac{\partial \rho}{\partial T} \\[2mm] \theta u_z & 0 & 0 & \rho & u_z \dfrac{\partial \rho}{\partial T} \\[2mm] \theta h - \delta & \rho u_x & \rho u_y & \rho u_z & h \dfrac{\partial \rho}{\partial T} + \rho c_p \end{bmatrix}. \tag{11.34}$$

The derivative of density with respect to temperature is taken at constant pressure. For an ideal gas, one obtains

$$\left(\frac{\partial \rho}{\partial T} \right)_p = -\frac{p}{RT} \tag{11.35}$$

and δ is set equal to one, while for an incompressible fluid, both quantities are set equal to zero. The most important parameter is θ, which is defined as:

$$\theta = \frac{1}{U_r^2} - \frac{\partial \rho / \partial T}{\rho c_p}, \tag{11.36}$$

where U_r stands for a reference velocity. It is selected such that the eigenvalues of the system with respect to convection and diffusion time scales remain well conditioned, i.e., they are rescaled to eliminate the stiffness of the equations. This is achieved by limiting U_r in such a way that its value nowhere falls below the local convection or diffusion velocity:

$$U_r = \max\left(|\mathbf{v}|, \, v/\Delta x, \, \epsilon \sqrt{\delta p / \rho} \right), \tag{11.37}$$

and the third limiting value is chosen for reasons of numerical stability (especially because of stagnation point/line zones; ϵ is typically set to 10^{-3}). Δx is the local length scale for diffusion based on grid spacing. For compressible flows, U_r is additionally limited by the local speed of sound c.

A finite volume method is used, with spatial discretization of second order. The cell-center gradients, which are used to interpolate the variables to cell-face centroids, are limited in order to avoid oscillations; the method suggested by Barth and Jespersen (1989) is used.

The preconditioning of the time derivative destroys the time accuracy and the method is thus applicable only to steady-state problems, where only the final solution is of interest, i.e., when the time derivative becomes equal to zero and the preconditioning does no harm. If a time-accurate solution for a transient problem is required, one has to use dual time stepping. Then, we begin with the system of governing equations for unsteady flow. They each have a derivative term with respect to "physical time"; to those is added a preconditioned time derivative with respect to a pseudo-time. For every physical time step, several steps in pseudo-time are taken until the solution stops changing (i.e., a steady state in pseudo-time is achieved, the preconditioned time derivative goes to zero, and the original equations are recovered).

The steps in pseudo-time correspond to outer iterations in the sequential pressure-correction scheme described earlier. For the integration in pseudo-time, the first-order implicit Euler scheme is used, because it allows large time steps and the accuracy in pseudo-time is not required. When solving transient flow problems, the size of physical time steps needs to be selected according to accuracy requirements; the schemes used are typically of second order, e.g., implicit backward scheme with three time levels or the Crank-Nicolson scheme.

Steady-state incompressible flow problems like those presented in Chaps. 8 and 9 can be solved very efficiently using this coupled solver if the step in pseudo-time is selected such that the Courant-number is very large (between 1000 and 10000); if smaller time steps are prescribed, the method is not very efficient. The commercial codes usually offer default values of Courant number between 1 and 10, which is seldom optimal. The problem is that the maximum usable value of Courant number (which usually provides the highest efficiency) is problem-dependent and can vary by several orders of magnitude.

The efficiency of the above coupled solver (both for incompressible and compressible flows) depends strongly on the use of an algebraic multigrid method (see Sect. 6) for solving the linearized coupled equations. More details and some illustrative application examples can be found in Weiss et al. (1999).

11.4 Comments on Applications

The compressible flow literature is rich with applications and with papers that bring together, for example, higher-order methods, large-eddy simulation, and wall modeling, etc. We comment here upon some of these.

One of the earliest works that attacked large-eddy simulation of compressible flow and a scalar is Moin et al. (1991). In addition to using a dynamic SGS model (see Sect. 10.3.3.3), they recast the governing equations in terms of Favre-filtered (density-weighted) variables. As Bilger (1975) points out, this averaging "makes the continuity equation exact and eliminates double correlations involving density fluctuations from the turbulent fluxes." A Farve-filtered variable is defined as

$$\bar{u}_i = \frac{\overline{\rho u_i}}{\bar{\rho}} , \tag{11.38}$$

where the overbar signifies either RANS or LES averaging. The resulting governing equations look very much like the incompressible RANS or LES equations, except for the presence of the averaged density and the continuity equation with a time derivative. Standard solution methods are applicable and the results for decaying isotropic turbulence and channel flow were excellent. Garnier et al. (2009),

Sect. 2.3.6 discusses Favre filtering (noting that most authors have used this change of variables), while Moin et al. (1991) present the development of the momentum and scalar equations in detail.

An area of active research is the prediction of noise from jet engine exhausts. This introduces acoustic propagation and new constraints on discretizations and boundary conditions. Bodony and Lele (2008) review the prediction of jet noise using large-eddy simulation, but improvements have been made since. Brès et al. (2017) apply LES to supersonic jets using an unstructured-grid code. Housman et al. (2017) use an overset grid (see Sect. 9.1.3) with hybrid RANS/LES models to study jet noise as part of a focused effort to develop a quiet supersonic business jet. Brehm et al. (2017) use implicit LES (see Sect. 10.3.2) and a modified sixth-order shock-capturing WENO scheme (Sect. 11.3) to study noise generation by impinging supersonic jets as part of NASA project on engine-noise-shielding to reduce community noise. Overview articles on prediction of flow-generated sound and numerical methods for high-speed flows can be found in Wang et al. (2016) and Pirozzoli (2011).

Finally, we call attention to Le Bras et al. (2017) who examine wall-bounded compressible flows, combining higher-order numerical schemes (e.g., sixth-order compact schemes; see Sect. 3.3.3), an eddy viscosity-based SGS model, and a wall model (using Reichardt 1951 (velocity) and Kader 1981 (temperature) analytical laws; cf. Sect. 10.3.5.5).

The NASA Turbulence Modeling Resource (NASA TMR 2019) provides guidance on implementing turbulence models into the compressible RANS equations.

Chapter 12
Efficiency and Accuracy Improvement

12.1 Introduction

12.1.1 Grid and Flow Feature Resolution

This chapter addresses computational efficiency and accuracy from the viewpoint of the numerical methods. As an introduction, we take a brief look at the accuracy of numerical solutions in the context of the flow physics, i.e., does the numerical solution represent the flow physics accurately? It is important to have a clear understanding of the goal of a simulation; in particular, precisely what flow physics is one trying to represent or educe? Knowledge of the physics (often achieved from theory, observation of the flow, dimensional analysis, etc.) helps us define the grid resolution needed. In Chap. 10, we showcased the simulation of flow about a sphere. It was clear that the drag coefficient for a given configuration was not sensitive to the grid resolution, but adding a trip wire had a large effect (reducing the drag by over 50 percent). On the other hand, the details of the small eddies near flow separation changed as the grid was refined for the smooth sphere. If we were concerned about heat transfer, for example, the local changes in the flow there might well have been of interest. Clearly, we need to understand the success of the simulation in capturing the flow physics of interest.

The predicted flow physics can be strongly influenced by the grid resolution. Rayleigh-Benard convection (fluid constrained between two horizontal plates is heated from below) offers an example. In that flow, the critical Rayleigh number denotes the initial instability leading to a transition from molecular conduction to convection and its wavelength is about twice the spacing between the plates in an experiment or direct numerical simulation. However, in a simulation, if the grid resolution is on the order of the plate spacing, the initial instability cannot be properly represented; indeed, using linear stability theory to estimate that initial wavelength, we would likely select a grid spacing on the order of a tenth of the wavelength. Zhou et al. (2014) studied the convective atmospheric boundary layer and demonstrated that the size of initial instability structures and the "critical" turbulent Rayleigh

© Springer Nature Switzerland AG 2020

J. H. Ferziger et al., *Computational Methods for Fluid Dynamics*,
https://doi.org/10.1007/978-3-319-99693-6_12

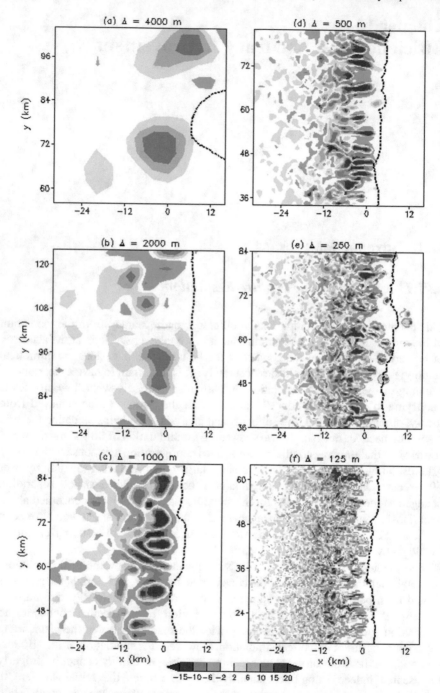

Fig. 12.1 Color map of vertical velocity (m/s) 5 km AGL (above ground level) and 6 h into a weather-front simulation showing grid spacings ranging (Δ) from 4 km to 125 m. Dashed contour marks the surface gust front. Courtesy of George Bryan, National Center for Atmospheric Research

number depend on the "grid spacing rather than the natural state of the flow." In effect, the flow physics that we see in (or can interpret from) a simulation depends on how well the grid resolves the real flow phenomena; as long as the resolution is insufficient, what we see is different from what we would see in reality (nature or an experiment).

A convincing demonstration of the effect of grid size on predicted flow physics was presented by Bryan (2007) and Bryan et al. (2003). They reported on simulations of deep-moist-convection flow across a storm weather front using a non-hydrostatic compressible code (the same solver as used in the WRF (Weather Research and Forecasting) model) in LES mode (cf. Sect. 10.3.3); the domain was 128 km along the front, 512 km across it and 18 km high. For a wind shift of 20 m/s across the front, they presented the results in Fig. 12.1 for the vertical velocity at 5 km above ground level and 6 h into the simulation. The results are startling in that the number and size of updrafts and downdrafts (i.e., cloud forms) is strongly dependent on grid resolution until one reaches 125 m; Bryan shows with energy spectrum plots (Fig. 12.2) that the flow is reasonably resolved at 125 m. In that figure, it is again startling to note that for larger grid spacing the size of the most energetic eddies is driven by the grid size rather than the flow physics; indeed, the most energetic updraft size scales as six times the grid size asymptotically with increasing grid size! A back-of-the-envelope analysis for this case, assuming a 2-km diameter for the convective cells in this situation (based on field data), leads to a suggested horizontal

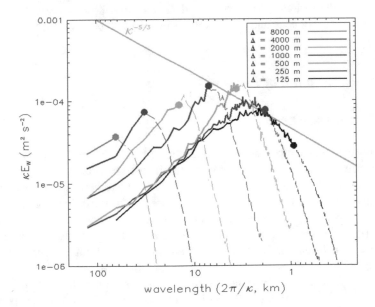

Fig. 12.2 Vertical velocity spectra at 5 km AGL in a weather-front simulation at grid spacings (Δ) ranging from 8 km to 125 m. Here κ is wavenumber. Dashed lines terminate approximately at grid spacing wavelength. Courtesy George Bryan, National Center for Atmospheric Research

grid resolution of O(100–200 m) for a resolved LES, which is consistent with the detailed grid-resolution study of Bryan; see also Matheou and Chung (2014).

In sum, refining and improving a grid for a simulation usually leads to improved accuracy. It may, however, lead to a radical change in the simulated flow physics as well. An understanding of the expected flow physics can be a guide to a correct simulation.

12.1.2   Organization

The best measure of the efficiency of a solution method is the computational effort required to achieve the desired accuracy. There are several methods for improving the efficiency and accuracy of CFD methods; we shall present three that are general enough to be applied to any of the solution schemes described in previous chapters.

12.2   Error Analysis and Estimation

The various types of errors which are unavoidable in the numerical solution of fluid flow problems have been briefly discussed in Sect. 2.5.7. Here we give a more detailed discussion of the various types of error and discuss how these can be estimated and eliminated. Issues of code and model validation will also be addressed.

12.2.1   Description of Errors

12.2.1.1   Modeling Errors

Fluid flow and related processes are usually described by integral or partial differential equations that represent basic conservation laws. The equations may be considered a *mathematical model* of the problem. Although the Navier–Stokes equations can be considered exact, solving them is impossible for most flows of engineering interest. Turbulence places huge demands on computer resources if it is to be simulated directly; other phenomena like combustion, multi-phase flow, chemical processes, etc., are difficult to describe exactly and inevitably require the introduction of modeling approximations. Newton's and Fourier's laws are themselves only models, although they are solidly based on experimental observations for many fluids.

Even when the underlying mathematical model is nearly exact, some properties of the fluid may not be exactly known. All fluid properties depend strongly on temperature, species concentration and, possibly, pressure; this dependence is often ignored, introducing additional modeling errors (e.g., the use of the Boussinesq approximation for natural convection, the neglect of compressibility effects in low Mach-number flows, etc.).

The equations require initial and boundary conditions. These are often difficult to specify exactly. In other cases, one is forced to approximate them for various reasons. Often what should be an infinite solution domain is taken as finite and artificial boundary conditions are applied. We often have to make assumptions about the flow at the inlet to the solution domain as well as at the lateral and outlet boundaries. Thus, even when the governing equations are exact, approximations made at the boundaries may affect the solution.

Finally, the geometry may be difficult to represent exactly; often we have to neglect details for which it is difficult to generate grids. Many codes that use structured or block-structured grids cannot be applied to very complicated problems without simplifying the geometry.

Thus, even if we were able to solve the equations and specified boundary conditions exactly, the result will not describe the flow exactly due to the errors in the model assumptions. We therefore define the *modeling error* as the difference between the real flow and the exact solution of the mathematical model for the specified geometry, fluid properties, and initial and boundary conditions.

12.2.1.2 Discretization Errors

Furthermore, we are seldom able to solve the governing equations exactly. Every numerical method produces *approximate solutions*, because various approximations have to be made to obtain an algebraic system of equations that can be solved on computer. For example, in FV methods one has to employ appropriate approximations for surface and volume integrals, variable values at intermediate locations, and time integrals. Obviously, the smaller the spatial and temporal discrete elements, the more accurate these approximations become. Using better approximations can also increase the accuracy; however, this is not a trivial matter as more accurate approximations are more difficult to program, need more computing time and storage, and may be difficult to apply to complex geometry. Usually, one selects the approximations prior to writing a code so the spatial and temporal grid resolution are the only parameters at user's disposal to control the accuracy.

The same approximation may be very accurate in one part of the flow but inaccurate elsewhere. Uniform spacing (either in space or in time) is seldom optimal, because the flow may vary strongly locally in both space and time; where the changes in variables are small, the errors will also be small. Thus, with the same number of discrete elements and the same approximations, the errors in the results may differ by an order of magnitude or more. Because the computational effort is proportional to the number of discrete elements, their proper distribution and size is essential for computational efficiency (the cost of achieving the prescribed accuracy).

We define the *discretization error* as the difference between the exact solution of the governing equations and the exact solution of the discrete approximation.

12.2.1.3   Iteration Errors

The discretization process normally produces a coupled set of non-linear algebraic equations. These are usually linearized and the linearized equations are also solved by an iterative method because direct solution is usually too expensive.

Any iteration process has to be stopped at some stage. We must therefore define a *convergence criterion* to decide when to stop the process. Usually, iteration is continued until the level of residuals has been reduced by a particular amount; this can be shown to be equivalent to reducing the error by a similar amount.

Even if the solution process is convergent and we iterate long enough, we never obtain the *exact* solution of the discretized equations; round-off errors due to finite arithmetic precision of the computer will provide a lower bound on the error. Fortunately, round-off error does not become an issue until the solution error becomes close to the arithmetic precision of computer and that is far more accuracy than is usually necessary.

We define the *iteration error* as the difference between the exact and the iterative solutions of the discretized equations. Although this kind of error has nothing to do with discretization itself, the effort required to reduce the error to a given size grows as the number of discrete elements is increased. It is therefore essential to choose an optimum level of iteration error—one that is small enough compared to the other errors (which could not be assessed otherwise) but not smaller (because the cost would be larger than necessary).

12.2.1.4   Programming and User Errors

It is often said that all computer codes have bugs—which is probably true. It is the responsibility of the code developer to try to eliminate them; an issue that we shall discuss here. It is difficult to locate programming errors by studying the code—a better approach is to devise test problems in which errors caused by bugs might show up. Results of test calculations must be carefully examined before applying the code to routine applications. One should check that the code converges at the expected rate, that the errors decrease with the number of discrete elements in the expected way, and that the solution agrees with accepted solutions produced either analytically or by another code.

A critical part of the code is the implementation of boundary conditions. The results must be checked to see if the boundary condition applied is really satisfied; it is not unusual to find that they are not. Perić (1993) discussed one such problem associated with adiabatic boundaries in a natural convection flow. Another common source of problems is the inconsistency in approximations of terms that are closely coupled; for example, in a stationary bubble the pressure drop across the free surface must be balanced by the surface tension. Simple flows for which analytical solutions are known are very useful for the verification of computer codes. For example, a code using moving grids can be examined by moving the interior grid while keeping the boundaries fixed and using stationary fluid as the initial condition; the fluid should remain stationary and should not be affected by the grid movement.

The accuracy of a solution depends not only on the discretization method and the code but also on the user of the code; it is easy to obtain bad results even with a good code! Although most user mistakes lead to errors which fall into one of the above three categories, it is important to distinguish between systematic errors, which are inherently present in the method, and the avoidable errors, which are due to bugs in the code or inappropriate or improper use of the code.

Many user errors are due to incorrect input data; often the error is found only after many computations have been carried out—and sometimes it is never found! Frequent errors are due to geometry scaling or parameter selection, when dimensionless form of the equations is used. Another kind of user error is due to a poor numerical grid (an inadequate distribution of grid points can increase the errors by an order of magnitude or more—or prevent one from getting a solution at all).

12.2.2 Estimation of Errors

Every numerical solution contains errors; the important thing is to know how big the errors are, and whether their level is acceptable in the particular application. The acceptable level of error can vary enormously. What may be an acceptable error in an optimization study in the early design stage of a new product, where only qualitative analysis and the response of the system to design changes is important, could be catastrophic in another application.

It is, thus, as important to know how good the solution is for the particular application, as it is to obtain the solution in the first place. Especially when using commercial codes, the user should concentrate on a careful analysis of the results and on estimation of the errors, as far as possible. This may be a great burden for a beginner, but an experienced CFD practitioner will do this routinely.

Error analysis should be done in an order reversed from the order in which the errors were introduced above. That is, one should begin by estimating the iteration error (which can be done within a single calculation), then the discretization error (which requires a minimum of two calculations on different grids) and, finally, the modeling error (which requires reference data and possibly many calculations). Each of these should be an order of magnitude smaller than the one it precedes or the estimation of the later errors will not be sufficiently accurate.

12.2.2.1 Estimation of Iteration Errors

Because the Navier–Stokes equations are non-linear, we have two iteration loops (see Fig. 7.6): *inner iterations* while solving the linearized (and possibly de-coupled) equation systems for a particular variable, and *outer iterations* for updating the coefficients of the linear equation systems and the right-hand side.

Knowing when to stop the iteration process is crucial from the point of view of computational efficiency. For the inner iterations, there is no point in iterating too much because the matrix coefficients and the right-hand side need to be updated many times before the non-linear, coupled equation system is properly solved. In most cases reducing the residual level one order of magnitude before coefficient update is enough; iterating longer would not reduce the number of required outer iterations and thus only lead to longer computing time. On the other hand, if inner iterations are stopped too soon, more outer iterations will be required, thus again increasing the computing effort. The optimum is, as usual, problem-dependent.

It is more critical to control the outer iterations: the discretized non-linear equations are properly solved when an update of matrix coefficients and right-hand side leads to a negligible change in solution. As a rule of thumb, the outer iteration errors (sometimes also called convergence errors) should be at least an order of magnitude lower than discretization errors. There is no point in iterating to the round-off level; for most engineering applications, relative accuracy (error compared to a reference value) of the three to four significant digits in any variable is more than sufficient.

There are a number of ways of estimating these errors; Ferziger and Perić (1996) analyzed three of them in detail; see also Sect. 5.7. It can be shown that the rate of reduction of error is the same as the rate at which the residual and the difference between successive iterates are reduced, except in the initial stage of iteration. This was demonstrated in Fig. 8.9: the curves for the norm of the residual, the norm of difference between successive iterates, the estimated error, and the actual iteration error are all parallel after some iterations. Note that, for outer iterations, the residual computed using the current solution of linearized equations and the updated matrix coefficients and right-hand side is relevant (i.e., the residual computed at the beginning of a new inner iteration loop). The relevant difference in solutions is obtained by subtracting the values from the last inner iteration of two successive loops.

Therefore, if one knows the error level at the start of computation (which is the solution itself if one starts with zero fields and somewhat lower if a rough but reasonable guess is made), then one can be confident that the error will fall 2–3 orders of magnitude if the norm of residuals (or of differences between two iterates) has fallen 3–4 orders of magnitude. This would mean that the first two or three most significant digits will not change in further iterations, and thus that the solution is accurate within 0.01–0.1%.

The above statements apply to the solution of steady-state problems. When solving unsteady problems, the estimation of iteration errors is a bit more complicated. In the case of explicit methods, one only needs to ensure that the pressure or pressure-correction equation is solved to a sufficiently tight tolerance, to ensure that the mass-conservation equation is adequately satisfied; reduction of residuals by three orders of magnitude usually suffices. In the case of implicit methods one may not need to require such a stringent criterion for outer iterations if the time step is very small (as would be the case in LES-simulations), because the solution is not changing much from one time step to another and reducing the residual levels three orders of magnitude within each time step may thus be an overkill. In such a case, 3–5 outer iterations may be enough to update the non-linearity and coupling effects. It is advisable in any new application area to test the effects of varying the convergence criteria in order to make sure that iteration errors are small enough.

A common error is to look at the magnitude of the differences between successive iterates and stop computations when they do not differ by more than a certain small number. However, the difference can be small because the iterations are slowly converging while the iteration error may be enormous. In order to estimate the magnitude of the error, one has to properly normalize the difference between successive iterates; when the convergence is slow, the normalization factor becomes large (see Sect. 5.7). On the other hand, requiring that the norm of differences falls three to four orders of magnitude is usually a safe criterion. Because the linear equation solvers in most CFD methods require the computation of residuals, the simplest practice is to monitor their norm (the sum of absolute values or square root of the sum of squares).

On a coarse grid, where the discretization errors are large, one can allow larger iteration errors; tighter tolerance is required for fine grids. This is automatically taken into account if the convergence criterion is based on the sum of residuals and not on the average residual per node, because the sum grows with growing number of nodes and thus tightens the convergence criterion.

When a new code is developed, or a new feature is added, one has to demonstrate beyond reasonable doubt that the solution process does converge until the residuals reach the round-off level. Very often, the lack of such convergence indicates that errors are present, especially in the implementation of boundary conditions. Sometimes, the limit is below the threshold at which the convergence is declared and the problem may not be noticed. In other cases, the procedure may stop converging (or even diverge) much earlier. Once all new features have been thoroughly tested, one can return to the usual convergence criteria.

Also, if one tries to obtain a steady solution for a problem which is inherently unsteady (e.g., flow around a circular cylinder at a Reynolds number for which the von Karman vortex street is present), iterations will not converge. Because each iteration can be interpreted as a pseudo-time step (see Sect. 7.2.2.2), it is likely that the process will not diverge, but that the residuals oscillate indefinitely. This often happens if the geometry is symmetric and the steady symmetric solution is unstable (e.g., diffusers or sudden expansions; steady solutions—both laminar and Reynolds-averaged—are often asymmetric, with a larger separation region on one side). One can check whether this is the problem by reducing the Reynolds number or computing the flow for one half of the geometry, using a symmetry boundary condition—or by performing a transient computation. Especially in complex geometries, the flow may be locally unsteady in a small part of the solution domain (e.g., behind a mirror of a car). In such a case, residuals may drop below the usual convergence level but if one tried to reduce them further, they would start oscillating at some stage. Often the instability is very weak and if the computation is continued as unsteady, the integral quantities (like forces, moments, total heat flux, etc.) may not change visibly in time—and yet the steady-state computation will not converge.

Figures 12.3 and 12.4 show an example of problems that may occur when one tries to compute a steady-state turbulent flow around a wall-mounted obstacle. The residuals oscillate around the same (too high) value without any sign of reduction

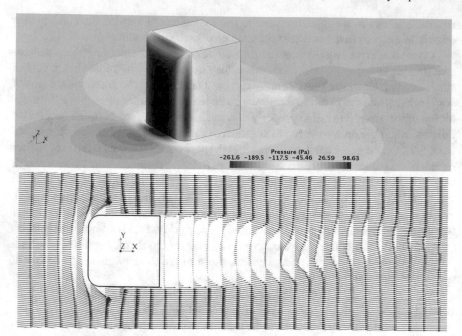

Fig. 12.3 Simulation of flow around a wall-mounted obstacle: instantaneous pressure distribution on obstacle and bottom wall (upper) and velocity vectors in the section parallel to bottom wall at obstacle mid-height (lower) from transient simulation

when a steady-state computation is attempted. When we switch to a transient simulation after 2000 iterations, the residuals at the start of each new time step remain at a high level and do not show a tendency to reduce, but within each time step, the outer iterations converge nicely: with only 5 outer iterations per time step, the residuals in the momentum equations drop more than two orders of magnitude. Obviously, the flow does not have a steady-state solution because the wake behind the obstacle is unsteady, as can be seen from the velocity vectors (they are not symmetric).

The drag and lift forces on obstacle from steady-state and transient simulation are shown in Figs. 12.5 and 12.6. In the steady-state computation, the drag coefficient oscillates around a value which is much lower than in the case of the transient simulation. The lift force oscillates around zero in both cases, but the amplitude is almost twice as large in the case of the steady-state computation. The reason for showing these plots is to warn users of CFD-codes: if the residuals in a steady-state computation start oscillating at a level higher than the usual convergence criterion, one does not have a solution of the governing equations and should not try to interpret the pictures from visualization of variables or to average oscillating forces. Only by switching to a transient simulation can one obtain, at the end of each time step, a valid solution of the governing equations which can be physically interpreted. Oscillations in forces or other integral parameters can now be both averaged (e.g., to evaluate the mean drag or heat transfer coefficient) or otherwise processed (e.g., to obtain frequency of oscillation, rms-value of fluctuation around mean, etc.).

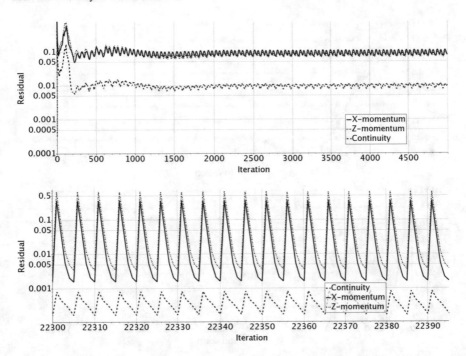

Fig. 12.4 Simulation of flow around a wall-mounted obstacle: variation of residuals in a steady-state computation (upper) and in a transient simulation (lower)

It is not uncommon that convergence problems are encountered when one refines the grid or switches from a lower to a higher-order discretization scheme. The reason is that, when the unsteadiness of the flow is weak, discretization errors may introduce sufficient damping (e.g., numerical diffusion of the first-order upwind scheme) and iterations may converge to a steady state. Flow unsteadiness is often associated with separation and small separation zones (e.g., on the suction side of an airfoil) may only show up once the grid is sufficiently refined. In any case, if outer iterations within each time step of a transient computation converge and in a steady-state computation residuals oscillate, the flow is inherently unsteady and should be computed as such. If outer iterations do not converge within each time step of a transient simulation, the cause can be (i) too large time step, (ii) too high under-relaxation factors, or (iii) mistakes in the set-up of the simulation (grid quality, boundary conditions, fluid properties, etc.).

12.2.2.2 Estimation of Discretization Errors

Discretization errors can only be estimated if solutions on systematically refined grids are compared; see Sects. 3.11.1.2 and 3.9 for more details. As noted earlier, these errors are due to the use of approximations for the various terms in the equations

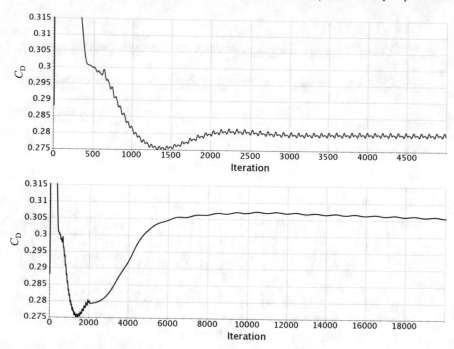

Fig. 12.5 Simulation of flow around a wall-mounted obstacle: variation of drag coefficient with iterations in a steady-state computation (upper) and in a transient simulation (lower)

and the boundary conditions. For problems with smooth solutions, the quality of an approximation is described in terms of its *order*, which relates the *truncation error* of the approximation to the grid spacing to a certain power; if the truncation error of a spatial derivative is proportional to say $(\Delta x)^p$, we say that the approximation is of pth order. The order is not a direct measure of the *magnitude* of the error; it indicates how the error changes when the spacing is changed. Approximations of the same order may have errors on a given grid which differ as much as an order of magnitude; also, an approximation of a lower order may have a smaller error for a particular grid than one of higher order. However, as the spacing becomes smaller, the higher-order approximation will certainly become more accurate.

It is easy to find the order of many approximations using Taylor series expansion. On the other hand, different approximations may be used for different terms, so the order of the solution method as a whole may not be obvious (it is usually of the order of the least accurate approximation of a significant term in the equation). Also, errors in the implementation of the algorithm in the computer code may yield a different order than expected. It is therefore important to check the order of the method for each class of problems using the actual code.

The best way to analyze discretization errors on structured grids is to halve the spacing in each direction. However, this is not always possible; in 3D, this leads to an eight-fold increase in the number of nodes. Thus, the third grid has 64 times as many

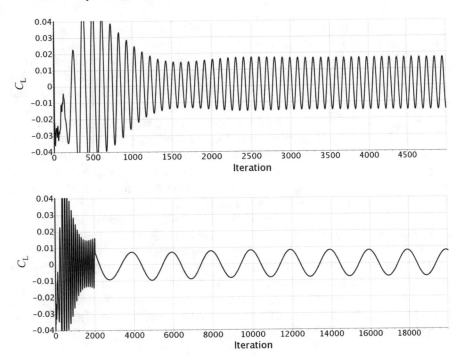

Fig. 12.6 Simulation of flow around a wall-mounted obstacle: variation of lift coefficient with iterations in a steady-state computation (upper) and in a transient simulation (lower)

points as the first one, and we may not be able to afford another refinement level. On the other hand, the errors are usually not uniformly distributed, so there is no point in refining the whole grid. Furthermore, when unstructured grids with arbitrary control volumes or elements are used, there are no local coordinate directions and the elements are refined in a different manner.

What is important is that the refinement is *substantial* and *systematic*. Increasing the number of nodes in one direction from say 112 to 128 is not very useful, except in an academic problem with uniform error distribution and a uniform grid; the refined grid should have at least 50% more nodes in each direction than the original grid (or, the grid spacing should be reduced by at least a factor of 1.5). Systematic refinement means that the grid topology and relative spatial density of grid points should remain comparable on all grid levels. A different distribution of grid points may lead to substantial changes in discretization errors without changing the number of nodes. An example is shown in Fig. 8.10: the results for ψ_{max} obtained on a non-uniform grid, which is finer near walls, are an order of magnitude more accurate than those obtained on a uniform grid with the same number of nodes. Both solutions converge to the same grid-independent solution with the same order (second), but the errors differ in magnitude by a factor of 10 or more (for grids with the same number of nodes)! Other examples with similar conclusions were shown in Figs. 6.4, 6.6, and 8.17.

The above example stresses the importance of good grid design. For practical engineering applications, grid generation is the most time-consuming task; it is often difficult to generate any grid, let alone a grid of high quality. A good grid should be as nearly orthogonal as possible (note that orthogonality has a different meaning in different methods; in a FV method, the angle between the cell-face normal and the line connecting neighboring cell centers is what counts—a tetrahedral grid may be orthogonal in this sense). It should be dense where large truncation errors are expected—hence the grid designer should know something about the solution (as suggested in Sect. 12.1.1). This is the most important criterion and is best met by using an unstructured grid with local refinement. Other criteria of quality depend on the method used (grid smoothness, aspect and expansion ratios, etc.). More details about grid quality measures will be given in Sect. 12.3.

The simplest means of estimation of discretization errors is based on Richardson extrapolation and assumes that calculations can be done on grids sufficiently fine that monotone convergence is obtained. (If this is not the case, it is likely that the error is larger than one would like.) The method is therefore only accurate when the two finest grids are both *fine enough* and the order of error reduction is known. The order may be computed from the results on three consecutive grids from the following formula provided that all three are fine enough in the above sense (see Roache 1994; and Ferziger and Perić 1996, for more details):

$$p = \frac{\log\left(\dfrac{\phi_{rh} - \phi_{r^2 h}}{\phi_h - \phi_{rh}}\right)}{\log r} , \tag{12.1}$$

where r is the factor by which the grid density was increased ($r = 2$ if the spacing is halved), and ϕ_h denotes the solution on a grid with an average spacing h. The discretization error is then estimated as (see Sect. 3.9 for details):

$$\epsilon_h \approx \frac{\phi_h - \phi_{rh}}{r^p - 1} . \tag{12.2}$$

Thus, when the spacing is halved, the error in the solution on one grid is equal to one third of the difference between the solutions on that and the preceding grid for a second-order method; for a first-order method, the error is equal to the aforementioned difference.

For the example from Fig. 8.10, Richardson extrapolation applied to both uniform and non-uniform grid leads to the same estimate of the grid-independent solution within five significant digits, although the errors in the solutions are an order of magnitude different.

Note that the error can be computed for integral quantities (drag, lift, etc.) as well as for field values but the order of convergence may not be the same for all quantities. It is usually equal to the theoretical order (e.g., second) for problems with smooth solutions, e.g., laminar flows. When complicated models (for turbulence, combustion, two-phase flow, etc.) or schemes in which switches or limiters are used, the definition of order may be difficult. However, it is not absolutely necessary to compute quantities like the order p or what Roache (1994) calls the *grid convergence index*; it is sufficient to show the change in the computed quantity of interest for a series of grids (preferably three). If the change is monotonic and the difference decreases with grid refinement, one can easily estimate where the grid-independent solution lies. Of course, one should use the Richardson extrapolation to estimate the grid-independent solution whenever possible.

Note also that the grid refinement need not extend over the whole domain. If the estimate indicates that the error is much smaller in some regions than elsewhere, local refinement can be used. This is particularly true for flows around bodies, where high resolution is needed only in the vicinity of the body and in the wake. Methods using local grid refinement strategies are always more efficient than those which require refinement in the whole domain or grid block. With unstructured grids, already the coarsest grid usually involves local refinement. However, care is needed; if the grid is not refined where the error sources are large (i.e., large truncation errors), the effect of refinement may not be substantial because the errors are subject to the same transport processes (convection and diffusion) as the variables themselves.

Two causes for difficulties when estimating discretization errors deserve to be mentioned. One is associated with wall functions which are often used for computing turbulent flows; see Sect. 10.3.5.5. In this case one does not specify a unique boundary condition at the wall but rather links the wall shear stress to the velocity at the center of the cell next to wall and the assumption that this location is within the logarithmic range of the boundary layer. The log-law is usually not strictly valid in flows subject to complicated wall shapes (i.e., when significant gradients are present in both wall-normal and wall-tangential direction). When the grid is refined, the location of the cell-centroid next to a wall moves and thus the boundary condition in momentum equations also effectively changes; this often leads to variation in integral quantities which do not follow the behavior expected from the discretization scheme used. This is especially true if the grid is refined so that the centroid of the cell next to wall falls into the buffer layer ($5 < n^+ < 30$). If wall functions have to be used, one should ensure that computational points next to the wall remain within the logarithmic range in all grids ($n^+ > 30$); an alternative is to keep the thickness of the first prism layer next to the wall fixed for all grids, so that n^+ at near-wall cells remains the same and the grid immediately at the wall is refined only in the tangential direction.

When the boundary layer is resolved by the grid (so-called "low-Re" approach, i.e., $n^+ \approx 1$ at near-wall cell centers), the no-slip condition is used in momentum equations, which makes the boundary condition unique. In this case the variation in integral quantities (forces, moments, heat transfer etc.) is usually more favorable for error estimation using Richardson extrapolation. However, one has to ensure that $n^+ < 2$ for all near-wall cells on all grids. Using the so-called "all-y^+"-version of wall functions is better than using the "high-Re"-version if n^+-values are in the intermediate range, but it is still better to ensure that, when wall functions are used, the n^+ lies above 30 over most of the wall surface in any grid.

The other problem is associated with the resolution of geometry features by the grids used, especially wall curvature. A typical example is the leading edge of turbo-

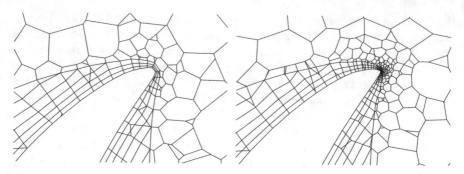

Fig. 12.7 Resolution of propeller blade leading edge in a computational grid: Too coarse grid in wall-tangential direction (left), leading to rough corner at the leading edge, and an improvement by local refinement (right)

machine blades (e.g., ship propellers, fans, water or gas turbine blades, wind or tidal turbine blades, etc.). The curvature is often so high that the distance between locations with maximum and minimum pressure is very short, which requires fine grid not only in wall-normal, but also in wall-tangential direction. However, automatic grid generation tools in commercial software may not resolve the leading edge curvature adequately unless special attention is paid during the grid generation process (e.g., by specifying the grid spacing along leading edge, or requiring that the curvature is resolved by a certain number of points on a circle). This can lead to the leading edge becoming sharp rather than rounded on coarse grids. In such a case, grid refinement effectively leads to the change in solution domain geometry, which again makes the variation of some quantities not following the expectations and thus making the use of Richardson extrapolation difficult. Accordingly, it is important that the initial grid already represents the geometry of the solution domain as well as possible. An example is shown in Fig. 12.7.

12.2.2.3   Estimation of Modeling Errors

Modeling errors are the most difficult ones to estimate; to do so, we need data on the real flow. In most cases, data are not available. Therefore, modeling errors are usually estimated only for some test cases, for which detailed and accurate experimental data are available, or for which accurate simulation data exist (e.g., large-eddy or direct numerical simulation data). In any case, before one can compare a computation with experiment, the iteration and discretization errors should be analyzed and shown to be small enough. In some cases the modeling and discretization errors cancel each other, so that results on a coarse grid may agree better with experimental data than ones obtained on a finer grid. Experimental data should therefore not be used to verify the code; one must use systematic analysis of the results. Only when it is proven beyond reasonable doubt that the results do converge towards a grid-independent solution and that the discretization errors are small enough, can one proceed with comparison of numerical solution and experimental data.

It is important to note that the grid-dependence study, which is used to estimate discretization errors before trying to quantify modeling errors, must be conducted very carefully. If the grid is not designed properly, some features may not show up in any of the grids used, and a small difference between solutions obtained on different grids might lead to the conclusion that discretization errors are small. The difference between the simulation result on the finest grid and experiment would then be wrongly interpreted as a modeling error. An example will be shown in Sect. 13.8: tip-vortex cavitation is not captured unless the grid is locally highly refined within the space occupied by the tip vortex, because otherwise the low pressure in vortex core is under-estimated. If all grids used in the grid-dependence study do not resolve the tip vortex well enough to capture the low pressure in the core accurately, the thrust and torque may still appear to be well converged, but tip-vortex cavitation will be missing. Making the cavitation model responsible for the missing tip-vortex cavitation is unfair because the result is quite good when the grid is adequately locally refined. Similar problems are also encountered in other situations, where, e.g., the turbulence model is blamed for discrepancies between simulation and experiment while a good deal of it may be due to an inadequate grid. Everybody knows that the grid needs to be fine in wall-normal direction to capture the boundary layer features, but curved walls, shear layers, swirl or secondary flows often require local tangential-direction refinements in areas which may not be obvious to an unexperienced user of commercial software.

It is important to bear in mind that the experimental data are also only approximate, and that the measurement and data processing errors can be significant. They may also contain significant systematic errors. However, they are indispensable for the validation of models. One should compare computational results only with experimental data of high accuracy. Careful analysis of the experimental data is essential, if they are to be used for validation purposes.

One should also note that modeling errors differ for different quantities; for example, computed pressure drag may agree well with the measured value, but the computed friction drag may be substantially in error. Mean velocity profiles are sometimes well predicted, while the turbulence quantities may be under- or over-predicted by a factor of two. It is important to compare results with a variety of quantities in order to assure that the model really is accurate.

12.2.2.4 Detection of Programming and User Errors

A kind of error that is difficult to quantify is the programming error. These may be simple "bugs" (typing errors that do not prevent the code from compiling) or serious algorithmic errors. The analysis of iteration and discretization errors usually helps the developer find them, but some may be so consistent that they remain undiscovered for years (if ever), especially when there are no exact reference solutions to compare.

A critical analysis of results is essential for the discovery of potential user errors; it is therefore crucial that the user have solid knowledge of fluid dynamics in general and of the problem to be solved in particular (cf. Sect. 12.1.1). Even if the CFD code that is being used has been validated on other flows, the user can make errors in setting-up the simulation so that the results may be significantly in error (e.g., due to errors in geometry representation, in boundary conditions, in flow parameters, etc.). User errors may be difficult to spot (e.g., when an error in scaling is made and the computed flow corresponds to a different Reynolds number than anticipated); the results should therefore be critically evaluated, if possible also by someone other than the person who performed the computation.

12.2.3 Recommended Practice for CFD Uncertainty Analysis

One should distinguish between *validation* of a newly developed CFD code (or new features added to an existed code) and *validation* of an established code for a particular problem.

12.2.3.1 Validation of a CFD Code

Any new code or added feature should undergo systematic analysis with the aim of assessing the discretization errors (both spatial and temporal), of defining convergence criteria in order to assure small iteration errors, and of eliminating as many 'bugs' as possible. For this purpose one has to select a set of test cases representative of the range of problems solvable by the code, and for which sufficiently accurate solutions (analytical or numerical) are available. Because one wants to assure that the equations are correctly solved for the specified boundary conditions, experimental data are not the best way to measure the quality of numerical solutions. Reference solutions are needed to locate errors in the algorithm or programming which may pass the tests associated with estimation of iteration and discretization errors. Note that one should design the grid such that situations are avoided in which some terms become zero, because in that case implementation errors may not show up; thus, even when using a Cartesian grid in a simple geometry, it is useful to rotate the solution domain such that grid lines are not aligned with Cartesian coordinates (avoiding that two of the three surface vector components at cell faces are zero). Also, one has to ensure that the solution is independent of the orientation of the solution domain relative to the coordinate system orientation.

One should first analyze the approximations used in the discretization to determine the order of convergence of solutions towards a grid (or time step) independent solution. This is the lowest-order truncation error in the significant terms in the equations (but note that not all terms are equally important—their importance depends on the problem). In some cases approximations of lower order may be used at a boundary than in the interior without reducing the overall order. An example is the use of one-sided first-order approximations for diffusion fluxes at boundaries while second-order central differences are used in the interior; the overall convergence is second order. However, this may not be true if low-order approximations are used with Neumann-type boundary conditions.

Iteration errors should be analyzed next; as a first step, one should make a calculation in which iterations are continued until their level is reduced to the double-precision round-off level (this requires at least 12 orders of magnitude reduction of the residual). A test case, which has a known steady solution, must be selected. Otherwise, iterations may stop converging at some stage because iterations can be interpreted as pseudo-time steps and the natural instability of the flow may not allow a steady solution. An example is the case of flow around a circular cylinder at Reynolds numbers above 50. Once an accurate solution of discretized equations on the given grid is available, one can compare it with solutions at intermediate stages, thus evaluating the iteration error. The errors can be compared with estimates, or their reduction can be related to the reduction of the residual or the difference between successive iterates, as discussed above. This should help to establish convergence criteria (both for inner iterations, i.e., for linear equation solver, and for outer iterations, i.e., solution of the non-linear equations).

Discretization errors should be analyzed by comparing solutions on a sequence of systematically refined grids and time steps. Systematic refinement is easy for structured or block-structured grids: one creates, e.g., three grids of different sizes. For unstructured grids, this task is not as straightforward, but one can create grids with similar distributions of relative grid sizes but different absolute sizes. Systematic refinement is crucial in regions of high truncation errors, which act as sources of discretization errors, which are both convected and diffused in the same way as the dependent variables themselves. As a rule of thumb, the grid must be fine and systematically refined where the second and higher-order derivatives of the solution are large. This is typically near walls and in shear layers and wakes.

Solutions with sufficiently small iteration errors should be obtained on at least three grids and compared; both the order of convergence and the discretization error can be estimated in this way if the grids are fine enough so that monotonic convergence prevails. If this is not the case, further refinement is necessary. If the computed order is not the expected one, errors have been made in discretization or programming and must be sorted out. The estimated discretization error should be compared with the required accuracy.

This procedure must be repeated for a number of test cases similar to the applications for which the code is intended in order to try to root out as many error sources as possible. Only when a systematic analysis of the results produced by the code has been made and grid and time-step independent solutions (in the sense that the discretization errors have been reliably estimated and are small enough) have been obtained, should one compare the solutions with analytical or other reference solutions. This is the final check for programming or algorithmic errors. Comparing solutions obtained on one grid with reference solutions is not meaningful, because often some quantities may accidentally agree well or some errors may cancel out.

Code validation says nothing about the accuracy with which the numerically accurate solutions represent real flows. No matter which turbulence (or other) model we use, we have to be sure that we are solving the equations that incorporate the models correctly. Comparisons of solutions obtained by different groups using the same grid and the same turbulence model but different codes often show larger differences than when one group uses the same code but different turbulence models (this was the conclusion reached at many workshops in the 1990s). The models are often differently implemented, the boundary conditions differently treated, etc. This is a difficult problem for which no satisfactory solution has been found. These differences may be due to differences in implementation but—if the models used are really identical, the implementation is correct and errors have been evaluated and eliminated—every code should produce the same result and the differences should disappear. This is why we have stressed the need for validation and error evaluation.

12.2.3.2   Validation of CFD Results

Validation of CFD results includes the analysis of discretization and modeling errors; one can assume that a validated code is used with appropriate convergence criteria, so that iteration errors can be excluded.

One of the most important factors which affects the accuracy of CFD results is the quality of the numerical grid. Note that even a poor grid, if refined enough, should produce the correct solution; it will just cost more. Furthermore, even the best code may produce poor results on a bad and insufficiently refined grid, and a code based on simpler and less accurate approximations may produce excellent results if the grid is tuned for the problem being solved. (However, this is often a matter of getting the various errors to cancel each other.) Discretization errors may be reduced by a proper distribution of grid points; see Fig. 8.10.

Many commercial codes have been made sufficiently robust that they run on any grid the user might provide. However, robustness is usually achieved at the expense of accuracy (for example, by using upwind approximations). A careless user may not pay much attention to grid quality and thereby obtain inaccurate solutions with little effort. The effort invested in grid generation, error estimation, and optimization should be related to the desired level of accuracy of the solutions. If only qualitative features of the flow are sought, a quick job may be acceptable, but for quantitatively accurate results at reasonable cost, high grid quality is required.

Comparison with experimental data requires that the experimental uncertainty be known. It is best to compare only fully converged results (ones from which iteration and discretization error have been removed to a high degree) with experimental data, because this is the only way that the effect of a model can be assessed. Experimental uncertainty bars usually extend on both sides of the reported value. If discretization errors are not small compared to the experimental uncertainty, nothing can be learned about the value of the models used. Estimation of modeling errors is the most difficult task in CFD.

In many cases the exact boundary conditions are not known and one has to make assumptions. Examples are far-field conditions for flows around bodies and inlet turbulence properties. In such a case, it is essential to vary the critical parameter (location of far-field boundary, turbulence quantities) over a substantial range to estimate the sensitivity of the solution to this factor. Often it is possible to obtain good agreement for reasonable values of the parameters but, unless the experimental data provide them, this amounts to little more than sophisticated curve fitting. That is why it is essential to choose experimental data that provide all of the necessary quantities and to discuss the importance of taking such data with the people who make the measurements. Some turbulence models are very sensitive to inlet and free-stream turbulence levels, leading to substantial changes in results for relatively minor variation in the parameters.

If a number of variants of the same geometry are to be studied, one can often rely on a validation performed for a typical representative case. It is reasonable to assume that the same grid resolution and the same model will produce discretization and modeling errors of the same order as those in the test case. While this is true in many cases, it may not always be so and care is needed. Changes in geometry may lead to the appearance of new flow phenomena (separation, secondary flow, instability etc.), which the model used may not capture. Thus, the modeling error may increase dramatically from one case to another although the change in geometry may be minor (e.g., the computation of flow around an engine valve may be accurate within 3% for one valve opening and qualitatively wrong for a slightly smaller opening; see Lilek et al. (1991) for a more detailed description).

12.2.3.3   General Suggestions

The definition of rigid rules for validation of CFD codes and results is difficult and sometimes impractical. While use of Richardson extrapolation is recommended wherever possible for estimating discretization errors, it may be difficult to obtain conclusive answers on all questions (e.g., the order may turn out to be different for different quantities). As we noted in Sect. 10.3.3.7 assessing the simulation quality of LES is a challenge; Sullivan and Patton (2011) present a detailed example of LES quality assessment.

Many journals and professional organizations have developed their own rules and guidelines for assessing and quantifying the uncertainty in CFD solutions; examples are:

- The American Society of Mechanical Engineers (ASME) has developed a standard for verification and validation (https://www.asme.org/products/codes-standards/v-v-20-2009-standards-verification-validation) and organizes at regular intervals conferences on this topic (https://event.asme.org/V-V) [Celik et al. 2008 summarizes these procedures from an ASME perspective];
- The American Institute of Aeronautics and Astronautics (AIAA) has also developed a standard for verification and validation of CFD simulations (Guide: Guide for the Verification and Validation of Computational Fluid Dynamics Simulations (https://doi.org/10.2514/4.472855; AIAA G-077-1998(2002))
- The International Towing Tank Conference (ITTC) has published a set of guidelines for using CFD in maritime engineering (see, e.g., https://ittc.info/media/4184/75-03-01-01.pdf)

Richardson extrapolation is the major ingredient in all error estimation procedures, but many of such guidelines go further and assess the uncertainty in the broader sense. We shall not go into details of those guidelines but recommend that they should be followed when working in the particular application area.

When several types of models are employed (for turbulence, two-phase flow, free-surface effects, etc.) it may be difficult to separate different effects from one another. The most important steps in any quantitative CFD analysis can, however, be summarized as:

- Generate a grid of appropriate structure and fineness (locally refined in regions of rapid variation of the flow and wall curvature).
- Refine the grid systematically (unstructured grids may be selectively refined: where the errors are small, refinement is not needed).
- Compute the flow on at least three grids and compare the solutions (making sure that the iteration errors are small); if the convergence is not monotonic, refine the grid again. Estimate the discretization error on the finest grid.
- If available, compare numerical solutions with reference data to estimate the modeling errors.

Any reasonable estimate of numerical errors is better than none; and because numerical solutions are always *approximate solutions*, one has to question their accuracy *all of the time*.

Many educational organizations offer specialized courses or conduct research on uncertainty quantification in general and CFD in particular. Examples are the UQLab at Stanford University (http://web.stanford.edu/group/uq/) or the Lecture Series at the von Karman Institute (VKI Lecture Series STO-AVT-236 on Uncertainty Quantification in Computational Fluid Dynamics). The number of publications on uncertainty quantification in CFD is also rapidly increasing; recent examples are a book edited by Bijl et al. (2013) and a paper by Rakhimov et al. (2018), among others.

12.3   Grid Quality and Optimization

Discretization errors are always reduced when a grid is refined, but a reliable estimation of these errors requires a grid refinement study for each new application. Optimization of a grid with a given number of grid points can reduce the discretization errors by as much (or more) than systematic refinement of a non-optimal grid. It is therefore important to pay attention to grid quality.

Grid optimization is aimed at improving the accuracy of approximations to surface and volume integrals. This depends on the discretization method used; in this section, we discuss grid features which affect the accuracy of the methods described in this book.

To obtain convection fluxes with maximum accuracy using linear interpolation and/or the midpoint rule, the line connecting two neighboring CV centers should

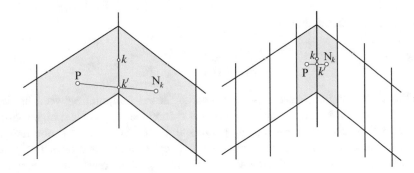

Fig. 12.8 An example of poor grid quality due to a large distance between k and k' (left) and the improvement through local grid refinement (right)

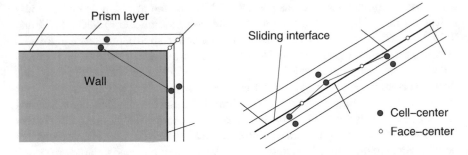

Fig. 12.9 Examples of high grid non-orthogonality when prism layers wrap around sharp corners (left) and at a non-conformal interface (right)

pass through the center of the common face. In certain cases, especially when a block-structured grid is used, situations like the one shown in Fig. 12.8 are unavoidable. Most automatic grid generators create grids of this kind at protruding corners, because they usually create layers of hexahedra or prisms at boundaries, as shown in Fig. 12.9. High non-orthogonality results also at non-conformal interfaces (e.g., in block-structured grids or at sliding interfaces) if cells parallel to interface are thin; such a situation is also depicted in Fig. 12.9. To improve the accuracy without adaptation one should locally refine the grid as shown in Fig. 12.8. This reduces the distance between cell-face center k and the point at which the straight line connecting nodes C and N_k passes through the cell face, k'. The distance between these two points, relative to the size of the cell face (e.g., $\sqrt{S_k}$) is a measure of the grid quality. Cells in which this distance is too large should be refined until the distance between k' and k is reduced to an acceptable level. At non-conformal block interfaces, cells should have a similar size on both sides of the interface and the aspect ratio should not be too large, in order to limit the non-orthogonality to an acceptable level.

Maximum accuracy for the diffusion flux is obtained when the line connecting the neighboring CV centers is orthogonal to the cell face and passes through the cell-face center. Orthogonality increases the accuracy of the central-difference approximation to the derivative in the direction of cell-face normal:

$$\left(\frac{\partial \phi}{\partial n}\right)_{k'} \approx \frac{\phi_{N_k} - \phi_C}{(\mathbf{r}_{N_k} - \mathbf{r}_C) \cdot \mathbf{n}} . \tag{12.3}$$

This approximation is second-order accurate at the midpoint between the two cell centers when the line connecting them is orthogonal to the face; higher-order approximations can be obtained using polynomial fits even when k' is not midway between the nodes. If the non-orthogonality is not negligible, estimation of the normal derivative requires the use of many nodes. This may lead to convergence problems.

If k' is not the cell-face center, the assumption that the value at k' represents the mean value over the cell face is no longer second-order accurate. Although corrections or alternative approximations are possible, most general-purpose CFD codes use simple approximations such as (12.3) and the accuracy is substantially reduced if the grid properties are unfavorable. We have shown in Sects. 9.7.1 and 9.7.2 how second order can be restored; higher-order approximations can also be obtained for arbitrary grids, but this increases the complexity of the code and reduces its robustness. Also, higher-order methods may lead to less accurate approximations than lower-order methods on relatively coarse grids with poor properties.

In most finite-volume methods, it is not important that the grid lines be orthogonal at CV corners; only the angle between the line connecting neighboring CV centers and the cell-face normal matters (see angle θ in Fig. 12.10). A tetrahedral grid can be orthogonal in this sense. An angle θ that is far from $0°$ can lead to large errors and convergence problems and should be avoided. In the situation shown in Fig. 12.8, the line connecting the neighboring CV-centers is nearly orthogonal to the cell face, so that the gradient at k' is accurately computed but, due to the large distance between k' and k, the accuracy of the flux integrated over the surface is poor.

Other kinds of undesirable distortions of CVs may be encountered. Two are depicted in Fig. 12.11. In one case, the upper face of a regular hexahedral CV is rotated around its normal, warping the adjacent faces. In the other case, the top face is sheared in its own plane. Both features are undesirable and should be avoided if at all possible. Warping is especially problematic when thin prismatic cells are present at a curved wall. The location of the cell centroid may then fall outside cell, which can lead to serious problems. The resolution is to either increase the thickness of prism layers, or to refine the grid in wall-tangential direction, or both.

Fig. 12.10 An example of grid non-orthogonality, measured by the angle θ between the face normal and the line connecting the cell centers on either side

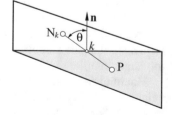

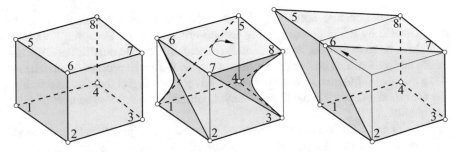

Fig. 12.11 An example of poor grid quality due to cell warping (middle) and distortion (right)

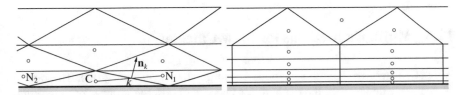

Fig. 12.12 An example of poor-quality triangular or tetrahedral grids (left) and a remedy by adding prism layers near boundary (right)

Grids made of triangles in 2D and tetrahedra in 3D are often used when solving the Euler-equations. When they are used to solve the Navier–Stokes equations, problems can be encountered if tetrahedra are of poor quality. One problematic situation is encountered in corners between two boundaries; it is possible that some tetrahedra near the edge where two boundaries meet have only 2 or possibly even only one neighbor (other cell faces lying in solution domain boundaries). It is then not possible to accurately compute the gradient (i.e., the derivatives in three coordinate directions) using only the data from immediate neighbors. If such cells are present in the grid and the usual discretization is used, it is likely that variables at those cells will oscillate and one may not be able to obtain a converged solution. The usual resolution is to generate prism layers along boundaries (especially along walls, where high gradients in wall-normal direction are present—and also possibly in tangential direction, if wall is curved). This ensures that next-to-boundary cells have at least 4 neighbors and also solves the problem of high grid non-orthogonality which results when tetrahedra are made too flat near wall, as shown in Fig. 12.12. Especially for viscous and turbulent flows, prism layers near walls are a must if the core grid is triangular or tetrahedral.

Another problematic situation which can be encountered with flat, distorted tetrahedra is when all neighbor cell centroids fall nearly into a plane, making it difficult to approximate the gradient component in the direction normal to this plane. This also usually leads to variable oscillation and convergence problems. The cure is to enforce the Delaunay-criterion (see Sect. 9.2.2) during generation of tetrahedral grids, and to create prisms near walls where cells need to be flat.

If the computational nodes are placed at the CV centroids, volume integrals approximated by the midpoint rule are second-order accurate. However, CVs may sometimes be so deformed, that the centroid is actually situated outside the CV. This should be avoided. The grid generator should inspect the grid it produces and indicate to the user that problematic cells exist unless it is able to correct them automatically. The user should then try to modify the control parameters in order to achieve a better grid quality.

Some of these problems can be avoided by subdividing problematic cells (and, possibly, some surrounding cells). Unfortunately, in some cases the only solution is the generation of a new grid.

12.4   Multigrid Methods for Flow Calculation

Almost all iterative solution methods converge slower on finer grids. The rate of convergence depends on the method; for many methods, the number of outer iterations to obtain a converged solution is linearly proportional to the number of nodes in one coordinate direction. This behavior is related to the fact that information travels a limited distance per iteration and, for convergence, information has to travel back and forth across the domain several times. Multigrid methods, for which the required number of iterations is independent of the number of grid points, have received a lot of attention in the 1990s (see Brandt 1984; and Briggs et al. 2000). It has been demonstrated by many authors, including the present ones, that solution of the Navier–Stokes equations by multigrid methods is very efficient. Experience with a wide variety of steady-state laminar and turbulent flows shows a tremendous reduction in computing effort resulting from implementation of the multigrid idea (see the review paper by Wesseling 1990). We give here a brief summary of a version of the method used by the authors; many other variants are possible, see proceedings of the international conferences devoted to multigrid methods in CFD, e.g., McCormick 1987; and Hackbusch and Trottenberg (1991).

In Chap. 5 we presented a multigrid method for solving linear systems of equations efficiently. We saw there that the multigrid method uses a hierarchy of grids; in the simplest case, the coarse ones are subsets of the fine ones. It is ideal for solving the Poisson-like pressure or pressure-correction equation when fractional-step or other explicit time-stepping methods are applied to unsteady flows because accurate solution of the pressure equation is required; this is often done in LES and DNS of flows in complex geometry. On the other hand, when implicit methods are used, the linear equations need not be solved very accurately at each iteration; reduction of the residual level by one order of magnitude suffices and can usually be achieved with a few iterations of one of the basic solvers such as ILU or CG. More accurate solution will not reduce the number of outer iterations but may increase the computing time. Thus, if the multigrid method is applied only to the solution of linear equation systems in an implicit solution method, the achievable acceleration is limited.

For steady flow problems, we have seen that implicit solution methods are preferred and acceleration of the outer iterations is very important. Fortunately, the multigrid method can be applied to outer iterations as well. The sequence of operations that constitute one outer iteration is then considered as the 'smoother' in accord with multigrid terminology.

In a multigrid version of a finite volume method for steady flows on a structured grid, each coarse grid CV is composed of four CVs of the next finer grid in 2D and eight in 3D. The coarsest grid is usually generated first and the solution process starts by solving the problem on it. Each CV is then subdivided in finer CVs. After a converged solution is found on the coarsest grid, it is interpolated to the next finer grid to provide the starting solution. Then a two-grid procedure is begun. The process is repeated until the finest grid is reached and a solution on it is obtained. As noted earlier, this strategy is called the *full multigrid* procedure (FMG). After m outer iterations on the grid with spacing h, the short-wavelength error components have been removed and the intermediate solution satisfies the following equation:

$$A_h^m \boldsymbol{\phi}_h^m - \mathbf{Q}_h^m = \boldsymbol{\rho}_h^m , \qquad (12.4)$$

where $\boldsymbol{\rho}_h^m$ is the residual vector after the mth iteration. The solution process is now transferred to the next coarser grid whose spacing is $2h$. As noted earlier, both the cost of an iteration and the convergence rate are much more favorable on the coarse grid, giving the method its efficiency.

The equations solved on the coarse grid should be smoothed versions of the fine grid equations. With a careful choice of definitions, one can assure that the equations solved appear identical to the ones solved earlier on that grid, i.e., the coefficient matrix is the same. However, the equations now contain an additional source term:

$$\hat{A}_{2h} \hat{\boldsymbol{\phi}}_{2h} - \hat{\mathbf{Q}}_{2h} = \tilde{A}_{2h} \tilde{\boldsymbol{\phi}}_{2h} - \tilde{\mathbf{Q}}_{2h} - \tilde{\boldsymbol{\rho}}_{2h} . \qquad (12.5)$$

If set to zero, the left-hand side of Eq. (12.5) would represent the coarse grid equations. The right-hand side contains the correction that assures that the solution is a smoothed fine grid solution rather than the coarse grid solution itself. The additional terms are obtained by smoothing ('restricting') of the fine grid solution and residual; they remain constant during the iterations on the coarse grid. The initial values of all terms on the left-hand side of the above equation are the corresponding terms on the right-hand side. If the fine-grid residual is zero, the solution will be $\hat{\boldsymbol{\phi}}_{2h} = \tilde{\boldsymbol{\phi}}_{2h}$.

Only if the residual on the fine grid is non-zero, will the coarse-grid approximation change from its initial value (because the problem is non-linear, the coefficient matrix and the source term also change, which is why these terms carry a ˆ symbol). Once the solution on the coarse grid is obtained (within a certain tolerance), the correction

$$\boldsymbol{\phi}' = \hat{\boldsymbol{\phi}}_{2h} - \tilde{\boldsymbol{\phi}}_{2h} \qquad (12.6)$$

is transferred by interpolation ('prolonged') to the fine grid and added to the existing solution $\boldsymbol{\phi}_h^m$. With this correction, much of the low-frequency error in the solution

Fig. 12.13 Transfer of
variables from fine to coarse
grid and vice-versa

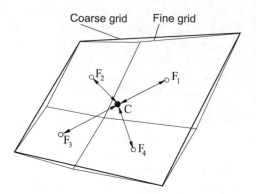

on the fine grid is removed, saving a lot of iterations on the fine grid. This process is
continued until the solution on fine grid is converged. Richardson extrapolation may
then be used to obtain an improved starting guess for the next finer grid, and a three
level V-cycle is initiated, and so on.

For structured grids, simple bilinear (in 2D) or trilinear (in 3D) interpolation is
usually used to transfer variable values from fine to coarse grids and corrections from
coarse to fine grids. Although more complex interpolation techniques can and have
been used, in most cases this simple technique is quite adequate.

Another way to transfer a variable from one grid to another is to compute the
gradient of that variable at the CV centers of the grid on which it was calculated
(coarse or fine). An efficient way to calculate gradients at the centers of arbitrary
CVs using Gauss theorem was described in Chap. 9. It is then easy to calculate
the variable value anywhere nearby using this gradient (this corresponds to linear
interpolation). For the case shown in Fig. 12.13, we can calculate the coarse-grid
variable value at node C by averaging the values calculated using the fine-grid CV
gradients:

$$\phi_C = \frac{1}{N_f} \sum_{i=1}^{N_f} [\phi_{F_i} + (\nabla \phi)_{F_i} \cdot (\mathbf{r}_C - \mathbf{r}_{F_i})] , \qquad (12.7)$$

where N_f is the number of fine-grid CVs in one coarse-grid CV (on structured grids,
four in 2D and eight in 3D). The coarse CV does not need to know which fine CVs
belong to it—it only needs to know how many there are. On the other hand, each
fine-grid CV (child) knows to which coarse grid CV (parent) it belongs; it has only
one parent.

Similarly, the coarse-grid correction is easily transferred to the fine grid. One
calculates the gradient of the correction at the coarse-grid CV center; the correction
at the fine-grid nodes which lie within this CV are calculated from:

$$\phi'_{F_i} = \phi'_C + (\nabla \phi')_C \cdot (\mathbf{r}_{F_i} - \mathbf{r}_C) . \qquad (12.8)$$

This interpolation is more accurate than simple injection of the coarse CV correction to all fine CVs within it (this can also be done, but smoothing of the correction is necessary after prolongation). On structured grids, one can easily implement other kinds of polynomial interpolation.

In FV methods, the conservation property can be used to transfer the mass fluxes and residuals from the fine to the coarse grid. In 2D, the coarse grid CV is made up of four fine grid CVs and the coarse grid equation should be the sum of its daughter fine grid CVs equations. The residuals are thus simply summed over fine grid CVs, and the initial mass flux at coarse CV faces is the sum of the mass fluxes at the fine CV faces. During calculations on coarse grids, the mass fluxes are not calculated using restricted velocities but are *corrected* using velocity corrections, (the former would be less accurate and the correction is smoother than the variable itself). For the generic variables we may write:

$$\tilde{\phi}_{k-1} = I_k^{k-1}\phi_k^m \quad \text{and} \quad \phi_k^{m+1} = \phi_k^m + I_{k-1}^k\phi_{k-1}', \tag{12.9}$$

where I_k^{k-1} is the operator describing the transfer from fine to coarse grid and I_{k-1}^k is the operator describing the transfer from coarse to fine grid; Eqs. (12.7) and (12.8) provide examples of these operators.

The treatment of the pressure terms in the momentum equations deserves special mention. Because initially $\hat{p} = \tilde{p}$ and the pressure terms are linear, we may work with the difference $p' = \hat{p} - \tilde{p}$. Then we do not need to restrict the pressure from fine to coarse grids. Note that this is not the pressure correction p' of SIMPLE and related algorithms; it is a correction of the finer grid pressure and is based on the velocity corrections $u_i' = \hat{u}_i - \tilde{u}_i$. As already mentioned, the initial coarse grid mass fluxes, $\tilde{m}$, are obtained by summing the corresponding fine grid mass fluxes. These change only if the velocities change; we assume that the fine grid mass fluxes are mass-conserving at the beginning of the multigrid cycle; if not, the mass imbalance can be included in the pressure-correction equation on the coarse grid.

It is important to take care that the implementation of boundary conditions also fulfills the consistency requirements. For example, if a symmetry boundary condition is implemented by setting the boundary value equal to the value at the near-boundary node, the restriction operator cannot calculate the boundary value of $\tilde{\phi}$ by interpolating the fine grid boundary values; it must calculate $\tilde{\phi}$ at all inner nodes and then apply the boundary condition to it, i.e., set $\tilde{\phi}$ at the symmetry boundary equal to $\tilde{\phi}$ at the near-boundary nodes. If the boundary condition is applied to $\hat{\phi}$, then ϕ' would not be the same at the boundary and next-to-boundary nodes, and a gradient of ϕ' would be passed to the finer grid. Then the solution on the fine grid cannot be converged beyond a certain limit. A similar situation can occur due to inconsistencies in treating other boundary conditions, but we shall not list all of the possibilities here. It is important to assure that the iteration errors can be reduced to machine accuracy (even though this criterion will not be used when the code is put into production); if this is not possible, something is wrong!

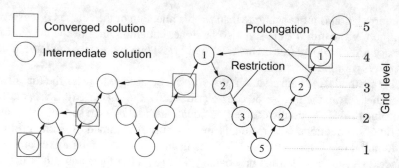

Fig. 12.14 Schematic presentation of FMG scheme using V-cycles, showing typical numbers of outer iterations at different stages in one cycle

Other strategies (e.g., W-cycles) may be used for cycling between the grids. Efficiency may be improved by basing the decision to switch from one grid to another on the rate of convergence. The simplest choice is the V-cycle described above with a fixed number of iterations on each grid level. The behavior of the FMG method for the V-cycle with typical numbers of iterations at each level is schematically shown in Fig. 12.14. The optimum choice of parameters is problem dependent, but their effect on performance is not as dramatic as for the single-grid method. Details of multigrid methods can be found in the book by Hackbusch (2003).

The multigrid method can be applied to unstructured grids as well as structured grids. In FV methods, one usually joins fine grid CVs to produce coarse grid CVs; the number of fine CVs per coarse CV may differ, depending on the shape of CVs (tetrahedra, pyramids, prisms, hexahedra etc.). The multigrid idea can even be used if the coarse and fine grids are not related by systematic refinement or coarsening—the grids may be arbitrary; it is only important that the solution domain and boundary conditions be the same on all levels, and that one grid is substantially coarser than the other (otherwise computational efficiency is not improved). The restriction and prolongation operators are then based on generic interpolation methods; such multigrid methods are called *algebraic multigrid methods* (see e.g., Raw 1995; and Weiss et al. 1999).

For computing unsteady flows with implicit methods and small time steps, the outer iterations usually converge very rapidly (residual reduction by an order of magnitude per outer iteration), so multigrid acceleration is not necessary. The biggest savings are achieved for fully elliptic (diffusion-dominated) problems, and the smallest for convection-dominated problems (Euler equations). Typical acceleration factors for steady-state problems range from 10 to 100 when five grid levels are used. An example is given below.

When computing turbulent flows with the k-ε turbulence model, interpolation may produce negative values of k and/or ε in the early cycles of a multigrid method; the corrections then have to be limited to maintain positivity. In problems with variable properties, the properties may vary by several orders of magnitude within the solution domain. This strong non-linear coupling of the equations may cause the multigrid

Fig. 12.15 Number of outer
iterations on the finest grid in
a multigrid method as a
function of the
under-relaxation factor α_u
for the lid-driven cavity flow
at Re = 1000

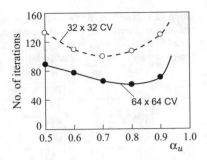

method to become unstable. It may be better to update some quantities (e.g., the
turbulent viscosity in the k-ϵ turbulence model) only on the finest grid and keep
them constant within a multigrid cycle.

Under-relaxation factors are relatively unimportant in multigrid methods for laminar flows; the methods are less sensitive to these parameters than the single grid
method. In Fig. 12.15 we show the dependence of the number of required outer iterations to solve the lid-driven cavity problem at Re = 10,000 on the under-relaxation
factor for velocity (using optimum under-relaxation of pressure correction, see Sect.
8.4) for two grids. The number of iterations varies by about 30% in the range of α_u
between 0.5 and 0.9, while for the single grid method, the variation was by a factor
of five to seven (higher for finer grids; see Fig. 8.14). However, for turbulent flows,
heat transfer, etc., under-relaxation may affect the multigrid method substantially.

The full multigrid method described above provides solutions on all grids. The
cost of solving on all of the coarse grids is about 30% of the cost of the finest
grid solution. If the solution process were started on the finest grid with zero fields,
more total effort is needed than if the FMG approach is used. The savings resulting
from more accurate initial fields usually outweigh the cost needed to obtain them.
In addition, having solutions on a series of grids allows evaluation of discretization
errors, as discussed in Sect. 3.9, and provides a basis for grid refinement. The grid
refinement process may be stopped when the desired accuracy is achieved. Also,
Richardson extrapolation may be used.

The multigrid approach to accelerating outer iterations described above can be
applied to any solution method for the Navier–Stokes equations. Vanka (1986)
applied it to the point-coupled solution method; Hutchinson and Raithby (1986)
and Hutchinson et al. (1988) use it with a line-coupled solution technique. Methods
of the SIMPLE type and fractional-step methods are also well suited for multigrid
acceleration; see, e.g., Hortmann et al. (1990), Lilek et al. (1997a) and Thompson
and Ferziger (1989). The role of the *smoother* is now taken by the basic algorithm
(e.g., SIMPLE); the linear equation solver plays a minor role.

In Table 12.1 we compare the numbers of outer iterations needed to solve the 2D
lid-driven cavity flow problem at Reynolds numbers Re = 1000 and Re = 1000 using
different solution strategies. SG denotes the single-grid method with a zero initial
field. PG denotes the prolongation scheme, in which the solution from the next

Table 12.1 Numbers of iterations and computing times required by various versions of the solution method to reduce the L_1 residual norm by four orders of magnitude when solving the lid-driven 2D cavity flow problem ($\alpha_u = 0.8$, $\alpha_p = 0.2$, non-uniform grid; for PG and FMG, CPU-times include computing times on all coarser grids)

Re	Grid	No. outer Iter.				CPU-time			
		SG	PG	MG	FMG	SG	PG	MG	FMG
100	8^2	58	58	58	58	0.3	0.3	0.3	0.3
	16^2	61	51	47	45	0.9	1.2	1.4	1.5
	32^2	156	99	41	41	9.1	7.0	4.0	5.0
	64^2	555	256	40	40	140.8	71.1	13.0	16.9
	128^2	2119	620	40	40	2141.9	702.6	50.9	66.5
	256^2	–	–	40	40	–	–	242.2	293.8
1000	8^2	124	124	124	124	0.5	0.5	0.5	0.5
	16^2	156	162	123	132	2.2	2.5	2.8	2.9
	32^2	250	288	132	132	14.0	19.2	11.2	13.8
	64^2	433	400	93	73	97.0	120.7	32.0	38.5
	128^2	1352	725	83	41	1383.4	851.1	121.5	92.4
	256^2	–	–	83	31	–	–	512.9	278.8

coarser grid is used to provide the initial field. MG denotes the multigrid method using V-cycles, with the finest grid having zero initial fields. Finally, FMG denotes the multigrid method described above, which can be considered a combination of the PG and MG schemes.

The results show that, for the Re = 1000 case, MG and FMG need about the same number of outer iterations on the finest grid level—a quality initial guess does not save much. For the single-grid scheme, the savings are substantial: on the 128 × 128 CV grid, the number of iterations is reduced by a factor of 3.5. The multigrid method reduces the number of iterations on that grid by a factor of 15; this factor increases as the grid is refined. The number of iterations remains constant in the MG and FMG methods from the third grid onwards, while it increases by a factor of four in SG and by a factor of 2.5 in the PG scheme.

For high Reynolds number flows, the situation changes a bit. SG needs fewer iterations than at Re = 1000, except on coarse grids, on which the use of CDS slows convergence. PG reduces the number of iterations by less than a factor of two. MG needs about twice as many iterations as it did for Re = 100. However, FMG becomes more efficient as the grid is refined—the number of iterations required is actually lower on a 256 × 256 CV grid than for Re = 1000. This is typical behavior of the multigrid method applied to the Navier–Stokes equations. The FMG approach is usually the most efficient one.

Results similar to those presented in Table 12.1 are also obtained for the 3D cavity flow; see Lilek et al. (1997a) for details.

In Fig. 12.16 the reduction of residual norm (the sum of absolute values of the residual over all CVs) of the turbulent kinetic energy k are shown for the computation

Fig. 12.16 Reduction of
residual norm for the
turbulent kinetic energy k in
a calculation of flow in a
tube bundle segment; the
finest grid had 176 × 48 CV,
five levels were used (from
Lilek 1995)

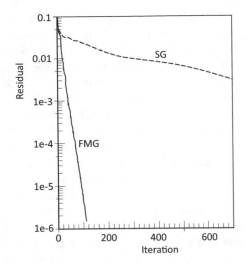

of turbulent flow in a segment of a tube bundle made with the $k - \epsilon$ model. The curves are typical for the MG and SG methods. In practical applications, reduction of residuals by three to four orders of magnitude usually suffices. Here the residuals were reduced more than necessary, to show that the rate of convergence does not deteriorate. The saving in computing time varies from application to application: it is lower for convection-dominated than for diffusion-dominated flows.

12.5 Adaptive Mesh Refinement (AMR)

12.5.1 Motivation for Adaptive Mesh Refinement

Issues of accuracy have plagued computational fluid dynamics from its inception. There are many published results with significant errors. Tests intended to determine model validity have sometimes proven inconclusive because the numerical errors were greater than the effects of the model. Comparative studies in which the same problem was solved by different groups using different codes revealed that the differences between solutions obtained using different codes with the same models are often larger than the differences between solutions obtained using the same code and different models (Bradshaw et al. 1994; Rodi et al. 1995). These differences can only be due to numerical errors or user mistakes, if the models *are* really the same (it is not unusual that supposedly same models turn out to be different due to different interpretation, implementation, or boundary treatment). For valid comparisons of models, it is critical that errors be estimated and reduced.

The first essential is a method of estimating errors; the Richardson method given earlier is a good choice. A method of estimating the error without the need to perform calculations on two grids is to compare the fluxes through CV faces that result from the discretization method employed in the solution and a more accurate (higher-order) method. This is not as accurate as the Richardson method, but it does serve the purpose of indicating where the errors are large. Because the higher-order approximation is usually more complex, it is used only to calculate the fluxes after a converged solution with the base method has been obtained. For example, one can fit a polynomial of degree three through the cell centers on either side of a particular face and use the variable values and gradients at these two cell centers to find the coefficients of the polynomial (see Sect. 4.4.4 for an example). Assume that, if this approximation had been used, the exact solution Φ would have been obtained instead of the solution ϕ that resulted from the usual approximations. The difference between fluxes computed using the cubic (F_k^Φ) and linear fits (F_k^ϕ) should be added to the discretized equation as an additional source term to recover the "exact" solution. If we define the discretization error ϵ^d and the source term τ (which is often called *tau-error*) as follows:

$$\epsilon^d = \Phi - \phi \quad \text{and} \quad \tau_P = \sum_k (F_k^\Phi - F_k^\phi) , \tag{12.10}$$

we obtain the following link between an estimate of the discretization error and the tau-error:

$$A_P \epsilon_P^d + \sum_k A_k \epsilon_k^d = \tau_P . \tag{12.11}$$

Instead of solving this equation system for ϵ^d, it is often sufficient to simply normalize the τ_P by A_P and use this quantity as an estimate of the discretization error; this corresponds to performing one Jacobi iteration on the system of Eqs. (12.11) starting with zero initial values. The reason is that the above analysis is only approximate and the computed quantity is rather an *indication* than *estimation* of the discretization error. For more details and examples of application of this method of error estimation, see Muzaferija and Gosman (1997). The key point is that we now have a strategy for defining error pointwise throughout the flow domain. This information can be used to adjust the grid (i.e., the mesh) to make the error levels more even. Essentially, if the error estimate at a particular grid point is larger than a prescribed level, that grid cell is labeled for refinement.

12.5.2   The AMR Strategy

From the error estimates, a map of the grid cells labeled for refinement can be constructed. To provide a buffer zone, the boundaries of the refinement region typically are extended by some margin, which should be a function of the local mesh size; the width of two to four cells is usually a sensible choice.

Block-structured grids require that refinement be performed block-wise; non-matching interface capability is required if not all blocks are refined. For unstructured grids, local refinement can be cell-wise. Otherwise, cells to be refined may be clustered and new blocks of refined grid defined, as will be described later.

The objective is to make the error everywhere smaller than some tolerance δ, either in terms of the absolute error, $\|\epsilon\|$, or the relative error, $\|\epsilon/\phi_{ref}\|$, where ϕ_{ref} is the representative variable value used for normalization. This can be accomplished by using methods of differing accuracy, an approach commonly used in ordinary differential equation solvers, but this is rarely done. One can also refine the grid everywhere but this is wasteful. A more flexible choice is to refine the grid *locally* where the errors are large. Experienced users of CFD codes may generate grids that are fine where necessary and coarse elsewhere, so that they yield a nearly uniform distribution of discretization error. This is especially important when the geometry contains small but important protrusions, e.g., mirrors on cars, appendages on ships and other vessels, small inlets and outlets on walls of large chambers etc. However, in complex geometries it is difficult to do local refinements "by hand"; in transient flows, regions which need refinement may also change with time. Thus, a method of automatic, adaptive mesh refinement is essential if the desired accuracy is to be achieved with minimum effort. For example, the Combo package of C++ codes (Adams et al. 2015) supports block-structured AMR applications. The AMR is discussed in a number of papers, beginning with Berger and Oliger (1984); examples include Skamarock et al. (1989) and Thompson and Thompson and Ferziger (1989).

In some flow problems, it is obvious where the grid needs to be refined so the automation of adaptive refinement is easy. Examples are compressible flows with shocks (fine grid is needed around shocks, e.g., flow around airfoils and in turbomachines etc), free-surface flows (fine grid is needed to resolve the free surface, e.g., jet break-up, rising bubbles, water entry of rigid bodies, ships in waves, etc.), flows with cavitation etc. An example of grid adaptation to a shock is shown in Figs. 12.17 and 12.18; examples of adaptation to free surface and cavitation zone will be presented in the next chapter. A suitable error indicator or estimator would show other regions where the initial grid is not fine enough. Sometimes error estimation may indicate that errors in one variable are high in one zone but other variables may require grid refinement elsewhere; one may need to compromise or apply weighting to each variable and error level. This indicates that the issue of deciding where to refine the grid and to which level is far from trivial; it is for this reason that commercial CFD-codes do not offer adaptive local grid refinement as a standard feature yet, but prototypes have been presented and it is clear that future versions will include it.

Some authors perform calculations on the refined portion of the grid only, using boundary conditions taken from the coarse grid solution at the refinement interface. This is called the *passive* method because the solution on the unrefined part of the grid is not re-computed (see Berger and Oliger 1984) This feature makes the method inappropriate for elliptic problems in which a change in conditions in any region may affect the solution everywhere. Methods that allow the influence of the refined grid solution to spread over the whole domain are called *active* methods. Such methods have been developed by Caruso et al. (1985) and Muzaferija (1994), among others.

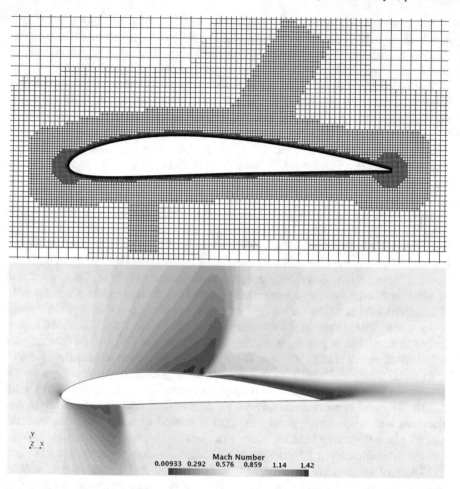

Fig. 12.17 Initial grid (upper) and computed Mach-number contours (lower) for the turbulent flow around an airfoil at Ma = 0.8

One kind of active method (e.g., Caruso et al. 1985) proceeds exactly as the passive method with the important difference that the procedure is not complete when the fine grid solution has been computed. Rather, it is necessary to compute a new coarse grid solution; this solution is not the one that would be computed on the coarse grid covering the entire domain but a smoothed version of the fine grid solution. To see what is needed, suppose that:

$$\mathcal{L}_h(\phi_h) = Q_h \tag{12.12}$$

is a discretization of the problem on a grid of size h; $\mathcal{L}$ represents the operator. To force the solution to be a smoothed version of the fine-grid solution in the region which has been refined we replace the coarse-grid problem by:

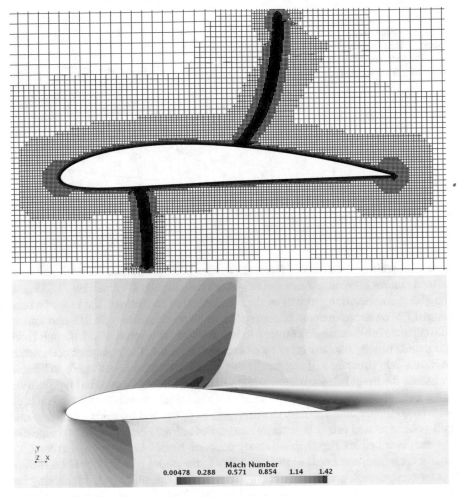

Fig. 12.18 Shock-adapted grid (upper) and computed Mach-number contours (lower) for the turbulent flow around an airfoil at Ma $= 0.8$

$$\mathcal{L}_{2h}(\phi_{2h}) = \begin{cases} \mathcal{L}_{2h}(\tilde{\phi}_h), & \text{in the refined region;} \\ Q_{2h}, & \text{in the remainder of the domain,} \end{cases} \qquad (12.13)$$

where $\tilde{\phi}_h$ is the smoothed fine-grid solution (i.e., its representation on the coarse grid). The solution is then iterated between the coarse and fine grids until the iteration error is small enough; about four iterations usually suffice. Because the solution on each grid does not need to be iterated to final tolerance each time, this method costs only a little more than the passive method.

In another kind of active method (Muzaferija 1994; Muzaferija and Gosman 1997) the grids are combined into a single global grid, including the refined grid as well as the non-refined part of the original grid. This requires a solution method that allows

Fig. 12.19 A non-refined
CV at the refinement
interface: it has six faces (c_1,
...,c_6) in common with six
neighbor CVs (N_1, ..., N_6)

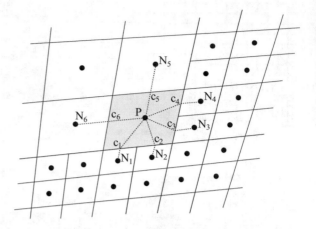

CVs with arbitrary numbers of faces. CVs at an interface between refined and non-refined regions have more faces and neighbors than regular CVs; see Fig. 12.19. For the global conservation property of the FV method to be retained, the face of a non-refined CV on the refinement boundary has to be treated as two (in 2D, and four in 3D) separate sub-faces, each common to two CVs. In the discretization, the sub-faces are treated exactly like any other face between two CVs. The computer code needs to have a data structure that can handle this situation and the solver needs to be able to handle the irregular matrix structure that results. Conjugate gradient type solvers are a good choice; multigrid solvers with Gauss-Seidel smoothers can also be used with some limitations. The data structure can be optimized by storing cell-face and cell-volume related values in separate arrays. For a simple discretization scheme like the one described in Chap. 9, this is easily done: each cell face is common to two CVs so, for each face, one needs to store pointers to the nodes of neighbor CVs, surface vector components, and the matrix coefficients. The computer code for solving the flow problem is then the same for locally refined and standard grids; only the pre-processor needs adaptation to enable it to handle the data for locally refined grids. This is akin to the treatment of non-conformal grid block interfaces, see Sect. 9.6.1.

If Chimera grids are used, the code needs no changes, but the interpolation coefficients and the nodes involved in interpolation need be redefined after each refinement.

As many levels of grid refinement as necessary can be used; usually at least three levels are required but as many as eight have been used. The advantage of the adaptive mesh method is that, because the finest grid occupies only a small part of the domain, the total number of grid points is relatively small so both the cost of computation and the memory requirements are reduced enormously. Furthermore, it can be designed so that the user need not be an expert grid designer. Especially for flows around bluff bodies such as cars, airplanes, and ships, in which very fine grids are needed near the body and in the wake but coarse grids can be used elsewhere, local grid refinement is essential for accurate and efficient simulation. An example of a user-defined cell-wise local grid refinement was presented in Sect. 10.3.4.2.

Finally, these methods combine very well with the multigrid method. The nested grids can be regarded as the ones used in multigrid; the only important difference is that, because the coarse grids provide enough accuracy where refinement was not necessary, the finest grids do not cover the entire domain. In the non-refined region, equations to be solved remain the same for both levels. Because the largest cost of the multigrid method is due to iterations on the finest grid, the savings can be very large, especially in 3D. For further details, see Thompson and Ferziger (1989) and Muzaferija (1994).

12.6 Parallel Computing in CFD

At the time when the first edition of this book appeared in 1996, workstations were single-processor computers. It was clear already then that further increases in computing power would require multiple processors, i.e., parallel computers. Nowadays (around 2020) all computers have multiple processing units (called cores); workstations usually have around 36 cores and 120 gigabytes of memory, with a trend to larger numbers. Multiple workstations can be connected to form clusters, which may have thousands of cores and terabytes of memory. The advantage of such clusters over classical vector supercomputers is scalability. They also use standard chips and are therefore cheaper to produce. However, algorithms designed for traditional serial processing may not run efficiently on parallel computers.

If parallelization is performed at the loop level (as is the case with auto-parallelizing compilers), Amdahl's law, which essentially says that the speed is determined by the least efficient part of the code, comes into play. To achieve high efficiency, the portion of the code that cannot be parallelized has to be very small.

A better approach is to subdivide the solution domain into sub-domains and assign each sub-domain to one processor. In this case the same code runs on all processors, on its own set of data. Because each processor also needs some data that resides in other sub-domains, exchange of data between processors and/or storage overlap is necessary.

Explicit schemes are relatively easy to parallelize, because all operations are performed on data from preceding time steps. It is only necessary to exchange the data at the interface regions between neighboring sub-domains after each step is completed. The sequence of operations and the results are identical on one and many processors. The most difficult part of the problem is usually the solution of the elliptic Poisson-like equation for the pressure (see, however, Sullivan and Patton 2008, who use a 2-D x-y plane decomposition to solve an incompressible flow problem for the planetary boundary layer).

Implicit methods are more difficult to parallelize. While calculation of the coefficient matrix and the source vector uses only 'old' data and can be efficiently performed in parallel, solution of the linear equation systems is not easy to parallelize. For example, Gauss elimination, in which each computation requires the result of the previous one, is very difficult to perform on parallel machines. Some other solvers can be parallelized and perform the same sequence of operations on n processors as on a single one, but they are either not efficient or the communication overhead is very large. We shall describe two examples.

12.6.1 Parallelization of Iterative Solvers for Linear Equation Systems

The *red-black Gauss-Seidel* method is well suited for parallel processing. It was briefly described in Sect. 5.3.8 and consists of performing Jacobi iterations on two sets of points in an alternating manner. In 2D, the nodes are colored as on a checkerboard; thus, for a five point computational molecule in 2D, Jacobi iteration applied to a red point calculates the new value using data only from black neighbor nodes, and vice versa. The convergence properties of this solver are exactly those of the Gauss-Seidel method, which gave the method its name.

Computation of new values on either set of nodes can be performed in parallel; all that is needed is the result of the previous step. The result is exactly the same as on a single processor. Communication between processors working on neighbor sub-domains takes place twice per iteration—after each set of data is updated. This local communication can be overlapped with computation of the new values. This solver is suitable only when used in conjunction with a multigrid method, as it is rather inefficient on its own.

ILU-type methods (e.g., the SIP-method presented in Sect. 5.3.4) are recursive, making parallelization less straightforward. In the SIP-algorithm, the elements of the L and U matrices, Eqs. (5.41), depend on the elements at the W and S nodes. One cannot start the calculation of the coefficients on a sub-domain, other than the one in the southwest corner, before data are obtained from its neighbors. In 2D, the best strategy is to subdivide the domain into vertical stripes i.e., use a 1D processor topology. Computation of L and U matrices and iteration can then be performed fairly efficiently in parallel (see Bastian and Horton 1989). The processor for sub-domain 1 needs no data from other processors and can start immediately; it proceeds along its bottom or southernmost line. After it has calculated the elements for the rightmost node, it can pass those values to the processor for sub-domain 2. While the first processor starts calculation on its next line, the second one can compute on its bottom line. All n processors are busy when the first one reaches the n-th line from bottom. When the first processor reaches the top boundary, it has to wait until the last processor, which is n lines behind, is finished; see Fig. 12.20. In the iteration scheme, two passes are needed. The first is done in the manner just described while the second is essentially its mirror image.

The algorithm is as follows:

```
for j = 2 to N_j - 1 do:
    receive U_E(i_s - 1, j), U_N(i_s - 1, j) from west neighbor;
    for i = i_s to i_e do:
```

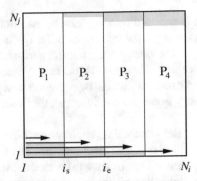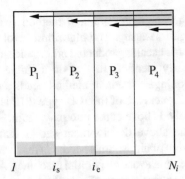

Fig. 12.20 Parallel processing in the SIP solver in the forward loop (left) and in the backward loop (right); shaded are regions of uneven load

```
        calculate U_E(i, j), L_W(i, j), U_N(i, j), L_S(i, j), L_P(i, j);
      end i;
      send U_E(i_e, j), U_N(i_e, j) to east neighbor;
    end j;

    for m = 1 to M do:
      for j = 2 to N_j - 1 do:
        receive R(i_s - 1, j) from west neighbor;
        for i = i_s to i_e do:
          calculate ρ(i, j), R(i, j);
        end i;
        send R(i_e, j) to east neighbor;
      end j;

      for j = N_j - 1 to 2 step -1 do:
        receive δ(i_e + 1, j) from east neighbor;
        for i = i_e to i_s step -1 do:
          calculate δ(i, j);
          update variable;
        end i;
        send δ(i_s, j) to west neighbor;
      end j;
    end m.
```

The problem is that this parallelization technique requires a lot of (fine grain) communication and there are idle times at the beginning and end of each iteration; these reduce the efficiency. Also, the approach is limited to structured grids. Bastian and Horton (1989) obtained good efficiency on transputer-based machines, which had a favorable ratio of communication to computation speed. With a less favorable ratio, the method would be less efficient.

The conjugate gradient method (without preconditioning) can be parallelized straightforwardly. The algorithm involves some global communication (gathering of partial scalar products and broadcasting of the final value), but the performance is nearly identical to that on a single processor. However, to be really efficient, the conjugate gradient method needs a good pre-conditioner. Because the best pre-conditioners are of the ILU-type (SIP is a very good pre-conditioner), the problems described above come into play again.

The above development shows that parallel computing environments require redesign of algorithms. Methods that are excellent on serial machines may be almost impossible to use on parallel machines. Also, new standards have to be used in assessing the effectiveness of a method. Good parallelization of implicit methods requires modification of the solution algorithm. The performance in terms of the number of numerical operations may be poorer than on a serial computer, but if the load carried by the processors is equalized and the communication overhead and computing time are properly matched, the modified method may be more efficient overall.

12.6.2 Domain Decomposition in Space

Parallelization of implicit methods is usually based on data parallelism or *domain decomposition*, which can be performed both in space and time. In spatial domain decomposition, the solution domain is divided into a certain number of sub-domains; this is similar to block-structuring of grids. In block-structuring the process is governed by the geometry of the solution domain, while, in domain decomposition, the objective is to maximize efficiency by giving each processor the same amount of work to do. Each sub-domain is assigned to one processor but more than one grid block may be handled by one processor; if so, we may consider all of them as one logical sub-domain.

As already noted, one has to modify the iteration procedure for parallel machines. The usual approach is to split the global coefficient matrix A into a system of diagonal blocks A_{ii}, which contain the elements connecting the nodes that belong to the ith sub-domain, and off-diagonal blocks or coupling matrices A_{ij} ($i \neq j$), which represent the interaction of blocks i and j. For example, if a square 2D solution domain is split into four sub-domains and the CVs are numbered so that the members of each sub-domain have consecutive indexes, the matrix has the structure shown in Fig. 12.21; a five-point molecule discretization is used in this illustration. The method described below is applicable to schemes using larger computational molecules; in this case, the coupling matrices are larger.

For efficiency, the iterative solver for the inner iterations should have as little data dependency (data provided by the neighbors) as possible; data dependency may result in long communication and/or idle times. Therefore, the global iteration matrix is selected so that the blocks are de-coupled, i.e., $M_{ij} = 0$ for $i \neq j$. The iteration scheme on sub-domain i is then:

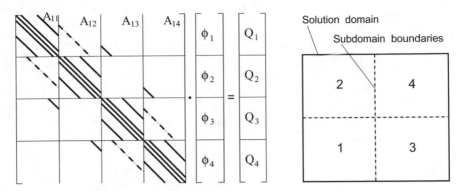

Fig. 12.21 Structure of the global coefficient matrix when a square 2D solution domain is subdivided into 4 sub-domains

$$M_{ii}\boldsymbol{\phi}_i^m = \mathbf{Q}_i^{m-1} - (A_{ii} - M_{ii})\boldsymbol{\phi}_i^{m-1} - \sum_j A_{ij}\boldsymbol{\phi}_j^{m-1} \quad (j \neq i) . \qquad (12.14)$$

Thus, data from neighbor sub-domains is taken from the previous inner iteration and treated as known; it is updated after each inner iteration. The SIP solver is easily adapted to this method. Each diagonal block matrix M_{ii} is decomposed into L and U matrices in the normal way; the global iteration matrix $M = LU$ is not the one found in the single processor case. After one iteration is performed on each sub-domain, one has to exchange the updated values of the unknown ϕ^m so that the residual ρ^m can be calculated at nodes near sub-domain boundaries.

When SIP solver is parallelized in this way, the performance deteriorates as the number of processors becomes large; the number of iterations may double when the number of processors is increased from one to 100. However, if the inner iterations do not have to be converged very tightly, as is the case for implicit schemes described in Chaps. 7 and 8, parallel SIP can be quite efficient because SIP tends to reduce the error rapidly in the first few iterations. Especially if the multigrid method is used to speed-up the outer iterations, the total efficiency is quite high (80–90%; see Schreck and Perić 1993; and Lilek et al. 1995, for examples). A 2D flow prediction code parallelized in this way is available via Internet; see appendix for details.

Conjugate gradient based methods can also be parallelized using the above approach. Below we present a pseudo-code for the preconditioned CG solver. It was found (Seidl et al. 1996) that the best performance is achieved by performing two pre-conditioner sweeps per CG iteration, on either single- or multi-processors. Results of solving a Poisson equation with Neumann boundary conditions, which simulates the pressure-correction equation in CFD applications, are shown in Fig. 12.22. With one pre-conditioner sweep per CG iteration, the number of iterations required for convergence increases with the number of processors. However, with two or more pre-conditioner sweeps per CG iteration, the number of iterations remains nearly constant. However, in different applications one may obtain different behavior.

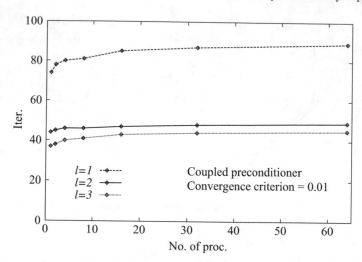

Fig. 12.22 Number of iterations in the ICCG solver as a function of the number of processors (uniform grid with 64^3 CVs, Poisson equation with Neumann boundary conditions, LC after each pre-conditioner sweep, l sweeps per CG iteration, residual norm reduced two orders of magnitude; from Seidl 1997)

- Initialize by setting: $k = 0$, $\boldsymbol{\phi}^0 = \boldsymbol{\phi}_{\text{in}}$, $\boldsymbol{\rho}^0 = Q - A\boldsymbol{\phi}_{\text{in}}$, $\mathbf{p}^0 = \mathbf{0}$, $s_0 = 10^{30}$
- Advance the counter: $k = k + 1$
- On each sub-domain, solve the system: $M\mathbf{z}^k = \boldsymbol{\rho}^{k-1}$
   LC: exchange $\mathbf{z}^k$ along interfaces
- Calculate: $s^k = \boldsymbol{\rho}^{k-1} \cdot \mathbf{z}^k$
   GC: gather and scatter s^k
   $\beta^k = s^k / s^{k-1}$
   $\mathbf{p}^k = \mathbf{z}^k + \beta^k \mathbf{p}^{k-1}$
   LC: exchange $\mathbf{p}^k$ along interfaces
   $\alpha^k = s_k / (\mathbf{p}^k \cdot A\mathbf{p}^k)$
   GC: gather and scatter α^k
   $\boldsymbol{\phi}^k = \boldsymbol{\phi}^{k-1} + \alpha^k \mathbf{p}^k$
   $\boldsymbol{\rho}^k = \boldsymbol{\rho}^{k-1} - \alpha^k A\mathbf{p}^k$
- Repeat until convergence.

To update the right-hand side of Eq. (12.14), data from neighbor blocks is necessary. In the example of Fig. 12.21, processor 1 needs data from processors 2 and 3. On parallel computers with shared memory, this data is directly accessible by the processor. When computers with distributed memory are used (which is true for most clusters), communication between processors is necessary. Each processor then needs to store data from one or more layers of cells on the other side of the interface. It is important to distinguish *local* (LC) and *global* (GC) communication.

Local communication takes place between processors operating on neighboring blocks. It can take place simultaneously between pairs of processors; an example is the communication within inner iterations in the problem considered above. GC means gathering of some information from all blocks in a 'master' processor and broadcasting some information back to the other processors. An example is the computation of the norm of the residual by gathering of the residuals from the processors and broadcasting the result of the convergence check. There are communication libraries which can be used for this purpose; nowadays, the Message-Passing Interface (MPI; see https://www.mpi-forum.org/docs/) is the de facto standard, but others like PVM (Sunderam 1990) or TCGMSG (Harrison 1991) have been used in the past and may still be available. This makes the codes portable—there is usually no need to adapt the CFD-code when using it on different parallel computers.

If the number of cells or computational points allocated to each processor (i.e., the load per processor) remains the same as the grid is refined (which means that more processors are used), the ratio of local communication time to computing time will remain the same. We say that LC is fully scalable. However, the GC-time increases when the number of processors increases, independent of the load per processor. The global communication time will eventually become larger than the computing time as the number of processors is increased. Therefore, GC is not scalable and is the limiting factor in massive parallelism. Methods of measuring the efficiency are discussed below.

12.6.3 Domain Decomposition in Time

Implicit methods are usually used for solving steady flow problems. Although one is tempted to think that these methods are not well suited to parallel computing, they can be effectively parallelized by using domain decomposition in time as well as in space. This means that several processors simultaneously perform work on the same sub-domain for different time steps. This technique was first proposed by Hackbusch (1984).

Because none of the equations needs to be solved accurately within an outer iteration, one can also treat the 'old' variables (i.e., those from previous time steps) in the discretized equation as unknowns. For a two-time-level scheme the equations for the solution at time step n can be written:

$$A^n \phi^n + B^n \phi^{n-1} = Q^n .$$ (12.15)

Because we are considering implicit schemes, the matrix and source vector may depend on the new solution, which is why they carry the index n. The simplest iterative scheme for solving simultaneously for several time steps is to de-couple the equations for each time step and use existing values of the variables (i.e., those from the previous outer iteration) where necessary. This allows one to start the calculation for the next time step as soon as the first estimate for the solution at the current time step is available, i.e., after one outer iteration is performed. The extra source term containing the information from the previous time step(s) is updated after each outer

Fig. 12.23 Structure of the
global coefficient matrix
when four time time steps
are calculated in parallel

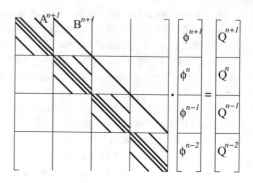

iteration, rather than being held constant as in serial processing. When the processor
k, working on time level t_n, is performing its mth outer iteration, the processor $k - 1$,
working on time level t_{n-1}, is performing its $(m + 1)$th outer iteration. The equation
system to be solved by processor k in the mth outer iteration is then:

$$(A^n \boldsymbol{\phi}^n)_k^m = (\mathbf{Q}^n)_k^{m-1} - (B^n \boldsymbol{\phi}^{n-1})_{k-1}^m . \tag{12.16}$$

The processors need to exchange data only once per outer iteration, i.e., the linear
equation solver is not affected. Of course, much more data is transferred each time
than in the method based on domain decomposition in space. If the number of time
steps treated in parallel is not larger than the number of outer iterations per time step,
using the lagged old values does not cause a significant increase in computational
effort per time step. If too many time steps are computed in parallel, the number of
required outer iterations per time step will grow and the efficiency will diminish.
On the last time step in a parallel sequence, the term $(B^n \boldsymbol{\phi}^{n-1})_{k-1}$ is included in the
source term, as it does not change within iterations.

Figure 12.23 shows the structure of the matrix for a two-time-level scheme with
simultaneous solution on four time steps. Time-parallel solution methods for CFD
problems have been used by Burmeister and Horton (1991), Horton (1991), and Seidl
et al. (1996), among others. The method can also be applied to multilevel schemes;
in that case the processors have to send and receive data from more than one time
level.

12.6.4 Efficiency of Parallel Computing

The analysis of the performance of parallel programs is usually measured by the
speed-up factor and *efficiency* defined by:

$$S_n = \frac{T_s}{T_n} , \quad E_n^{tot} = \frac{T_s}{n \, T_n} . \tag{12.17}$$

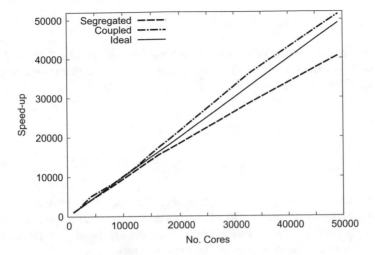

Fig. 12.24 Speed-up as a function of the number of used processor units (cores) in a computation of turbulent flow around a Le Mans racing car, using segregated and coupled solution algorithms in the Simcenter STAR-CCM+ commercial code and a grid consisting of 1.02 billion cells

Here T_s is the execution time for the best serial algorithm on a single processor and T_n is the execution time for the parallelized algorithm using n processors. In general $T_s \neq T_1$, as the best serial algorithm may be different from the best parallel algorithm; one should not base the efficiency on the performance of the parallel algorithm executed on a single processor.

The speed-up is usually less than n (the ideal value), so the efficiency is usually less than 1 (or 100%). However, when solving coupled non-linear equations, it may turn out that solution on two or four processors is more efficient than on 1 processor so, in principle, efficiencies higher than 100% are possible (the increase is often due to the better use of cash memory when a smaller problem is solved by one processor). An example is shown in Fig. 12.24.

Although not necessary, the processors are usually synchronized at the start of each iteration. Because the duration of one iteration is dictated by the processor with the largest number of CVs, other processors experience some idle time. Delays may also be due to different boundary conditions in different sub-domains, different numbers of neighbors, or more complicated communication.

The computing time T_s may be expressed as:

$$T_s = N^{cv} \tau i_s \,, \tag{12.18}$$

where N^{cv} is the total number of CVs, τ is the time per floating point operation and i_s is the number of floating point operations per CV required to reach convergence. For a parallel algorithm executed on n processors, the total execution time consists of computing and communication time:

$$T_n = T_n^{\text{calc}} + T_n^{\text{com}} = N_n^{\text{cv}} \tau i_n + T_n^{\text{com}} , \tag{12.19}$$

where N_n^{cv} is the number of CVs in the largest sub-domain and T_n^{com} is the total communication time during which calculation cannot take place. Inserting these expressions into the definition of the total efficiency yields:

$$E_n^{\text{tot}} = \frac{T_s}{n\,T_n} = \frac{N^{\text{cv}} \tau i_s}{n\,(N_n^{\text{cv}} \tau i_n + T_n^{\text{com}})} =$$
$$\frac{i_s}{i_n} \frac{1}{1 + T_n^{\text{com}} / T_n^{\text{calc}}} \frac{N^{\text{cv}}}{n\,N_n^{\text{cv}}} = E_n^{\text{num}} E_n^{\text{par}} E_n^{\text{lb}} . \tag{12.20}$$

This equation is not exact, because the number of floating point operations per CV is not constant (due to branching in the algorithm and the fact that boundary conditions affect only some CVs). However, it is adequate to identify the major factors affecting the total efficiency. The meanings of these factors are:

- E_n^{num}—The *numerical efficiency* accounts for the effect of the change in the number of operations per grid node required to reach convergence due to modification of the algorithm to allow parallelization;
- E_n^{par}—The *parallel efficiency* accounts for the time spent on communication during which computation cannot take place;
- E_n^{lb}—The *load balancing efficiency* accounts for the effect of some processors being idle due to uneven load.

When the parallelization is performed in both time and space, the overall efficiency is equal to the product of time and space efficiencies.

The total efficiency is easily determined by measuring the computing time necessary to obtain the converged solution. The parallel efficiency cannot be measured directly, because the number of inner iterations is not the same for all outer iterations (unless it is fixed by the user). However, if we execute a certain number of outer iterations with a fixed number of inner iterations per outer iteration on 1 and n processors, the numerical efficiency is unity and the total efficiency is then the product of the parallel and load balancing efficiencies. If the load balancing efficiency is reduced to unity by making all sub-domains equal, we obtain the parallel efficiency. Some computers have tools which allow operation counts to be performed; then the numerical efficiency can be directly measured.

For both space and time domain decomposition, all three efficiencies are usually reduced as the number of processors is increased for a given grid. This decrease is both non-linear and problem-dependent. The parallel efficiency is especially affected, because the time for LC is almost constant, the time for GC increases, and the computing time per processor decreases due to reduced sub-domain size. For time parallelization, the time for GC increases while the LC and computing times remain the same when more time steps are calculated in parallel for the same problem size. However, the numerical efficiency will decrease disproportionately if the number of processors is increased beyond a certain limit (which depends on the required number of outer iterations per time step). Optimization of the load balancing is difficult in general, especially if the grid is unstructured and local refinement is employed. There are algorithms for optimization, but they may take more time than the flow computation!

Parallel efficiency can be expressed as a function of three main parameters:

- set-up time for data transfer (called *latency time*);
- data transfer rate (usually expressed in Mbytes/s);
- computing time per floating point operation (usually expressed in Mflops).

For a given algorithm and communication pattern, one can create a model equation to express the parallel efficiency as a function of these parameters and the domain topology. Schreck and Perić (1993) presented such a model and showed that the parallel efficiency can be fairly well predicted. One can also model the numerical efficiency as a function of alternatives in the solution algorithm, the choice of solver and the coupling of the sub-domains. However, empirical input based on experience with similar flow problems is necessary, because the behavior of the algorithm is problem-dependent. These models are useful if the solution algorithm allows alternative communication patterns; one can choose the one most suitable for the computer used. For example, one can exchange data after each inner iteration, after every second inner iteration, or only after each outer iteration. One can employ one, two, or more pre-conditioner iterations per conjugate gradient iteration; the pre-conditioner iterations may include local communication after each step or only at the end. These options affect both the numerical and parallel efficiency; a trade-off is necessary to find an optimum.

The frequency of communication is obviously an important factor in parallel computing. Coupled solution methods (which treat velocities, pressure and temperature as a single vector of unknowns) perform five times less data exchanges (LC) than segregated methods, which solve for each variable in turn. In addition, coupled solution methods usually need to perform substantially more computing operations per iteration than segregated solvers. Although the amount of exchanged data is five times larger, the saved latency times and the more favorable ratio of communication to computing time leads to coupled solvers having in general higher efficiency than segregated solvers. It also happens more frequently that a coupled solver shows super-linear performance, as seen in Fig. 12.24. In this example, because the total number of cells was over one billion, computation could not be performed using a single core; the lowest number of processors used was 1024 (about 1 million cells per core). However, calculation on 48 times as many cores (with about 20,000 cells per core) was more than 50 times faster using the coupled solver and about 40 times faster using the segregated solver.

Combined space and time parallelization is more efficient than pure spatial parallelization because, for a given problem size, the efficiency goes down as the number of processors increases. Table 12.2 shows results of the computation of the unsteady 3D flow in a cubic cavity with an oscillating lid at a maximum Reynolds number of 10^4, using a $32 \times 32 \times 32$ CV grid and a time step of $\Delta t = T/200$, where T is the period of the lid oscillation. When sixteen processors were used with four time steps

Table 12.2 Numerical efficiency for various domain decompositions in space and time for the calculation of 3D cavity flow with an oscillating lid

Decomposition in space and time $x \times y \times z \times t$	Mean number of outer iterations per time step	Mean number of inner iterations per time step	Numerical efficiency (in %)
$1 \times 1 \times 1 \times 1$	11.3	359	100
$1 \times 2 \times 2 \times 1$	11.6	417	90
$1 \times 4 \times 4 \times 1$	11.3	635	68
$1 \times 1 \times 1 \times 2$	11.3	348	102
$1 \times 1 \times 1 \times 4$	11.5	333	104
$1 \times 1 \times 1 \times 8$	14.8	332	93
$1 \times 1 \times 1 \times 12$	21.2	341	76
$1 \times 2 \times 2 \times 4$	11.5	373	97

calculated in parallel and the space domain decomposed into four sub-domains, the total numerical efficiency was 97%. If all processors are used solely for spatial or temporal decomposition, the numerical efficiency drops below 70%.

Communication between processors halts computation on many machines. However, if the communication and computation could take place simultaneously (which is possible on some new parallel computers), many parts of the solution algorithm could be rearranged to take advantage of this. For example, while LC takes place in the solver, one can do computation in the interior of the sub-domain, where data from neighbor sub-domains is not required. With time parallelism, one can assemble the new coefficient and source matrices while LC is taking place. Even the GC in a conjugate gradient solver, which appears to hinder execution, can be overlapped with computation if the algorithm is rearranged as suggested by Demmel et al. (1993). Convergence checking in steady-state computations can be skipped in the early stage or the convergence criterion can be rearranged to monitor the residual level at previous iterations—one can decide to stop iterating according to projection based on residual reduction rate.

Perić and Schreck (1995) analyzed the possibilities of overlapping communication and computation in more detail and found that it can significantly improve parallel efficiency. New hard- and software are likely to allow concurrency of computation and communication, so one can expect that parallel efficiency can be optimized. One of the main concerns for the developers of parallel implicit CFD algorithms is numerical efficiency. It is essential that the parallel algorithm not need many more computing operations than the serial algorithm for the same accuracy.

Results show that parallel computing can be efficiently used in CFD. The use of workstation clusters is especially useful with this respect, as they are available to almost all users and big problems are not solved all the time. All computers (PCs, workstations and mainframes) are nowadays multiprocessor machines; it is therefore essential to have parallel processing in mind when developing new solution methods.

12.6.5 Graphics Processing Units (GPUs) and Parallel Processing

Graphics processors (GPUs) are designed for interactive gaming, but have proven to be effective in solution of fluid flow problems. Because of their different design a separate language (Compute Unified Device Architecture—CUDA) has been developed and a literature has grown in this domain. While GPUs offer impressive speed when used as single processors ($\sim 20\times$ over a CPU) in a workstation, for example, much greater acceleration of computations has come in multi-GPU workstations with hundreds of cores.

To date many applications are from projects which use local software packages to link the GPU framework; however, more general software exists. Here, we only seek to alert readers to this opportunity and to provide some references. Khajeh-Saeed and Perot (2013) did DNS of turbulence using GPU-accelerated supercomputers, demonstrating how the GPU, the MPI and the supercomputer can be linked and optimized. We note that the basic algorithms of the numerical method are not changed, much of the effort going then to the coding in CUDA and the linkage and optimization work. Schalkwijk et al. (2012a) present a meteorological example of simulation of turbulent clouds on a desktop PC with a linked GPU. Their sidebar "Porting to the GPU" is particularly instructive. Often applications take a slow part of a current code and implement a GPU solver for that portion alone and thereby achieve significant speed ups (e.g., Williams et al. 2016, who did a GPU version of a sparse linear matrix pressure solver for unstructured grids). Some commercial codes can also link to the GPU to speed up appropriate parts of the computations.

Performing computations of complex flows (involving modeling of turbulence, combustion, multiphase flows, etc.) on adaptive unstructured grids using both CPUs and GPUs is a challenge, but there is no way around it if high efficiency is to be obtained. None of the approaches developed for simpler flow problems and structured grids is alone optimal for complex problems—one will have to develop smarter parallelization concepts that dynamically adapt to the problem at hand. Both the solution algorithm and the data structure may need to be re-organized to facilitate an efficient use of both CPUs and GPUs in such applications. For example, some loops over faces or volumes may need to be broken-up into multiple loops and some data may need to be stored twice (and copied between particular steps) to minimize data dependency and allow concurrency of computation and communication. It is expected that a significant research effort will be put into this area in the coming decade.

Chapter 13
Special Topics

13.1 Introduction

Fluid flows may include a broad range of additional physical phenomena that take the subject far beyond the single-phase non-reacting flows that have been the focus of this work up to this point. Many types of physical processes may occur in flowing fluids. Each of these may interact with the flow to produce an amazing range of new phenomena. Almost all of these processes occur in important applications. Computational methods have been applied to them with varying degrees of success.

The simplest element that can be added to a flow is a scalar quantity such as the concentration of a soluble chemical species or temperature. The case in which the presence of the scalar quantity does not affect the properties of the fluid has already been treated in earlier chapters; in such a case, we speak of a passive scalar. In a more complex case, the density and viscosity of the fluid may be modified by the presence of the scalar and we have an active scalar. In a simple example, the fluid properties are functions of temperature or the concentration of the species. This field is known as heat and mass transfer.

In other cases, the presence of a dissolved scalar or the physical nature of the fluid itself causes the fluid to behave in way that the stress is not related to the strain rate by the simple Newtonian relationship (Eq. 1.9). In some fluids, the viscosity becomes a function of the instantaneous strain rate and we speak of shear-thinning or shear-thickening non-Newtonian fluids. In more complex fluids, the stress is determined by an additional set of non-linear partial differential equations. We then say that the fluid is viscoelastic. Many polymeric materials, including biological ones, exhibit this kind of behavior, giving rise to unexpected flow phenomena. This is the field of non-Newtonian fluid mechanics.

Flows may contain various kinds of interfaces. These may be due to the presence of a solid body in the fluid. In simple cases of this kind, it is possible to transform to a coordinate system moving with the body and the problem is reduced to one of the kind treated earlier, albeit in a complex geometry. In other problems, there may be bodies that move with respect to each other and there is no choice but to introduce

© Springer Nature Switzerland AG 2020
J. H. Ferziger et al., *Computational Methods for Fluid Dynamics*,
https://doi.org/10.1007/978-3-319-99693-6_13

a moving coordinate system. A particularly important and difficult case of this kind
is one in which the surface is deformable. Surfaces of bodies of liquid are examples
of this type.

In still other flows, multiple phases may coexist. All of the possible combinations
are of importance. The solid-gas case includes such phenomena as dust in the atmosphere, fluidized beds, and gas flow through a porous medium. In the solid-liquid
category are slurries (in which the liquid is the continuous phase), again, porous
medium flows. Gas-liquid flows include sprays (in which the gas phase is continuous) and bubbly flows (in which the reverse is true). Finally, there may be three-phase
flows. Each of these cases has many sub-categories.

Chemical reaction may take place in flows and again, there are many individual cases. When the reacting species are dilute, the reaction rates may be assumed
constant (they may, however, depend on temperature) and the reacting species are
essentially passive scalars with respect to their effect on the flow. Examples of this
kind are pollutant species in the atmosphere or the ocean. Another kind of reaction
involves major species and releases a large amount of energy. This is the case of
combustion. Still another example is that of airflow at high speeds; compressibility
effects may lead to large temperature increases and the possibility of dissociation or
ionization of the gas.

Geophysics and astrophysics also require the solution of the equations of fluid
motion. Other than plasma effects (discussed below), the new elements in these
flows are the enormous scales compared to engineering flows. In meteorology and
oceanography, rotation and stratification have a great influence on the flow behavior.

Finally, we mention that in plasmas (ionized fluids), electromagnetic effects play
an important role. In this field, the equations of fluid motion have to be solved along
with the equations of electro-magnetism (the Maxwell equations) and the number of
phenomena and special cases is enormous.

In the remainder of this chapter, we shall describe methods for dealing with
some, but not all, of these difficulties. We should point out that each of the topics
mentioned above is an important sub-specialty of fluid mechanics and has a large
literature devoted to it; references to textbooks in each area are given below. It is
impossible to do justice to each of these topics in the space available here.

13.2 Heat and Mass Transfer

Of the three mechanisms of heat transfer—conduction, radiation, and convection—
usually presented in courses on the subject, the last is most closely connected with
fluid mechanics. The link is so strong that convective heat transfer may be regarded
as a sub-area of fluid mechanics.

Steady heat conduction is described by Laplace's equation (or equations very similar to it) while unsteady conduction is governed by the heat equation; these equations
are readily solved by methods presented in Chaps. 3, 4 and 6. A complication arises
when the properties are temperature-dependent. In such a case, the properties are

calculated using the temperature in the current iteration, the temperature is updated, and the process is repeated. Convergence is usually nearly as rapid as in the fixed-property case.

Radiation involving solid surfaces has little connection with fluid mechanics (except in problems with multiple active mechanisms of heat transfer). There are interesting problems (for example, flows in rocket nozzles and combustors) in which both fluid mechanics and radiative heat transfer in the gas are important. The combination also occurs in astrophysical applications and in meteorology. We shall not deal with this type of problem here; however, see Sect. 13.7.

In laminar convective heat transfer, the dominant processes are *advection* (which we previously called convection!) in the streamwise direction and conduction in the direction normal to the flow. When the flow is turbulent, much of the role played by conduction in laminar flows is taken over by the turbulence and is represented by a turbulence model; these models are discussed in Chap. 10. In either case, interest generally centers on exchange of thermal energy with solid surfaces.

If the temperature differences are small (less than 5 K in water or 10 K in air) and the Reynolds number is high, the variations of the fluid properties are not important and the temperature behaves as a passive scalar. Problems of this sort can be treated by methods described earlier in this book. Because the temperature is a passive scalar in this case, it can be computed after the computation of the velocity field has been completely converged, making the task much simpler. In the case in which the flow is driven by density differences, the latter must be taken into consideration. This can be done with the aid of the Boussinesq approximation described below.

Another important special case is that of heat transfer occurring in flows past bodies of smooth shape. In flows of this type, one can first compute the potential flow around the body and then use the pressure distribution obtained as input to a boundary layer code for the prediction of heat transfer. If the boundary layer does not separate from the body, it is possible to compute these flows using the boundary-layer simplification of the Navier–Stokes equations (see, for example, Kays and Crawford 1978; or Cebeci and Bradshaw 1984). The boundary-layer equations are parabolic and can be solved in a matter of seconds (for the 2D case) or a minute or so (for the 3D case) on a modern workstation or personal computer. Methods for computing these flows have not been covered in detail in this work (but the general principles are found in Chaps. 3–8); the interested reader can find them in the works by Cebeci and Bradshaw (1984) and Patankar and Spalding (1977).

In the general case, temperature variations are significant. They affect the flow in two ways. The first is through the variation of the transport properties with temperature. These can be very large and must be taken into account but are not difficult to handle numerically. The important issue is that the energy and momentum equations are now coupled and must be solved simultaneously. Fortunately, the coupling is not usually so strong as to prevent solution of the equations in sequential fashion. On each outer iteration, the momentum equations are first solved using transport properties computed from the 'old' temperature field. The temperature field is updated after the solution of the momentum equations has been obtained for the new outer iteration and the properties are updated. This technique is very similar to the one for solving the momentum equations with a turbulence model described in Chap. 10.

Another effect of temperature variation is that density variation, interacting with gravity, produces a body force that may modify the flow considerably and may be the principal driving force in the flow. In the latter case, we talk of *buoyancy-driven* or *natural convection* flow. The relative importance of *forced* convection and buoyancy effects is measured by the ratio of the Grashof and Reynolds numbers. The former was defined in Eq. (1.33); if the ratio

$$\frac{\text{Gr}}{\text{Re}^2} = \frac{\text{Ra}}{\text{Pr}\,\text{Re}^2} \ll 1 \,,$$

the effects of natural convection may be ignored. In purely buoyancy-driven flows, if the density variations are small enough, it may be possible to ignore the density variations in all terms other than the body force in the vertical momentum equation. This is called the Boussinesq approximation and it allows the equations to be solved by methods that are essentially identical to those used for incompressible flow. An example was presented in Sect. 8.4.

Computation of flows in which buoyancy is important is usually made by methods of the type described above, i.e., iteration of the velocity field precedes iteration for the temperature and density fields. Because the coupling between the fields may be quite strong, this procedure may converge more slowly than in isothermal flows. Solution of the equations as a coupled system increases the convergence rate at the cost of increased complexity of programming and storage requirements; see Galpin and Raithby (1986) for an example. The strength of the coupling also depends on the Prandtl number. It is stronger for fluids with high Prandtl numbers; for these fluids, the coupled solution approach yields much faster convergence than the sequential approach. This statement, however, applies only to computations of steady-state flows. At high Rayleigh-numbers buoyancy-driven flows become unsteady and eventually turbulent even if boundary conditions are steady; unsteadiness may also be caused by time-varying boundary conditions. When time-accurate solutions are required, the time step must be adequately small; in that case, the solution does not change much from one time step to another and segregated solution methods (like the one described above) may actually be computationally more efficient than the coupled solver. This is because an implicit solution method will need only few outer iterations per time step; a coupled solver requires more computing time per iteration and is only beneficial if it can significantly reduce the required number of iterations, which is usually true only for steady or weakly transient flows. With sufficiently small time steps, even non-iterative time-advancing methods, like the fractional-step method described in Chap. 7, can be used; see Armfield and Street (2005) for an example.

In some applications, heat conduction in a solid needs to be considered along with convection in an adjacent fluid. Problems of this kind are called *conjugate heat transfer* problems and need to be solved by iterating between the equations describing the two types of heat transfer. It is also possible to solve the energy equation across

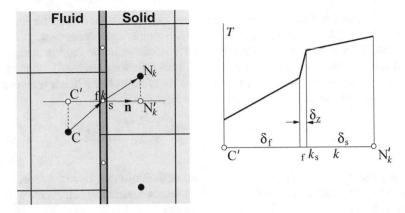

Fig. 13.1 On the discretization of diffusion flux at a solid-fluid interface with a coating layer, assuming non-conformal grids at the interface

both fluid and solid domain simultaneously; we describe below discretization steps which require special attention and also account for coatings or baffles which are too thin to be resolved by the grid but which may represent a significant resistance to heat transfer.

Solving the energy equation in fluid and solid separately and updating boundary conditions for each continuum once per outer iteration is usually not efficient; it is desirable to solve the energy equation simultaneously in all continua. This requires that the heat flux at solid-fluid interfaces is expressed as a function of temperatures in adjacent cell-centers in fluid and solid, without relying on the knowledge of the temperature at the interface itself. We describe below how this can be achieved. Consider the situation shown in Fig. 13.1. If one assumes that on the fluid side the boundary layer is resolved, the temperature profile along a line normal to the cell face k would look like shown on the right-hand side of Fig. 13.1. For the sake of generality, it is assumed that the grids in solid and fluid do not match at the interface; the mesh is drawn as Cartesian to point out that even then the situation requires a special treatment; the same approach applies if grids are not orthogonal to the interface.

The heat flux per unit area through the face k common to cells centered around nodes C and N_k, see Fig. 13.1, can be expressed as:

$$q_k = \lambda_f \left(\frac{\partial T}{\partial n}\right)_{k_f} = \lambda_z \left(\frac{\partial T}{\partial n}\right)\bigg|_{k_f}^{k_s} = \lambda_s \left(\frac{\partial T}{\partial n}\right)_{k_s}, \qquad (13.1)$$

where T is the temperature, λ is the heat conductivity, n is the direction of coordinate normal to cell face, and indexes "f", "z", and "s" denote fluid side, coating, and solid side of the interface, respectively (see Fig. 13.1). The coating is usually very thin so the conductivity λ_z can be assumed constant, as well as the temperature gradient across the layer. For fluid and solid, heat conductivities should be taken for interface temperatures T_{k_f} and T_{k_s}, respectively, if they are temperature-dependent.

Immediately next to a wall, the temperature variation in the fluid is always linear; if the viscous layer is not resolved (i.e., wall functions are used), an effective conductivity has to be used in approximations that follow. For a linear temperature distribution, the discrete approximation of the heat flux is:

$$q_k = \lambda_f \frac{T_{k_f} - T_{C'}}{\delta_f} = \lambda_z \frac{T_{k_s} - T_{k_f}}{\delta_z} = \lambda_s \frac{T_{N'_k} - T_{k_s}}{\delta_s} . \tag{13.2}$$

The points C' and N'_k lie on the face normal n; their distance from the face, δ_f and δ_s, respectively, represent the projection of the vector connecting cell center with the face center, i.e.:

$$\delta_f = (\mathbf{r}_k - \mathbf{r}_C) \cdot \mathbf{n}_k \quad \text{and} \quad \delta_s = (\mathbf{r}_{N_k} - \mathbf{r}_k) \cdot \mathbf{n}_k . \tag{13.3}$$

By introducing resistance coefficients α as follows:

$$\alpha_f = \frac{\lambda_f}{\delta_f} , \quad \alpha_z = \frac{\lambda_z}{\delta_z} , \quad \alpha_s = \frac{\lambda_s}{\delta_s} , \tag{13.4}$$

the expression (13.2) can be re-written as:

$$q_k = \alpha_f(T_{k_f} - T_{C'}) = \alpha_z(T_{k_s} - T_{k_f}) = \alpha_s(T_{N'_k} - T_{k_s}) . \tag{13.5}$$

From this expression interface temperatures T_{k_f} and T_{k_s} can be eliminated to yield:

$$q_k = \frac{T_{N'_k} - T_{C'}}{\dfrac{1}{\alpha_f} + \dfrac{1}{\alpha_z} + \dfrac{1}{\alpha_s}} . \tag{13.6}$$

By introducing an effective resistance coefficient and an effective conductivity:

$$\alpha_{\text{eff}} = \frac{1}{\dfrac{1}{\alpha_f} + \dfrac{1}{\alpha_z} + \dfrac{1}{\alpha_s}} = \frac{\lambda_{\text{eff}}}{\delta_f + \delta_z + \delta_s} , \tag{13.7}$$

the heat flux can be conveniently expressed as:

$$q_k = \alpha_{\text{eff}}(T_{N'_k} - T_{C'}) = \lambda_{\text{eff}} \frac{T_{N'_k} - T_{C'}}{\delta_f + \delta_z + \delta_s} . \tag{13.8}$$

If the grid is non-orthogonal or non-conformal, temperatures at auxiliary nodes C' and N'_k have to be expressed through nodal values by a suitable interpolation. The following approximation is consistent with other approximations used in the discretization process and is second-order accurate:

$$T_{C'} = T_C + (\nabla T)_C \cdot (\mathbf{r}_{C'} - \mathbf{r}_C) \quad \text{and} \quad T_{N'_k} = T_{N_k} + (\nabla T)_{N_k} \cdot (\mathbf{r}_{N'_k} - \mathbf{r}_{N_k}) . \quad (13.9)$$

The coordinates of auxiliary nodes C' and N'_k are easily obtained as (see Eq. (13.3) and Fig. 13.1):

$$\mathbf{r}_{C'} = \mathbf{r}_k - \delta_f \mathbf{n}_k \quad \text{and} \quad \mathbf{r}_{N'_k} = \mathbf{r}_k + \delta_s \mathbf{n}_k . \quad (13.10)$$

Finally, the heat flux per unit area can be expressed as (see Eq. (13.8)):

$$q_k = \alpha_{\text{eff}}(T_{N_k} - T_C) + \underline{\alpha_{\text{eff}} \left[(\nabla T)_{N_k} \cdot (\mathbf{r}_{C'} - \mathbf{r}_C) - (\nabla T)_C \cdot (\mathbf{r}_{N'_k} - \mathbf{r}_{N_k}) \right]} .$$
$$(13.11)$$

In order to obtain the total heat flux through the cell face, one needs to multiply q_k by the face area, S_k. The underlined term can be treated as a deferred correction, i.e., computed using values from previous iteration. The first term on the right-hand side is treated implicitly, i.e., it contributes to the coefficients of the matrix equation.

The underlined term is small compared to the principal term if the distance between cell centers and auxiliary nodes is small compared to the distance to cell face, i.e., when the non-orthogonality is moderate. It vanishes when the face normal through face center passes through cell centers. In the case of severe non-orthogonality, problems of two kinds can result: (i) unphysical solutions (over- or undershoots) or (ii) convergence problems. In such a case one could limit the underlined term (or even set it equal to zero as the last resort); this reduces the order of approximation but can help to avoid problems if grid quality cannot be improved. For example, by neglecting the underlined term altogether, one effectively sets $T_{C'} = T_C$, which corresponds to assuming constant temperature within the cell (a first-order approximation).

Note that Eq. (13.8) resembles the expression by which the heat transfer from a wall to environment is often described,

$$q_{\text{wall}} = \alpha(T_{\text{wall}} - T_\infty) , \quad (13.12)$$

where q_{wall} is the heat flux at wall, T_{wall} is wall temperature, T_∞ is the temperature of the environment and α is the so-called *heat transfer coefficient*; experimental data is often presented in this form. Many users of CFD-codes want to compute and visualize the heat transfer coefficient at walls in internal flows, e.g., on turbine blades or hot engine surfaces. This is tricky because the heat transfer coefficient is not a uniquely defined quantity—only wall heat flux is unique (and this is the quantity one should be looking at). While it is clear that T_{wall} should be the local wall temperature, it is not clear which reference temperature for the environment should be used in internal flows. Thus, α and T_∞ must always come in pairs. In some commercial codes, several different versions of heat transfer coefficient can be extracted from simulation results; some change considerably when the grid is refined and one therefore has to be careful when comparing solutions.

Fig. 13.2 Geometry of the test case showing all solid and fluid regions in which the energy equation is solved simultaneously

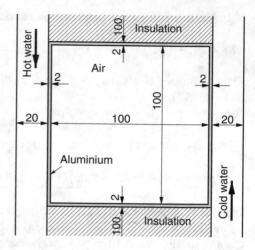

We now present an example in which natural convection of air in an enclosure is of main interest. We discussed a similar problem in Sect. 8.4.1, but there we only solved the energy equation in the fluid domain and prescribed the temperature at side walls while using adiabatic conditions at top and bottom boundaries. Here, we are taking into account the solid structure around the cavity: the insulation above the top and below the bottom walls, as well as the hot and cold water flowing in side channels with the aim of keeping nearly constant wall temperatures, as shown in Fig. 13.2. In an experiment, it is difficult to achieve both constant wall temperature and zero heat flux (corresponding to the adiabatic boundary condition); such boundary conditions are always an approximation and the error is often unknown. Whenever possible, one should try to specify approximate boundary conditions as far from the region of interest as possible.

In this 2D example, air occupies a cavity with dimensions 100×100 mm; it is enclosed by 2 mm thick aluminum walls. The left wall is heated by hot water flowing downward with a mean velocity of 10 m/s in a channel 20 mm wide, with a temperature of 310 K. On the right-hand side, cold water at 300 K is flowing upward with the mean velocity of 10 m/s, in a channel 20 mm wide. The Rayleigh number in the cavity filled with air is around 1.5×10^6, meaning that the air flow is laminar. The Reynolds number in the water channels is 200,000, meaning that the water flow is turbulent; it is simulated using a k-ε turbulence model. It is assumed that water loses no heat to the outer side walls; this is an approximation but because the flow rate is rather high, the heat passed on to the air cavity will not be affected even if some heat escapes through the other wall. It is also assumed that no heat is lost at the top and bottom boundaries of the two insulation layers; because they are 100 mm thick and the heat conductivity of the material is very low (0.036 W/mK, compared to 237 W/mK for the aluminum frame), the effect of this approximation is also expected to be small. There are no boundary conditions imposed on the solution domain for the air flow—the temperatures driving the flow are part of the conjugate heat transfer solution; only the velocity and temperature at the two water inlets are imposed.

Fig. 13.3 Flow pattern in
fluid domains (upper) and
temperature contours in air,
aluminum frame and
insulation (lower)

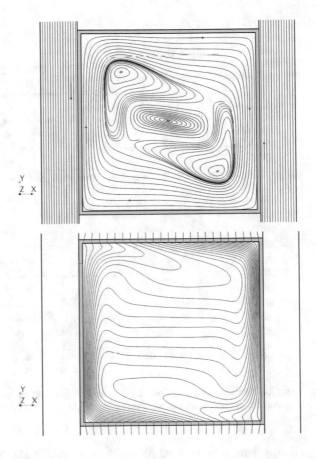

For the sake of simplicity of grid generation, a uniform grid across all regions
was used, with a spacing of 0.5 mm. This is just about adequate near walls in fluid
zones for demonstration purposes, but the grid is too fine in the insulation regions.
This will be obvious from the results presented below. The iterative solution method
follows the steps outlined before for the computation of steady-state flows: on each
outer iteration, the temperature is updated first by solving the energy equation across
all regions, using the velocity fields in fluid regions from the previous outer iteration.
Then, one step of the SIMPLE algorithm (i.e., solving the momentum equations and
the pressure-correction equation in turn) is performed in each fluid region, using the
updated temperature to determine fluid properties of air. In the forced flow of water,
properties were held constant because the temperature varies little, while air was
treated as an ideal gas. The Boussinesq-approximation was not used: density was
updated at each outer iteration in all terms using the equation of state, while viscosity variation with temperature was accounted for by using a polynomial expression.
Under-relaxation parameters were 0.8 for velocities, k and ε, 0.3 for pressure and
0.99 for temperature. About 500 outer iterations were performed; residual levels for

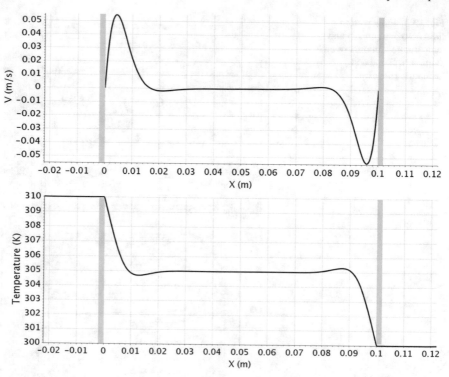

Fig. 13.4 Profiles along a horizontal cut at mid-height: vertical velocity component in air (upper) and temperature across all materials (lower); gray bars denote cavity walls

all equations were reduced by 6 orders of magnitude. The linear upwind scheme (2nd-order; see Sect. 4.4.5) was used for convection and central differences for diffusion terms. A detailed report with simulation file is available for download; see the appendix for details.

Figure 13.3 shows flow patterns in fluid domains and temperature contours. Air rises up the hot wall, turns then toward the cold wall where it gives away the heat collected from the hot wall, and comes back along the bottom wall. In the central part of the cavity, the air is almost stagnant, as can be seen from velocity profile shown in Fig. 13.4. Isotherms are also almost horizontal in the central part of the cavity (stably-stratified zone). Isotherms in the air approach the aluminum frame separating air from insulation at sharp angles, indicating a significant heat transfer between air and metal frame; on the other side, between metal and insulation, isotherms are almost orthogonal to the interface, indicating very low heat exchange between these two materials. The heat exchanged between air and aluminum at the upper and lower interface is primarily conducted along the frame. In insulation, the temperature varies—as expected—linearly from hot to cold wall; irregular variation is only visible near cavity corners. Both velocity and temperature have high gradients near hot and cold walls, requiring a fine grid in the wall-normal direction.

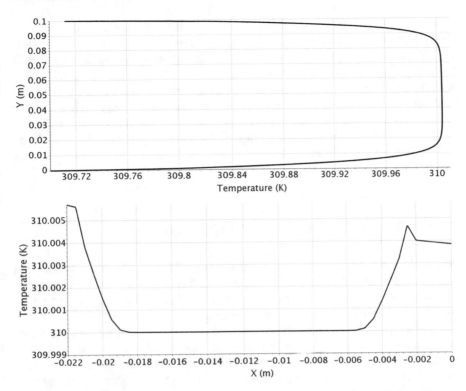

Fig. 13.5 Temperature profile along the hot wall on air side (upper) and along a horizontal cut at mid-height in the hot water and aluminum wall (lower)

It is interesting to see whether the temperature is constant along the hot and cold walls. Figure 13.5 shows temperature variation along the hot wall on the air side. Not only is it not constant—in the central part it is higher than the hot-water temperature (310 K)! A bug in the code, one might think; however, a closer look at the temperature variation across the hot-water channel reveals that water heats up a bit along both channel walls, while in the central part it has a constant temperature of 310 K, as was specified at the inlet. This is due to the fact that the energy equation solved contains the source term due to viscous heat generation; when water flows at 10 m/s in a channel 20 mm wide, a measurable heat gain is produced. This is why all experiments in which fluid recirculates through the test section need to have a heat exchanger to take the generated heat away and keep the working temperature constant. Thus, temperature is slightly higher than 310 K over ca. 65% of the hot wall, while it is ca. 0.3 K lower near the corners. With the temperature difference between hot and cold walls being 10 K, the deviation from constant temperature is significant. Therefore, whenever possible, one should include as much of the experimental setup in the simulation as possible.

In this example air flow was laminar; when the flow of interest is turbulent, it is desirable to resolve the near-wall layer (i.e., use the low-Re wall treatment). If this is not affordable, than one should ensure that the first computational point near wall is in the log-range, so that wall functions are appropriate. Although so-called "all-y^+ wall functions" are available in most commercial codes, they are only helpful if the grid over a small part of wall surface is in the intermediate range. When the grid is such that y^+ is between say 5 and 20 over most of wall surface, the results are likely to be poor. This needs to be accounted for when performing grid-dependence studies: if finer grids fall into the buffer range, the difference between solutions on consecutive grids is likely to increase rather than decrease!

13.3   Flows with Variable Fluid Properties

Although we have dealt mainly with incompressible flows, the density, viscosity, and other fluid properties have been kept inside the differential operators. This allows the discretization and solution methods presented in the preceding chapters to be used to solve problems with variable fluid properties.

The variation in fluid properties is usually caused by temperature variation; pressure variation also affects the change of density. This kind of variation was considered in Chap. 11, where we dealt with compressible flows. However, there are many cases in which the pressure does not change substantially, but the temperature and/or concentration of solutes can cause large variation in fluid properties. Examples are gas flows at reduced pressure, flows in liquid metals (crystal growth, solidification and melting problems, etc.), and environmental flows of fluids stratified by dissolved salt.

Variations in density, viscosity, Prandtl number, and specific heat increase the non-linearity of the equations. The sequential solution methods can be applied to these flows in much the same way they are applied to flows with variable temperature. One recalculates the fluid properties after each outer iteration and treats them as known during the next outer iteration. If the property variation is significant, the convergence may be slowed considerably. For steady flows, the multigrid method can result in a substantial speed-up; see Durst et al. (1992) for an example of application to metalorganic chemical vapor deposition problems, and Kadinski and Perić (1996) for application to problems involving thermal radiation.

For transient flows, especially when time steps are relatively small, the sequential solution method described earlier is likely to work well because the changes in solution from one time step to the next are not large. Outer iterations within a time step are usually required anyway to update deferred corrections (e.g., those due to grid non-orthogonality) and non-linear terms (convection fluxes, fluid properties, contributions from turbulence models etc.). One may need stronger under-relaxation and more outer iterations per time step when the problem becomes stiff.

Flows in the atmosphere and the oceans are special examples of variable density flows; they are discussed later.

13.4 Moving Grids

In many application areas the solution domain changes with time due to the movement of boundaries. The movement is determined either by external effects (as in piston-driven flows) or by calculation as part of the solution (for example, in the case of a floating or flying body). In either case, the grid has to move to accommodate the changing boundary. If the coordinate system remains fixed and the Cartesian velocity components are used, the only change in the conservation equations is the appearance of the relative velocity in convection terms; see Sect. 1.2. We describe here briefly how the equations for a moving grid system can be derived.

First consider the one dimensional continuity equation:

$$\frac{\partial \rho}{\partial t} + \frac{\partial (\rho v)}{\partial x} = 0 \ . \tag{13.13}$$

By integrating this equation over a control volume whose boundaries move with time, i.e., from $x_1(t)$ to $x_2(t)$, we get:

$$\int_{x_1(t)}^{x_2(t)} \frac{\partial \rho}{\partial t} \, dx + \int_{x_1(t)}^{x_2(t)} \frac{\partial (\rho v)}{\partial x} \, dx = 0 \ . \tag{13.14}$$

The second term causes no problems. The first requires the use of Leibniz's rule and, as a result, Eq. (13.14) becomes:

$$\frac{d}{dt} \int_{x_1(t)}^{x_2(t)} \rho \, dx - \left[\rho_2 \frac{dx_2}{dt} - \rho_1 \frac{dx_1}{dt} \right] + \rho_2 v_2 - \rho_1 v_1 = 0 \ . \tag{13.15}$$

The derivative dx/dt represents the velocity with which the grid (CV-surface) moves; we denote it by v_s. The terms in square brackets have therefore a form similar to the last two terms involving fluid velocity, so we can rewrite Eq. (13.14) as:

$$\frac{d}{dt} \int_{x_1(t)}^{x_2(t)} \rho \, dx + \int_{x_1(t)}^{x_2(t)} \frac{\partial}{\partial x} [\rho (v - v_s)] \, dx = 0 \ . \tag{13.16}$$

When the boundary moves with fluid velocity, i.e., $v_s = v$, the second integral becomes zero and we have the Lagrangian mass conservation equation, $dm/dt = 0$.

The three-dimensional version of Eq. (13.15) (obtained using the 3D version of Leibniz's rule) gives:

$$\frac{d}{dt} \int_V \rho \, dV - \int_S \rho \frac{d\mathbf{r}}{dt} \cdot \mathbf{n} \, dS + \int_S \rho \mathbf{v} \cdot \mathbf{n} \, dS = 0 \ , \tag{13.17}$$

or, in the notation used above:

$$\frac{d}{dt} \int_V \rho \, dV + \int_S \rho(\mathbf{v} - \mathbf{v}_s) \cdot \mathbf{n} \, dS = 0 . \tag{13.18}$$

In Sect. 1.2 we noted that the conservation laws can be transformed from the control mass to control volume form by using Eq. (1.4); this also leads to the above mass conservation equation. The same approach may be applied to any transport equation.

The integral form of the conservation equation for the ith momentum component takes the following form when the CV-surface moves with velocity $\mathbf{v}_s$:

$$\frac{d}{dt} \int_V \rho u_i \, dV + \int_S \rho u_i (\mathbf{v} - \mathbf{v}_s) \cdot \mathbf{n} \, dS = \int_S (\tau_{ij} \mathbf{i}_j - p \mathbf{i}_i) \cdot \mathbf{n} \, dS + \int_V b_i \, dV . \tag{13.19}$$

Conservation equations for scalar quantities are easily derived from the corresponding equations for a fixed CV by replacing the velocity vector in the convection term with the relative velocity $\mathbf{v} - \mathbf{v}_s$.

Obviously, if the boundary moves with the same velocity as the fluid, the mass flux through the CV face will be zero. If this is true for all CV faces, then the same fluid remains within the CV and it becomes a *control mass*; we then have the Lagrangian description of fluid motion. On the other hand, if the CV does not move, the equations are those dealt with earlier. The time derivative in the above equations has a different meaning in fixed and moving grids, although it is approximated in the same way. If the CV does not move, the time derivative represents the local change of the conserved quantity at a fixed location and is denoted by $\partial \phi / \partial t$; when the CV moves, we use $d\phi / dt$ to denote the change of ϕ in time at a location which moves in space. In the above-mentioned extreme of a CV whose surface moves exactly with fluid velocity, the time derivative becomes the total (material) derivative because the CV contains the same fluid all the time and thus represents the control mass. This change of meaning of the time derivative is accounted for by the convection fluxes, which also change depending on CV-motion.

When the motion of the grid is known as a function of time, solution of the Navier–Stokes equations poses no new problems: we simply calculate the convection fluxes (e.g., the mass fluxes) using the relative velocity components at the cell faces. However, when the cell faces move, conservation of mass (and all other conserved quantities) is not necessarily ensured if the grid velocities are used to calculate the mass fluxes. For example, consider the continuity equation with implicit Euler time integration; for the sake of simplicity we assume that the CV is rectangular and that the fluid is incompressible and moves at constant velocity. Figure 13.6 shows the relative sizes of the CV at the old and new time levels. We also assume that the grid lines (CV faces) move with constant, but different velocities, so that the size of the CV grows with time.

Fig. 13.6 A rectangular control volume whose size increases with time due to a difference in the grid velocities at its boundaries

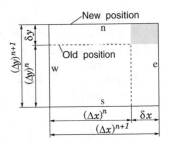

The discretized continuity equation for the CV shown in Fig. 13.6 with the implicit Euler scheme reads:

$$\frac{\rho\,[(\Delta V)^{n+1} - (\Delta V)^n]}{\Delta t} + \rho\,[(u - u_s)_e - (u - u_s)_w]^{n+1}(\Delta y)^{n+1} +$$

$$\rho\,[(v - v_s)_n - (v - v_s)_s]^{n+1}(\Delta x)^{n+1} = 0 \,, \qquad (13.20)$$

where u and v are the Cartesian velocity components. Because we assume that the fluid moves with a constant velocity, the contribution of fluid velocity in the above equation cancels out—only the difference in grid velocities remains:

$$\frac{\rho}{\Delta t}[(\Delta V)^{n+1} - (\Delta V)^n] - \rho(u_{s,e} - u_{s,w})(\Delta y)^{n+1} -$$

$$\rho(v_{s,n} - v_{s,s})(\Delta x)^{n+1} = 0 \,. \qquad (13.21)$$

Under the assumptions made above, the difference in grid velocities at the opposite CV sides can be expressed as (see Fig. 13.6):

$$u_{s,e} - u_{s,w} = \frac{\delta x}{\Delta t} \,, \qquad v_{b,n} - v_{s,s} = \frac{\delta y}{\Delta t} \,. \qquad (13.22)$$

By substituting these expressions into Eq. (13.21) and noting that $(\Delta V)^{n+1} = (\Delta x \,\Delta y)^{n+1}$ and $(\Delta V)^n = [(\Delta x)^{n+1} - \delta x][(\Delta y)^{n+1} - \delta y]$, we find that the discretized mass conservation equation is not satisfied—there is a mass source

$$\delta \dot m = \frac{\rho\,\delta x\,\delta y}{\Delta t} = \rho(u_{s,e} - u_{s,w})(v_{s,n} - v_{s,s})\,\Delta t \,. \qquad (13.23)$$

The same error (with opposite sign) is obtained with the explicit Euler scheme. For constant grid velocities it is proportional to the time step size, i.e., it is a first-order discretization error. One might think that this is not a problem, because the scheme is only first-order accurate in time; however, artificial mass sources may accumulate with time and cause serious problems. The error disappears if only one set of grid lines moves, or if the grid velocities are equal at opposite CV sides.

Under the above assumptions, both the Crank–Nicolson and three-time-level implicit scheme satisfy the continuity equation exactly. More generally, when the fluid and/or grid velocities are not constant, these schemes can also produce artificial mass sources.

Mass conservation can be obtained by enforcing the so-called *space conservation law* (SCL) which can be thought of as the continuity equation in the limit of zero fluid velocity:

$$\frac{d}{dt} \int_V dV - \int_S v_s \cdot n \, dS = 0 \,. \tag{13.24}$$

This equation describes the conservation of space when the CV changes its shape and/or position with time.

Why is it important to obey the SCL can be seen by considering the mass-conservation equation (13.18) for a fluid of constant density; it can then be written as:

$$\frac{d}{dt} \int_V dV - \int_S v_s \cdot n \, dS + \int_S v \cdot n \, dS = 0 \,. \tag{13.25}$$

The first two terms represent the SCL and add up to zero, cf. Eq. (13.24); thus, for fluids with constant density, the mass conservation equation reduces to

$$\int_S v \cdot n \, dS = 0 \quad \text{or} \quad \nabla \cdot v = 0 \,. \tag{13.26}$$

It is therefore important to ensure that the above two terms cancel out in the discretized equations as well (i.e., the sum of volume fluxes through CV faces due to their movement must equal the rate of change of volume); otherwise, artificial mass sources are introduced and they may accumulate in time and spoil the solution, as demonstrated by Demirdžić and Perić (1988).

In what follows the implicit three-time-level scheme for time integration and the SIMPLE algorithm are used for illustration. The expressions for the Crank-Nicholson scheme are easily derived, as well as those for the implicit Euler scheme; implementation in the implicit fractional-step method is also straightforward. The use of a first-order scheme for time integration in the case of a transient flow with moving boundaries makes sense only if we know that the flow is developing towards a steady-state solution, as is the case for some floating body problems (ships moving in a calm water; the initial position is that of both vessel and fluid being at rest, but at constant speed both trim and sinkage of the vessel will change due to generated waves). For spatial integration we use the midpoint rule and central-difference schemes.

The discretized SCL equation can be cast into the following form (see Eq. (6.25)):

$$\frac{3 \, (\Delta V)^{n+1} - 4 \, (\Delta V)^n + (\Delta V)^{n-1}}{2 \, \Delta t} = \left[\sum_k (v_s \cdot S)_k \right]^{n+1} , \tag{13.27}$$

Fig. 13.7 A typical 2D CV at two time steps and the volume swept by a cell face

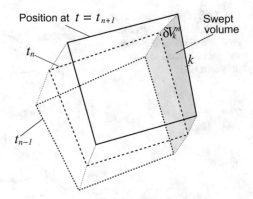

where the summation is over all faces of the CV. Note that the difference between CV volumes at consecutive time levels can be expressed as the sum of volumes δV_k swept by each CV face when moving from its old to the new position, see Fig. 13.7, i.e.:

$$(\Delta V)^{n+1} - (\Delta V)^n = \sum_k \delta V_k^n . \tag{13.28}$$

When this expression is introduced in Eq. (13.27), the following expression is obtained:

$$\frac{3\sum_k \delta V_k^n - \sum_k \delta V_k^{n-1}}{2\,\Delta t} = \left[\sum_k (\mathbf{v}_s \cdot \mathbf{S})_k\right]^{n+1} . \tag{13.29}$$

Although the above equation could be satisfied under different circumstances, it is reasonable to assume that the corresponding parts of the sum on the left- and the right-hand side should be equal (i.e., the contributions from each face are equal on the two sides of the equation). Under this assumption one finds that the SCL is satisfied identically if the volume fluxes through cell faces are defined as:

$$\dot{V}_k^{n+1} = [(\mathbf{v}_s \cdot \mathbf{S})_k]^{n+1} \approx \frac{3\,\delta V_k^n - \delta V_k^{n-1}}{2\,\Delta t} . \tag{13.30}$$

Therefore, the volumes swept by each face over one time step, δV_k, are calculated from the grid position at two time levels and used to calculate volume fluxes $\dot{V}_k^{n+1}$; there is then no need to define explicitly the velocity of the CV-surface, $\mathbf{v}_s$.

The mass flux through one cell face can now be calculated as (see Eq. (13.18)):

$$\dot{m}_k^{n+1} = \left(\int_{S_k} \rho\mathbf{v} \cdot \mathbf{n}\,dS - \int_{S_k} \rho\mathbf{v}_s \cdot \mathbf{n}\,dS\right)^{n+1} \approx (\rho v_i S^i)_k^{n+1} - (\rho_k \dot{V}_k)^{n+1} . \tag{13.31}$$

Here $(v_i)_k$ and $(S^i)_k$ stand for the Cartesian components of the fluid velocity vector $\mathbf{v}$ and surface vector $S\mathbf{n}$ at the face k.

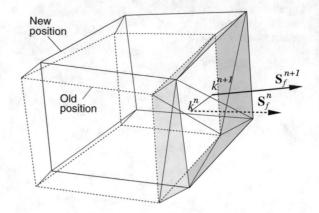

Fig. 13.8 On the calculation of a volume swept by a cell face of a 3D CV; shaded are surfaces common to neighbor CVs

The discretized mass conservation equation which should be satisfied (to within a certain tolerance) at each SIMPLE iteration reads:

$$\frac{3\,(\rho\,\Delta V)^{n+1} - 4\,(\rho\,\Delta V)^{n} + (\rho\,\Delta V)^{n-1}}{2\,\Delta t} + \sum_{k} \dot{m}_{k}^{n+1} = 0\,. \tag{13.32}$$

Within the transient SIMPLE algorithm, values for the new time level t_{n+1} are approached via outer iterations. An approximation to the new mass flux is computed at each outer iteration according to Eq. (13.31), using prevailing density and velocities obtained by solving the momentum equations. Mass fluxes are then corrected to satisfy the mass conservation equation by applying a correction to cell-face velocity (which is proportional to the gradient of pressure correction), and—in the case of a compressible flow—a correction to the cell-face density (which is directly proportional to pressure correction). These corrections follow the procedure described in Sects. 7.2.2 and 11.2.1 and will not be repeated here.

In 3D, one has to be careful in calculating the volumes swept by cell faces. Because the cell face edges may turn, the calculation of the swept volume requires triangulation of the shaded surfaces in Fig. 13.8. The volume can then be calculated using approach described in Sect. 9.6.4. However, as the shaded surfaces are common to two CVs, one has to ensure that they are triangulated in the same way for both CVs to assure space conservation.

The pressure-correction equation resulting from the mass conservation equation has the same form as in the case of fixed grids (for both compressible and incompressible flows), except for the time-dependent term. If the grid position at the new time level is known (e.g., in piston-driven flows, flows around rotating machinery, etc.), the volume fluxes through cell faces $\dot{V}_{k}^{n+1}$ do not depend on the outer iteration counter and are computed once at the beginning of each new time step. However, in the case of mesh adaptation to the position of a solid structure (fluid-structure interaction) or to an interface (e.g., in interface-tracking treatment of free surfaces), the volume fluxes need to be corrected during outer iterations together with other corrections; see Demirdžić and Perić (1990) for more details and an example.

In the case of compressible flow, where the rate of change of mass within the solution domain is non-zero, one cannot formally prove that a particular implementation of SCL guarantees conservation, as was described above for incompressible flows. The reason is that densities appearing on the left-hand side of Eq. (13.32) are cell-center values, while on the right-hand side (in cell-face mass fluxes), face values of densities are used. However, the three-time-levels scheme is second-order in time, so any temporal discretization error (which affects the rate of change) will reduce by a factor of four when the time step is halved. The additional errors due to mesh motion are of the same order as discretization errors for fixed grids and reduce at the same rate when time step is refined.

One can show that, for a compressible flow in a closed system (without inlet and outlet) with varying shape and volume of the solution domain, the mass is conserved when the above-described SCL discretization is used. This is demonstrated below for the three-time-level scheme.

From the discretized mass conservation equation (13.32), one obtains the total mass in the solution domain by summing over all CVs. The mass fluxes through all internal cell faces cancel out in this summation; there remains only the sum of mass fluxes through boundary faces (note that mass fluxes contain contributions from both fluid velocity and boundary motion, see Eq. (13.31); thus, mass fluxes are zero at an impermeable wall, irrespective of whether it moves or not):

$$3M^{n+1} - 4M^n + M^{n-1} = -\sum_B \dot{m}_B , \qquad (13.33)$$

where $\dot{m}_B$ are mass fluxes through boundary cell faces and

$$M^n = \sum_{\text{all cells}} (\rho \Delta V)^n \qquad (13.34)$$

represents the mass in the solution domain at time t_n.

In the case of a closed system without inflow or outflow, the sum of mass fluxes through boundary cell faces is equal to zero. From Eq. (13.33) it follows then:

$$M^{n+1} = \frac{4M^n - M^{n-1}}{3} . \qquad (13.35)$$

If at the beginning of simulation the mass was the same at two consecutive time levels, i.e., if $M^n = M^{n-1} = M^0$, it follows from the above equation that also $M^{n+1} = M^0$, i.e., the mass will be conserved within the solution domain, irrespective of mesh motion.

In some implicit time-integration methods (so-called *fully-implicit methods*), in which fluxes and source terms are computed only at the newest time level, grid motion can be ignored everywhere except near boundaries. Examples of such methods are the implicit Euler scheme and the three-time-level scheme, see Chap. 6. Because the fluxes are computed at time level t_{n+1}, we do not need to know where the grid was (or what shape the CVs had) at the previous time level t_n: instead of Eq. (13.19) we can use the usual equation for a space-fixed CV:

$$\frac{\partial}{\partial t} \int_V \rho u_i \, dV + \int_S \rho u_i \mathbf{v} \cdot \mathbf{n} \, dS = \int_S (\tau_{ij} \mathbf{i}_j - p \, \mathbf{i}_i) \cdot \mathbf{n} \, dS + \int_V b_i \, dV \; . \quad (13.36)$$

The two equations differ in the definition of the rate-of-change and convection terms: for a space-fixed CV convection fluxes are computed using fluid velocity, and time derivative represents the local rate of change at a fixed point in space (e.g., CV-center). On the other hand, for a moving CV convection fluxes are computed using relative velocity between fluid and CV-surface, and the time derivative expresses the rate of change in a volume whose location changes.

Because solutions from previous time steps are not needed to compute surface and volume integrals, the grid can not only move but may also change its topology, i.e., both the number of CVs and their shape can change from one time step to another. The only term in which the old solution appears is the unsteady term, which requires that volume integrals over the *new* CV of some *old* quantities have to be approximated. If midpoint rule is used for this purpose, all we need to do is to interpolate the old solutions to the locations of the new CV-centers. One possibility is to compute gradient vectors at the center of each old CV and then, for each new CV-center, find the nearest center of an old CV and use linear interpolation to obtain the old value at the new CV-center:

$$\phi_{C_{\text{new}}}^{\text{old}} = \phi_{C_{\text{old}}}^{\text{old}} + (\nabla \phi)_{C_{\text{old}}}^{\text{old}} \cdot (\mathbf{r}_{C_{\text{new}}} - \mathbf{r}_{C_{\text{old}}}) \; . \quad (13.37)$$

Near moving boundaries we have to account for the fact that the boundary moved during the time step and either displaced fluid or made space to be filled by fluid. For small motions this can be taken into account by prescribing mass sources or sinks in the near-boundary CVs; these are computed in the same way as one would compute inlet or outlet mass flux through the boundary face using the velocity with which the boundary moves as fluid velocity. A problem can arise if the CV moves more than its width in the direction of motion in one time step, because the center of a new CV may lie outside the old mesh. Thus, for grids which are fine near moving walls it may be desirable to use a moving grid and equations based on moving control volumes in the near-wall region, while away from walls the grid motion may be ignored, allowing for the grid to be re-generated if its properties deteriorate due to excessive deformation. One example of a method of the above kind can be found in the thesis by Hadžić (2005); he also performed comparison of this method and the one using equations for moving control volumes and found good agreement between results of both methods and experimental data for a piston-driven flow in a pipe with sudden expansion.

Many engineering applications require the use of moving grids. However, different problems require different solution methods. An important example is rotor-stator interaction which is common to turbomachinery and mixers: one part of the grid is attached to the stator and does not move, while another part is attached to the rotor and moves with it. The interface between the moving and fixed grids is usually a combination of cylindrical and planar surfaces. If grids match at the interface at the initial time, one can allow the rotating part of the grid to move while keeping the boundary points "glued" to the fixed grid, until the deformation becomes substantial (45° angles should not be exceeded); then, the boundary points "leap" one cell ahead and stay glued to the new location for a while. This kind of "clicking" grid has been used in these applications in combination with regular, structured grids.

Another possibility is to let the moving grid "slide" along the interface without deformation. In this case the grids do not match at the interface, so some CVs have more neighbors than others. However, this situation is completely analogous to that encountered in block-structured grids with non-conformal interfaces and can be handled by the methods described in Sect. 9.6.1; the only difference is that the cell connectivity changes with time and has to be re-established after each time step. This approach is more flexible than the one described above; the grids can be of different kinds and/or fineness, and the interface can be an arbitrary surface. This approach can also be applied to flows around bodies passing each other, entering a tunnel, or moving in an enclosure with a known trajectory. All commercial codes offer this feature, which is used to simulate flows involving rotor-stator interaction in many kinds of turbo-machines, propeller rotation on a ship, etc.

The third approach is to use overlapping (Chimera) grids. Again, one grid is attached to the fixed part of the domain and the other to the moving body. This approach can be used even if the trajectory of the moving body is not known in advance, when it is very complicated, or when the surrounding domain is of a complex shape (e.g., when a sliding interface cannot be constructed because paths of moving parts intersect). The fixed grid may cover the whole "environment" in which the body is moving. The overlap region changes with time and the relationship between the grids needs to be re-established after each time step. Except for difficulties in ensuring exact conservation (as discussed in Sect. 9.1.3) there are almost no limitations on the applicability of this approach (one can even account for the contact of bodies approaching each other).

As noted above, the same equations and discretization methods apply to both the fixed and moving grids, the only difference being that on the fixed grid, the grid velocity v_s is obviously zero. Sometimes it may be advantageous to use different coordinate systems on the two domains; for example, one may use Cartesian velocity components in one part and polar components on the other grid. This is possible provided: (i) one adds the body forces due to frame acceleration and (ii) one transforms the vector components from one system to another at the interface or in the overlap region. Both of these are easy to do in principle but the programming may be tedious.

An example of the use of overlapping grids to couple the solutions on a fixed grid attached to ship hull and on a grid rotating with propeller is shown in Fig. 13.9. The ship is equipped with POD-drives which can be rotated around a vertical axis; the thrust direction can thus be changed and the vessel does not need a rudder. In this

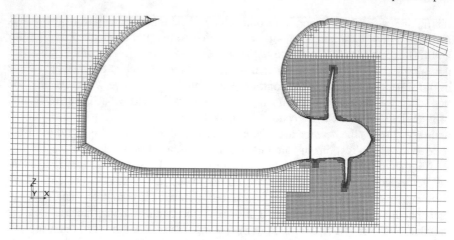

Fig. 13.9 Longitudinal section through the propeller of a full-scale ship equipped with so-called POD-drives, showing overlapping of a rotating grid attached to the propeller and a fixed grid attached to the hull

study the POD was fixed and only the propeller is rotating at a fixed rate; the ship speed is varied until resistance matches propeller thrust. In this example the fixed grid is twice as coarse in the zone where the propeller is inserted as the outer part of the rotating grid; a larger mismatch in cell size is undesirable, while nearly equal cell size in both grids would be optimal. Space limitations do not allow us to discuss details of the particular study, but in addition to showing the overlapping grids in two sections in Fig. 13.9, we also show velocity and pressure contours in Fig. 13.10. Here we want to draw the attention to contour lines: in the zone where active cells of both grids overlap, each contour is represented twice: once in cells of the fixed grid, and once in cells of the rotating grid. Inevitably, the lines will not be identical but if everything is done correctly, the miss-match will be small. In this particular case, both pressure and velocity contours match well. A more detailed report about this simulation can be downloaded from the book web site; see the appendix for details.

A simpler approach is to use the Cartesian velocity components but different reference frames in the fixed and moving region; this is often done if the flow is steady in the moving (e.g., rotating) region when viewed in the reference frame moving with the part. The equations in the moving region then need to be extended by the extra terms resulting from the motion of the reference frame. This approach cannot account for the interaction between moving and fixed parts (e.g., blade-passing effects in turbo-machinery) and thus introduces an additional modeling error. This error is small if the flow is steady and axisymmetric in both reference frames near the interface, but may be large if the gap between moving and fixed parts is small. Such approaches were commonly used in the past in conjunction with eddy-viscosity types of turbulence models and steady-state analysis, in order to save computing time; however, because the interaction between moving and fixed parts is often an important element of the analysis, one can expect that in future moving grids will be used in most applications of this kind.

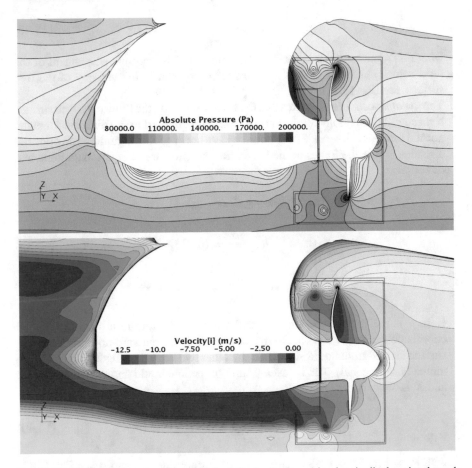

Fig. 13.10 Contours of pressure (upper) and axial velocity (lower) in a longitudinal section through the propeller, computed using overlapping grids shown in Fig. 13.9

13.5   Free-Surface Flows

Flows with free surfaces are an especially difficult class of flows with moving boundaries. The position of the boundary is known only at the initial time; its location at later times has to be determined as part of the solution. To accomplish this, the SCL and the boundary conditions at the free surface must be used.

In the most common case, the free surface is an air-water boundary but other liquid-gas surfaces occur, as do liquid-liquid interfaces. If phase change at the free surface can be neglected, the following boundary conditions apply:

- The *kinematic condition* requires that the free surface be a sharp interface separating the two fluids that allows no flow through it, i.e.:

$$[(\mathbf{v} - \mathbf{v}_s) \cdot \mathbf{n}]_{fs} = 0 \quad \text{or} \quad \dot{m}_{fs} = 0 , \tag{13.38}$$

where 'fs' denotes the free surface. This states that the normal component of the fluid velocity at the surface is the normal component of the velocity of the free surface, see Eq. (13.18).

- The *dynamic condition* requires that the forces acting on the fluid at the free surface be in equilibrium (momentum conservation at the free surface). This means that the normal forces on either side of the free surface are of equal magnitude and opposite direction, while the forces in the tangential direction are of equal magnitude and direction:

$$(\mathbf{n} \cdot \mathsf{T})_l \cdot \mathbf{n} + \sigma K = -(\mathbf{n} \cdot \mathsf{T})_g \cdot \mathbf{n} ,$$

$$(\mathbf{n} \cdot \mathsf{T})_l \cdot \mathbf{t} - \frac{\partial \sigma}{\partial t} = (\mathbf{n} \cdot \mathsf{T})_g \cdot \mathbf{t} , \tag{13.39}$$

$$(\mathbf{n} \cdot \mathsf{T})_l \cdot \mathbf{s} - \frac{\partial \sigma}{\partial s} = \mathbf{n} \cdot \mathsf{T})_g \cdot \mathbf{s} .$$

Here σ is the surface tension, $\mathbf{n}$, $\mathbf{t}$ and $\mathbf{s}$ are unit vectors in a local orthogonal coordinate system (n, t, s) at the free surface (n is the outward normal to the free surface viewed from liquid side while the other two lie in the tangent plane and are mutually orthogonal). The indexes 'l' and 'g' denote liquid and gas, respectively, and K is the curvature of the free surface,

$$K = \frac{1}{R_t} + \frac{1}{R_s} , \tag{13.40}$$

with R_t and R_s being radii of curvature along coordinates t and s, see Fig. 13.11. The surface tension σ is the force per unit length of a surface element and acts tangential to the free surface; in Fig. 13.11, the magnitude of the force $\mathbf{f}_\sigma$ due to surface tension is $f_\sigma = \sigma \, dl$. For an infinitesimally small surface element dS, the tangential components of the surface tension forces cancel out when $\sigma =$const., and the normal component can be expressed as a local force that results in a pressure jump across the surface, as in Eq. (13.39).

Surface tension is a thermodynamic property of a liquid that depends on the temperature and other state variables such as chemical composition and surface cleanliness. If the temperature differences are small, the temperature dependence of σ can be linearized so that $\partial \sigma / \partial T$ is constant; it is usually negative. When the temperature varies substantially along the free surface, the gradient in surface tension results in a shear force that causes fluid to move from the hot region to the cold region. This phenomenon is called *Marangoni* or *capillary convection* and its importance is characterized by the dimensionless Marangoni number:

Fig. 13.11 On the description of boundary conditions at the free surface

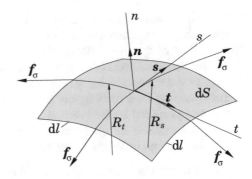

$$\mathrm{Ma} = -\frac{\partial \sigma}{\partial T} \frac{\Delta T \, L}{\mu \kappa} \,, \tag{13.41}$$

where ΔT is the bulk temperature difference across the domain, L is a characteristic length of the surface and κ is the thermal diffusivity.

In many applications the shear stress at the free surface can be neglected (e.g., ocean waves with no appreciable wind or waves generated by a large ship). The normal stress and the effect of the surface tension are also often neglected; in this case the dynamic boundary condition reduces to $p_1 = p_g$.

The implementation of these boundary conditions is not as trivial as it would appear. If the position of the free surface were known, there would be little problem. The mass flux can be set to zero on cell faces lying on the free surface, and the forces acting on the cell face from outside can be calculated; if surface tension is neglected, only the pressure force remains. The problem is that the location of the free surface must be computed as part of the solution and is not usually known in advance. One can therefore directly implement only one of the boundary conditions at the free surface; the other must be used to locate the surface. Location of the free surface must be done iteratively, greatly increasing the complexity of the task. Often only the flow on one side of the interface (typically, the liquid flow) is of interest; however, in many applications flow of both fluids on either side of the interface must be computed simultaneously (e.g., flow of gas bubbles moving in liquid or of liquid drops moving in gas).

Many methods have been used to find the shape of the free surface. They can be classified into two major groups:

- Methods which treat the free surface as a sharp interface whose motion is followed *(interface-tracking methods)*. In this type of methods, boundary-fitted grids are used and advanced each time the free surface is moved. In explicit methods, which must use small time steps, the problems associated with grid movement that were discussed earlier are often ignored. As an example, Hodges and Street (1999) presented a FV fractional-step method with a moving boundary-orthogonal grid for LES of free-surface flows.

- Methods which do not define the interface as a boundary *(interface-capturing methods)*. The computation is performed on a fixed grid, which covers fluids on both sides of the free surface. The shape of the free surface is determined by computing the fraction of each near-interface cell that is partially filled. This can be achieved by introducing massless particles at the free surface at the initial time and following their motion; this is called the *Marker-and-Cell* or MAC scheme that was proposed by Harlow and Welsh (1965). Alternatively, one can solve a transport equation for the fraction of the cell occupied by the liquid phase (the *Volume-of-Fluid* or VOF scheme, Hirt and Nichols 1981).

There are also hybrid methods. All of these methods can also be applied to some kinds of two-phase flows as will be discussed in the following section.

13.5.1 *Interface-Capturing Methods*

The MAC-scheme is attractive because it can treat complex phenomena like wave breaking. However, the computing effort is large, especially in three dimensions because, in addition to solving the equations governing fluid flow, one has to follow the motion of a large number of marker particles.

In VOF-methods, in addition to the conservation equations for mass and momentum, one solves an equation for the filled fraction of each control volume, c, so that $c = 1$ in filled CVs and $c = 0$ in empty CVs. From the continuity equation, one can show that the evolution of c is governed by the transport equation:

$$\frac{\partial c}{\partial t} + \nabla \cdot (c\mathbf{v}) = 0 \quad \text{or} \quad \frac{\partial}{\partial t} \int_V c \, dV + \int_S c\mathbf{v} \cdot \mathbf{n} \, dS = 0 . \tag{13.42}$$

In incompressible flows this equation is invariant with respect to interchange of c and $1 - c$; for this to be assured in the numerical method, mass conservation has to be strictly enforced.

This approach is more efficient than the MAC scheme and can be applied to complex free surface shapes including breaking waves. However, the free surface profile is not sharply defined; it is usually smeared over one or more cells (similar to shocks in compressible flows). Local grid refinement is important for accurate resolution of the free surface. The refinement criterion is simple: cells with $0 < c < 1$ need to be refined. A method of this kind, called the marker and micro-cell method, has been developed by Raad and his colleagues (see, for example Chen et al. 1997).

There are several variants of the above approach. In the original VOF-method (Hirt and Nichols 1981) Eq. (13.42) is solved in the whole domain to find the location of the free surface; the mass and momentum conservation equations are solved for the liquid phase only. The method can calculate flows with overturning free surfaces, but the gas enclosed by the liquid phase will not feel buoyancy effects and will therefore behave in an unrealistic manner.

Kawamura and Miyata (1994) used Eq. (13.42) to calculate the distribution of the *density function* (the product of the density and the volume fraction c) and to locate the free surface, which is the contour with $c = 0.5$. The computation of the motion of the liquid *and* gas flows are done separately. The free surface is treated as a boundary at which the kinematic and dynamic boundary conditions are applied. Cells which become irregular due to being cut by the free surface require special treatment (variable values are extrapolated to nodal locations lying on the other side of the interface). The method was used to calculate flows around ships and submerged bodies.

Alternatively, one can treat fluids on both sides of the interface as a single fluid whose properties vary in space according to the volume fraction of each phase, i.e.:

$$\rho = \rho_1 c + \rho_2 (1 - c) , \quad \mu = \mu_1 c + \mu_2 (1 - c) , \tag{13.43}$$

where subscripts 1 and 2 denote the two fluids (e.g., liquid and gas). In this case, the interface is not treated as a boundary so no boundary conditions need to be prescribed on it. The interface is simply the location where the fluid properties change abruptly. However, solution of Eq. (13.42) implies that the kinematic condition is satisfied, and the dynamic condition is also implicitly taken into account. If surface tension is significant at the free surface, its effects can be accounted for as a body force.

The surface-tension force acts only in the region of interface, i.e., in partially filled cells. Because in full or empty cells the gradient of c is zero, the normal component of the surface tension force can be expressed as (*continuum surface force* approach, Brackbill et al. 1992):

$$\mathbf{F}_{fs} = \int_V \sigma \kappa \, \nabla c \, dV . \tag{13.44}$$

However, there are problems when the surface-tension effects become dominant, such as in the case of droplets or bubbles whose diameter is of the order of 1 mm and which move with very low velocity. In this case, there are two very large terms in the momentum equations (the pressure term and the body force representing the surface-tension effects) which have to balance each other; they are the only non-zero terms if the bubble or droplet is stationary. Due to the fact that curvature of the interface also depends on c (note that the gradient of c at interface points in the direction of normal to interface),

$$\kappa = \nabla \cdot \mathbf{n} = -\nabla \cdot \left(\frac{\nabla c}{|\nabla c|} \right) , \tag{13.45}$$

it is difficult to ensure on an arbitrary 3D grid that the two terms are identical (one is linear in p and the other is non-linear in c), so their difference may cause the so called *parasitic currents*. These can be avoided by using special discretization methods in 2D (see Scardovelli and Zaleski 1999, for some examples of such special methods; see also the analysis of Harvie et al. 2006). We do not know at present of a method that eliminates the problem completely for unstructured arbitrary grids in 3D.

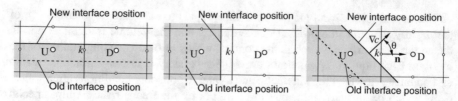

Fig. 13.12 Possible orientations of free surface relative to cell face k at two consecutive time steps

The critical issue in this type of methods is the discretization of convection term in Eq. (13.42). Low-order schemes (like the first-order accurate upwind method) smear the interface and introduce artificial mixing of the two fluids, so higher-order schemes are preferred. Because c must satisfy the condition

$$0 \le c \le 1 \,,$$

it is important to ensure that the method does not generate overshoots or undershoots. Fortunately, it is possible to derive schemes which both keep the interface sharp and produce monotone profiles of c across it; see Leonard (1997) for some examples of monotone schemes and Lafaurie et al. (1994), Ubbink (1997) or Muzaferija and Perić (1999) for methods specifically designed for interface-capturing in free surface flows. Many more variants of interface-capturing or hybrid methods have been published during the past two decades; we shall mention some of them further below. We give here only a brief explanation of the HRIC-scheme (High-Resolution Interface-Capturing) by Muzaferija and Perić (1999). The convected value of c at a cell face k is expressed as a blend of upwind and downwind values:

$$c_k = \gamma_k \, c_U + (1 - \gamma_k)c_D \,, \tag{13.46}$$

where subscripts U and D denote cell centers on the upstream and downstream side of the face k, see Fig. 13.12. Second-order central differencing scheme is obtained if $\gamma_k = 0.5$; otherwise, the scheme is formally first-order accurate. However, the aim here is not an accurate interpolation to the cell-face center of a smoothly varying function (as is the case when computing the convected value of velocity or other scalar variables) but preserving the sharp interface between two immiscible fluids; therefore, the order of approximation is not as important as in other transport equations. In the HRIC-scheme, the blending factor γ_k is determined as a function of three properties of the solution:

- Variation of c across interface, based on values of c at two upstream and one downstream cell center;
- Local CFL-number, which estimates how far will the interface move during one time step;
- Angle θ between normal to the interface (defined by the gradient of c) and cell-face normal (see Fig. 13.12).

The location of the interface is not explicitly computed, but it is assumed that the value $c = 0.5$ represents the free surface. If the interface is orthogonal to the cell face ($\nabla c \cdot \mathbf{n} = 0$), than both upwind and downwind approximation give the same value so it is not important which value is assigned to γ_k. On the other hand, if the interface is parallel to the cell face (the vectors ∇c and $\mathbf{n}$ are co-linear), the CFL-number plays the most important role: if the interface is not likely to reach the cell face during the current time step, than the fluid crossing interface is the one on downstream side of it, so downwind approximation should be used. In case of an arbitrary orientation of the interface, all three factors may be important; see Muzaferija and Perić (1999) for details.

Because fluid properties may differ by several orders of magnitude across an interface, care needs to be taken when using interpolation to compute cell-face values of some quantities; pressure is the most prominent example. In the case of purely hydrostatic variation of pressure, the slope is proportional to density and it thus changes abruptly at the interface. In order to correctly compute pressure forces acting on the control volume, one should not interpolate across interface but rather use one-sided extrapolation; see, e.g., Vukčević et al. (2017) for a detailed description of one suitable approach.

An example of the capabilities of interface-capturing methods is shown in Fig. 13.13, which illustrates the solution to the 'dam-break' problem, a standard test case for methods for computing free-surface flows. The barrier holding back the liquid is suddenly removed, leaving a free vertical water face. As water moves to the right along the floor, it hits an obstacle, flows over it and hits the opposite wall. The confined air escapes upwards due to buoyancy as water falls to the floor on the other side of obstacle. Numerical results compare well with experiments by Koshizuka et al. (1995). This example demonstrates the importance of computing the flow of both liquid and gas phase. If the gas phase was ignored, liquid would fall down without feeling any resistance from trapped air, leading to a significantly different motion. This is even more important in the next example, where surface tension forces are also important.

Buoyancy effects are always important when gas is trapped in liquid or vice-versa; surface tension effects are only significant if high curvature of the free surface is present or if surface tension coefficient varies along free surface due to temperature or concentration gradient. An example where both buoyancy and surface tension are important is the rising of small gas bubbles. Figure 13.14 shows one such situation: air flows into a pipe with 40 mm diameter at one end with a velocity of 1 mm/s (the other end being closed) and from there through a connecting pipe with a diameter of 5 mm into a larger container filled with liquid 45 mm deep. Liquid density is 1500 kg/m^3 and viscosity is 1 Pa·s, while air density is 1.18 kg/m^3 and viscosity is 0.0000185 Pa·s. The surface tension coefficient is constant, $\sigma = 0.074$ N/m. This large variation in fluid properties is a big challenge for the numerical method if a sharp interface is to be obtained. Here the HRIC-scheme (High-Resolution Interface-Capturing) of Muzaferija and Perić (1999) was used and it typically leads to only one cell having a liquid volume fraction between 0 and 1 when the interface is supposed to be sharp. In order to better resolve bubble curvature, adaptive grid refinement was

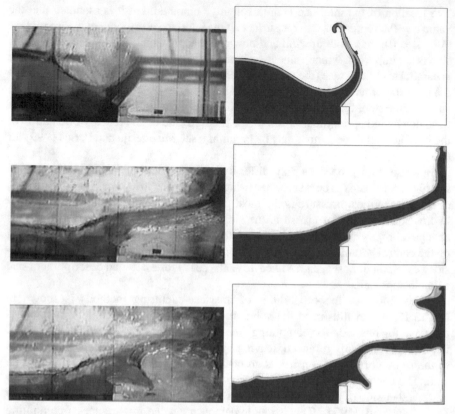

Fig. 13.13 Comparison of experimental visualization (left) and numerical prediction (right) of collapsing water column flow over an obstacle (experiments by Koshizuka et al. 1995; prediction by Muzaferija and Perić 1999)

used; the refinement criterion was the presence of the interface (i.e., volume fraction being between 0.01 and 0.99) and additional two cell layers around interface were also refined. Once the interface reaches the end of the refined zone, grid refinement and coarsening needs to be re-done; thus, grid adaptation was performed every tenth time step. This causes additional work for re-distribution of cells between processors when parallel computing is used (which is nowadays the rule rather than exception). Because second-order time discretization was used, the interface was not allowed to move more than one third of the cell in each time step, in order to avoid over- and under-shoots (which would normally be obtained if a parabola is passed through a step). Without adaptive mesh refinement one would need to have the finest grid level everywhere where the free surface might be present at any time, which would increase the cell count (and thus the computing time) by a factor of 10. A more detailed report on this simulation is available from the book web site; see the appendix for details.

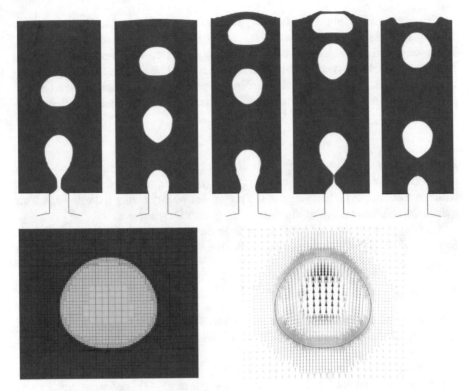

Fig. 13.14 Simulation of motion and flow around rising bubbles using the HRIC-scheme: a sequence of bubbles rising and bursting through the free surface (upper); distribution of vapor volume fraction and the grid adapted to the bubble free surface (lower left); bubble contour and velocity vectors inside and outside of the bubble (lower right)

Another class of interface-capturing methods is based on the *level-set formulation*, introduced by Osher and Sethian (1988). The surface is defined as the one on which a level-set function $\phi = 0$. Other values of this function have no significance and to make it a smooth function, ϕ is typically initialized as the signed distance from the interface, i.e., its value at any point is the distance from the nearest point on the surface and its sign is positive on one side and negative on the other. This function is then allowed to evolve as a solution of the transport equation:

$$\frac{\partial \phi}{\partial t} + \nabla \cdot (\phi \mathbf{v}) = 0 , \tag{13.47}$$

where v is the local fluid velocity and, at any time, the surface on which $\phi = 0$ is the interface. If the function ϕ becomes too complicated, it can be re-initialized in the manner described above. As in VOF-like methods, fluid properties are determined by the local value of ϕ but here, only the sign of ϕ is important.

The apparent advantage of this approach relative to VOF-methods is that ϕ varies smoothly across the interface while the volume fraction c is discontinuous there. However, when solving for the volume fraction c, its step-wise variation across the interface is usually not maintained—the step is smeared by the numerical approximation. As a result, the fluid properties experience a smooth change across the interface. In level-set methods, the step-wise variation of the properties is maintained, because we define

$$\rho = \rho_l \ \text{if} \ \ \phi < 0 \,, \quad \rho = \rho_g \ \text{if} \ \ \phi > 0 \,.$$

However, this usually causes problems when computing viscous flows so one needs to introduce a region of some finite thickness ϵ (usually one to three cells wide) over which a smooth but rapid change of the properties occurs across the interface. This makes the solution becoming similar to that from VOF-methods.

As noted above, the computed ϕ needs to be re-initialized every now and then. Sussman et al. (1994) proposed that this be done by solving the following equation:

$$\frac{\partial \phi}{\partial \tau} = \text{sgn}(\phi_0)(1 - |\nabla \phi|) \,, \tag{13.48}$$

until steady state is reached. This guarantees that ϕ has the same sign and zero level as ϕ_0 and fulfills the condition that $|\nabla \phi| = 1$, making it similar to a signed distance function.

Because ϕ does not explicitly occur in any of the conservation equations, the original level-set method did not exactly conserve mass. Mass conservation can be enforced by making the right-hand side of Eq. (13.48) a function of the local mass imbalance $\Delta \dot{m}$ as was done by Zhang et al. (1998). The more frequently one solves this equation, the fewer iterations are needed to reach steady state; of course, frequent solution of this equation increases the computational cost so there is a trade-off.

Many level-set methods have been proposed; they differ in the choices for the various steps. Zhang et al. (1998) describe one such method, which they applied to bubble-merging and mold-filling, including melt solidification. They used a FV-method on structured, non-orthogonal grids to solve the conservation equations, and a FD-method for the level-set equation. An ENO-scheme was used to discretize the convection term in the latter. Enright et al. (2002) describe a Lagrangian marker particle and level-set method for better interface capture.

Another version of this method is used to study flame propagation. In this case, the flame propagates relative to the fluid and this introduces the possibility that the surface will develop cusps, locations at which the surface normal is discontinuous. This will be discussed further below.

More details on level-set methods can be found in the books by Sethian (1996) and Osher and Fedkiw (2003); see also Smiljanovski et al. (1997) and Reinecke et al. (1999) for examples of similar approaches used for flame tracking.

Interface-capturing or VOF-methods are most widely used approaches to computing free surface flows; they are available in all major commercial and public codes and have been successfully applied to simulate water entry and slamming of bodies onto liquid surface, flow around ships and submerged bodies, primary jet break-up, bubble-collapse near wall, droplet-wall interaction etc. Some examples of application will be briefly described further below.

13.5.2 Interface-Tracking Methods

In the calculation of flows around submerged bodies, many authors linearize about the unperturbed free surface. This requires introduction of a *height function*, which is the free surface elevation relative to its unperturbed state:

$$z = H(x, y, t) . \tag{13.49}$$

The kinematic boundary condition (13.38) then becomes the following equation describing the local change of the height H:

$$\frac{\partial H}{\partial t} = u_z - u_x \frac{\partial H}{\partial x} - u_y \frac{\partial H}{\partial y} . \tag{13.50}$$

This equation can be integrated in time using the methods described in Chap. 6. The fluid velocity at the free surface is obtained either by extrapolation from the interior or by using the dynamic boundary condition (13.39).

This approach is usually used in conjunction with structured grids and explicit time integration. Many authors used a FV method for the flow calculation and a FD method for the height equation and enforced both boundary conditions at the free surface only at the converged steady state (see e.g., Farmer et al. 1994).

Hino (1992) used a FV method with the enforcement of the SCL, thus satisfying all conditions at each time step and ensuring volume conservation. Similar methods were developed by Raithby et al. (1995), Thé et al. (1994) and Lilek (1995). One fully-conservative FV method of this type consists of the following steps:

- Solve the momentum equations using the specified pressure at the current free surface to obtain velocities u_i^*.
- Enforce local mass conservation in each CV by solving a pressure-correction equation, with a zero pressure correction boundary condition at the current free surface (see Sect. 11.2.4). Mass is conserved both globally and in each CV, but the prescribed pressure at the free surface produces a velocity correction there, so that mass fluxes through the free surface are non-zero.
- Correct the position of the free surface to enforce the kinematic boundary condition. Each free-surface cell face is moved so that the volume flux due to its movement compensates the flux obtained in the previous step.
- Iterate until no further adjustment is necessary and both the continuity equation and momentum equations are satisfied.
- Advance to the next time step.

The critical issue for the efficiency and stability of the method is the algorithm for the movement of the free surface. The problem is that there is only one discrete equation per free-surface cell face but a larger number of grid nodes that have to be moved. Correct treatment of the intersections of the free surface with other boundaries (inlet, outlet, symmetry, walls) is essential if wave reflection and/or instability is to be avoided. We shall briefly describe one such method. Only two-time-level schemes are considered, but the approach can be extended.

The mass flux through a moving free-surface cell face is (see Eq. (13.31)):

$$\dot{m}_{fs} = \int_{S_{fs}} \rho \mathbf{v} \cdot \mathbf{n}\, dS - \int_{S_{fs}} \mathbf{v}_s \cdot \mathbf{n}\, dS \approx \rho (\mathbf{v} \cdot \mathbf{n})^\tau_{fs} S^\tau_{fs} - \rho \dot{V}_{fs} \ . \qquad (13.51)$$

The superscript τ denotes the time ($t_n < t_\tau < t_{n+1}$) at which the quantity is calculated; for the implicit Euler scheme, $t_\tau = t_{n+1}$, while for the Crank–Nicolson scheme, $t_\tau = \frac{1}{2}(t_n + t_{n+1})$.

The mass fluxes obtained from the pressure-correction equation with prescribed pressure at the free surface are non-zero; we compensate by displacing the free surface, i.e.:

$$\dot{m}_{fs} + \rho \dot{V}'_{fs} = 0 \ . \qquad (13.52)$$

From this equation we obtain the volume of fluid $\dot{V}'_{fs}$ which has to flow into or out of the CV due to free-surface motion. We need to obtain the coordinates of the CV vertexes that lie on the free surface from this equation. This has to be done carefully and thus requires special attention. Because there is a single volumetric flow rate for each cell face but a greater number of CV vertexes, there are more unknowns than equations.

Thé et al. (1994) suggested using staggered CVs in the layer adjacent to the free surface, but only in the continuity (pressure-correction) equation. The method was applied to several problems in 2D and showed good performance. However, it requires substantial adaptation of the solution method, especially in 3D; see Thé et al. (1994) for more details.

Another possibility is to define the CVs under the free surface not by vertexes but by cell-face centers; the vertexes are then defined by interpolating cell-face center locations, as shown in Fig. 13.15 for a 2D structured grid. The volume swept by the free-surface cell face is then:

Fig. 13.15 Control volumes under the free surface, whose vertexes lie on the free surface and are defined by the coordinates of the cell face centers (open symbols); the volume swept by the cell face during the time step is shaded

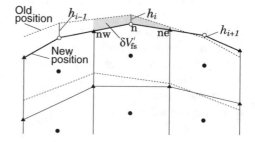

$$\delta V'_{\text{fs}} = \frac{1}{2} \Delta x \, (h_{\text{nw}} + 2 \, h_{\text{n}} + h_{\text{ne}}) \,, \tag{13.53}$$

where h is the distance the free-surface markers move during one time step; $h_{\text{n}} = h_i$ while h_{nw} and h_{ne} are obtained by linearly interpolating h_i and h_{i-1} or h_i and h_{i+1}. By expressing h_{nw}, h_{n} and h_{ne} in terms of h_i, h_{i-1}, h_{i+1} and inserting the above expression into Eq. (13.52), we obtain a system of equations for the locations of the cell-face centers, h_i. In 2D, the system is tridiagonal and can be solved directly by the TDMA method of Sect. 5.2.3. In 3D, the system is block-tridiagonal and is best solved by one of the iterative solvers presented in Chap. 5. 'Boundary conditions' have to be specified at the CV vertexes at the edges of the free surface. If the boundary is not allowed to move, $h = 0$. If the edge of the free surface is allowed to move, e.g., for an open system, the boundary condition should be of the non-reflective or 'wave-transmissive' type that does not cause wave reflection; the condition (10.4) is one appropriate possibility.

This approach was applied by Lilek (1995) to 2D and 3D problems on structured grids. When the lateral bounding surface has an irregular shape (e.g., a ship hull), the expressions for the volume $\delta V'_{\text{fs}}$ become complicated and require iterative solution on each outer iteration.

Muzaferija and Perić (1997) suggested a simpler approach. They noted that it is not necessary to calculate the swept volume from the geometry of the cell vertexes on the free surface; it can be obtained from Eq. (13.52). The displacement of free-surface markers located above the cell-face center is defined by the height h, obtained from the known volume and cell-face area. The new vertex locations are then computed by interpolating h; the resulting swept volume is not exact and iterative correction is necessary. The method is suitable for implicit schemes, for which outer iterations are required at each time step anyway. The 'old' and 'new' locations shown in Fig. 13.15 are now the values for the current and preceding outer iterations; each outer iteration corrects the swept volume according to Eq. (13.52). At the end of each time step, when the outer iterations converge, all corrections are zero. For a detailed description of this approach and its implementation on arbitrary unstructured 3D-grids, see Muzaferija and Perić (1997, 1999).

Flows with free surfaces, such as open channel flows, flows around ships, etc., are characterized by the Froude number:

$$\text{Fr} = \frac{v}{v_{\text{w}}} = \frac{v}{\sqrt{gL}} \,, \tag{13.54}$$

where g is the gravity acceleration, v is the reference velocity, L is the reference length; $\sqrt{gL}$ is the velocity of a wave of length L in deep water. When $\text{Fr} > 1$, the fluid velocity is greater than the wave speed and the flow is said to be *supercritical* and waves cannot travel upstream (as is the case with pressure waves in a supersonic compressible flow). When $\text{Fr} < 1$, waves can travel in all directions. If the method of calculating the free surface shape is not properly implemented, disturbances in the form of small waves may be generated and it may not be possible to obtain a steady

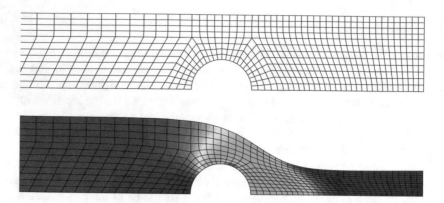

Fig. 13.16 Initial (upper) and final (lower) coarsest grid used in the interface-tracking method to compute the critical flow over a semi-cylinder (the lower plot also shows the dynamic pressure distribution; two-dimensional model)

solution. A method which does not generate waves where they physically should not be (e.g., in front of ships) is said to satisfy the *radiation condition*. Another approach is described in Sect. 13.6.

We show here one example in which both interface-tracking and interface-capturing methods were used and compared. A turbulent flow over a semi-cylinder placed onto channel bottom with the upstream depth-based Froude number lower than unity has been studied. The flow may turn supercritical while passing over the cylinder. In a critical flow, the Froude number becomes equal to unity above cylinder and the flow undergoes a transition from subcritical (Fr < 1) upstream to supercritical (Fr > 1) downstream of cylinder. The upstream water level and the velocity cannot be both independently set; one of these quantities has to adapt to the critical conditions (increasing the flow rate leads to an increase in upstream water depth). Here the case with the ratio of upstream water level to cylinder radius of 2.3 has been selected and the inlet velocity has been set according to results by Forbes (1988) to 0.275 m/s. Figure 13.16 shows a coarse mesh used in the interface-tracking method in its initial and final shape. The problem with this approach is that the mesh needs to be moved and adapted to the shape of the free surface, and this is not easy to automate for arbitrary situations. When the surface mesh points can be simply moved say in vertical direction, the solution is relatively easy; however, the mesh inside the solution domain has to be adapted to the surface mesh motion and this can be complicated for unstructured meshes and large deformation. In this particular case a block-structured grid was used; it has been designed so that the cells retain reasonable quality throughout the solution process. An algebraic smoothing method was used to adapt the interior grid to the motion of free surface. The grid undergoes a large deformation compared to its initial shape above and downstream of the cylinder.

The same flow has been computed using the interface-capturing approach and a trimmed Cartesian grid with prism layers along walls. Figure 13.17 shows the

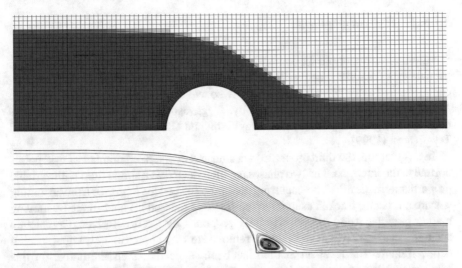

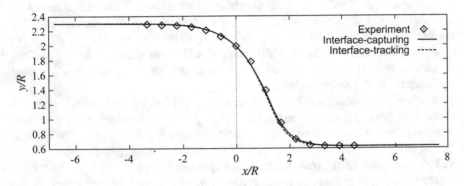

Fig. 13.17 The medium grid showing also volume fraction of water and the interface location (upper) and the predicted flow pattern (lower) obtained using the interface-capturing method to compute the critical flow over a semi-cylinder

Fig. 13.18 Comparison of free-surface shape predicted using interface-tracking and interface-capturing methods with experimental data of Forbes (1988)

grid and the computed distribution of volume fraction. Figure 13.18 shows free-surface profiles computed on the finest mesh used with both methods, compared with experimental data of Forbes (1988). The agreement is very good; actually, even with the coarse mesh shown in Fig. 13.16, the free-surface shape is predicted relatively well. In the case of interface-capturing method, the zone in which volume fraction of water varies between unity and zero is about one cell wide, as can be seen from Fig. 13.17; In this case too the location of the iso-surface of volume fraction 0.5 does not change much when grid fineness is varied. However, in other applications, e.g., when waves need to be captured, the mesh-dependence of the results can be large.

13.5.3   Hybrid Methods

Finally, there are methods for computing two-phase flows that do not fall into either of the categories described above. These methods borrow elements from both interface-capturing and interface-tracking methods so we shall call them hybrid methods. Among these is a method developed by Tryggvason and his colleagues that has been applied to bubbly flows, see Tryggvason and Unverdi (1990) and Bunner and Tryggvason (1999).

In this method, the fluid properties are smeared over a fixed number of grid points normal to the interface. The two phases are then treated as a single fluid with variable properties as in the interface-capturing methods. To keep the interface from becoming smeared, it is also tracked as in interface-tracking methods. This is done by moving marker particles using the velocity field generated by the flow solver. To maintain accuracy, marker particles are added or removed to keep approximately equal spacing between them. The level-set method has been suggested as an alternative for this purpose; see the particle level set method of Enright et al. (2002) or Osher and Fedkiw (2003). After each time step, the properties are re-computed.

Tryggvason and his colleagues have computed a number of flows with this method including some containing hundreds of bubbles of water vapor in water. Phase change, surface tension and merging and splitting of the bubbles can all be treated with this method.

Similar hybrid methods, in which both an additional equation for the volume fraction of one phase and the tracking of the interface are used, have been reported by Scardovelli and Zaleski (1999).

Another hybrid method is known under acronyms PLIC (or VOF-PLIC), meaning piece-wise linear interface construction (Youngs 1982). An equation for volume fraction of one phase is solved as in the interface-capturing method described above, but the interface is in each cell geometrically reconstructed using line segments in 2D and plane surfaces in 3D, such that the cell is split into two parts corresponding to the volume fraction of phases and the direction of normal to the interface (given by the gradient of volume fraction). A review of similar methods is presented by Rider and Kothe (1998). Some special methods have been developed specifically for Cartesian grids (see e.g., Weymouth and Yue 2010 and Qin et al. 2015). Combinations of VOF and level-set methods have also been presented; see e.g., Sussman (2003). Special attention is required when dealing with compressible phases; see e.g., Johnsen and Ham (2012) and Beig and Johnsen (2015).

Methods similar to those described above have been used to produce accurate simulations of detailed free-surface phenomena like air entrainment and mixing in breaking waves (e.g., Deike et al. 2016) and a hydraulic jump (Mortazavi et al. 2016) using DNS. A large number of publications is available in the literature dealing with jet break-up, rising bubbles, bubble collapse near a wall, etc. These methods have been used in conjunction with both LES and RANS modeling of turbulence.

In many applications the interface between two fluids is not sharp; an example is a breaking wave or a hydraulic jump, where a region exists in which the water and air form a foamy mixture. In such a case it would be necessary to add a model for the mixing of the two fluids, similar to the models of turbulent transport in single-phase flows. Such models can be developed using DNS-data as obtained by Deike et al. (2016). Special treatment is also required when computing bubble growth due to heat addition during boiling, as well as in some other flows with free surfaces; it is beyond the scope of this book to go into details of each particular application, but most of the methods in use are related to one of the methods described above.

13.6 Solution Forcing

Instead of trying to construct a numerical scheme which does not reflect waves at boundaries, one can apply *solution forcing* to achieve the same goal. In this approach one blends the solution of the Navier–Stokes (or RANS) equations with a reference solution which is usually either a trivial solution (such as uniform flow or a flat free surface), a theoretical solution, or solution obtained using another method (e.g., one based on potential flow theory) or different boundary conditions (e.g., wave propagation in an infinite domain without any obstacles). The blending is achieved by adding a source term to the discretized momentum equations (and possibly also to the equation for volume fraction of one phase, if an interface-capturing method is used to compute a free-surface flow) of the following form:

$$A_P \phi_P + \sum_k A_k \phi_k = Q_P - \gamma \left[\rho V (\phi - \phi^*) \right]_P , \qquad (13.55)$$

where ϕ_P^* stands for the reference solution at the cell centroid. The source term needs to be blended in gradually, which is accomplished by varying the forcing coefficient γ between zero (at the start of the forcing zone) to its maximum value (at the boundary of the solution domain). It is usually desirable to use a variation which asymptotically tends to zero at the start of blending zone, e.g., an exponential or $\cos^2$-variation. Note that γ is not a dimensionless coefficient; its dimension is [1/s] and the optimum value is problem-dependent.

This generic approach has been used by many authors for various purposes and can be found in literature under different names, depending on application area (relaxation zone, damping zone, sponge layer etc.); we shall not go into details of various variants. The key issue is the choice of the maximum value of the forcing coefficient γ. In most publications, a trial-and-error approach was used to determine the optimum value of this parameter. Recently, Perić and Abdel-Maksoud (2018) published a method which allows determination of a parameter range which leads to reflection coefficients below target value when free-surface waves are to be damped towards outlet boundary. The optimum value depends on wavelength and width of the forcing zone; higher values apply to shorter wavelengths. For free surface waves, values between 1 (very long ocean waves) and 100 (short wavelengths in laboratory experiments) and forcing zones one to two wavelengths wide appear to lead to good results.

If one only wants to force waves towards a flat free surface, it is enough to gradually force the vertical velocity component toward zero. However, one can also force all velocity components, e.g., towards a 5th-order Stokes wave (see e.g., Fenton 1985) if one wants to impose this wave at the inlet boundary. The advantage of using forcing over say one wavelength to just specifying theoretical values for velocities and volume fraction at inlet is that the former will avoid reflection of upstream-traveling disturbances, while the latter will not. Also, if, for example, in a study of flow around a ship side boundaries are treated as symmetry planes, one can force the lateral velocity component to zero, thus avoiding wave reflection from these boundaries. An example of using forcing to avoid wave reflection at boundaries when studying flow around ship in waves will be shown in Sect. 13.10.1.

The same approach can also be used to damp acoustic pressure waves in compressible flows (both in gases and liquids) to avoid reflection from boundaries; the theory of Perić and Abdel-Maksoud (2018) can be used to determine optimal values of γ also in this case. For example, in order to damp acoustic pressure fluctuation at 500 Hz in water, the optimum value of γ is around 8,000 when the forcing zone is two wavelengths wide.

One can also use the forcing approach to avoid disturbances at downstream boundaries in DNS and LES simulations, if the flow carries eddies which make it difficult to specify an appropriate boundary condition which does not reflect disturbances. By forcing the flow towards, e.g., a constant velocity in streamwise direction and zero velocities in cross-flow directions (which would be appropriate in a study of flow around a body in free stream), one can achieve a smooth transition from turbulent flow to a uniform flow for which a simple outlet or specified pressure boundary condition is sufficient.

13.7 Meteorological and Oceanographic Applications

The atmosphere and the oceans are the sites of the largest scale flows on Earth. The velocities may be tens of metres per second and the length scales are enormous so the Reynolds numbers are huge. Due to the very large aspect ratios of these flows (thousands of kilometres horizontally and a few kilometres vertically for both the atmosphere and the ocean), the large-scale flow is almost two-dimensional (although vertical motions are important) while the flow at the small scales is three-dimensional. The Earth's rotation is a major force on the large scales but is less significant at the small scales. Stratification or a stable variation of density is important, mainly at the smaller scales. The forces and phenomena that play dominant roles are different on different scales.

Also, one needs predictions on different time scales. In the case of greatest interest to the public, one wants to predict the state of the atmosphere or ocean for a relatively short time in the future. In weather forecasting, the time scale is a few days while in the ocean, which changes more slowly, the scale is a few weeks to a few months. In either case, a method that is accurate in time is required. At the other extreme are climate studies, which require prediction of the average state of the atmosphere over a relatively long time period. In this case, the short term time behavior can be averaged out and the time accuracy requirement relaxed; however, it is essential to model the ocean as well as the atmosphere in this case. Depending on the resolution required in a simulation (e.g., kilometres or metres), typically the methods are divided as to how turbulence is treated. If the 3D state of the flow is important, large-eddy simulations are often carried out. On the other hand, for flows which are dominantly 2D, but with vertical mixing, RANS approaches with various styles of turbulence modeling are used.

Computations are done on a wide range of different length scales. The smallest region of interest is the atmospheric boundary layer or ocean mixed layer that has dimensions of hundreds of metres. The next scale may be called the basin scale and consists of a city and its surroundings. On the regional- or meso-scale, one considers a domain that is a significant part of a continent or ocean. Finally there are the continental (or ocean) and global scales. In each case, computational resources dictate the number of grid points that can be used and thus the grid size. Phenomena that occur on smaller scales must be represented by an approximate model. Even on the smallest scales of interest, the size of the regions over which averaging must be done is obviously much larger than in engineering flows. Consequently, the models used to represent the smaller scales are much more important than in the engineering large-eddy simulations discussed in Chap. 10.

The fact that significant structures cannot be resolved in simulations at the largest scales requires that calculations be performed at a number of different scales; on each scale, the aim is to study phenomena particular to that scale. Meteorologists distinguish four to ten scales on which simulations are performed (depending on who does the counting). As one might expect, the literature on this subject is vast and we cannot even begin to cover all of what has been done.

As already noted, on the largest scales, atmospheric and oceanic flows are essentially two-dimensional (although there are important influences of vertical motion). In simulations of the global atmosphere or an entire ocean basin, the capacity of current computers requires the grid size in the horizontal directions be between 10 and 100 km. As a result, in these types of simulations, significant structures such as fronts (zones that exist between masses of fluid of different properties) have to be treated by approximate models to render their thickness sufficiently large that the grid can resolve them. Models of this kind are very difficult to construct and are a major source of error in predictions.

Three-dimensional motion is important only on the smallest scales of atmospheric or oceanographic flows. It is also important to note that, despite the high Reynolds numbers, only the portion of the atmosphere closest to the surface is turbulent; this is the atmospheric boundary layer and it usually occupies a region about 1–3 km thick. Above the boundary layer, the atmosphere is stratified and remains laminar. Similarly, only the top layer of the ocean is turbulent; it is 100–300 m thick and is called the mixed layer. Modeling these layers is important because it is within them that the atmosphere and ocean interact and their impact on large-scale behavior is very important.

The numerical methods used in these simulations vary somewhat with the scale on which the simulation is performed. For simulations at the smallest atmospheric or oceanic scales, e.g., in the atmospheric or oceanic boundary layers, one may use methods similar to those used in large-eddy simulation of engineering flows (cf. Chap. 10). For example, Sullivan et al. (2016) studied the nighttime stably-stratified atmospheric BL using a nonhydrostatic code at extreme resolution (0.39 m); this LES code uses a spectral formulation in the horizontal and finite differences in the vertical with RK3 time advancement. An ocean application of a FV nonhydrostatic LES code by Skyllingstad and Samelson (2012) at 3 m resolution allowed examination of frontal instabilities and turbulent mixing in the ocean surface boundary layer. A regional ocean code is the Regional Ocean Modeling System (ROMS; https://www.myroms.org) which solves the hydrostatic flow equations and other coupled models. It is a FV formulation using a predictor (Leapfrog; Sect. 6.3.1.2) and corrector (Adams–Moulton; Sect. 6.2.2) time-step scheme. Here, hydrostatic means that the vertical accelerations are small and the vertical pressure gradient and gravitational forces are in balance.

For simulations not involving clouds, standard codes are adequate, but when clouds and their associated liquid water and water vapor along with ice, etc. are present, it is necessary to add a microphysics package to handle the moist processes (see, Morrison and Pinto 2005). This adds a significant number of additional partial differential equations to solve, but does not otherwise change the numerics significantly. The CM1 code (http://www2.mmm.ucar.edu/people/bryan/cm1/) is a straightforward implementation of such a system, being a nonhydrostatic FD code for compressible fluids with RK3 time advancement. Also, like many atmospheric codes it uses 5th-order or 6th-order accurate advection schemes. Because LES plays an ever more important role in atmospheric modeling, Shi et al. (2018b, 2018a) have explored the accuracy of the subfilter-scale and subgrid-scale models used in a cloud-resolving LES code such as CM1 (see also Khani and Porté-Agel 2017).

As another example, Schalkwijk et al. (2012a, 2015) demonstrated the generation of turbulent clouds using graphics processing units (GPU) and the Dutch Atmospheric Large-Eddy Simulation (DALES) code (Heus et al. 2010), which is a nonhydrostatic Boussinseq FD formulation on an Arakawa C-grid with RK3 time advancement. The code solves seven main prognostic variables (i.e., those advanced in time by PDE). Generally, applications aimed at high resolution and/or eddy/cloud resolution are nonhydrostatic and LES, while global atmospheric and ocean models are hydrostatic and RANS; see, e.g., Washington and Parkinson (2005) for a view of climate modeling. At the global scale, finite-volume methods are used but a spectral method specifically designed for the surface of a sphere is more common. This method uses spherical harmonics as the basis functions.

In choosing a time-advancement method, one must take into account the need for accuracy, but it is also important to note that wave phenomena play a significant role in both meteorology and oceanography. The large weather systems that are familiar from weather maps and satellite photographs may be regarded as very large scale traveling waves. The numerical method should not amplify or dissipate them. For this reason, it was quite common to use the leapfrog method in these fields (Sect. 6.3.1.2).

This method is second-order accurate and neutrally stable for waves. Unfortunately, it is also unconditionally unstable (it amplifies exponentially decaying solutions) so it must be stabilized. For this reason, Runge–Kutta methods (Sect. 6.2.3) have generally replaced it. In particular, the third-order accurate-in-time RK3 schemes are now commonly used (Sect. 6.2.3).

13.8 Multiphase Flows

Engineering applications often involve multiphase flows; examples are solid particles carried by gas or liquid flows (fluidized beds, dusty gases, and slurries), gas bubbles in liquid (bubbly fluids and boilers) or liquid droplets in gas (sprays), etc. A further complication is that multiphase flows often occur in combustion systems. In many combustors, liquid fuel or powdered coal is injected as a spray. In others, coal is burned in a fluidized bed. Finally, multiphase flows with phase change (cavitation, boiling, condensation, melting, solidification) are also often encountered in engineering.

The methods described in Sect. 13.5 may be applied to some types of two-phase flows, especially those in which both phases are fluids. In these cases, the interface between the two fluids is treated explicitly as described above. Some of the methods were specifically designed for this type of flow. However, the computational cost associated with the treatment of interfaces limits these methods to flows in which the interfacial area is relatively small.

There are several other approaches to computing two-phase flows. The carrier or continuous phase fluid is always treated by an Eulerian approach, but the dispersed phase may be handled by either a Lagrangian or an Eulerian method.

The Lagrangian approach is often used when the mass loading of the dispersed phase is not very large and the particles of the dispersed phase are small; dusty gases and some fuel sprays are examples of flows to which this method might be applied. In this approach, the dispersed phase is represented by a finite number of particles or drops whose motion is computed in a Lagrangian manner. The number of particles whose motion is tracked is usually much smaller than the actual number in the fluid. Each computational particle then represents a number (or packet) of actual particles; these are called packet methods. If phase change and combustion are not present and the loading is light, the effect of the dispersed phase on the carrier flow can be neglected and the latter can be computed first. Particles are then injected and their trajectories are computed using the pre-computed velocity field of the background fluid. (This approach is also used for flow visualization; one uses massless point particles and follows their motion to create streaklines.) This method requires interpolation of the velocity field to the particle location; the interpolation scheme needs to be at least as accurate as the methods used for time advancement. Accuracy also requires that the time step be chosen so that particles do not cross more than one cell in one time step.

When the mass loading of the dispersed phase is substantial, the influence of particles on the fluid motion has to be taken into account. If a packet method is used, the computation of particle trajectories and fluid flow must be done simultaneously and iteration is needed; each particle contributes momentum (and energy and mass) to the fluid in the cell in which it is located. Interaction between particles (collision, agglomeration, and splitting) and between particles and walls needs to be modeled. For these exchanges, correlations based on experiment have been used but the uncertainties may be rather large. These issues require another book to be described in any detail; see the book by Crowe et al. (1998) for a description of most widely used methods.

For large mass loadings and when phase change takes place, an Eulerian approach (the *two-fluid model*) is applied to both phases. In this case, both phases are treated as continua with separate velocity and temperature fields; the two phases interact via exchange terms analogous to those used in the mixed Eulerian–Lagrangian approach. A function defines how much of each cell is occupied by each phase. The principles of two-fluid models are described in detail by Ishii (1975) and Ishii and Hibiki (2011); see also Crowe et al. (1998) for a description of some methods for gas-particle and gas-droplet flows. The methods used to compute these flows are similar to those described earlier in this book, except for the addition of the interaction terms and boundary conditions and, of course, twice as many equations need to be solved. The equation system is also much stiffer than in the case of single-phase flows, which requires stronger under-relaxation and smaller time steps.

Cavitation is an important phenomenon which falls into the class of two-phase flows and requires specialized models for its prediction. The most widely used approach is to use a homogeneous model of a multiphase flows, i.e., there is no slip between phases (they share the same velocity, pressure and temperature fields) and the flow of an effective fluid with variable properties is computed. The distribution of each phase is determined by the volume fraction of vapor c_v for which an equation needs to be solved:

$$\frac{\partial}{\partial t} \int_V c_v \, dV + \int_S c_v \mathbf{v} \cdot \mathbf{n} \, dS = \int_V q_v \, dV \ . \tag{13.56}$$

The similarity with interface-capturing method for free-surface flows is obvious, with two important distinctions: (i) cavitation does not necessarily lead to a sharp interface between phases on the grid scale, so one does not need special discretization of convection term (methods used for other scalar variables can be used), and (ii) the equation for vapor volume fraction contains a source term q_v to model the growth and collapse of vapor bubbles. The derivation of the source term for the cavitation model is usually based on the Rayleigh–Plesset equation, which describes the dynamics of a single vapor bubble:

$$R \frac{d^2 R}{dt^2} + \frac{3}{2} \left(\frac{dR}{dt} \right)^2 = \frac{p_s - p}{\rho_l} - \frac{2\sigma}{\rho_l R} - 4 \frac{\mu_l}{\rho_l R} \frac{dR}{dt} \ . \tag{13.57}$$

Here R is bubble radius, t is time, p_s is the saturation pressure for given temperature, p is the local pressure in the surrounding liquid, σ is the surface tension coefficient and ρ_l is the liquid density.

The inertial term is often neglected, because its inclusion in the modeling increases the complexity significantly without substantial benefits in most applications. Also, surface tension effects are only important for the beginning of bubble growth (it sets the limit on the size of seed bubbles which can grow—smaller bubbles are prevented from expanding by the surface tension). The simplified Rayleigh–Plesset equation (in which the inertial, surface tension and viscous terms are neglected) is strictly speaking only meaningful for the bubble growth phase, where it represents an asymptotic solution; the quadratic equation cannot be solved when the surrounding pressure is larger than saturation pressure, because the right-hand side is then negative. The solution to this problem adopted by most authors is to simply take the square root from the absolute value of pressure difference and apply its sign to the result:

$$\frac{\mathrm{d}R}{\mathrm{d}t} = \mathrm{sgn}(p_s - p)\sqrt{\frac{2}{3}\frac{|p_s - p|}{\rho_l}}\,, \tag{13.58}$$

In spite of this deficiency, reasonably good results are obtained with models based on this approximation in most practical applications.

Because with this simplification the rate of bubble radius variation only depends on the local pressure, one does not need to track bubbles explicitly; for each bubble present in a control volume, irrespective of its diameter and the previous history of its motion, the growth rate is only a function of the local pressure in fluid. There are several cavitation models in literature which use Eq. (13.58) to determine the source term in the equation for vapor volume fraction; we describe here only briefly the most widely used model proposed by Sauer (2000); see Zwart et al. (2004) for another similar model.

The model assumes that spherical seed bubbles of an initial radius R_0 are present and uniformly distributed in liquid, as characterized by the number of bubbles per unit volume of liquid, n_0. The number of bubbles in a control volume is thus determined by the amount of liquid. The above model parameters are related to liquid "quality": it is well known that pure liquids (without any dissolved or free gas or solid particles) can withstand very high tensile stresses. Most technical liquids have impurities and the parameter n_0 is usually taken to be of the order of 10^{12}. By filtration and degassing one can reduce or even avoid cavitation in experiments; in simulation, this corresponds to lowering n_0 by several orders of magnitude.

Under the above assumptions, the number of bubbles within a control volume at any time is equal to

$$N = n_0 c_l V\,, \tag{13.59}$$

where c_l is the volume fraction of liquid within control volume V. Obviously, $c_l + c_v = 1$, with c_v being the volume fraction of vapor. The total vapor volume equals

$$V_v = NV_b = N\frac{4}{3}\pi R^3 , \tag{13.60}$$

where V_b is the volume of one bubble with R being the local bubble radius. The vapor volume fraction c_v can now be expressed in terms of n_0, R, and volume fraction of cavitating liquid, c_l:

$$c_v = \frac{NV_b}{V} = \frac{4}{3}\pi R^3 n_0 c_l . \tag{13.61}$$

The local bubble radius can be computed from this expression when the volume fraction of vapor is known:

$$R = \left(\frac{3c_v}{4\pi n_0 c_l}\right)^{1/3} . \tag{13.62}$$

One now needs to define the rate of vapor production or consumption in Eq. (13.56) using the above modeling framework. Obviously, the vapor production results in bubble growth and vice versa, so the rate of bubble radius change is the key parameter. Another parameter is the amount of liquid in the control volume that can cavitate.

The vapor bubbles are moving with the flow; the rate at which vapor is created at any instant of time can thus be approximated by the rate at which volume of bubbles present in the control volume at the given time is changing:

$$q_v \approx \frac{N}{V}\frac{dV_b}{dt} = n_0 c_l \frac{\partial V_b}{\partial R}\frac{dR}{dt} = n_0 c_l 4\pi R^2 \frac{dR}{dt} . \tag{13.63}$$

The rate at which bubble radius grows can be computed from Eq. (13.58), which completes the cavitation model. A more elaborate derivation of the model can be found in Sauer (2000) and Schnerr and Sauer (2001). This model is available in most commercial and public CFD-codes and has been successfully applied to study cavitating flows around ship propellers, in pumps and turbines, fuel injectors and other devices. Some authors multiply the source term with another parameter which has a different value for the growth and the collapse phase (positive or negative source term).

We use an example of flow around a propeller under the open-water test condition, (i.e., the propeller is operating in a uniform flow, rotating at a fixed, prescribed rate) to demonstrate both the performance of the above-described cavitation model and several aspects of error estimation and the interaction of errors from different sources. Under open-water conditions, one can compute the flow around a single propeller blade, using periodic boundary conditions in the circumferential direction and the rotating frame of reference. This reduces computational cost and allows the use of much finer grids than would be possible if the ship and rudder were present (in which case the whole propeller would have had to be included in the simulation and the grid attached to propeller would have to rotate with it, as shown in Figs. 13.9 and 13.10). The propeller is the one used as a test case at the Symposium on Marine Propulsors in 2011 and 2015 (www.marinepropulsors.com); the operating conditions are from the cavitating test case (Propeller VP1304 from SVA Potsdam, test case 2.3.1; see the

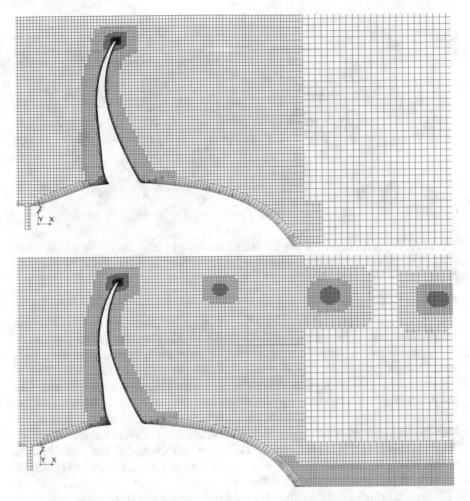

Fig. 13.19 Computational grid in a longitudinal section through propeller: without (upper) and with a special local refinement for tip-vortex capturing (lower)

workshop proceedings for a detailed description of the test case and the towing-tank report; Heinke 2011).

Using systematically refined grids without special local refinement for tip-vortex capturing (see upper plot in Fig. 13.19) leads to cavitation patterns and values of thrust and torque which do not vary much on the two finest grids (the difference being of the order of 0.1%), leading to two false conclusions: (i) the discretization errors appear to be very small, (ii) the modeling errors appear to be very large, because one does not see any cavitation in the tip vortex range (see left picture in Fig. 13.20). Tip vortexes can be identified in these solutions using iso-surfaces of vorticity magnitude, as shown in the right picture of Fig. 13.20. The vorticity distribution can therefore be used to guide local grid refinement within the tip vortex, such that in

Fig. 13.20 Computed iso-surface of vapor volume fraction 0.05 using a grid without local refinement for tip-vortex capturing and k-ω turbulence model (left) and an iso-surface of vorticity magnitude from the same solution (right)

the vortex core the cell size is small enough to resolve the rapid variation of pressure and velocity there (smaller than 1/1000th of propeller diameter), as shown in one longitudinal section through the grid in Fig. 13.19. This leads to the appearance of tip vortex cavitation, but in the unsteady RANS simulation, the cavitation zone ends much sooner than the local grid refinement ends, as shown in Fig. 13.21. This is due to the over-estimation of turbulent viscosity within the tip vortex, which smears the peaks in pressure and velocity profiles, causing pressure to rise above saturation level too soon. By switching to LES, the cavitating zone lasts to the very end of the local grid refinement, which is also shown in Fig. 13.21. Turbulent viscosity from the subgrid-scale model is much lower than that from RANS-model. The LES-solution corresponds very well to the experimentally observed pattern shown in Fig. 13.22. While both the mean shape and position of tip and hub vortex remain stable in both LES and experiment, one can also observe fluctuations around the mean shape.

Although the finest grid in this study was relatively fine (ca. 15 million cells for a single propeller blade), no significant improvement in RANS-solution could be observed compared to the solution on the next coarser grid. This is one of the major drawbacks of this modeling approach: once discretization errors become smaller than modeling errors, further grid refinement brings no benefits. With LES, grid refinement not only reduces discretization errors but also the part of the turbulence spectrum that needs to be modeled; at a sufficiently fine grid level, it turns essentially into a DNS and turbulent viscosity becomes negligible. In this example, LES on a coarse grid produces pulsating tip-vortex cavitation: it periodically varies between extending to the end of local grid refinement and shrinking to the extent similar to the RANS-solution. However, once the grid was further refined (around 5 million cells), the cavitation zone became stable. The next refinement exposed finer structures in both tip and hub vortex, as expected. A separate report with more details about this simulation is available on the book web-site; see the appendix for details.

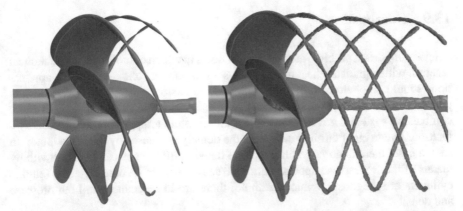

Fig. 13.21 Computed iso-surface of vapor volume fraction 0.05 using a grid with local refinement for tip-vortex capturing and k-ω turbulence model (left) and using the same grid but LES-simulation (right)

Fig. 13.22 Photographs of cavitation pattern observed in an experiment at two time instants, showing stable and fluctuating features of tip and hub vortex cavitation (source: SVA Potsdam)

This example also shows that one has to be careful when assessing both discretization and modeling errors. It is not enough to just monitor some integral quantities like thrust and torque, unless one is only interested in them; local flow features may still be far from converged when integral quantities stop changing significantly, especially in the wake of bodies (due to a limited upstream influence at high Reynolds numbers). In this particular case, the solution is much more sensitive to the local grid refinement and turbulence model than to the cavitation model. The latter performs surprisingly well, taking into account that the relation between bubble growth rate and pressure given by Eq. (13.58) is a very crude approximation of the Rayleigh–Plesset equation.

13.9   Combustion

Another important problem area deals with flows in which combustion, i.e., chemical reaction with significant heat release, plays an important role. Some of the applications should be obvious to the reader. Some combustors operate at nearly constant pressure so the principal effect of heat release is reduction of the density. In many combustion systems, it is not unusual for the absolute temperature to increase by a factor of five to eight through the flame; the density decreases by the same factor. In such a case, there is no possibility that the density differences can be dealt with by means of the Boussinesq approximation discussed earlier. In other systems (engine cylinders are the most common example) there are large changes in both pressure and density.

It is possible to do direct numerical simulation of turbulent combusting flows but only for very simple cases. It is important to note that the speed of travel of a flame relative to the gas is rarely greater than 1 m/s (explosions or detonations are an exception). This speed is much lower than the speed of sound in the gas; usually, the fluid velocities are also well below the sound speed. Then the Mach number is much less than unity and we have the strange situation of a flow with large temperature and density changes that is essentially incompressible.

It is possible to compute combusting flows by solving the compressible equations of motion. This has been done (see Poinsot et al. 1991). The problem is that, as we have pointed out earlier, most methods designed for compressible flows become very inefficient when applied to low speed flows, raising the cost of a simulation. For this reason, these simulations are very expensive; this is especially so when the chemistry is simple. However, when more realistic (and therefore more complex) chemistry is included, the range of time scales associated with the chemical reaction is almost always very large and this dictates that small time steps be used. In other words, the equations are very stiff. In this case, the penalty for using compressible flow methods may be largely eliminated.

An alternative approach is to introduce a low-Mach-number approximation (McMurtry et al. 1986). One starts with the equations describing a compressible flow and assumes that all of the quantities to be computed may be expressed as power series in the Mach number. This is a non-singular perturbation theory so no special care is required. The results are, however, somewhat surprising. To lowest (zeroth) order, the momentum equations reduce to the statement that the pressure $p^{(0)}$ is constant everywhere. This is the thermodynamic pressure and the density and temperature of the gas are related by its equation of state. The continuity equation has the compressible (variable density) form, which is no surprise. At the next order, the momentum equations in their usual form are recovered but they contain only the gradient of the first-order pressure, $p^{(1)}$, which is essentially the dynamic head found in the incompressible equations. These equations resemble the incompressible Navier–Stokes equations and can be solved by methods that have been given in this book.

In the theory of combustion, two idealized cases are distinguished. In the first, the reactants are completely mixed before any reaction takes place and we have the case of premixed flames. Internal combustion engines are close to this limit. In premixed combustion, the reaction zone or flame propagates relative to the fluid at the laminar flame speed. In the other case, the reactants mix and react at the same time and one speaks of non-premixed combustion. These two cases are quite different and are treated separately. Of course, there are many situations that are not close to either limit; they are called partially premixed. For a complete treatment of the theory of combustion, the reader is advised to consult the well-known work by Williams (1985).

The key parameter in reacting flows is the ratio of the flow time scale to the chemical time scale; it is known as the Damköhler number, Da. When the Damköhler number is very large, chemical reaction is so fast that it takes place almost instantaneously after the reactants have mixed. In this limit, flames are very thin and the flow is said to be mixing-dominated. Indeed, if the effects of heat release can be ignored, it is possible to treat the limit $Da \rightarrow \infty$ as one involving a passive scalar and the methods discussed at the beginning of this chapter can be used.

For the calculation of practical combustors, in which the flow is almost always turbulent, it is necessary to rely on solution of the Reynolds-averaged Navier–Stokes (RANS) equations. This approach and the turbulence models that need to be used in conjunction with it for non-reacting flows were described in Chap. 10. When combustion is present, it is necessary to solve additional equations that describe the concentrations of the reacting species and to include models that allow one to compute the reaction rate. We shall describe some of these models in the remainder of this section.

The most obvious approach, that of Reynolds-averaging the equations for a reacting flow, does not work. The reason is that chemical reaction rates are very strong functions of temperature. For example, the reaction rate between species A and species B might be given by:

$$R_{AB} = K e^{-E_a/RT} Y_A Y_B , \qquad (13.64)$$

where E_a is called the activation energy, R is the gas constant, and Y_A and Y_B are the concentrations of the two species. The presence of the Aarhenius factor $e^{(-E_a/RT)}$ is what makes the problem difficult. It varies so rapidly with temperature that replacing T with its Reynolds-averaged value produces large errors.

In a high Damköhler number non-premixed turbulent flame, the reaction takes place in thin wrinkled flame zones. For this case, several approaches have been used, of which we will briefly describe two. Despite the significant difference in philosophy and appearance between them, they are more alike than it would seem.

In the first approach, one takes the point of view that, because mixing is the slower process, the reaction rate is determined by how fast it takes place. In that case, the rate of reaction between two species A and B is given by an expression of the form:

$$R_{AB} = \frac{Y_A Y_B}{\tau} ,$$ (13.65)

where τ is the time scale for mixing. For example, if the k-ε model is used, $\tau = k/\varepsilon$. In the k-ω model, $\tau = 1/\omega$. A number of models of this kind have been proposed, perhaps the best known of which is the eddy break-up model of Spalding (1978). Models of this type are in common use for the prediction of the performance of industrial furnaces.

Another type of model for non-premixed combustion is the laminar flamelet model. Under stagnant conditions, a non-premixed flame would slowly decay as its thickness increases with time. To prevent this from happening, there must be a compressive strain on the flame. The state of the flame is determined by this strain rate or, its more commonly used surrogate, the rate of scalar dissipation, χ. Then it is assumed that the local structure of a flame is determined by just a few parameters; at minimum, one needs the local concentrations of the reactants and the scalar dissipation rate. The data on flame structure is tabulated. Then the volumetric reaction rate is computed as the product of the reaction rate obtained by table look-up and the flame area per unit volume. A number of versions of the equation for the flame area have been given. We shall not present one here but it should suffice to say that these models contain terms describing the increase of the flame area by stretching of the flame and destruction of flame area.

For premixed flames, which propagate relative to the flow, the equivalent of the flamelet model is a kind of level-set method. If the flame is assumed to be the location where some variable $G = 0$, then G satisfies the equation:

$$\frac{\partial G}{\partial t} + u_j \frac{\partial G}{\partial x_j} = S_L |\nabla G| ,$$ (13.66)

where S_L is the laminar flame speed. One can show that the rate of consumption of reactants is $S_L |\nabla G|$, thus completing the model. In more complex versions of the model, the flame speed may be a function of the local strain rate just as it depends on the scalar dissipation rate in non-premixed flames.

Finally, note that there are many effects that are very difficult to include in any combustion model. Among these are ignition (the initiation of a flame) and extinction (the destruction of a flame). Models for turbulent combustion are undergoing rapid development at the present time and no snapshot of the field can remain current for very long. The reader interested in this subject should consult the book by Peters (2000).

13.10 Fluid-Structure Interaction

Fluids always exert forces onto submerged structures (even without a flow), but so far we assumed that solid walls are rigid. By *fluid-structure interaction* (FSI) we mean motion or deformation of solid bodies exposed to fluid flow. The flow is thus affected by the motion of solid walls and a coupled simulation of both motions is necessary.

The simplest form of FSI is the motion of flying or floating rigid bodies; the next level of complexity is introduced when the solid structure also deforms due to flow-induced forces. We discuss briefly how these phenomena can be simulated and show some illustrative examples.

13.10.1 Floating and Flying Bodies

If the flying or floating body can be considered rigid, its motion can be simulated by solving ordinary differential equations for the translation of center of mass and rotation around it:

$$\frac{d(m\mathbf{v})}{dt} = \mathbf{f} \, , \tag{13.67}$$

$$\frac{d(M\boldsymbol{\omega})}{dt} = \mathbf{m} \, , \tag{13.68}$$

where m is the mass of the body, $\mathbf{v}$ is the velocity of its center of mass, $\mathbf{f}$ is the resultant force acting on the body, M is the tensor of moments of inertia of the body with respect to the global (inertial) coordinate system, $\boldsymbol{\omega}$ is the angular velocity of the body and $\mathbf{m}$ is the resultant moment acting on the body. Because during motion the body continuously changes its orientation with respect to the global coordinate system, one would have to recompute its moments of inertia at each time step; for all but the simplest body shapes, this would be quite impractical. For this reason, one usually solves the equation of angular motion in a modified form, using a coordinate system fixed to the body at its center of mass, with respect to which the moments of inertia do not change as the body moves (see, e.g., Shabana 2013):

$$M_b \frac{d\boldsymbol{\omega}}{dt} + \boldsymbol{\omega} \times M_b \boldsymbol{\omega} = \mathbf{m} \, , \tag{13.69}$$

where M_b stands for the tensor of moments of inertia in the body-fixed (moving) coordinate system.

The forces acting on the body always include gravity force and flow-induced pressure (normal to body surface) and shear force (tangential to body surface); in addition, there may be external forces which are either independent of the flow (e.g., a propulsion force generated by a motor inside body) or depending on flow and body motion (e.g., a force resulting from a mooring line). Except for the gravity force (which, by definition, acts at the center of mass and thus creates no moment), all other forces in general produce a moment which influences the rotational motion of the body.

Note that—contrary to equations of motion for solid particles in a Lagrangian-modeling of particle-laden two-phase flows—we do not need to include in **f** drag, lift, virtual mass or other forces which account for special effects resulting from flow-body interaction. These forces are used to *model* the fluid-particle interaction in the Lagrangian approach because the grid then does not resolve the flow around an individual particle. Here, the interaction between body and flow is taken into account directly at the fluid-solid interface; and thus, all effects are included in the pressure and shear force.

Equations (13.67) and (13.69) represent a system of six ODE for three components of each velocity vector (linear and angular); one therefore says that a moving body has *six degrees of freedom* (6 DoF) if its motion is not constrained in any way. These equations can be solved using methods described in Chap. 6 to obtain **v** and ω at the new time level; in a coupled simulation of flow and body motion, it is common to use the same method for time-advancement in both sets of equations. In order to obtain the new position of the body center of mass and its new orientation, one has to integrate also the following set of equations (which are just the definitions of **v** and ω):

$$\frac{d\mathbf{r}}{dt} = \mathbf{v} \quad \text{and} \quad \frac{d\mathbf{\Omega}}{dt} = \omega, \tag{13.70}$$

where **r** is the vector defining the position of the center of mass of the body and $\mathbf{\Omega}$ is the vector defining the orientation of the body with respect to the body-fixed coordinate system.

Because flow-induced forces affect the body motion and the change of position and orientation of the body affect the flow, the two problems are strongly coupled (both ways). It is therefore necessary to solve the two sets of momentum equations in a coupled way. Because the flow around moving bodies is always unsteady and time steps are not too large, sequential solution methods like the one depicted in Fig. 7.6 are used. It is then easy to include the solution of equations of motion for rigid bodies inside the outer iteration loop. One first estimates the flow field at the new time step while the body is still at the position computed in the previous time step. Then estimated forces acting on the body are applied to determine an estimate of the new body position (under-relaxation may be used in both cases). The grid in the fluid region is then adapted to the newly estimated body position and the next outer loop is started. These iterations continue until neither the forces resulting from estimated flow at the new time step nor the estimated position of the body change more than a prescribed tolerance. Even if a non-iterative time-advancing method is used to compute the fluid flow (like PISO or fractional-step method described in Sects. 7.2.1 and 7.2.2), it is still desirable to introduce the outer iteration loop in order to implicitly couple the solutions for fluid flow and body motion, especially if light bodies are moving in a heavy fluid.

Consider as an example a light body with density 1 submerged in liquid with density 1000; let us assume that the body is fixed to a bottom wall by a rope and at the start of simulation, both the fluid and the body are at rest. The forces acting on the body in vertical direction are the gravity force $-\rho_b V g$, the buoyancy force

$\rho_l V g$, and the restraint force in the rope, which is equal to the difference between the buoyancy and the gravity force. Here ρ_b is body density, ρ_l is liquid density, V is body volume and g is the magnitude of gravity component in vertical direction. If we now cut the rope and let the body loose, it will start moving upward due to the fact that the buoyancy force is larger than the gravity force. In the case of an explicit solution method, the fluid would stay at rest in the first time step because the body is still at its old position. Now the forces acting on the body are just the gravity and the buoyancy force, leading to an astronomical body acceleration of $999g$ (see Eq. (13.67)):

$$\rho_b V \frac{dv}{dt} = (\rho_l - \rho_b) V g \quad \Rightarrow \quad \frac{dv}{dt} = \frac{(\rho_l - \rho_b)g}{\rho_b} = 999g \ .$$

This would lead to body moving too far in the first time step; the flow would then in the second time step react with an excessively large resistance force in the downward direction. The second time step for body motion would result in reversed direction of motion, because the fluid resistance is larger than buoyancy force, and so on; the result is an oscillatory divergence. With an implicit coupling and under-relaxation to both fluid flow and body motion, one achieves after a few outer iterations a situation with a balance between computed body acceleration on the one side and the sum of gravity, buoyancy and resistance force on the other side; this balanced equation delivers a reasonable body motion.

The largest challenge in this kind of simulations is the adaptation of the computational grid in the fluid to the body motion. Contrary to problems with a prescribed body motion, where the grid needs to be adapted only once at the beginning of each new time step, now the adaptation needs to be performed in each outer iteration. This increases the demand for an efficient solution. Two approaches are most widely used: overlapping grids and grid morphing. In the first approach, one grid is attached to the body and moves with it without any deformation. The grid quality remains the same all the time, and there is no change in discretization errors around body due to variation of grid quality, as is the case with alternative approaches. The downside is that the interpolation stencils for the coupling of solutions on the overlapping and background grid change and need to be updated in each outer iteration.

Mesh morphing relies on either solution of additional (partial-differential or algebraic) equations or on some kind of algebraic smoothing of coordinates of grid vertexes. One approach is to consider the fluid domain as a pseudo-solid (with favorable properties); the motion of boundary vertexes as a result of body motion is imposed as a Dirichlet boundary condition for the displacement and by solving the momentum equations for the pseudo-solid, this displacement propagates into the body of the fluid domain. By varying the properties of the pseudo-solid, one can achieve that, e.g., the prism layer zone near solid walls moves almost as a rigid body, while the grid deforms further away from boundary. However, the initial grid can be deformed by morphing only for moderate body motion (e.g., motion of a ship in waves); extreme motions may lead to grid distortions which make it unusable. One can, in principle, halt the simulation when the grid quality becomes poor, generate a new grid for given body position, interpolate current and one or more old solutions onto the new cell centroids, and then continue the simulation by treating the interpolated solution as the initial condition.

We now present two illustrative examples of coupled simulation of flow and motion of floating and flying bodies. In the first example we want to predict added resistance when a ship moves in head waves at constant speed. In reality, ships are propelled with a (nearly) constant thrust and the speed of forward motion drops when waves are present compared to speed at calm water; however, experiments in a towing tank are often performed by towing a ship model at constant speed. The main questions to be answered are: (i) by how much is the mean resistance increased due to the presence of waves compared to resistance in calm water, and (ii) what speed can the vessel achieve in waves with the available power on board?

When waves are present, the grid design has to change a bit: in a calm water, we can coarsen the grid in all directions far away from hull, especially ahead and sideways, but with waves, we have to maintain a nearly constant number of cells per wavelength in the wave propagation direction and also per wave height in the vertical direction throughout the domain. Changes in grid resolution (especially when they are abrupt and lead to a too coarse resolution of the wave profile) can cause disturbances which then propagate further throughout the domain. In Fig. 13.23 the minimum resolution of wavelength is maintained, while the grid is coarsened in the spanwise direction where the incoming long-crested wave is not disturbed. The hull is a model of a container ship (the so-called KCS—KRISO Container Ship; this is a ship which has never been built but whose model has been expensively tested in many towing tanks), ca. 7.5 m long. The wavelength is equal to ship length, and the solution domain is about 3.3 wavelengths long and just above one wavelength wide; only half of the flow domain is included in the simulation, because the propeller is not present and the flow is assumed to be symmetric about the ship symmetry plane. The ship is allowed to heave and pitch; all other motions are suppressed. Computation is performed in a coordinate system attached to ship hull and moving with it at constant speed (here: 1.2 m/s). The wave height is 0.2 m and the wave period is 2.184 s; it is modeled as a Stokes' 5th-order wave using the solution presented by Fenton (1985). Ship motion was accounted for by using overlapping grids, and the free surface is captured using the HRIC-scheme (Muzaferija and Perić 1999). More detailed information about this simulation can be found on this book's web page; see the appendix for details.

In order to keep the solution domain as small as possible and avoid reflection of waves from boundaries, the forcing method presented in Sect. 13.6 is used: both the velocity and the volume fraction of water are forced towards the theoretical solution of Fenton (1985) over a distance of 5 m from inlet, side and outlet boundary, using $\cos^2$ variation of the forcing coefficient, from 0 at the start of forcing to 10 at the boundary. The ship-induced disturbance of incoming waves propagates in all directions, but it gradually vanishes in the forcing zone and the solutions of RANS-equations transition smoothly into the theoretical solution, see Fig. 13.23. This kind of forcing on all boundaries can only be applied when the reference solution sufficiently well satisfies the Navier–Stokes equations; otherwise, disturbances within the forcing zone would appear due to the mismatch between the wave propagation by the Navier–Stokes

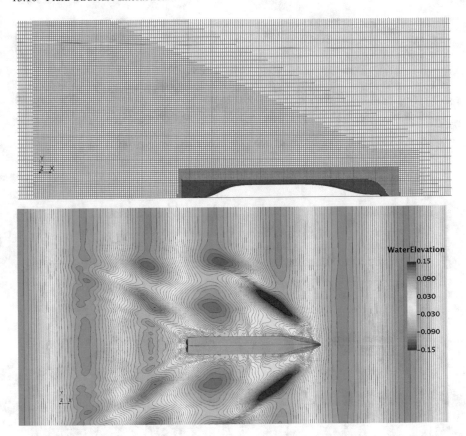

Fig. 13.23 Computational grid in the undisturbed free surface around a ship model (upper) and the computed instantaneous free surface elevation (lower)

equations within the solution domain and the theoretical solution (this would be the case with theoretical solutions based on linear theory, e.g., long-crested irregular wave models based on superposition of linear waves). The Stokes' 5th-order theory is a very accurate wave model and the solution of Navier–Stokes equations—with a sufficient resolution of wavelength and wave height by the numerical grid—matches this theory quite well.

Figure 13.24 shows the ship at two phases of its motion through waves: when its bow dives deep into the crest of the incoming wave, and when it comes completely out of the water as the wave crest moves towards the mid-ship. The forces acting on the ship and its motions are shown in Fig. 13.25. While the shear force remains nearly constant, the pressure force varies by a large amplitude around the mean value; the amplitude of oscillation is 5 times larger than the mean value, which leads to resistance being negative over more than a third of the period! The ship is pitching by $\pm\,3°$ and heaving by $\pm\,5.5$ cm; the period of motion is shorter than the wave period—1.62 s versus 2.184 s. This is due to the fact that the ship moves towards

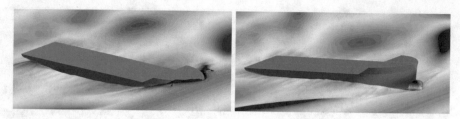

Fig. 13.24   Ship motion in waves: diving into a wave crest (left) and bow out of water shortly before crest reaches midship (right)

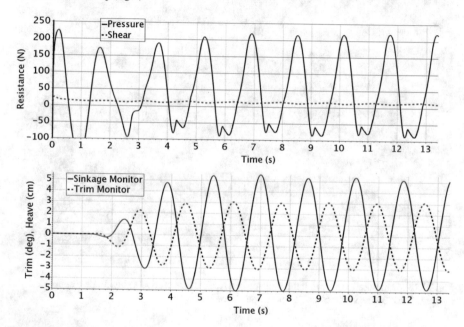

Fig. 13.25   Predicted time-variation of shear and pressure drag components (upper) and of ship heave and pitching motion (lower)

the waves, which increases the encounter frequency. After about 4 wave periods, the solution becomes nearly periodic.

The second example deals with a simulation of lifeboat water entry. When a lifeboat is released from an offshore platform or a ship, it first flies through air, then dives into water, re-surfaces and continues to float and move with its own propulsion as an ordinary boat. The key questions here are: (i) what deceleration will be experienced by people inside the lifeboat when it hits water, (ii) to what loads will the lifeboat structure be exposed during water entry, and (iii) when and where will the lifeboat re-surface upon diving into water? The answers to these questions are critical for the survival of people inside the lifeboat: (i) a human body cannot sustain too high a deceleration over too long a period; (ii) too high loads during water entry may damage the lifeboat structure and, thus, threaten the lives of people inside it; (iii) if the lifeboat dives too deep and comes out of water too late or at a wrong location, it may be hit by falling objects or collide with platform or ship.

The major problem is that the answers to all of the questions depend on too many parameters, e.g.: drop height, wave propagation direction, wavelength and wave height, the point on wave profile at which lifeboat enters water, wind direction and speed, etc. All of these factors influence each other, so the number of combinations to look at is enormous. Many effects cannot be realized in model experiments, so the interest in simulation has become very high lately. One of the first studies of lifeboat water entry and a comparison of simulation results with limited experimental data was presented by Mørch (2008, 2009) showed that the maximum pressure on the lifeboat surface during water entry can vary by a factor of 4, depending on where the boat hits the wave, for the same wave propagation direction, wavelength and wave height.

Simulations of lifeboat water entry are easiest done using overlapping grids: one creates a grid in a domain obtained by, e.g., subtracting the lifeboat body from a cylinder, which moves as a rigid body with the boat, and a separate grid for the environment, which is adapted to resolving wave propagation and which may include the platform, ship or any other objects. Usually, the release of a lifeboat and its initial motion through air are simulated using simpler methods; a detailed simulation using CFD starts at some distance above water, with orientation, linear and angular velocities taken from the other simulation method and imposed as initial conditions. Because the lifeboat is very heavy, air flow around it will quickly adapt to its motion without introducing a significant disturbance (resistance in air is relatively low compared to boat's weight and initial momentum). Figure 13.26 presents two photographs from an experimental study of lifeboat entry into calm water, showing clearly the complexity of the resulting two-phase flow around the boat. Fortunately, one does not need to resolve all the flow features (like thin water sheets, their transition to droplets or the foamy zone in the wake after re-surfacing) in order to provide reasonably good answers to the questions posed above.

Figure 13.27 shows overlapping grids used to compute the flow around the lifeboat shown in Fig. 13.26, under the same conditions as in the experiment. Ideally, the

Fig. 13.26 Pictures from an experimental study of lifeboat water entry (from Mørch et al. 2008)

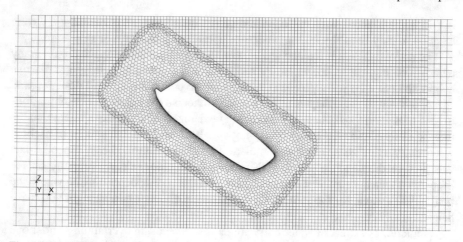

Fig. 13.27 Overlapping grids used to simulate the water entry of a lifeboat

background grid should, every now and then, be refined ahead and coarsened behind the lifeboat as it moves forward, but that feature was not available in the code when the simulation was performed. For this reason the background grid was refined to a cell size similar to that of cells in the outer layers of the overlapping grid attached to the lifeboat in a larger zone within which the lifeboat is expected to move. It is in addition refined around free surface in order to capture its deformation during water entry and re-surfacing of the lifeboat. In this example, the lifeboat is about 15 m long, the drop height was 36 m and the launching angle was 35°. More details about this and similar simulations can be found in Mørch et al. (2008) and (2009).

During the free fall through air, the lifeboat accelerates nearly with gravity acceleration, g, because the resistance in air is not high. However, when the lifeboat hits water surface, it experiences a sudden increase in resistance to its motion, which leads to a significant deceleration. Figure 13.28 shows variation of acceleration relative to gravity, evaluated at two locations within lifeboat: one in the front and one in the rear part (zero means the acceleration equals gravity). Within a very short period, deceleration in front reaches $5g$; at the same time, the rear part of the lifeboat experiences acceleration of $3g$, because, as the bow hits water, the boat starts rotating around it. Within about 0.3 s, the acceleration of the rear part changes from $3g$ in gravity direction to $6g$ in the opposite direction, which makes in total $9g$ deceleration. This is already a critical value for a normal human, but in some experiments decelerations of up to $30g$ were measured! Even if the boat structure remained intact, nobody inside would have survived!!

As can be seen from Fig. 13.28, the agreement between simulation and experiment is quite good, in spite of a relatively coarse grid used in the simulation (about 1 million cells for half a boat; because the boat was only free to move in x and z direction and rotate around y-axes) and the water was calm, symmetry conditions were applied. While the agreement is not perfect in all details, the major features of the water entry and re-surfacing are well captured. The second deceleration at 5 s comes about

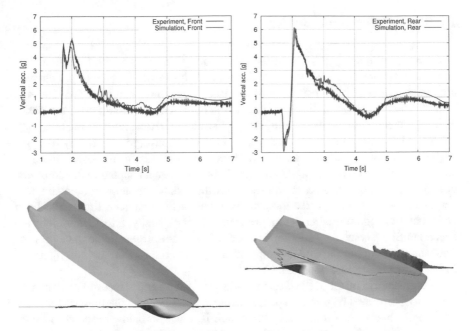

Fig. 13.28 Predicted time variation of acceleration in the front and rear part of the lifeboat (upper) and pressure on the lifeboat surface and the free-surface deformation at two time instances (lower)

because, after re-surfacing, the boat first jumps out of water and then lands again, leading to a deceleration of about $1g$.

One might expect that the highest pressure on the boat surface would be recorded at the bow when it hits the water, but that is not the case. As can be seen from Fig. 13.28, the pressure is highest at the intersection of the undisturbed free surface and the boat surface, and because the angle between the boat axis and the free surface reduces during the entry phase, the pressure increases along the boat from the bow toward the stern during this time. The highest pressure is obtained at the location just before the stern curvature of boat bottom starts; at that location, the deadrise angle is the smallest. In this examples pressures above 5 bar are measured; much higher pressures can be obtained under different conditions.

As the boat dives into water, an air cavity is formed behind it, because the body entering water displaces liquid to the side. As the body moves forward, liquid comes back and closes the cavity; this creates a big splash on the surface, but also high pressure loads on the rear part of the boat. Tregde (2015) investigated the behavior of the air cavity and found that it was essential to account for the effects of air compressibility in such simulations; treating the gas phase as incompressible led to a significant under-estimation of pressure loads, while the comparison with experiments was pretty good if compressibility was accounted for. Berchiche et al. (2015) presented an extensive validation study in which CFD-simulations of free-fall lifeboat launches into regular waves were compared with experimental data.

13.10.2 Deformable Bodies

The next level in complexity is reached if the body interacting with fluid flow may deform under flow-induced forces. If the solution has a steady state (e.g., when an airplane is cruising at a constant speed in calm air, its wings bend upwards and stay steady in that position, until turbulence is encountered), one can perform a series of independent simulations of fluid flow and structural deformation. One first computes fluid flow for undeformed body, until it (nearly) converges to a steady state; the computed pressure and shear forces are passed on as loads for the computation of structural deformation. The grid in the fluid domain is then adapted to the deformed body and the computation is continued, until the forces acting on body again converge. The process is repeated until the changes in flow-induced forces and body deformation become smaller than a prescribed limit. Usually, 5 to 10 iterations are needed until the converged solution is obtained. The convergence tolerance for both flow and structural deformation computation can be relaxed initially and tightened as the converged state is approached.

In the scenario just described, the simulation of flow and structural deformation is usually performed by different people using different codes; the flow is computed using an in-house or commercial code which is, most probably, based on a finite-volume method like those described in this book, while structural deformation is most likely computed using a commercial finite-element code. The two simulation teams and codes communicate via file exchange: the flow team writes out a file with forces acting on each wall boundary face, while the structural team sends back the new body shape or displacement of a set of boundary points. Because the grids are usually very different in fluid and in solid, interpolation (also called "mapping") of solutions is necessary. Very often, even the geometries in the two codes are not the same. For example, for the fluid flow analysis, the skin of an airplane wing is the relevant wall boundary; however, the skin is hardly relevant for the structural analysis—it only takes the fluid load but the structurally relevant body is a frame under the skin. Thus, the loads computed by simulating fluid flow along the skin have to be passed on to the frame, and the deformation of the frame (displacements at a set of discrete locations) needs to be mapped to the vertexes of fluid grid at the skin surface. Accordingly, the coupling is not as trivial as it may sound.

When the problem is transient, a much tighter coupling between flow and structural computation is needed. At minimum the exchange of forces and displacements needs to occur once per time step. However, as already mentioned for the simulation of the motion of rigid bodies, an explicit coupling—in which the flow solver completes one time step using frozen body shape and then the structural solver computes deformation using frozen flow-induced forces—may not always converge. Ideally, the exchange should happen after every outer iteration in each time step, if a tight (implicit) coupling is to be achieved. This is difficult to accomplish when using two different codes, unless they are designed for such co-simulation (in which one code leads and the other follows); the authors know of only one set of commercial codes that can communicate in this way. They can run concurrently and exchange data via memory sockets, so that it is not necessary to write any exchange files.

An ideal situation would be the computation of both fluid flow and structural deformation using the same code. For a while it seemed that finite volume methods could be suitable for both tasks. Demirdžić and Muzaferija (1994, 1995) presented a finite volume method based on arbitrary polyhedral control volumes and second-order discretization for both fluid flow and structural deformation; a number of similar methods were published subsequently. Demirdžić et al. (1997) demonstrated that the computation of structural deformation using finite volume method can be greatly accelerated using the multigrid method.

However, it turned out that second-order discretization, which is typically used to compute fluid flow (midpoint rule for integral approximations, linear interpolation and central differences for gradients), is not sufficient for some structural problems—in particular for deformation of thin structures under certain conditions (a cantilever is a typical example). Demirdžić (2016) demonstrated that a fourth-order 2D finite volume method (based on Simpson's rule approximation of surface integrals, interpolation by cubic polynomials and 4th-order central differences) is both efficient and accurate when solving the cantilever problem. However, developing a high-order finite volume method for arbitrary polyhedra is not a simple task and may not bring advantages compared with well-established finite element methods for structural analysis; the latter do not use the same equations for all structures but rather use different elements (based on different theories) for shells, plates, membranes, beams and generic volume elements.

The most recent trend in commercial codes is to integrate the finite volume flow solver and finite element structural solver in the same code; in this case, a tight coupling without any external communication can be achieved at outer iteration level, allowing for a simultaneous computation of both flow and structural deformation. The only problem with this approach is that now the same person needs to know how to set-up both simulations, or the setting-up of coupled simulations has to be performed by a team of two people—one being an expert for fluid flow and the other being expert for the structural side. Once the simulation is properly set up, a single code user (who might be the third person) can easily perform parametric studies, because for this one only needs to change the geometry or input data.

The field of fluid-structure interaction is vast; we cannot but touch on some illustrative examples. In engineering applications, one is usually interested in predicting situations which might lead to *resonance*. If fluid flow produces oscillating loads on the structure at a frequency which is close to the structure's resonant frequency, the structure may start to oscillate with increasing amplitudes, leading eventually to failure. Some structures may have multiple resonant frequencies for different modes of deformation (e.g., bending and torsion). Either the flow or the structure needs to be modified to obtain a sufficient gap between frequencies of pronounced flow-induced force oscillations and natural frequencies of the structure.

We describe here briefly an example of a coupled simulation of fluid flow and the motion of a deformable structure shown in Fig. 13.29; more details are available in a separate report on the Internet; see the appendix for details. The test case was designed to be used for validation of computational methods and is described in detail in Gomes et al. (2011) and Gomes and Lienhart (2010). A thin, 50 mm long

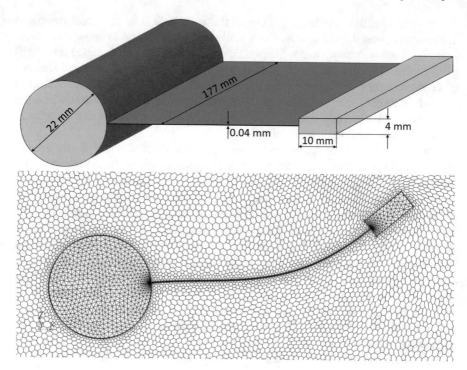

Fig. 13.29 A sketch of the assembly (upper) and the grid in solid structure and the surrounding fluid (lower)

and 0.04 mm thick sheet of stainless steal is attached to a rigid cylinder with a diameter of 22 mm; at the end of the sheet, a rectangular rigid body 10 mm long and 4 mm wide is attached. The cylinder can rotate around its axis. The spanwise dimension of the assembly was 177 mm; it was placed at the center of a test section in a tunnel with a rectangular cross-section with dimensions 240 mm × 180 mm such that the assembly spans the whole tunnel width, thus leading to a nominally 2D flow. A very viscous fluid (dynamic viscosity $\mu = 0.1722$ Pa·s, density $\rho = 1050$ kg/m^3) flowing with a velocity of 1.08 m/s was used; the flow was laminar.

The simulation was carried out using the commercial code Simcenter STAR-CCM+, which includes finite-element routines that compute the motion of rigid parts and the deformation of the flexible sheet, while the flow is computed using the usual finite-volume method. Outer iterations are performed in the same manner as is done for fluid flow, using the SIMPLE algorithm; the loop is only extended by adding one iteration on the structural side. Fluid forces acting on the structure and structural displacements are updated after each outer iteration. Figure 13.29 shows the grid used; it was polyhedral in the fluid, tetrahedral in the rigid solid parts and hexahedral in the flexible sheet (4 layers). The time step was 1 ms.

Figure 13.30 shows the shape and position of the structure at the two extreme displacements of the tail body, together with vectors in the fluid in the vicinity of

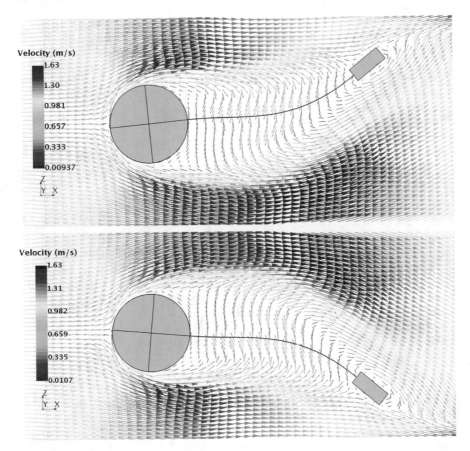

Fig. 13.30 Predicted deformation of the structure and velocity vectors around it at the time when the tail body is at its largest displacement on either side

the structure. The simulation was started with the structure aligned with the flow; over time, vortex shedding on both sides of the cylinder leads to a swinging motion of the tail body. Figure 13.31 shows this development by plotting the rotation of the cylinder and the sideways displacement of the tail versus time. The disturbance grows exponentially, but after 2 s a periodic state is established. The period of oscillation and the amplitudes are in a reasonably good agreement with experimental data published by Gomes and Lienhart (2010).

Practical examples of unsteady fluid-structure interaction are: flow around wind turbine blades (see, Sect. 10.3.4.3), flow around sails of sailboats, flow through heart valves,[1] flow around propellers made of composite materials, etc. Flow-induced vibration of structures can also produce noise, which is another reason for studying

---

[1] Gilmanov et al. (2015) report on FSI for several problems, including a heart valve, using a fractional-step LES.

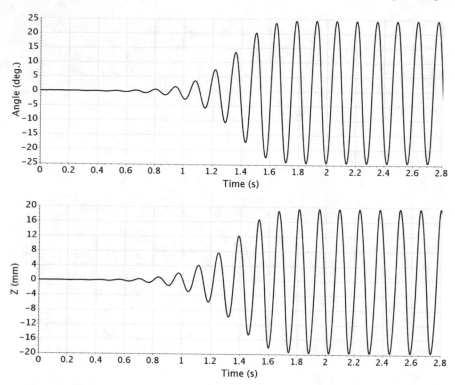

Fig. 13.31 Predicted development of cylinder rotation (upper) and vertical displacement of the tail (lower)

fluid-structure interaction. If the structure significantly deforms in operation (like airplane wings), it is important to know how large the deformation is, because one has to take it into account in the manufacturing process. For example, we can use optimization software coupled to a flow solver to find the optimal shape of an airplane wing, but because this should be the shape under the operation load, we then need to find the undeformed shape to be manufactured, such that, when the wing deforms under flow-induced forces, it attains the desired optimal shape.

Appendix
Supplementary Information

A.1 List of Computer Codes and How to Access Them

A number of computer codes embodying some of the methods described in this book can be obtained by readers via the Internet. These codes may be useful as they stand, but they can also serve as the starting point for further development. They will be updated from time to time, and new codes may be added. All computer codes can be accessed from the download-center of web-site www.cfd-peric.de.

Included are codes used to solve the one- and two-dimensional generic conservation equation; these were used to do the examples in Chaps. 3, 4 and 6. Several schemes for discretization of the convection and diffusion terms and time integration are used in these codes. They can be used to study features of the schemes, including convergence and discretization errors and the relative efficiency of the solvers. They can also be used as the basis for student assignments; students could, for example, be asked to modify the discretization scheme and/or boundary conditions.

Several solvers are given in the initial package including:

- TDMA solver for 1D problems;
- Line-by-line TDMA solver for 2D problems (five-point molecule);
- Stone's ILU solver (SIP) for 2D and 3D problems (five- and seven-point molecules; the 3D version is also given in vectorized form);
- Conjugate gradient solver preconditioned by the Incomplete Cholesky method (ICCG) for symmetric matrices in 2- and 3-D (five- and seven-point molecules);
- A modified SIP solver for a nine-point molecule in 2D;
- CGSTAB solver for non-symmetric matrices and 3D problems;
- Multigrid solver for 2D problems using Gauss-Seidel, SIP, and ICCG as smoothers.

Finally, there are several codes for solving fluid flow and heat transfer problems. The source codes of the following are included:

- A code for generating Cartesian 2D grids;
- A code for generating non-orthogonal structured 2D grids;

© Springer Nature Switzerland AG 2020
J. H. Ferziger et al., *Computational Methods for Fluid Dynamics*,
https://doi.org/10.1007/978-3-319-99693-6

- A code for post-processing 2D data on Cartesian and non-orthogonal grids, which can plot the grid, velocity vectors, profiles of any quantity on lines of $x =$ const. or $y =$ const., and contours of any quantity in black and white or color (the output is a postscript file);
- A FV code for a Cartesian 2D grid with the staggered variable arrangement, for steady problems;
- A FV code using Cartesian 2D grids with the colocated variable arrangement, for steady or unsteady problems;
- A FV code using Cartesian 3D grids and the colocated variable arrangement, for steady and unsteady problems, with multigrid applied to the outer iterations;
- A FV code using boundary-fitted non-orthogonal 2D grids and the colocated variable arrangement, for laminar steady or unsteady flows (including moving grids);
- Versions of the above code that include k–ε and k–ω turbulence models with wall functions and a version that does not use wall functions;
- A multigrid version of the above code for laminar flows (multigrid applied to outer iterations).

The codes are programmed in standard FORTRAN77 and have been tested on many computers. For the larger codes there are also explanatory files in the directory; many comment lines are included in each code, including suggestions how to adapt them to 3D problems on unstructured grids.

Finally, the main directory contains a file named errata; in it, errors which might be found will be documented (we hope that this file will be very small, if not empty).

A.2   Extended Reports on Example Simulations

In Chaps. 9–13, we have briefly presented results from several test cases; space did not allow a detailed description of all relevant data. The download-center includes more detailed reports in pdf-format (in most cases the simulation file used to obtain the results is also provided) which readers can download. The simulations are performed using the commercial software Simcenter STAR-CCM+ from Siemens. The simulation files can be used to vary the parameters (geometry, fluid properties, boundary conditions, turbulence model etc.) and to study the grid or time-step dependence of solutions.

For some test cases, animations of flow behavior are also provided.

A.3   Other Free CFD Codes

One can find on the Internet free versions of both CFD and grid generation codes. The most prominent example is OpenFOAM, which we mention here because it is widely used and has a large community base. It is officially released every six months and includes many built-in models to compute incompressible and compressible flow with many additional physics models, like multiphase flow, combustion, and solid stress analysis. One can use it as is, but it is also suitable as the basis for further development of either numerical techniques or physics models. More information is available at www.openfoam.com.

References

Abe, K., Jang, Y.-J., & Leschziner, M. A. (2003). An investigation of wall-anisotropy expressions and length-scale equations for non-linear eddy-viscosity models. *International Journal of Heat Fluid Flow, 24*, 181–198.

Abgrall, R. (1994). On essentially non-oscillatory schemes on unstructured meshes: Analysis and implementation. *Journal of Computational Physics, 114*, 45–58.

Achenbach, E. (1972). Experiments on the flow past spheres at very high Reynolds numbers. *Journal of Fluid Mechanics, 54*, 565–575.

Adams, M., Colell, P., Graves, D. T., Johnson, J. N., Johansen, H. S., Keen, N. D. et al. (2015). Chombo software package for AMR application design document (Technical Report). Berkeley, CA: Lawrence Berkeley National Laboratory, Applied Numerical Algorithms Group, Computational Research Division.

Amezcua, J., Kalnay, E.,& Williams, P. D. (2011). The effects of the RAW filter on the climatology and forecast skill of the SPEEDY model. *Monthly Weather Review, 13*, 608–619.

Arcilla, A. S., Häuser, J., Eiseman, P. R., & Thompson, J. F. (Eds.). (1991). *Numerical grid generation in computational fluid dynamics and related fields*. Amsterdam: North- Holland.

Aris, R. (1990). *Vectors, tensors and the basic equations of fluid mechanics*. New York: Dover Publications.

Armfield, S. (1991). Finite difference solutions of the Navier-Stokes equations on staggered and non-staggered grids. *Computers Fluids, 20*, 1–17.

Armfield, S. (1994). Ellipticity, accuracy, and convergence of the discrete Navier-Stokes equations. *Journal of Computational Physics, 114*, 176–184.

Armfield, S., & Street, R. (1999). The fractional-step method for the Navier-Stokes equations on staggered grids: The accuracy of three variations. *Journal of Computational Physics, 153*, 660–665.

Armfield, S., & Street, R. (2000). Fractional-step methods for the Navier-Stokes equations on non-staggered grids. *ANZIAM Journal, 42*(E), C134–C156.

Armfield, S., & Street, R. (2002). An analysis and comparison of the time accuracy of fractionalstep methods for the Navier-Stokes equations on staggered grids. *International Journal for Numerical Methods in Fluids, 38*, 255–282.

Armfield, S., & Street, R. (2003). The pressure accuracy of fractional-step methods for the Navier-Stokes equations on staggered grids. *ANZIAM Journal, 44*(E), C20–C39.

Armfield, S.,& Street, R. (2004). Modified fractional-step methods for the Navier-Stokes equations. *ANZIAM Journal, 45*(E), C364–C377.

© Springer Nature Switzerland AG 2020
J. H. Ferziger et al., *Computational Methods for Fluid Dynamics*,
https://doi.org/10.1007/978-3-319-99693-6

Armfield, S., & Street, R. (2005). A comparison of staggered and non-staggered grid Navier-Stokes solutions for the 8:1 cavity natural convection flow. *ANZIAM Journal, 46*(E), C918–C934.

Armfield, S., Williamson, N., Kirkpatrick, M., & Street, R. (2010). A divergence free fractionalstep method for the Navier-Stokes equations on non-staggered grids. *ANZIAM Journal, 51*(E), C654–C667.

Aspden, A., Nikiforakis, N., Dalziel, S., & Bell, J. (2008). Analysis of implicit LES methods. *Communications in Applied Mathematics and Computational Science, 3*, 103–126.

Asselin, R. (1972). Frequency filter for time integration. *Monthly Weather Review, 100*, 487–490.

Bakić, V. (2002). *Experimental investigation of turbulent flows around a sphere* (Ph.D. Dissertation). Germany: Technical University of Hamburg-Harburg.

Baliga, B. R. (1997). Control-volume finite element method for fluid flow and heat transfer. In W. J. Minkowycz & E. M. Sparrow (Eds.), *Advances in numerical heat transfer* (Vol. 1, pp. 97–135). New York: Taylor and Francis.

Baliga, B. R., & Patankar, S. V. (1983). A control-volume finite element method for two-dimensional fluid flow and heat transfer. *Numerical Heat Transfer, 6*, 245–261.

Bardina, J., Ferziger, J. H., & Reynolds, W. C. (1980). Improved subgrid models for large-eddy simulation. In *13th Fluid and Plasma Dynamics Conference (AIAA Paper 80-1357)*.

Barth, T. J., & Jespersen, D. C. (1989). The design and application of upwind schemes on unstructured meshes. In *27th Aerospace Science Meeting (AIAA Paper 89-0366)*.

Bastian, P., & Horton, G. (1989). Parallelization of robust multi-grid methods: ILU factorization and frequency decomposition method. In W. Hackbusch & R. Rannacher (Eds.), *Notes on numerical fluid mechanics* (Vol. 30, pp. 24–36). Braunschweig: Vieweg.

Beam, R. M., & Warming, R. F. (1976). An implicit finite-difference algorithm for hyperbolic systems in conservation-law form. *J. Comput. Phys, 22*, 87–110.

Beam, R. M., & Warming, R. F. (1978). An implicit factored scheme for the compressible navier-stokes equations. *AIAA Journal, 16*, 393–402.

Beare, R. J., MacVean, M. K., Holtslag, A. A. M., Cuxart, J., Esau, I., Golaz, J.-C., et al. (2006). An intercomparison of large-eddy simulations of the stable boundary layer. *Boundary-Layer Meteorology, 118*, 247–272.

Beig, S. A., & Johnsen, E. (2015). Maintaining interface equilibrium conditions in compressible multiphase flows using interface capturing. *Journal of Computational Physics, 302*, 548–566.

Berchiche, N., Östman, A., Hermundstad, O. A., & Reinholdtsen, S.-A. (2015). Experimental validation of CFD simulations of free-fall lifeboat launches in regular waves. *Ship Technology Research, 62*, 148–158.

Berger, M. J., & Oliger, J. (1984). Adaptive mesh refinement for hyperbolic partial differential equations. *Journal of Computational Physics, 53*, 484–512.

Bermejo-Moreno, I., Pullin, D. I., & Horiuti, K. (2009). Geometry of enstrophy and dissipation, grid resolution effects and proximity issues in turbulence. *Journal of Fluid Mechanics, 620*, 121–166.

Bewley, T., Moin, P., & Temam, R. (1994). Optimal control of turbulent channel flows. In *Active control of vibration and noise* (Vol. DE 75, pp. 221–227). New York: American Society of Mechanical Engineers, Design Engineering Division.

Bhagatwala, A., & Lele, S. K. (2011). Interaction of a Taylor blast wave with isotropic turbulence. *Physics of Fluids, 23*, 035103.

Bhaskaran, R., & Lele, S. K. (2010). Large-eddy simulation of free-stream turbulence effects on heat transfer to a high-pressure turbine cascade. *Journal of Turbulence, 11*, N6.

Bijl, H., Lucor, D., Mishra, S., & Schwab, C. (2013). *Uncertainty quantification in computational fluid dynamics*. Switzerland: Springer International Publishing.

Bilger, R. W. (1975). A note on Favre averaging in variable density flows. *Combustion Science and Technology, 11*, 215–217.

Billard, F., Laurence, D., & Osman, K. (2015). Adaptive wall functions for an elliptic blending eddy-viscosity model applicable to any mesh topology. *Flow, Turbulence and Combustion, 94*, 817–842.

Bini, D. A., & Meini, D. (2009). The cyclic reduction algorithm: from Poisson equation to stochastic processes and beyond. *Numerical Algorithms, 51*, 23–60.

Bird, R. B., Stewart, W. E., & Lightfoot, E. N. (2006). *Transport phenomena* (Revised 2 ed.). New York: Wiley.

Bird, R. B., & Wiest, J. M. (1995). Constitutive equations for polymeric liquids. *Annual Review of Fluid Mechanics, 27*, 169–193.

Blumberg, A. F., & Mellor, G. L. (1987). A description of a three-dimensional coastal ocean circulation model. In N. S. Heaps (Ed.) *Three-dimensional Coastal Ocean models, Coastal and Estuarine Science* (Vol. 4, pp. 1–16). Washington, D. C.: AGU.

Bodony, D. J., & Lele, S. K. (2008). Current status of jet noise predictions using large-eddy simulation. *AIAA J., 46*, 364–380.

Boris, J. P., & Book, D. L. (1973). Flux-corrected transport. i. SHASTA, a fluid transport algorithm that works. *Journal of Computational Physics, 11*, 38–69.

Bose, S. T., Moin, P., & You, D. (2010). Grid-independent large-eddy simulation using explicit filtering. *Physics of Fluids, 22*, 105103.

Boyd, J. P. (2001). *Chebyshev and Fourier spectral methods* (Revised 2 ed.). Mineola: Dover Publications.

Brackbill, J. U., Kothe, D. B., & Zemaach, C. (1992). A continuum method for modeling surface tension. *Journal of Computational Physics, 100*, 335–354.

Bradshaw, P., Launder, B. E., & Lumley, J. L. (1994). Collaborative testing of turbulence models. In K. N. Ghia, U. Ghia, & D. Goldstein (Eds.), *Advances in computational fluid mechanics* (Vol. 196). New York: ASME.

Brandt, A. (1984). Multigrid techniques: 1984 guide with applications to fluid dynamics. GMDStudien Nr. 85, Gesellschaft für Mathematik und Datenverarbeitung (GMD). Bonn, Germany (see also Multigrid Classics version at http://www.wisdom.weizmann.ac.il/~achi/).

Brazier, P. H. (1974). An optimum SOR procedure for the solution of elliptic partial differential equations with any domain or coefficient set. *Computer Methods in Applied Mechanics and Engineering, 3*, 335–347.

Brehm, C., Housman, J. A., Kiris, C. C., Barad, M. F., & Hutcheson, F. V. (2017). Four-jet impingement: Noise characteristics and simplified acoustic model. *International Journal of Heat and Fluid flow, 67*, 43–58.

Brès, G. A., Ham, F. E., Nichols, J. W., & Lele, S. K. (2017). Unstructured large-eddy simulations of supersonic jets. *AIAA Journal, 55*, 1164–1184.

Briggs, D. R., Ferziger, J. H., Koseff, J. R., & Monismith, S. G. (1996). Entrainment in a shear free mixing layer. *Journal of Fluid Mechanics, 310*, 215–241.

Briggs, W. L., Henson, V. E., & McCormick, S. F. (2000). *A multigrid tutorial* (2nd ed.). Philadelphia: Society for Industrial and Applied Mathematics (SIAM).

Brigham, E. O. (1988). *The fast Fourier transform and its applications*. Englewood Cliffs: Prentice Hall.

Bryan, G. H. (2007). A comparison of convection resolving-simulations with convection-permitting simulations. NSSL Colloquium, Norman, OK. http://www2.mmm.ucar.edu/people/-bryan/Presentations/bryan_2007_nssl_resolution.pdf.

Bryan, G. H., Wyngaard, J. C., & Fritsch, J. M. (2003). Resolution requirements for the simulation of deep moist convection. *Monthly Weather Review, 131*, 2394–2416.

Bückle, U., & Perić, M. (1992). Numerical simulation of buoyant and thermocapillary convection in a square cavity. *Numerical Heat Transfer, Part A (Applications), 21*, 101–121.

Bunner, B., & Tryggvason, G. (1999). Direct numerical simulations of three-dimensional bubbly flows. *Physics of Fluids, 11*, 1967–1969.

Burmeister, J., & Horton, G. (1991). Time-parallel solution of the Navier-Stokes equations. In *Proceedings of 3rd European Multigrid Conference*. Basel: Birkhäuser.

Butcher, J. C. (2008). *Numerical methods for ordinary differential equations*. Chichester: Wiley.

Calhoun, D. (2002). A cartesian grid method for solving the two-dimensional streamfunctionvorticity equations in irregular regions. *Journal of Computational Physics, 176*, 231–275.

Canuto, C., Hussaini, M. Y., Quarteroni, A.,& Zang, T. A. (2006). *Spectral methods: Fundamentals in single domains*. Berlin: Springer.

Canuto, C., Hussaini, M. Y., Quarteroni, A., & Zang, T. A. (2007). *Spectral methods: Evolution to complex geometries and applications to fluid dynamics*. Berlin: Springer.

Carati, D., Winckelmans, G. S., & Jeanmart, H. (2001). On the modelling of the subgrid-scale and filtered-scale stress tensors in large-eddy simulation. *Journal of Fluid Mechanics, 441*, 119–138.

Caretto, L. S., Gosman, A. D., Patankar, S. V., & Spalding, D. B. (1972). Two calculation procedures for steady three-dimensional flows with recirculation. In *Proceedings of the Third International Conference on Numerical Methods in Fluid Mechanics*. Paris.

Caruso, S. C., Ferziger, J. H., & Oliger, J. (1985). An adaptive grid method for incompressible flows (Technical Report No. TF-23). Stanford CA: Department of Mechanical Engineering, Stanford University.

Casulli, V., & Cattani, E. (1994). Stability, accuracy and efficiency of a semi-implicit method for three-dimensional shallow water flow. *Computers & Mathematics with Applications, 27*, 99–112.

Cebeci, T., & Bradshaw, P. (1984). *Physical and computational aspects of convective heat transfer*. New York: Springer.

Celik, I., Ghia, U., Roache, P. J., & Freitas, C. J. (2008). Procedure for estimation and reporting of uncertainty due to discretization in cfd applications. *Journal of Fluids Engineering, 130*, 078001.

Celik, I., Klein, M., & Janicka, J. (2009). Assessment measures for engineering LES applications. *Journal of Fluids Engineering, 131*, 031102.

Chang, W., Giraldo, F., & Perot, B. (2002). Analysis of an exact fractional-step method. *Journal of Computational Physics, 180*, 183–199.

Chen, C.-J., & Jaw, S.-Y. (1998). *Fundamentals of turbulence modeling*. Washington: Taylor & Francis.

Chen, S., Johnson, D. B., Raad, P. E., & Fadda, D. (1997). The surface marker and micro-cell method. *International Journal for Numerical Methods in Fluids, 25*, 749–778.

Chen, C., Zhu, J., Zheng, L., Ralph, E., & Budd, J. W. (2004a). A non-orthogonal primitive equation coastal ocean circulation model: Application to lake superior. *Journal of Great Lakes Research, 30*(Supplement 1), 41–54.

Chen, Y., Ludwig, F. L., & Street, R. L. (2004b). Stably stratified flows near a notched transverse ridge across the salt lake valley. *Journal of Applied Meteorology, 43*, 1308–1328.

Choi, H., & Moin, P. (1994). Effects of the computational time step on numerical solutions of turbulent flow. *Journal of Computational Physics, 113*, 1–4.

Choi, H., Moin, P., & Kim, J. (1994). Active turbulence control for drag reduction in wall-bounded flows. *Journal of Fluid Mechanics, 262*, 75–110.

Chorin, A. J. (1967). A numerical method for solving incompressible viscous flow problems. *Journal of Computational Physics, 2*, 12–26.

Chorin, A. J. (1968). Numerical solution of the Navier-Stokes equations. *Mathematics of Computation, 22*, 745–762.

Chow, F. K., & Moin, P. (2003). A further study of numerical errors in large-eddy simulations. *Journal of Computational Physics, 184*, 366–380.

Chow, F. K., & Street, R. L. (2009). Evaluation of turbulence closure models for large-eddy simulation over complex terrain: Flow over Askervein Hill. *Journal of Applied Meteorology and Climatology, 48*, 1050–1065.

Chow, F. K., Street, R. L., Xue, M., & Ferziger, J. (2005). Explicit filtering and reconstruction turbulence modeling for large-eddy simulation of neutral boundary layer flow. *Journal of the Atmospheric Sciences, 62*, 2058–2077.

Chow, F. K., Weigel, A. P., Street, R. L., Rotach, M. W., & Xue, M. (2006). High-resolution large-eddy simulations of flow in a steep Alpine valley. part i: Methodology, verification, and sensitivity experiments. *Journal of Applied Meteorology and Climatology, 45*, 63–86.

Colella, P. (1985). A direct Eulerian MUSCL scheme for gas dynamics. *SIAM Journal on Scientific and Statistical Computing, 6*, 104–117.

Colella, P. (1990). Multidimensional upwind methods for hyperbolic conservation laws. *Journol of Computational Physics*, *87*, 171–200.

Coelho, P., Pereira, J. C. F., & Carvalho, M. G. (1991). Calculation of laminar recirculating flows using a local non-staggered grid refinement system. *International Journal for Numerical Methods in Fluids*, *12*, 535–557.

Constantinescu, G., & Squires, K. (2004). Numerical investigations of flow over a sphere in the subcritical and supercritical regimes. *Physics Fluids*, *16*, 1449–1466.

Cooley, J. W., & Tukey, J. W. (1965). An algorithm for the machine calculation of complex Fourier series. *Mathematics of Computation*, *19*, 297–301.

Courant, R., Friedrichs, K., & Lewy, H. (1928). Über die partiellen Differenzengleichungen der mathematischen Physik. *Mathematische Annalen*, *100*, 32–74.

Crowe, C., Sommerfeld, M., & Tsuji, Y. (1998). *Multiphase flows with droplets and particles*. Boca Raton: CRC Press.

Deike, L., Melville, W. K., & Popinet, S. (2016). Air entrainment and bubble statistics in breaking waves. *Journal of Fluid Mechanics*, *801*, 91–129.

Demirdžić, I. (2015). On the discretization of the diffusion term in finite-volume continuum mechanics. *Numerical Heat Transfer*, *68*, 1–10.

Demirdžić, I. (2016). A fourth-order finite volume method for structural analysis. *Applied Mathematical Modelling*, *40*, 3104–3114.

Demirdžić, I., & Perić, M. (1988). Space conservation law in finite volume calculations of fluid flow. *International Journal for Numerical Methods in Fluids*, *8*, 1037–1050.

Demirdžić, I., & Perić, M. (1990). Finite volume method for prediction of fluid flow in arbitrarily shaped domains with moving boundaries. *International Journal for Numerical Methods in Fluids*, *10*, 771–790.

Demirdžić, I., & Muzaferija, S. (1994). Finite volume method for stress analysis in complex domains. *International Journal for Numerical Methods in Engineering*, *37*, 3751–3766.

Demirdžić, I., & Muzaferija, S. (1995). Numerical method for coupled fluid flow, heat transfer and stress analysis using unstructured moving meshes with cells of arbitrary topology. *Computer Methods in Applied Mechanics and Engineering*, *125*, 235–255.

Demirdžić, I., Muzaferija, S., & Peric, M. (1997). Benchmark solutions of some structural analysis problems using finite-volume method and multigrid acceleration. *International Journal for Numerical Methods in Engineering*, *40*, 1893–1908.

Demmel, J. W., Heath, M. T., & van der Vorst, H. A. (1993). Parallel numerical linear algebra. *Acta Numerica*, *2*, 111–197.

Demirdžić, I., & Lilek, V., & Peric, M. (1993). A colocated finite volume method for predicting flows at all speeds. *International Journal for Numerical Methods in Fluids*, *16*, 1029–1050.

Deng, G. B., Piquet, J., Queutey, P., & Visonneau, M. (1994). Incompressible flow calculations with a consistent physical interpolation finite volume approach. *Computers Fluids*, *23*, 1029–1047.

de Vahl Davis, G., & Mallinson, G. D. (1972). False diffusion in numerical fluid mechanics. Univ (Technical Report No. FM1). New South Wales, Australia: University of New South Wales, School of Mechanical Ind. Engineering

Donea, J., & Huerta, A. (2003). *Finite element methods for flow problems*. Chichester: Wiley. (available at Wiley Online Library).

Donea, J., Huerta, A., Ponthot, J.-P.,& Rodríguez-Ferran, A. (2004). Arbitrary Lagrangian-Eulerian methods. In *Encyclopedia of Computational Mechanics. Vol. 1: Fundamentals* (pp. 413–437).

Donea, J., Quartapelle, L., & Selmin, V. (1987). An analysis of time discretization in the finite element solution of hyperbolic problems. *Journal of Computational Physics*, *70*, 463–499.

Durbin, P. A. (1991). Near-wall turbulence closure modeling without 'damping functions'. *Theoretical and Computational Fluid Dynamics*, *3*, 1–13.

Durbin, P. A. (2002). A perspective on recent developments in RANS modeling. In W. Rodi & N. Furyo (Eds.), *Engineering urbulence modelling and experiments* (Vol. 5, pp. 3–16).

Durbin, P. A. (2009). Limiters and wall treatments in applied turbulence modeling. *Fluid Dynamics Research*, *41*, 012203.

Durbin, P. A., & Pettersson Reif, B. A. (2011). *Statistical theory and modeling for turbulent flows* (2nd ed.). Chichester: Wiley.

Durran, D. R. (2010). *Numerical methods for fluid dynamics with applications to geophysics* (2nd ed.). Berlin: Springer.

Durst, F., Kadinskii, L., Perić, M., & Schäfer, M. (1992). Numerical study of transport phenomena in MOCVD reactors using a finite volume multigrid solver. *Journal of Crystal Growth, 125*, 612–626.

El Khoury, G. K., Schlatter, P., Noorani, A., Fischer, P. F., Brethouwer, G., & Johansson, A. V. (2013). Direct numerical simulation of turbulent pipe flow at moderately high Reynolds numbers. *Flow, Turbulence and Combustion, 91*, 475–495.

Enright, D., Fedkiw, R., Ferziger, J. H., & Mitchell, I. (2002). A hybrid particle level set method for improved interface capturing. *Journal of Computational Physics, 183*, 83–116.

Enriquez, R. M. (2013). *Subgrid-scale turbulence modeling for improved large-eddy simulation of the atmospheric boundary layer (Ph.D. Dissertation)*. Stanford: Stanford University.

Enriquez, R. M., Chow, F. K., Street, R. L., & Ludwig, F. L. (2010). Examination of the linear algebraic subgrid-scale stress [LASS] model, combined with reconstruction of the subfilterscale stress, for large-eddy simulation of the neutral atmospheric boundary layer. In *19th Conference on Boundary Layers and Turbulence, AMS, Paper 3A*. (8 pages).

Erturk, E. (2009). Discussions on driven cavity flow. *International Journal for Numerical Methods in Fluids, 60*, 275–294.

Fadlun, E. A., Verzicco, R., Orlandi, P., & Mohd-Yusof, J. (2000). Combined immersed-boundary finite-difference methods for three-dimensional complex flowsimulations. *Journal of Computational Physics, 161*, 35–60.

Farmer, J., Martinelli, L., & Jameson, A. (1994). Fast multigrid method for solving incompressible hydrodynamic problems with free surfaces. *AIAA Journal, 32*, 1175–1182.

Fenton, J. D. (1985). A fifth-order Stokes theory for steady waves. *Journal of Waterway, Port, Coastal, Ocean Engineering, 111*, 216–234.

Ferziger, J. H. (1998). *Numerical methods for engineering application* (2nd ed.). New York: Wiley-Interscience.

Ferziger, J. H., & Perić, M. (1996). Further discussion of numerical errors in CFD. *International Journal for Numerical Methods in Fluids, 23*, 1–12.

Findikakis, A. N., & Street, R. L. (1979). An algebraic model for subgrid-scale turbulence in stratified flows. *Journal of the Atmospheric Sciences, 36*, 1934–1949.

Fletcher, C. A. J. (1991). *Computational techniques for fluid dynamics* (2nd ed., Vol. I & II). Berlin: Springer.

Fletcher, R. (1976). Conjugate gradient methods for indefinite systems. *Lecture Notes in Mathematics, 506*, 773–789.

Forbes, L. K. (1988). Critical free surface flow over a semicircular obstruction. *Journal of Engineering Mathematics, 22*, 3–13.

Fornberg, B. (1988). Generation of finite difference formulas on arbitrarily spaced grids. *Mathematics of Computation, 51*, 699–706.

Fox, D. G., & Orszag, S. A. (1973). Pseudospectral approximation to two-dimensional turbulence. *Journal of Computational Physics, 11*, 612–619.

Freitas, C. J., Street, R. L., Findikakis, A. N., & Koseff, J. R. (1985). Numerical simulation of three-dimensional flow in a cavity. *International Journal for Numerical Methods in Fluids, 5*, 561–575.

Frey, P. J., & George, P.-L. (2008). *Mesh generation: Application to finite elements* (2nd ed.). New Jersey: Wiley-ISTE.

Friedrich, O. (1998). Weighted essentially non-oscillatory schemes for the interpolation of mean values on unstructured grids. *Journal of Computational Physics, 144*, 194–212.

Fringer, O. B., Armfield, S. W., & Street, R. L. (2003). A nonstaggered curvilinear grid pressure correction method applied to interfacial waves. In *Second International Conference on Heat Transfer, Fluid Mechanics and Thermodynamics, HEFAT 2003, Paper FO1*. (6 pages).

Fringer, O. B., Armfield, S. W., & Street, R. L. (2005). Reducing numerical diffusion in interfacial gravity wave simulations. *International Journal for Numerical Methods in Fluids, 49*, 301–329.

Fringer, O. B., Gerritsen, M., & Street, R. L. (2006). An unstructured-grid, finite-volume, nonhydrostatic, parallel coastal ocean simulator. *Ocean Modelling, 14*, 139–173.

Fu, L., Hu, X. Y., & Adams, N. A. (2016). A family of high-order targeted ENO schemes for compressible-fluid simulations. *Journal of Computational Physics, 305*, 333–359.

Fu, L., Hu, X. Y., & Adams, N. A. (2017). Targeted ENO schemes with tailored resolution property for hyperbolic conservation laws. *Journal of Computational Physics, 349*, 97–121.

Galpin, P. F., & Raithby, G. D. (1986). Numerical solution of problems in incompressible fluid flow: Treatment of the temperature-velocity coupling. *Numerical Heat Transfer, 10*, 105–129.

Gamet, L., Ducros, F., Nicoud, F., & Poinsot, T. (1999). Compact finite difference schemes on nonuniform meshes. Application to direct numerical simulations of compressible flows. *International Journal for Numerical Methods in Fluids, 29*, 159–191.

Garnier, E., Adams, N., & Sagaut, P. (2009). *Large-eddy simulation for compressible flows*. Berlin: Springer.

Germano, M., Piomelli, U., Moin, P., & Cabot, W. H. (1991). A dynamic subgrid-scale eddy viscosity model. *Physics of Fluids A, 3*, 1760–1765.

Ghia, U., Ghia, K. N., & Shin, C. T. (1982). High-Re solutions for incompressible flow using the Navier-Stokes equations and a multigrid method. *Journal of Computational Physics, 48*, 387–411.

Gibbs, J. W. (1898). Fourier's series. *Nature, 59*, 200.

Gibbs, J. W. (1899). Fourier's series. *Nature, 59*, 606.

Gilmanov, A., Le, T. B., & Sotiropoulos, F. (2015). A numerical approach for simulating fluid structure interaction of flexible thin shells undergoing arbitrarily large deformations in complex domains. *Journal of Computational Physics, 300*, 814–843.

Glowinski, R., & Pironneau, O. (1992). Finite element methods for Navier-Stokes equations. *Annual Review of Fluid Mechanics, 24*, 167–204.

Göddeke, D., & Strzodka, R. (2011). Cyclic reduction tridiagonal solvers on GPUs applied to mixed-precision multigrid. *IEEE Transactions on Parallel and Distributed Systems, 22*, 22–32.

Golub, G. H., & van Loan, C. F. (1996). *Matrix computations* (3rd ed.). Baltimore: Johns Hopkins University Press.

Gomes, J. P., & Lienhart, H. (2010). Fluid-structure interaction-induced oscillation of flexible structures in laminar and turbulent flows. *Journal of Fluid Mechanics, 715*, 537–572.

Gomes, J. P., Yigit, S., Lienhart, H., & Schäfer, M. (2011). Experimental and numerical study on a laminar fluid-structure interaction reference test case. *Journal of Fluids Structures, 27*, 43–61.

Gottlieb, S., Mullen, J. S., & Ruuth, S. J. (2006). A fifth order flux implicit WENO method. *Journal of Scientific Computing, 27*, 271–287.

Gresho, P. M. (1990). On the theory of semi-implicit projection methods for viscous incompressible flow and its implementation via a finite element method that also introduces a nearly consistent mass matrix. Part 1: Theory. *International Journal for Numerical Methods in Fluids, 11*, 587–620.

Grinstein, F. F., Margolin, L. G., & Rider, W. J. (Eds.). (2007). *Implicit large-eddy simulation: Computing turbulent fluid dynamics*. Cambridge: Cambridge University Press.

Gullbrand, J., & Chow, F. K. (2003). The effect of numerical errors and turbulence models in large-eddy simulations of channel flow, with and without explicit filtering. *Journal of Fluid Mechanics, 495*, 323–341.

Gyllenram, W., & Nilsson, H. (2006). Very large-eddy simulation of draft tube flow. In *23rd IAHR Symposium Yokohama*. (10 pages).

Hackbusch, W. (1984). Parabolic multi-grid methods. In R. Glowinski & J.-R. Lions (Eds.), *Computing methods in applied sciences and engineering*. Amsterdam: North Holland.

Hackbusch, W. (2003). *Multi-grid methods and applications (2nd Printing)*. Berlin: Springer.

Hackbusch, W., & Trottenberg, U. (Eds.). (1991). In *Proceedings of Third European Multigrid Conference, International Series of Numerical Mathematics*. Basel: Birkhäuser.

Hadžić, H. (2005). Development and application of a finite volume method for the computation of flows around moving bodies on unstructured, overlapping grids (Ph.D. Dissertation). *Technische Universität*, Hamburg-Harburg.

Hadžić, I. (1999). Second-moment closure modelling of transitional and unsteady turbulent flows (Ph.D. Dissertation). Delft University of Technology.

Hageman, L. A., & Young, D. M. (2004). *Applied iterative methods*. Mineola: Dover Publications.

Ham, F., & Iaccarino, G. (2004). Energy conservation in collocated discretization schemes on unstructured grids. In *Annals Research Briefs*. Stanford, CA: Center for Turbulence Research.

Hanaoka, A. (2013). *An overset grid method coupling an orthogonal curvilinear grid solver and a Cartesian grid solver (Ph.D. Dissertation)*. Iowa City, IA: University of Iowa.

Hanjalić, K. (2002). One-point closure models for buoyancy-driven turbulent flows. *Annual Review of Fluid Mechanics, 34*, 321–347.

Hanjalić, K. (2004). Closure models for incompressible turbulent flows (Technical Report). Brussels, Belgium: Lecture Notes at the von Karman Institute for Fluid Dynamics.

Hanjalić, K., & Kenjereš, S. (2001). T-RANS simulation of deterministic eddy structure in flows driven by thermal buoyancy and Lorentz force. *Flow Turbulence and Combustion, 66*, 427–451.

Hanjalić, K., & Launder, B. E. (1976). Contribution towards a Reynolds-stress closure for low Reynolds number turbulence. *Journal of Fluid Mechanics, 74*, 593–610.

Hanjalić, K., & Launder, B. E. (1980). Sensitizing the dissipation equation to irrotational strains. *Journal of Fluids Engineering, 102*, 34–40.

Harlow, F. H., & Welsh, J. E. (1965). Numerical calculation of time dependent viscous Incompressible flow with free surface. *Physics of Fluids, 8*, 2182–2189.

Harrison, R. J. (1991). Portable tools and applications for parallel computers. *International Journal of Quantum Chemistry, 40*, 847–863.

Harten, A. (1983). High resolution schemes for hyperbolic conservation laws. *Journal of Computational Physics, 49*, 357–393.

Harvie, D. J. E., Davidson, M. R.,& Rudman, M. (2006). An analysis of parasitic current generation in Volume-of-Fluid simulations. *Applied Mathematical Modelling, 30*, 1056–1066.

Hatlee, S. C., & Wyngaard, J. C. (2007). Improved subfilter-scale models from the HATS field data. *Journal of the Atmospheric Sciences, 64*, 1694–1705.

Heil, M., Hazel, A. L.,& Boyle, J. (2008). Solvers for large-displacement fluid-structure interaction problems: segregated versus monolithic approaches. *Computational Mechanics, 43*, 91–101.

Heinke, H. J. (2011). Potsdam propeller test case (Technical Report No. 3753). Potsdam, Germany: SVA Potsdam.

Hess, J. L. (1990). Panel methods in computational fluid dynamics. *Annual Review of Fluid Mechanics, 22*, 255–274.

Heus, T., van Heerwaarden, C. C., Jonker, H. J. J., Siebesma, A. P. et al. (2010). Formulation of the dutch atmospheric large-eddy simulation (DALES) and overview of its applications. *Geoscientific Model Development, 3*, 415–444.

Hinatsu, M., & Ferziger, J. H. (1991). Numerical computation of unsteady incompressible flow in complex geometry using a composite multigrid technique. *International Journal for Numerical Methods in Fluids, 13*, 971–997.

Hino, T. (1992). Computation of viscous flows with free surface around an advancing ship. In *Proceedings of 2nd Osaka International Colloquium on Viscous Fluid Dynamics in Ship and Ocean Technology*. Osaka University.

Hirsch, C. (2007). *Numerical computation of internal and external flows* (2nd ed., Vol. I). Burlington: Butterworth-Heinemann (Elsevier).

Hirt, C. W., Amsden, A. A., & Cook, J. L. (1997). An arbitrary Lagrangean-Eulerian computing method for all flow speeds. *Journal of Computational Physics, 135*, 203–216. (Reprinted from 14, 1974, 227–253).

Hirt, C. W., & Harlow, F. H. (1967). A general corrective procedure for the numerical solution of initial-value problems. *Journal of Computational Physics, 2*, 114–119.

Hirt, C. W., & Nichols, B. D. (1981). Volume of fluid (VOF) method for dynamics of free boundaries. *Journal of Computational Physics, 39*, 201–221.

Hodges, B. R., & Street, R. L. (1999). On simulation of turbulent nonlinear free-surface flows. *Journal of Computational Physics, 151*, 425–457.

Hortmann, M., Perić, M., & Scheuerer, G. (1990). Finite volume multigrid prediction of laminar natural convection: Bench-mark solutions. *International Journal for Numerical Methods in Fluids, 11*, 189–207.

Horton, G. (1991). *Ein zeitparalleles Lösungsverfahren für die Navier-Stokes-Gleichungen (Ph.D. Dissertation)*. Erlangen-Nürnberg: Universität.

Housman, J. A., Stich, G.-D., & Kiris, C. C. (2017). Jet noise prediction using hybrid RANS/LES with structured overset grids. In *23rd AIAA/CEAS Aeroacoustics Conference, AIAA Paper 2017-3213*.

Hu, F. Q., Hussaini, M. Y., & Manthey, J. L. (1996). Low-dissipation and low-dispersion Runge-Kutta schemes for computational acoustics. *Journal of Computational Physics, 124*, 177–191.

Hubbard, B. J., & Chen, H. C. (1994). A Chimera scheme for incompressible viscous flows with applications to submarine hydrodynamics. In *25th AIAA Fluid Dynamics Conference (AIAA Paper 94-2210)*.

Hubbard, B. J., & Chen, H. C. (1995). Calculations of unsteady flows around bodies with relative motion using a Chimera RANS method. In *Proceedings of 10th ASCE Engineering Mechanics Conference*. Boulder, CO: University of Colorado at Boulder.

Hutchinson, B. R., Galpin, P. F., & Raithby, G. D. (1988). Application of additive correction multigrid to the coupled fluid flow equations. *Numerical Heat Transfer, 13*, 133–147.

Hutchinson, B. R., & Raithby, G. D. (1986). A multigrid method based on the additive correction strategy. *Numerical Heat Transfer, 9*, 511–537.

Hylla, E. A. (2013). *Eine Immersed Boundary Methode zur Simulation von Strömungen in komplexen und bewegten Geometrien (Ph.D. Dissertation)*. Berlin, Germany: Technische Universität Berlin.

Iaccarino, G., Ooi, A., Durbin, P. A., & Behnia, M. (2003). Reynolds averaged simulation of unsteady separated flow. *International Journal of Heat and Fluid Flow, 24*, 147–156.

Ishihara, T., Gotoh, T., & Kaneda, Y. (2009). Study of high-Reynolds number isotropic turbulence by direct numerical simulation. *Annual Review of Fluid Mechanics, 41*, 165–180.

Ishii, M. (1975). *Thermo-fluid dynamic theory of two-phase flow*. Paris: Eyrolles.

Ishii, M., & Hibiki, T. (2011). *Thermo-fluid dynamics of two-phase flow*. New York: Springer.

Issa, R. I. (1986). Solution of implicitly discretized fluid flow equations by operator-splitting. *Journal of Computational Physics, 62*, 40–65.

Issa, R. I., & Lockwood, F. C. (1977). On the prediction of two-dimensional supersonic viscous interaction near walls. *AIAA Jorunal, 15*, 182–188.

Ivanell, S., Sørensen, J., Mikkelsen, R., & Henningson, D. (2009). Analysis of numerically generated wake structures. *Wind Energy, 12*, 63–80.

Jacobson, M. Z., & Delucchi, M. A. (2009). A path to sustainable energy by 2030. *Scientific American, 301*, 58–65.

Jakirlic, S., & Jovanovic, J. (2010). On unified boundary conditions for improved predictions of near-wall turbulence. *Journal of Fluid Mechanics, 656*, 530–539.

Jakobsen, H. A. (2003). Numerical convection algorithms and their role in Eulerian CFD reactor simulations. *International Journal of Chemical Reactor Engineering, 1*, Art. A1, 15.

Johnsen, E., & Ham, F. (2012). Preventing numerical errors generated by interface-capturing schemes in compressible multi-material flows. *Journal of Computational Physics, 231*, 5705–5717.

Jovanović, J. (2004). *The statistical dynamics of turbulence*. Berlin: Springer.

Kader, B. A. (1981). Temperature and concentration profiles in fully turbulent boundary layers. *International Journal of Heat Mass Transfer, 24*, 1541–1544.

Kadinski, L., & Perić, M. (1996). Numerical study of grey-body surface radiation coupled with fluid flow for general geometries using a finite volume multigrid solver. *International Journal of Numerical Methods for Heat and Fluid Flow, 6*, 3–18.

Kaneda, Y., & Ishihara, T. (2006). High-resolution direct numerical simulation of turbulence. *Journal of Turbulence, 7*, N20.

Kang, S., Iaccarino, G., Ham, F., & Moin, P. (2009). Prediction of wall-pressure fluctuation in turbulent flows with an immersed boundary method. *Journal of Computational Physics, 228*, 6753–6772.

Karki, K. C., & Patankar, S. V. (1989). Pressure based calculation procedure for viscous flows at all speeds in arbitrary configurations. *AIAA Journal, 27*, 1167–1174.

Kawai, S., & Lele, S. K. (2010). Large-eddy simulation of jet mixing in supersonic crossflows. *AIAA Journal, 48*, 2063–2083.

Kawamura, T., & Miyata, H. (1994). Simulation of nonlinear ship flows by density-function method. *Journal of the Society of Naval Architects of Japan, 176*, 1–10.

Kays, W. M., & Crawford, M. E. (1978). *Convective heat and mass transfer*. New York: McGraw-Hill.

Kenjereš, S., & Hanjalić, K. (1999). Transient analysis of Rayleigh-Bénard convection with a RANS model. *International Journal of Heat and Fluid Flow, 20*, 329–340.

Kenjereš, S., & Hanjalic, K. (2002). Combined effects of terrain orography and thermal stratification on pollutant dispersion in a town valley: A T-RANS simulation. *Journal of Turbulence, 3*, N26.

Khajeh-Saeed, A., & Perot, J. B. (2013). Direct numerical simulation of turbulence using GPU accelerated supercomputers. *Journal of Computational Physics, 235*, 241–257.

Khani, S., & Porté-Agel, F. (2017). A modulated-gradient parameterization for the large-eddy simulation of the atmospheric boundary layer using the weather research and forecasting model. *Boundary-Layer Meteorology, 165*, 385–404.

Khosla, P. K., & Rubin, S. G. (1974). A diagonally dominant second-order accurate implicit scheme. *Computers Fluids, 2*, 207–209.

Kim, J. W. (2007). Optimised boundary compact finite difference schemes for computational aeroacoustics. *Journal of Computational Physics, 225*, 995–1019.

Kim, D., & Choi, H. (2000). A second-order time-accurate finite volume method for unsteady incompressible flow on hybrid unstructured grids. *Journal of Computational Physics, 162*, 411–428.

Kim, J., Kim, D., & Choi, H. (2001). An immersed-boundary finite-volume method for simulations of flow in complex geometries. *Journal of Computational Physics, 171*, 132–150.

Kim, J., & Moin, P. (1985). Application of a fractional-step method to incompressible Navier-Stokes equations. *Journal of Computational Physics, 59*, 308–323.

Kim, J., Moin, P., & Moser, R. D. (1987). Turbulence statistics in fully developed channel flow at low Reynolds number. *Journal of Fluid Mechanics, 177*, 133–166.

Kim, S., Kinnas, S. A., & Du, W. (2018). Panel method for ducted propellers with sharp trailing edge duct with fully aligned wake on blade and duct. *Journal of Marine Science and Engineering, 6*, 6030089.

Kirkpatrick, M. P., & Armfield, S. W. (2008). On the stability and performance of the projection- 3 method for the time integration of the Navier-Stokes equations. *ANZIAM Journal, 49*(EMAC2007), C559–C575.

Klemp, J. B., Skamarock, W. C., & Dudhia, J. (2007). Conservative split-explicit time integration methods for the compressible nonhydrostatic equations. *Monthly Weather Review, 135*, 2897–2913.

Knight, D. D. (2006). *Elements of numerical methods for compressible flows*. Cambridge: Cambridge University Press.

Konan, N. A., Simonin, O., & Squires, K. D. (2011). Detached-eddy simulations and particle Lagrangian tracking of horizontal rough wall turbulent channel flow. *Journal of Turbulence, 12*, N22.

Kordula, W., & Vinokur, M. (1983). Efficient computation of volume in flow predictions. *AIAA Journal, 21*, 917–918.

Koshizuka, S., Tamako, H., & Oka, Y. (1995). A particle method for incompressible viscous flow with fluid fragmentation. *Computational Fluid Dynamics Journal, 4*, 29–46.

Kosović, B. (1997). Subgrid-scale modelling for the large-eddy simulation of high-Reynoldsnumber boundary layers. *Journal of Fluid Mechanics, 336*, 151–182.

Kundu, P. K., & Cohen, I. M. (2008). *Fluid mechanics* (4th ed.). Burlington: Academic (Elsevier).

Kuzmin, D., Löhner, R., & Turek, S. (Eds.). (2012). *Flux-corrected transport: Principles, algorithms, and applications* (2nd ed.). Dordrecht: Springer.

Kwak, D., Chang, J. L. C., Shanks, S. P., & Chakravarthy, S. R. (1986). A three-dimensional incompressible Navier-Stokes flow solver using primitive variables. *AIAA Journal, 24*, 390–396.

Kwak, D., & Kiris, C. C. (2011). Artificial compressibility method. In *Computation of viscous incompressible flows*. Dordrecht: Springer.

Lafaurie, B., Nardone, C., Scardovelli, R., Zaleski, S., & Zanetti, G. (1994). Modelling merging and fragmentation in multiphase flows with SURFER. *Journal of Computational Physics, 113*, 134–147.

Lardeau, S. (2018). Consistent strain/stress lag eddy-viscosity model for hybrid RANS/LES. In Y. Hoarau, S. H. Peng, D. Schwamborn, & A. Revell (Eds.), *Progress in Hybrid RANS-LES Modelling* (pp. 39–51). Cham: Springer.

Launder, B. E., Reece, G. J., & Rodi, W. (1975). Progress in the development of a Reynolds-stress turbulence closure. *Journal of Fluid Mechanics, 68*, 537–566.

Launder, B. E., & Spalding, D. B. (1974). The numerical computation of turbulent flows. *Computer Methods in Applied Mechanics and Engineering, 3*, 269–289.

Le Bras, S., Deniau, H., Bogey, C., & Daviller, G. (2017). Development of compressible large-eddy simulations combining high-order schemes and wall modeling. *AIAA Journal, 55*, 1152–1163.

Leder, A. (1992). *Abgelöste Strömungen. Physikalische Grundlagen*. Wiesbaden, Germany: Vieweg.

Lee, M., & Moser, R. D. (2015). Direct numerical simulation of turbulent channel flow up to $Re_\tau \approx 5200$. *Journal of Fluid Mechanics, 774*, 395–415.

Leister, H.-J., & Perić, M. (1994). Vectorized strongly implicit solving procedure for seven-diagonal coefficient matrix. *International Journal of Numerical Methods for Heat and Fluid Flow, 4*, 159–172.

Lele, S. J. (1992). Compact finite difference schemes with spectral-like resolution. *Journal of Computational Physics, 3*, 16–42.

Leonard, B. P. (1979). A stable and accurate convection modelling procedure based on quadratic upstream interpolation. *Computer Methods in Applied Mechanics and Engineering, 19*, 59–98.

Leonard, B. P. (1997). Bounded higher-order upwind multidimensional finite-volume convection-diffusion algorithms. In W. J. Minkowycz & E. M. Sparrow (Eds.), *Advances in numerical heat transfer* (pp. 1–57). New York: Taylor and Francis.

Leonard, A.,& Wray, A. A. (1982). A new numerical method for the simulation of three dimensional flow in a pipe. In E. Krause (Ed.), *Eighth International Conference on Numerical Methods in Fluid Dynamics*. Lecture Notes in Physics (Vol. 170). Berlin: Springer.

Leonardi, S., Orlandi, P., Smalley, R. J., Djenidi, L., & Antonia, R. A. (2003). Direct numerical simulations of turbulent channel flow with transverse square bars on one wall. *Journal of Fluid Mechanics, 491*, 229–238.

Leschziner, M. A. (2010). Reynolds-averaged Navier-Stokes methods. In R. Blockley & W. Shyy (Eds.), *Encyclopedia of aerospace engineering* (pp. 1–13).

Lesieur, M. (2010). Two-point closure based on large-eddy simulations in turbulence, Part 2: Inhomogeneous cases. *Discrete & Continuous Dynamical Systems, 28*.

Lesieur, M. (2011). Two-point closure based on large-eddy simulations in turbulence, Part 1: Isotropic turbulence. *Discrete & Continuous Dynamical System Series S, 4*, 155–168.

Li, G., & Xing, Y. (1967). High order finite volume WENO schemes for the Euler equations under gravitational fields. *Journal of Computational Physics, 316*, 145–163.

Lilek, Ž. (1995). *Ein Finite-Volumen Verfahren zur Berechnung von inkompressiblen und kompressiblen Strömungen in komplexen Geometrien mit beweglichen Rändern und freien Oberflächen (Ph.D. Dissertation)*. Germany: University of Hamburg.

Lilek, Ž., Muzaferija, S., & Perić, M. (1997a). Efficiency and accuracy aspects of a full-multigrid SIMPLE algorithm for three-dimensional flows. *Numerical Heat Transfer, Part B, 31*, 23–42.

Lilek, Ž., Muzaferija, S., Perić, M., & Seidl, V. (1997b). An implicit finite-volume method using non-matching blocks of structured grid. *Numerical Heat Transfer, Part B, 32*, 385–401.

Lilek, Ž., Nadarajah, S., Peric, M., Tindal, M., & Yianneskis, M. (1991). Measurement and simulation of the flow around a poppet valve. In *Proceedings of 8th Symposium Turbulent Shear Flows* (pp. 13.2.1–13.2.6).

Lilek, Ž., & Perić, M. (1995). A fourth-order finite volume method with colocated variable arrangement. *Computers Fluids, 24*, 239–252.

Lilek, Ž., Schreck, E., & Perić, M. (1995). Parallelization of implicit methods for flow simulation. In S. G. Wagner (Ed.), *Notes on numerical fluid mechanics* (Vol. 50, pp. 135–146). Braunschweig: Vieweg.

Lilly, D. K. (1992). A proposed modification of the Germano subgrid-scale closure method. *Physics of Fluids A, 4*, 633–635.

Liu, X.-D., Osher, S., & Chan, T. (1994). Weighted essentially non-oscillatory schemes. *Journal of Computational Physics, 115*, 200–212.

Louda, P., Kozel, K., & Příhoda, J. (2008). Numerical solution of 2D and 3D viscous incompressible steady and unsteady flows using artificial compressibility method. *International Journal for Numerical Methods in Fluids, 56*, 1399–1407.

Lu, H., & Porté-Agel, F. (2011). Large-eddy simulation of a very large wind farm in a stable atmospheric boundary layer. *Physics of Fluids, 23*, 065101.

Ludwig, F. L., Chow, F. K., & Street, R. L. (2009). Effect of turbulence models and spatial resolution on resolved velocity structure and momentum fluxes in large-eddy simulations of neutral boundary layer flow. *Journal of Applied Meteorology and Climatology, 48*, 1161–1180.

Lumley, J. L. (1979). Computational modeling of turbulent flows. *Advances in Applied Mechanics, 18*, 123–176.

Lundquist, K. A., Chow, F. K., & Lundquist, J. K. (2012). An immersed boundary method enabling large-eddy simulations of flow over complex terrain in the WRF model. *Monthly Weather Review, 140*, 3936–3955.

MacCormack, R. W. (2003). The effect of viscosity in hypervelocity impact cratering. *Journal of Spacecraft and Rockets, 40*, 757–763. (Reprinted from AIAA Paper 69–354, 1969).

Mahesh, K. (1998). A family of high order finite difference schemes with good spectral resolution. *Journal of Computational Physics, 145*, 332–358.

Mahesh, K., Constantinescu, G., & Moin, P. (2004). A numerical method for large-eddy simulation in complex geometries. *Journal of Computational Physics, 197*, 215–240.

Malinen, M. (2012). The development of fully coupled simulation software by reusing segregated solvers. *Application of Parallel and Science Computing, Part 1, PARA 2010. LNCS, 7133*, 242–248.

Maliska, C. R., & Raithby, G. D. (1984). A method for computing three-dimensional flows using non-orthogonal boundary-fitted coordinates. *International Journal for Numerical Methods in Fluids, 4*, 518–537.

Manhart, M., & Wengle, H. (1994). Large-eddy simulation of turbulent boundary layer over a hemisphere. In P. Voke, L. Kleiser,& J. P. Chollet (Eds.), *Proceedings of 1st ERCOFTAC Workshop on Direct and Large Eddy Simulation* (pp. 299–310). Dordrecht: Kluwer Academic Publishers.

Marstorp, L., Brethouwer, G., Grundestam, O., & Johansson, A. V. (2009). Explicit algebraic subgrid stress models with application to rotating channel flow. *Journal of Fluid Mechanics, 639*, 403–432.

Mason, M. L., Putnam, L. E.,& Re, R. J. (1980). *The effect of throat contouring on two-dimensional converging-diverging nozzle at static conditions* (p. 1704). Paper No: NASA Techn.

Masson, C., Saabas, H. J., & Baliga, R. B. (1994). Co-located equal-order control-volume finite element method for two-dimensional axisymmetric incompressible fluid flow. *International Journal for Numerical Methods in Fluids, 18*, 1–26.

Matheou, G., & Chung, D. (2014). Large-eddy simulation of stratified turbulence. Part II: Application of the stretched-vortex model to the atmospheric boundary layer. *Journal of the Atmospheric Sciences, 71*, 4439–4460.

McCormick, S. F. (Ed.). (1987). *Multigrid methods*. Philadelphia: Society for Industrial and Applied Mathematics (SIAM).

McMurtry, P. A., Jou, W. H., Riley, J. J., & Metcalfe, R. W. (1986). Direct numerical simulations of a reacting mixing layer with chemical heat release. *AIAA Journal, 24*, 962–970.

Mellor, G. L., & Yamada, T. (1982). Development of a turbulence closure model for geophysical fluid problems. *Review of Geophysics, 20*, 851–875.

Meneveau, C., & Katz, J. (2000). Scale-invariance and turbulence models for large-eddy simulation. *Annual Review of Fluid Mechanics, 32*, 1–32.

Meneveau, C., Lund, T. S., & Cabot, W. H. (1996). A Lagrangian dynamic subgrid-scale model of turbulence. *Journal of Fluid Mechanics, 319*, 353–385.

Menter, F. R. (1994). Two-equation eddy-viscosity turbulence models for engineering applications. *AIAA Journal, 32*, 1598–1605.

Menter, F. R., Kuntz, M., & Langtry, R. (2003). Ten years of industrial experience with the SST turbulence model. In K. Hanjalić, Y. Nagano, & M. Tummers (Eds.), *Turbulence, heat and mass transfer* (Vol. 4, pp. 625–632). (Proceedings of 4th International Symposium on Turbulent for Heat and Mass Transfer, Begell House, Inc)

Mesinger, F., & Arakawa, A. (1976). *Numerical methods used in atmospheric models*. GARP Publications Series No. 17 (Vol. 1). Geneva: World Meteorological Organization

Meyers, J., Geurts, B. J., & Sagaut, P. (2007). A computational error-assessment of central finitevolume discretizations in large-eddy simulation using a Smagorinsky model. *Journal of Computational Physics, 227*, 156–173.

Michioka, T., & Chow, F. K. (2008). High-resolution large-eddy simulations of scalar transport in atmospheric boundary layer flow over complex terrain. *Journal of Applied Meteorology and Climatology, 47*, 3150–3169.

Mittal, R., & Iaccarino, G. (2005). Immersed boundary methods. *Annual Review of Fluid Mechanics, 37*, 239–261.

Moeng, C.-H. (1984). A large-eddy-simulation model for the study of planetary boundary-layer turbulence. *Journal of the Atmospheric Sciences, 41*, 2052–2062.

Moeng, C.-H., & Sullivan, P. P. (2015). Large-eddy simulation. *Encyclopedia of atmospheric sciences* (2nd ed., Vol. 4, pp. 232–240). Cambridge: Academic.

Moin, P. (2010). *Fundamentals of engineering numerical analysis* (2nd ed.). Cambridge: Cambridge University Press.

Moin, P., & Kim, J. (1982). Numerical investigation of turbulent channel flow. *Journal of Fluid Mechanics, 118*, 341–377.

Moin, P., & Mahesh, K. (1998). Direct numerical simulation: A tool in turbulence research. *Annual Review of Fluid Mechanics, 30*, 539–578.

Moin, P., Squires, K., Cabot, W., & Lee, S. (1991). A dynamic subgrid-scale model for compressible turbulence and scalar transport. *Physics of Fluids A, 3*, 2746–2757.

Mørch, H. J., Enger, S., Perić, M., & Schreck, E. (2008). Simulation of lifeboat launching under storm conditions. In *6th International Conference on CFD in Oil and Gas, Metallurgical and Process Industries*. Trondheim, Norway.

Mørch, H. J., Perić, M., Schreck, E., el Moctar, O., & Zorn, T. (2009). Simulation of flow and motion of lifeboats. In *ASME 28th International Conference on Ocean, Offshore and Arctic Engineering*. Honolulu, Hawaii.

Morrison, H., & Pinto, J. O. (2005). Intercomparison of bulk cloud microphysics schemes in mesoscale simulations of springtime arctic mixed-phase stratiform clouds. *Mon. Wea. Rev., 134*, 1880–1900.

Mortazavi, M., Le Chenadec, V., Moin, P., & Mani, A. (2016). Direct numerical simulation of a turbulent hydraulic jump: Turbulence statistics and air entrainment. *Journal of Fluid Mechanics, 797*, 60–94.

Moser, R. D., Moin, P., & Leonard, A. (1983). A spectral numerical method for the Navier-Stokes equations with applications to Taylor-Couette flow. *Journal of Computational Physics, 52*, 524–544.

Muzaferija, S. (1994). Adaptive finite volume method for flow predictions using unstructured meshes and multigrid approach (Ph.D. Dissertation). University of London.

Muzaferija, S., & Gosman, A. D. (1997). Finite-volume CFD procedure and adaptive error control strategy for grids of arbitrary topology. *Journal of Computational Physics, 138*, 766–787.

Muzaferija, S., & Perić, M. (1997). Computation of free-surface flows using finite volume method and moving grids. *Numerical Heat Transfer, Part B, 32*, 369–384.

Muzaferija, S., & Perić, M. (1999). Computation of free surface flows using interface-tracking and interface-capturing methods. In O. Mahrenholtz & M. Markiewicz (Eds.), *Nonlinear water wave interaction* (pp. 59–100). Southampton: WIT Press.

NASA CGTUM. (2010). Chimera grid tools user's manual, ver. 2.1. NASA Advanced Supercomputing Division. Retrieved from https://www.nas.nasa.gov/publications/software/-docs/chimera/index.html.

NASA TMR. (n.d.). Turbulence modeling resource. Langley Research Center. Retrieved from https://turbmodels.larc.nasa.gov/index.html.

NSF. (2006). *Simulation-based engineering science*. Retrieved from http://www.nsf.gov/-pubs/reports/sbes_final_report.pdf.

Oden, J. T. (2006). *Finite elements of non-linear continua*. Mineola: Dover Publications.

Orlandi, P., & Leonardi, S. (2008). Direct numerical simulation of three-dimensional turbulent rough channels: Parameterization and flow physics. *Journal of Fluid Mechanics, 606*, 399–415.

Osher, S., & Fedkiw, R. (2003). *Level set methods and dynamic implicit surfaces*. New York: Springer.

Osher, S., & Sethian, J. A. (1988). Fronts propagating with curvature-dependent speed: Algorithms based on Hamilton-Jacobi formulations. *Journal of Computational Physics, 79*, 12–49.

Pal, A., Sarkar, S., Posa, A., & Balaras, E. (2017). Direct numerical simulation of stratified flow past a sphere at a subcritical Reynolds number of 3700 and moderate Froude number. *Journal of Fluid Mechanics, 826*, 5–31.

Parrott, A. K., & Christie, M. A. (1986). FCT applied to the 2-D finite element solution of tracer transport by single phase flow in a porous medium. In K. W. Morton & M. J. Baines (Eds.), *Proceedings of ICFD-conference on Numerical Methods in Fluid Dynamics* (p. 609ff). Oxford: Oxford University Press.

Pascau, A. (2011). Cell face velocity alternatives in a structured colocated grid for the unsteady Navier-Stokes equations. *International Journal for Numerical Methods in Fluids, 65*, 812–833.

Patankar, S. V. (1980). *Numerical heat transfer and fluid flow*. New York: McGraw-Hill.

Patankar, S. V., & Spalding, D. B. (1972). A calculation procedure for heat, mass and momentum transfer in three-dimensional parabolic flows. *International Journal of Heat and Mass Transfer, 15*, 1787–1806.

Patankar, S. V., & Spalding, D. B. (1977). *Genmix: A general computer program for twodimensional parabolic phenomena*. Oxford: Pergamon Press.

Patel, V. C., Rodi, W., & Scheuerer, G. (1985). Turbulence models for near-wall and low-Reynolds number flows: A review. *AIAA Journal, 23*, 1308–1319.

Pekurovsky, D., Yeung, P. K., Donzis, D., Pfeiffer, W., & Chukkapallli, G. (2006). Scalability of a pseudospectral DNS turbulence code with 2D domain decomposition on Power41/Federation and Blue Gene systems. In *ScicomP12 and SP-XXL*. Boulder, CO: International Business Machines. Retrieved from http://www.spscicomp.org/ScicomP12/-Presentations/User/Pekurovsky.pdf.

Peller, N. (2010). *Numerische Simulation turbulenter Strömungen mit Immersed Boundaries (Ph.D. Dissertation)*. Fachgebiet Hydromechanik, Mitteilungen: Technische Universität München.

Perić, M. (1985). A finite volume method for the prediction of three-dimensional fluid flow in complex ducts (Ph.D. Dissertation). Imperial College, London.

Perić, M. (1987). Efficient semi-implicit solving algorithm for nine-diagonal coefficient matrix. *Numerical Heat Transfer, 11*, 251–279.

Perić, M. (1990). Analysis of pressure-velocity coupling on non-orthogonal grids. *Numerical Heat Transfer. Part B (Fundamentals), 17*, 63–82.

Perić, M. (1993). Natural convection in trapezoidal cavities. *Numerical Heat Transfer. Part A (Applications), 24,* 213–219.

Perić, M., & Schreck, E. (1995). Analysis of efficiency of implicit CFD methods on MIMD computers. In *Proceedings of Parallel CFD '95 Conference.*

Perić, R. (2019). *Minimierung unerwünschter Wellenreflexionen an den Gebietsrändern bei Strömungssimulationen mit Forcing Zones (Ph.D. Dissertation).* Germany: Technische Universität Hamburg.

Perić, R., & Abdel-Maksoud, M. (2018). Analytical prediction of reflection coefficients for wave absorbing layers in flow simulations of regular free-surface waves. *Ocean Engineering, 47,* 132–147.

Perktold, G., & K. and Rappitsch.,. (1995). Computer simulation of local blood flow and vessel mechanics in a compliant carotid artery bifurcation model. *J. Biomechanics, 28,* 845–856.

Perng, C. Y., & Street, R. L. (1991). A coupled multigrid-domain-splitting technique for simulating incompressible flows in geometrically complex domains. *International Journal for Numerical Methods in Fluids, 13,* 269–286.

Peskin, C. S. (1972). *Flow patterns around heart valves: A digital computer method for solving the equations of motion (Ph.D. Dissertation).* Albert Einstein College of Medicine, Yeshiva University.

Peskin, C. S. (2002). The immersed boundary method. *Acta Numerica, 11,* 479–517.

Peters, N. (2000). *Turbulent combustion.* Cambridge: Cambridge University Press.

Piomelli, U. (2008). Wall-layer models for large-eddy simulations. *Progress in Aerospace Sciences, 44,* 437–446.

Piomelli, U., & Balaras, E. (2002). Wall-layer models for large-eddy simulations. *Annual Review of Fluid Mechanics, 34,* 349–374.

Pirozzoli, S. (2011). Numerical methods for high-speed flows. *Annual Review of Fluid Mechanics, 43,* 163–194.

Poinsot, T., Veynante, D., & Candel, S. (1991). Quenching processes and premixed turbulent combustion diagrams. *Journal of Fluid Mechanics, 228,* 561–605.

Pope, S. B. (2000). *Turbulent flows.* Cambridge: Cambridge University Press.

Popovac, M., & Hanjalić, K. (2007). Compound wall treatment for RANS computation of complex turbulent flows and heat transfer. *Flow, Turbulence and Combustion, 78,* 177–202.

Porté-Agel, F., Meneveau, C., & Parlange, M. B. (2000). A scale-dependent dynamic model for large-eddy simulation: Application to a neutral atmospheric boundary layer. *Journal of Fluid Mechanics, 415,* 261–284.

Press, W. H., Teukolsky, S. A., Vettering, W. T., & Flannery, B. P. (2007). *Numerical recipes: The art of scientific computing* (3rd ed.). Cambridge: Cambridge University Press.

Pritchard, P. J. (2010). *Fox and McDonald's introduction to fluid mechanics* (8th ed.). New York: Wiley.

Purser, R. J. (2007). Accuracy considerations of time-splitting methods for models using two-timelevel schemes. *Monthly Weather Review, 135,* 1158–1164.

Qin, Z., Delaney, K., Riaz, A., & Balaras, E. (2015). Topology preserving advection of implicit interfaces on Cartesian grids. *Journal of Computational Physics, 290,* 219–238.

Raithby, G. D. (1976). Skew upstream differencing schemes for problems involving fluid flow. *Computer Methods in Applied Mechanics and Engineering, 9,* 153–164.

Raithby, G. D., & Schneider, G. E. (1979). Numerical solution of problems in incompressible fluid flow: Treatment of the velocity-pressure coupling. *Numerical Heat Transfer, 2,* 417–440.

Raithby, G. D., Xu, W.-X., & Stubley, G. D. (1995). Prediction of incompressible free surface flows with an element-based finite volume method. *Computational Fluid Dynamics Journal, 4,* 353–371.

Rakhimov, A. C., Visser, D. C., & Komen, E. M. J. (2018). Uncertainty quantification method for CFD applied to the turbulent mixing of two water layers. *Nuclear Engineering and Design, 333,* 1–15.

Ramachandran, S., & Wyngaard, J. C. (2010). Subfilter-scale modelling using transport equations: Large-eddy simulation of the moderately convective atmospheric boundary layer. *Boundary-Layer Meteorology*,. https://doi.org/10.1007/s10546-010-9571-3.

Rasam, A., Brethouwer, G., & Johansson, A. V. (2013). An explicit algebraic model for the subgrid-scale passive scalar flux. *Journal of Fluid Mechanics, 721*, 541–577.

Raw, M. J. (1985). *A new control-volume-based finite element procedure for the numerical solution of the fluid flow and scalar transport equations (Ph.D. Dissertation)*. Waterloo, Canada: University of Waterloo

Raw, M. J. (1995). A coupled algebraic multigrid method for the 3D Navier-Stokes equations. In W. Hackbusch & G. Wittum (Eds.), *Fast solvers for flow problems, notes on numerical fluid mechanics* (Vol. 49, pp. 204–215). Braunschweig: Vieweg.

Reichardt, H. (1951). Vollständige Darstellung der turbulenten Geschwindigkeitsverteilung in glatten Leitungen. *Z. Angew Mathematics and Mechanics, 31*, 208–219.

Reinecke, M., Hillebrandt, W., Niemeyer, J. C., Klein, R., & Gröbl, A. (1999). A new model for deflagration fronts in reactive fluids. *Astronomy and Astrophysics, 347*, 724–733.

Reynolds, O. (1895). On the dynamical theory of incompressible viscous fluids and the determination of the criterion. *Philosophical Transactions of the Royal Society London Series A, 186*, 123–164.

Rhie, C. M., & Chow, W. L. (1983). A numerical study of the turbulent flow past an isolated airfoil with trailing edge separation. *AIAA Journal, 21*, 1525–1532.

Riahi, H., Meldi, M., Favier, J., Serre, E., & Goncalves, E. (2018). A pressure-corrected immersed boundary method for the numerical simulation of compressible flows. *Journal of Computational Physics, 374*, 361–383.

Richardson, L. F. (1910). The approximate arithmetical solution by finite differences of physical problems involving differential equations with an application to the stresses in a masonry dam. *Philosophical Transactions of the Royal Society London Series A, 210*, 307–357.

Richtmyer, R. D., & Morton, K. W. (1967). *Difference methods for initial value problems*. New York: Wiley.

Rider, W. J. (2007). Effective subgrid modeling from the ILES simulation of compressible turbulence. *Journal of Fluids Engineering, 129*, 1493–1496.

Rider, W. J., & Kothe, D. B. (1998). Reconstructing volume tracking. *Journal of Computational Physics, 141*, 112–152.

Rizzi, A., & Viviand, H. (Eds.). (1981). *Numerical methods for the computation of inviscid transonic flows with shock waves.*, Notes on numerical fluid mechanics (Vol. 3) Braunschweig: Vieweg.

Roache, P. J. (1994). Perspective: A method for uniform reporting of grid refinement studies. *ASME Journal of Fluids Engineering, 116*, 405–413.

Rodi, W. (1976). A new algebraic relation for calculating the Reynolds stress. *ZAMM, 56*, T219–T221.

Rodi, W., Bonnin, J.-C., & Buchal, T. (Eds.). (1995). In*Proceedings of ERCOFTAC Workshop on Databases and Testing of Calculation Methods for Turbulent Flows, April 3–7*. Germany: University of Karlsruhe.

Rodi, W., Constantinescu, G., & Stoesser, T. (2013). *Large-eddy simulation in hydraulics*. London: Taylor & Francis.

Roe, P. L. (1986). Characteristic-based schemes for the Euler equations. *Annual Review of Fluid Mechanics, 18*, 337–365.

Rogallo, R. S. (1981). Numerical experiments in homogeneous turbulence (Technical Report No. 81315). Ames Research Center, CA: NASA.

Saad, Y. (2003). *Iterative methods for sparse linear systems* (2nd ed.). Philadelphia: Society for Industrial and Applied Mathematics (SIAM).

Saad, Y., & Schultz, M. H. (1986). GMRES: A generalized residual algorithm for solving nonsymmetric linear systems. *SIAM Journal on Scientific Statistical Computing, 7*, 856–869.

Sagaut, P. (2006). *Large-eddy simulation for incompressible flows: An introduction* (3rd ed.). Berlin: Springer.

Sani, R. L., Shen, J., Pironneau, O., & Gresho, P. M. (2006). Pressure boundary condition for the time-dependent incompressible Navier-Stokes equations. *International Journal for Numerical Methods in Fluids, 50,* 673–682.

Sauer, J. (2000). *Instationär kavitierende Strömungen - ein neues Modell, basierend auf Front Capturing (VoF) und Blasendynamik (Ph.D. Dissertation).* Germany: University of Karlsruhe.

Scardovelli, R., & Zaleski, S. (1999). Direct numerical simulation of free-surface and interfacial flow. *Annual Review of Fluid Mechanics, 31,* 567–603.

Schalkwijk, J., Griffith, E., Post, F. H., & Jonker, H. J. J. (2012). High-performance simulations of turbulent clouds on a desktop PC: Exploiting the GPU. *Bulletin of the American Meteorological Society, 93,* 307–314.

Schalkwijk, J., Jonker, H. J. J., Siebesma, A. P., & van Meijgaard, E. (2015). Weather forecasting using GPU-based large-eddy simulations. *Bulletin of the American Meteorological Society, 96,* 715–723.

Schlatter, P., & Örlü, R. (2010). Assessment of direct numerical simulation data of turbulent boundary layers. *Journal of Fluid Mechanics, 659,* 116–126.

Schneider, G. E., & Raw, M. J. (1987). Control-volume finite-element method for heat transfer and fluid flow using colocated variables. 1. Computational procedure. *Numerical Heat Transfer, 11,* 363–390.

Schneider, G. E., & Zedan, M. (1981). A modified strongly implicit procedure for the numerical solution of field problems. *Numerical Heat Transfer, 4,* 1–19.

Schnerr, G. H., & Sauer, J. (2001). Physical and numerical modeling of unsteady Cavitation dynamics. In *Fourth International Conference on Multiphase Flow. New Orleans, USA.*

Schreck, E., & Perić, M. (1993). Computation of fluid flow with a parallel multigrid solver. *International Journal for Numerical Methods in Fluids, 16,* 303–327.

Schumacher, J., Sreenivasan, K., & Yeung, P. (2005). Very fine structures in scalar mixing. *Journal of Fluid Mechanics, 531,* 113–122.

Sedov, L. (1971). *A course in continuum mechanics* (Vol. 1). Groningen: Wolters-Noordhoft Publishing.

Seidl, V. (1997). *Entwicklung und Anwendung eines parallelen Finite-Volumen-Verfahrens zur Strömungssimulation auf unstrukturierten Gittern mit lokaler Verfeinerung (Ph.D. Dissertation).* Germany: University of Hamburg.

Seidl, V., Muzaferija, S., & Perić, M. (1998). Parallel DNS with local grid refinement. *Applied Scientific Research, 59,* 379–394.

Seidl, V., Perić, M., & Schmidt, S. (1996). Space- and time-parallel Navier-Stokes solver for 3D block-adaptive Cartesian grids. In A. Ecer, J. Periaux, N. Satofuka, & S. Taylor (Eds.), *Parallel Computational Fluid Dynamics 1995: Implementations and results using parallel computers* (pp. 577–584). Amsterdam: North Holland, Elsevier.

Senocak, I., & Jacobsen, D. (2010). Acceleration of complex terrain wind predictions using many-core computing hardware. In *5th International Symposium on Computational Wind Engineering. (CWE2010) (Paper 498).* International Association for Wind Engineering.

Sethian, J. A. (1996). *Level set methods.* Cambridge: Cambridge University Press.

Shabana, A. A. (2013). *Dynamics of multibody systems* (4th ed.). New York: Cambridge University Press.

Shah, K. B., & Ferziger, J. H. (1995). Large-eddy simulations of flow past a cubic obstacle. In *Annual Research Briefs.* Stanford, CA: Center for Turbulence Research.

Shah, K. B., & Ferziger, J. H. (1997). A fluid mechanicians view of wind engineering: Large-eddy simulation of flow over a cubical obstacle. *Journal of Wind Engineering & Industrial Aerodynamics, 67 & 68,* 211–224.

Shen, J. (1993). A remark on the Projection-3 method. *International Journal for Numerical Methods in Fluids, 16,* 249–253.

Shewchuk, J. R. (1994). An introduction to the conjugate gradient method without the agonizing pain. Pitt., PA: School of Computer Science, Carnegie Mellon University. Retrieved from http://www.cs.cmu.edu/quake-papers/painless-conjugate-gradient.pdf.

Shi, X., Chow, F. K., Street, R. L., & Bryan, G. H. (2018a). An evaluation of LES turbulence models for scalar mixing in the stratocumulus-capped boundary layer. *Journal of the Atmospheric Sciences, 75*, 1499–1507.

Shi, X., Hagen, H. L., Chow, F. K., Bryan, G. H., & Street, R. L. (2018b). Large-eddy simulation of the stratocumulus-capped boundary layer with explicit filtering and reconstruction turbulence modeling. *J. Atmos. Sci., 75*, 611–637.

Shih, T.-H., & Liu, N.-S. (2009). A very-large-eddy simulation of the nonreacting flow in a single element lean direct injection combustor using PRNS with a nonlinear subscale model (Technical Report No. 2009-21564). Cleveland, OH: NASA Glenn Research Center.

Skamarock, W. C., & Klemp, J. B. (2008). A time-split nonhydrostatic atmospheric model for weather research and forecasting operations. *Journal of Computational Physics, 227*, 3465–3485.

Skamarock, W. C., Oliger, J., & Street, R. L. (1989). Adaptive grid refinement for numerical weather prediction. *Journal of Computational Physics, 80*, 27–60.

Skyllingstad, E. D., & Samelson, R. M. (2012). Baroclinic frontal instabilities and turbulent mixing in the surface boundary layer. Part I: Unforced simulations. *Journal of Physical Oceanography, 42*, 1701–1716.

Smagorinsky, J. (1963). General circulation experiments with the primitive equations. Part I: The basic experiment. *Monthly Weather Review, 91*, 99–164.

Smiljanovski, V., Moser, V., & Klein, R. (1997). A capturing-tracking hybrid scheme for deflagration discontinuities. *Combustion Theory and Modelling, 1*, 183–215.

Smits, A. J., & Marusic, I. (2013). Wall-bounded turbulence. *Physics Today, 66*, 25–30.

Smits, A. J., McKeon, B. J., & Marusic, I. (2011). High-Reynolds number wall turbulence. *Annual Review of Fluid Mechanics, 43*, 353–375.

Smolarkiewicz, P. K., & Margolin, L. G. (2007). Studies in geophysics. In F. Grinstein, L. Margolin, & W. Rider (Eds.), *Implicit large-eddy simulation: Computing turbulent fluid dynamics*. Cambridge: Cambridge University Press.

Smolarkiewicz, P. K., & Prusa, J. M. (2002). VLES modelling of geophysical fluids with nonoscillatory forward-in-time schemes. *International Journal for Numerical Methods in Fluids, 39*, 799–819.

Sonar, T. (1997). On the construction of essentially non-oscillatory finite volume approximations to hyperbolic conservation laws on general triangulations: Polynomial recovery, accuracy and stencil selection. *Computer Methods in Applied Mechanics and Engineering, 140*, 157.

Sonneveld, P. (1989). CGS, a fast Lanczos type solver for non-symmetric linear systems. *SIAM Journal on Scientific Computing, 10*, 36–52.

Spalart, P. R., & Allmaras, S. R. (1994). A one-equation turbulence model for aerodynamic flows. *La Recherche Aerospatiale, 1*, 5–21.

Spalart, P. R., Deck, S., Shur, M. L., Squires, K. D., Strelets, M. K., & Travin, A. (2006). A new version of detached-eddy simulation, resistant to ambiguous grid densities. *Theoretical and Computational Fluid Dynamics, 20*, 181–195.

Spalding, D. B. (1972). A novel finite-difference formulation for differential expressions involving both first and second derivatives. *International Journal for Numerical Methods in Fluids, 4*, 551–559.

Spalding, D. B. (1978). General theory of turbulent combustion. *Journal of the Energy, 2*, 16–23.

Spotz, W. (1998). Accuracy and performance of numerical wall boundary conditions for steady, 2D, incompressible streamfunction vorticity. *International Journal for Numerical Methods in Fluids, 28*, 737–757.

Spotz, W. F., & Carey, G. F. (1995). High-order compact scheme for the steady stream-function vorticity equations. *International Journal for Numerical Methods in Fluids, 38*, 3497–3512.

Sta. Maria, M., & Jacobson, M. (2009). Investigating the effect of large wind farms on the energy in the atmosphere. *Energies, 2*, 816–838.

Steger, J. L., & Warming, R. F. (1981). Flux vector splitting of the inviscid gas-dynamic equations with applications to finite difference methods. *Journal of Computational Physics, 40*, 263–293.

Stoll, R.,&Porté-Agel, F. (2006). Effect of roughness on surface boundary conditions for large-eddy simulation. *Boundary-Layer Meteorology, 118*, 169–187.

Stoll, R., & Porté-Agel, F. (2008). Large-eddy simulation of the stable atmospheric boundary layer using dynamic models with different averaging schemes. *Boundary-Layer Meteorology, 126*, 1–28.

Stolz, S., Adams, N. A., & Kleiser, L. (2001). An approximate deconvolution model for largeeddy simulation with application to incompressible wall-bounded flows. *Physics of Fluids, 13*, 997–1015.

Stone, H. L. (1968). Iterative solution of implicit approximations of multidimensional partial differential equations. *SIAM Journal on Numerical Analysis, 5*, 530–558.

Street, R. L. (1973). *Analysis and solution of partial differential equations*. Monterey: Brooks/Cole Publishing Co.

Street, R. L., Watters, G. Z., & Vennard, J. K. (1996). *Elementary fluid mechanics* (7th ed.). New York: Wiley.

Sullivan, P. P., Weil, J. C., Patton, E. G., Jonker, H. J. J., & Mironov, D. V. (2016). Turbulent winds and temperature fronts in large-eddy simulations of the stable atmospheric boundary layer. *Journal of the Atmospheric Sciences, 73*, 1815–1840.

Sullivan, P. P., Horst, T. W., Lenschow, D. H., Moeng, C.-H., & Weil, J. C. (2003). Structure of subfilter-scale fluxes in the atmospheric surface layer with application to large-eddy simulation modelling. *Journal of Fluid Mechanics, 482*, 101–139.

Sullivan, P. P., & Patton, E. G. (2008). A highly parallel algorithm for turbulence simulations in planetary boundary layers: Results with meshes up to 1024^3. In *18th Conference on Boundary Layers and Turbulence, AMS (pp. Paper 11B.5, 11). Stockholm, Sweden*.

Sullivan, P. P., & Patton, E. G. (2011). The effect of mesh resolution on convective boundary layer statistics and structures generated by large-eddy simulation. *Journal of the Atmospheric Sciences, 68*, 2395–2415.

Sunderam, V. S. (1990). PVM: A framework for parallel distributed computing. *Concurrency and computaton: Practice and Experience, 2*, 315–339.

Sussman, M. (2003). A second-order coupled level set and volume-of-fluid method for computing growth and collapse of vapor bubbles. *Journal of Computational Physics, 187*, 110–136.

Sussman, M., Smereka, P., & Osher, S. (1994). A level set approach for computing solutions to incompressible two-phase flow. *Journal of Computational Physics, 114*, 146–159.

Sweby, P. K. (1984). High resolution schemes using flux limiters for hyperbolic conservation laws. *SIAM Journal on Numerical Analysis, 21*, 995–1011.

Sweby, P. K. (1985). High resolution TVD schemes using flux limiters. *Large-scale computations in fluid mechanics*, Lecture notes in applied mathematics, Part 2 (Vol. 22, pp. 289–309). Providence: American Mathematical Society.

Taira, K., & Colonius, T. (2007). The immersed boundary method: A projection approach. *Journal of Computational Physics, 225*, 2118–2137.

Taneda, S. (1978). Visual observations of the flow past a sphere at Reynolds numbers between 10^4 and 10^6. *Journal of Fluid Mechanics, 85*, 187–192.

Tannehill, J. C., Anderson, D. A., & Pletcher, R. H. (1997). *Computational fluid mechanics and heat transfer* (Vol. 2). Penn: Taylor & Francis.

Tennekes, H., & Lumley, J. L. (1976). *A first course in turbulence*. Cambridge, MA: MIT Press.

Thé, J. L., Raithby, G. D., & Stubley, G. D. (1994). Surface-adaptive finite-volume method for solving free-surface flows. *Numerical Heat Transfer, Part B, 26*, 367–380.

Thibault, J. C., & Senocak, I. (2009). CUDA implementation of a Navier-Stokes solver on multi-GPU desktop platforms for incompressible flows. In *47th AIAA Aerospace Science Meeting, AIAA Paper 2009-758*.

Thompson, J. F., Warsi, Z. U. A., & Mastin, C. W. (1985). *Numerical grid generation - foundations and applications*. New York: Elsevier.

Thompson, M. C., & Ferziger, J. H. (1989). A multigrid adaptive method for incompressible flows. *Journal of Computational Physics, 82*, 94–121.

Tokuda, Y., Song, M.-H., Ueda, Y., Usui, A., Akita, T., Yoneyama, S., et al. (2008). Three-dimensional numerical simulation of blood flow in the aortic arch during cardiopulmonary bypass. *European Journal of Cardio-Thoracic Surgery, 33*, 164–167.

Tregde, V. (2015). Compressible air effects in CFD simulations of free fall lifeboat drop. In *ASME 34th International Conference on Ocean, Offshore and Arctic Engineering. St John's, Newfoundland, Canada.*

Truesdell, C. (1991). *A first course in rational continuum mechanics* (2nd ed., Vol. 1). Boston: Academic.

Tryggvason, G., & Unverdi, S. O. (1990). Computations of 3-dimensional Rayleigh-Taylor instability. *Physics of Fluids A, 2*, 656–659.

Tseng, Y. H., & Ferziger, J. H. (2003). A ghost-cell immersed boundary method for flow in complex geometry. *Journal of Computational Physics, 192*, 593–623.

Tu, J. Y., & Fuchs, L. (1992). Overlapping grids and multigrid methods for three-dimensional unsteady flow calculation in IC engines. *International Journal for Numerical Methods in Fluids, 15*, 693–714.

Tuković, Ž., Perić, M., & Jasak, H. (2018). Consistent second-order time-accurate non-iterative PISO algorithm. *Computers Fluids, 166*, 78–85.

Turkel, E. (1987). Preconditioned methods for solving the incompressible and low speed compressible equations. *Journal of Computational Physics, 72*, 277–298.

Ubbink, O. (1997). *Numerical prediction of two fluid systems with sharp interfaces (Ph.D. Dissertation).* London: University of London.

van der Vorst, H. A. (1992). BI-CGSTAB: a fast and smoothly converging variant of BI-CG for the solution of non-symmetric linear systems. *SIAM Journal on Scientific Computing, 13*, 631–644.

van der Vorst, H. A. (2002). Efficient and reliable iterative methods for linear systems. *Journal of Computational and Applied Mathematics, 149*, 251–265.

van der Vorst, H. A., & Sonneveld, P. (1990). CGSTAB, a more smoothly converging variant of CGS (Technical Report No. 90-50). Delft, NL: Delft University of Technology.

van der Wijngaart, R. F. (1990). *Composite-grid techniques and adaptive mesh refinement in computational fluid dynamics (Ph.D. Dissertation).* Stanford CA: Stanford University.

Van Doormal, J. P., & Raithby, G. D. (1984). Enhancements of the SIMPLE method for predicting incompressible fluid flows. *Numerical Heat Transfer, 7*, 147–163.

Van Doormal, J. P., Raithby, G. D., & McDonald, B. H. (1987). The segregated approach to predicting viscous compressible fluid flows. *ASME Journal of Turbomachinery, 109*, 268–277.

Vanka, S. P. (1986). Block-implicit multigrid solution of Navier-Stokes equations in primitive variables. *Journal of Computational Physics, 65*, 138–158.

Van Leer, B. (1977). Towards the ultimate conservative difference scheme. IV. A new approach to numerical convection. *Journal of Computational Physics, 23*, 276–299.

Van Leer, B. (1985). Upwind-difference methods for aerodynamic problems governed by the Euler equations. *Large-scale computations in fluid mechanics*, Lecture notes in Applied Mathematics, Part 2, (Vol. 22, 327–336). Providence: American Mathematical Society.

Viswanathan, A. K., Squires, K. D., & Forsythe, J. R. (2008). Detached-eddy simulation around a forebody with rotary motion. *AIAA Journal, 46*, 2191–2201.

Vukčević, V., Jasak, H., & Gatin, I. (2017). Implementation of the ghost fluid method for free surface flows in polyhedral finite volume framework. *Computers Fluids, 153*, 1–19.

Wakashima, S., & Saitoh, T. S. (2004). Benchmark solutions for natural convection in a cubic cavity using the high-order time-space method. *International Journal of Heat and Mass Transfer, 47*, 853–864.

Wallin, S., & Johansson, A. V. (2000). An explicit algebraic Reynolds stress model for incompressible and compressible turbulent flows. *Journal of Fluid Mechanics, 403*, 89–132.

Wang, R., Feng, H., & Huang, C. (2016). A new mapped weighted essentially non-oscillatory method using rational mapping function. *Journal of Scientific Computing, 67*, 540–580.

Warming, R. F., & Hyett, B. J. (1974). The modified equation approach to the stability and accuracy of finite-difference methods. *Journal of Computational Physics, 14*, 159–179.

Washington, W. M., & Parkinson, C. L. (2005). *An introduction to three-dimensional climate modeling* (2nd ed.). Sausalito: University Science Books.

Waterson, N. P., & Deconinck, H. (2007). Design principles for bounded higher-order convection schemes - a unified approach. *Journal of Computational Physics, 224*, 182–207.

Watkins, D. S. (2010). *Fundamentals of matrix computations* (3rd ed.). New York: Wiley- Interscience.

Wegner, B., Maltsev, A., Schneider, C., Sadiki, A., Dreizler, A., & Janicka, J. (2004). Assessment of unsteady RANS in predicting swirl flow instability based on LES and experiments. *International Journal of Heat and Fluid Flow, 25*, 528–536.

Weinan, E., & Liu, J.-G. (1997). Finite difference methods for 3D viscous incompressible flows in the vorticity - vector potential formulation on nonstaggered grids. *Journal of Computational Physics, 138*, 57–82.

Weiss, J. M., Maruszewski, J. P., & Smith, W. A. (1999). Implicit solution of preconditioned Navier-Stokes equations using algebraic multigrid. *AIAA Journal, 37*, 29–36.

Weiss, J. M., & Smith, W. A. (1995). Preconditioning applied to variable and constant density flows. *AIAA Journal, 33*, 2050–2057.

Wesseling, P. (1990). Multigrid methods in computational fluid dynamics. *ZAMM - Z. Angew Mathematics and Mechanics, 70*, T337–T347.

Weymouth, G., & Yue, D. K. P. (2010). Conservative volume-of-fluid method for free-surface simulations on Cartesian grids. *Journal of Computational Physics, 229*, 2853–2865.

White, F. M. (2010). *Fluid mechanics* (7th ed.). New York: McGraw Hill.

Wicker, L. J., & Skamarock, W. C. (2002). Time-splitting methods for elastic models using forward time schemes. *Monthly Weather Review, 130*, 2088–2097.

Wie, S. Y., Lee, J. H., Kwon, J. K., & Lee, D. J. (2010). Far-field boundary condition effects of CFD and free-wake coupling analysis for helicopter rotor. *Journal of Fluids Engineering, 132*, 84501-1-6.

Wikstrom, P. M., Wallin, S., & Johansson, A. V. (2000). Derivation and investigation of a new explicit algebraic model for the passive scalar flux. *Physics of Fluids, 12*, 688–702.

Wilcox, D. C. (2006). *Turbulence modeling for CFD* (3rd ed.). La Cañada: DCW Industries Inc.

Williams, F. A. (1985). *Combustion theory: The fundamental theory of chemically reacting flow systems*. Menlo Park: Benjamin-Cummings Publishing Co.

Williams, J., Sarofeen, C., Shan, H., & Conley, M. (2016). An accelerated iterative linear solver with GPUs and CFD calculations of unstructured grids. *Procedia Computer Science, 80*, 1291–1300.

Williams, P. D. (2009). A proposed modification to the Robert-Asselin time filter. *Monthly Weather Review, 137*, 2538–2546.

Wong, V. C., & Lilly, D. K. (1994). A comparison of two dynamic subgrid closure methods for turbulent thermal convection. *Physics of Fluids, 6*, 1016–1023.

Wosnik, M., Castillo, L., & George, W. K. (2000). A theory for turbulent pipe and channel flows. *Journal of Fluid Mechanics, 412*, 115–145.

Wrobel, L. C. (2002). The boundary element method. Vol. 1: Applications in thermo-fluids and acoustics. New York: Wiley.

Wu, X., & Moin, P. (2009). Direct numerical simulation of turbulence in a nominally zero-pressuregradient flat-plate boundary layer. *Journal of Fluid Mechanics, 630*, 5–41.

Wu, X., & Moin, P. (2011). Evidence for the persistence of hairpin forest in turbulent, zeropressure-gradient flat-plate boundary layers. In *7th International Symposium for Turbulence and Shear Flow Phenomena (TSFP-7), Paper 6A4P. Ottawa, Canada.*

Wyngaard, J. C. (2004). Toward numerical modeling in the "Terra Incognita". *Journal of the Atmospheric Sciences, 61*, 1816–1826.

Wyngaard, J. C. (2010). *Turbulence in the atmosphere*. Cambridge: Cambridge University Press.

Xing-Kaeding, Y. (2006). *Unified approach to ship seakeeping and maneuvering by a RANSE method* (Ph.D. Dissertation, TU Hamburg-Harburg. Hamburg). Retrieved from http://doku.b.tu-harburg.de/volltexte/2006/303/pdf/Xing-Kaeding-thesis.pdf.

Xue, M. (2000). High-order monotonic numerical diffusion and smoothing. *Monthly Weather Review*, *128*, 2853–2864.

Xue, M., Drogemeier, K. K., & Wong, V. (2000). The advanced regional prediction system (ARPS) - a multi-scale nonhydrostatic atmospheric simulation and prediction model. Part I: Model dynamics and verification. *Meteorology and Atmospheric Physics*, *75*, 161–193.

Yang, H. Q., & Przekwas, A. J. (1992). A comparative study of advanced shock-capturing schemes applied to Burger's equation. *Journal of Computational Physics*, *102*, 139–159.

Ye, T., Mittal, R., Udaykumar, H. S., & Shyy, W. (1999). An accurate Cartesian grid method for viscous incompressible flows with complex immersed boundaries. *Journal of Computational Physics*, *156*, 209–240.

Youngs, D. L. (1982). Time-dependent multi-material flow with large fluid distortion. In K. W. Morton & M. J. Baines (Eds.), *Numerical methods for fluid dynamics* (pp. 273–285). New York: Academic.

Zalesak, S. T. (1979). Fully multidimensional flux-corrected transport algorithms for fluids. *Journal of Computational Physics*, *31*, 335–362.

Zang, Y., & Street, R. L. (1995). A composite multigrid method for calculating unsteady incompressible flows in geometrically complex domains. *International Journal for Numerical Methods in Fluids*, *20*, 341–361.

Zang, Y., Street, R. L., & Koseff, J. R. (1993). A dynamic mixed subgrid-scale model and its application to turbulent recirculating flows. *Physics of Fluids A*, *5*, 3186–3196.

Zang, Y., Street, R. L., & Koseff, J. R. (1994). A non-staggered grid, fractional-step method for time-dependent incompressible Navier-Stokes equations in curvilinear coordinates. *Journal of Computational Physics*, *114*, 18–33.

Zhang, H., Zheng, L. L., Prasad, V., & Hou, T. Y. (1998). A curvilinear level set formulation for highly deformable free surface problems with application to solidification. *Numerical Heat Transfer*, *34*, 1–20.

Zhou, B., & Chow, F. K. (2011). Large-eddy simulation of the stable boundary layer with explicit filtering and reconstruction turbulence modeling. *Journal of the Atmospheric Sciences*, *68*, 2142–2155.

Zhou, B., Simon, J. S., & Chow, F. K. (2014). The convective boundary layer in the Terra Incognita. *Journal of the Atmospheric Sciences*, *71*, 2547–2563.

Zienkiewicz, O. C., Taylor, R. L., & Nithiarasu, P. (2005). *The finite element method for fluid dynamics* (6th ed.). Burlington: Butterworth- Heinemann (Elsevier).

Zwart, P. J., Gerber, G., & Belamri, T. (2004). A two-phase flow model for prediction of cavitation dynamics. In *Fifth International Conference on Multiphase Flow Yokohama, Japan*

Index

© Springer Nature Switzerland AG 2020
J. H. Ferziger et al., *Computational Methods for Fluid Dynamics*,
https://doi.org/10.1007/978-3-319-99693-6

Printed in the United States
By Bookmasters